PRINCIPLES AND APPLICATIONS OF AQUATIC CHEMISTRY

PRINCIPLES AND APPLICATIONS OF AQUATIC CHEMISTRY

FRANÇOIS M. M. MOREL

Massachusetts Institute of Technology
Cambridge, Massachusetts

JANET G. HERING

University of California, Los Angeles
Los Angeles, California

A Wiley–Interscience Publication

JOHN WILEY & SONS, INC.

New York / Chichester / Brisbane / Toronto / Singapore

Library of Congress Cataloging in Publication Data:

Morel, François, 1944–
    Principles and applications of aquatic chemistry / François M. M. Morel, Janet G. Hering.
        p.   cm.
    Rev. ed. of Principles of aquatic chemistry. c1983.
    "A Wiley–Interscience publication."
    Includes bibliographical references and index.
    ISBN 0-471-54896-0
    1. Water chemistry.   I. Hering, Janet G.   II. Morel, François, 1944–   Principles of aquatic chemistry.   III. Title.
GB855.M67   1993
551.48—dc20   92–18608

Printed in Mexico

20 19

To our families

CONTENTS

PREFACE

Over the past decade the subject of aquatic chemistry has evolved to take a more detailed mechanistic view of the reactions that control the chemistry of natural waters; it has also broadened to encompass the elemental cycles that determine the overall chemistry of the biosphere. This book reflects these developments and expands the scope of the original *Principles of Aquatic Chemistry*. Though it maintains a strong cartesian organization and expands the treatment of the chemical fundamentals, this book now emphasizes applications. Throughout we have introduced real-world examples to illustrate the practical utility of the methods and principles.

This teaching text is intended for a wide audience of advanced undergraduate and beginning graduate students in environmental science and engineering, earth sciences, and oceanography. General chemistry and calculus are the only prerequisites. In several chapters, supplementary material, such as a detailed derivation, is presented in separate sidebars. These and a few other sections may require somewhat greater mathematical or chemical sophistication: the necessary background in physical chemistry or differential equations can be supplied by ad hoc explanations tailored to the particular audience. The material in the book is more than can be covered in one trimester or even in one semester. Typically only the basic sections of each chapter will be covered in an introductory aquatic chemistry course. The advanced topics, supplemented by current literature, can serve as the basis of specialized courses.

The first three chapters have been completely revised. Chapter 1 begins with a short discussion of global geochemical cycles, examines the issue of mass conservation in chemical systems, and introduces the "tableau method" for setting up chemical problems. This approach is gaining broad acceptance as a pedagogical device[1,2] and is particularly helpful for initiating engineers to the joys of chemistry. Fundamentals of thermodynamics and kinetics are now covered in individual chapters (2 and 3), which also provide methods for solving practical equilibrium and kinetics problems. Chapter 4 on acid-base reactions, with a focus on the carbonate system, has been expanded to include the role of organic acids and gas exchange through ebullition. The updated coverage of precipitation/dissolution reactions in Chapter 5 has been complemented with an expanded discussion of kinetics. Fundamental chemical principles of coordination now introduce the subject of complexation in Chapter 6. A new approach to metal binding by humates echoes the treatment of organic acids in Chapter 4, and new discussions of kinetics and microbial uptake are provided. In Chapter 7, which covers redox reactions, the photochemistry and kinetics sections have been extensively updated. We have retained the didactic progression of Chapter 8, introducing adsorption as a surface complexation reaction before separating chemical and coulombic contributions to the free energies of sorption. Following the treatment of Dzombak and Morel[3] the presentation of the "surface complexation model" is now based on a diffuse layer calculation of the coulombic term and takes advantage of their compilation of coherent adsorption constants for hydrous ferric oxide.

Still there is no Chapter 9; no treatment of organic aquatic chemistry. Fortunately, this void has now been filled by the recently published text *Environmental Organic Chemistry*.[4]

As we enumerate those who have assisted us we cannot help but wonder if it is really our work after all; at least we can share the blame, widely: Beth Ahner, Tina Bartschat (humic acids), Pat Brady, Philippe van Capellen, David Dzombak, Liz Fechner (critical proofreading), Sarah Green (photochemistry), Jürg Hoigné (photochemistry), Bob Hudson, Jean LeCorre, Jenny Lee, George Luther, John MacFarlane (figures), Russ McDuff, Diane McKnight, Jim Morgan (kinetics), Brian Palenik, Lynn Roberts, Barbara Sulzberger (photochemistry), Paul Tratnyek (kinetics), Bernhard Wehrli, Don Yee.

The junior author is grateful to Werner Stumm for his encouragement and financial support during her postdoctoral stay at EAWAG and to her colleagues at UCLA for their understanding during the completion of this work.

The senior author acknowledges the inspiration of many generations of students in Aquatic Chemistry at MIT.

FRANÇOIS M. M. MOREL
JANET G. HERING

1. L. Sigg, W. Stumm, P. Behra, *Chimie des Milieux Aquatiques*, Masson, Paris, 1992.

2. L. Sigg and W. Stumm, *Aquatische Chemie*, Verlag der Fachvereine, Zürich, 1989.

3. D. A. Dzombak and F. M. M. Morel, *Surface Complexation Modeling*, Wiley–Interscience, New York, 1990.

4. R. P. Schwarzenbach, P. M. Gschwend, and D. M. Imboden, *Environmental Organic Chemistry*, Wiley–Interscience, New York, 1993.

PREFACE TO *PRINCIPLES OF AQUATIC CHEMISTRY*

Because the hydrologic cycle powers all others, and because life on earth is so intimately tied to water, aquatic chemistry is a central link between the cycles of elements at the surface of the earth and the workings of biological systems. Life adapts to its environment and modifies it. As noted by Alfred Redfield some 50 years ago, the composition of natural waters reflects both the geochemistry of the planet earth and the biochemical requirements of its inhabitants. These are delicately in tune, constantly modifying each other, linked by a few billion years of "co-evolution," and both ultimately responding to the same fundamental laws of chemistry.

This central position of aquatic chemistry in the natural sciences gives it an increasing popularity in science and engineering curricula; it also makes it a difficult topic to teach for it requires exploring some aspects of almost all sciences. This text is intended for first year graduate students and advanced undergraduates. I have assumed only a general college chemistry background and all incursions into geology, microbiology, oceanography, chemical kinetics, and other associated fields are kept strictly elementary. The first three chapters which deal with conservation of mass, thermodynamics and kinetics, and chemical equilibrium calculations, form effectively a separate entity. I intend them as a self-contained introduction to solution chemistry. In these three chapters, which provide the general background indispensible for the rest of the book, thermodynamics is the major stumbling block. Because typical student backgrounds differ widely, I have chosen a treatment of thermodynamics that is pragmatic and intuitive, to serve as a first introduction for some and as a review for others.

In the next five chapters of the text the subject of aquatic chemistry proper is developed following the accepted divisions of acid-base, precipitation-dissolution, coordination, redox, and surface chemistry. (A possible ninth chapter on organic aquatic chemistry never got written; I think it would require a text and a course of its own.) Chapters 5 and 6 focus more on chemical principles and less on the intricate realities of natural waters which are introduced more fully in the later chapters. At the end of each chapter the more advanced topics, such as kinetics, are treated in an elementary fashion. Sections that are either digressive or not essential to the overall comprehension of the text are indicated by three squares at the beginning and at the end.

□ □ □

The expression "aquatic chemistry" made its official entry into scientific language with the publication of the text by Stumm and Morgan in 1970. The title of this new text is meant in part as a general reference and acknowledgment to that classic book, as a plain expression of the debt I owe to its authors who started a scientific and pedagogic tradition.

My primary goal in writing this text was to satisfy the demands of a particular assemblage of students. For ten years I have taught aquatic chemistry to students from MIT engineering and science departments, from the MIT-WHOI Joint Program in Oceanography, and from Harvard College and the School of Public Health. During those years I developed class notes that are the basis for this text. Much has been added and reworked to be sure, but the skeleton of these original class notes remains, still reflecting a wistful and very Gallic ambition for organization and coherence.

From those notes also comes an emphasis on the geochemical and geobiological cycles of elements in natural waters. This central conceptual link between environmental chemists and oceanographers is used as a thread throughout the text. I have also tried to meet the students' demand for mathematical rigor without mathematical sophistication. The price sometimes is a cumbersomely detailed algebraic and numerical development. I hope to have at least generally resisted a natural propensity for reducing lively chemical concepts to cold mathematics.

While writing this text I often found myself squeezed between the lofty goals of pedagogy and the sheer boredom of a systematic and repetitive treatment. My pen attempted to escape and I had to fight bouts of lyricism, jocularity, and even obscurantism. (Such wonderful depth there is in the murky corners of the mind!) No help to students, of course, and most of it has been bravely edited away. Perhaps the scattered remains will entertain those who wander through; may they find the text more enjoyable where it is less orderly.

In the end, of course, this book is a compromise, a truce in a quixotic battle against the modern realities of academic life. Still, there is much satisfaction just in getting here and, as I sigh in relief, I must also thank all those who helped along the way. I shall mention only a few: W. Fish, P. M. Gschwend,

T. D. Waite, and O. Zafiriou who contributed some of their expertise to particular sections; R. E. McDuff, J. J. Morgan, and J. C. Westall who commented on the complete manuscript; R. Selman and B. A. Hanrahan who transformed my hieroglyphic handwriting into print; D. A. Dzombak and S. J. Tiffany who helped simplify and anglicize my prose and freed me of countless detailed tasks; S. W. Chisholm whose enlightening influence permeates the whole text. To these and many unnamed others I express my gratitude. Still, it is all the students in Aquatic Chemistry at MIT over the past ten years who deserve the greatest credit: they insisted on their right to understand. If this text helps clarify the subject in some way, the next generations of students will owe it to their elders, and I shall derive no small satisfaction from it.

FRANÇOIS M. M. MOREL

Lexington, Massachusetts
September 1983

PRINCIPLES AND APPLICATIONS OF AQUATIC CHEMISTRY

CHAPTER 1

CONSERVATION PRINCIPLES

A poet physicist once calculated how many atoms from Plato's body each of us has appropriated as our own—probably in excess of 10^{12}. Behind the vertigo of the multiple reincarnations and the comforting thought of all the great people each of us has been lies the profound and fundamental principle of conservation of mass.

The idea that the total mass of a closed system should remain constant regardless of internal changes is one of the great conceptual and experimental discoveries of the late eighteenth century. The equivalency of matter and energy discovered by Einstein in the twentieth century generalizes rather than invalidates this rule. In chemistry, where we break down a system into its constitutive elements, we hold that the total mass of an element is invariant through all its physical and chemical transformations. Nuclear reactions break that rule but, in so doing, help bring it into focus, as discussed later.

The solution to our Platonic riddle lies in the examination of global element cycles—chiefly that of carbon. One must consider the transformation and transfer of carbon over the whole earth's surface, including the air, water, some part of the earth, and, of course, the organisms. We must estimate how much carbon is respired into CO_2, how this CO_2 may be diluted and transferred among various reservoirs, and what other chemical species the carbon may be transformed into, including foodstuff. In the first section of this chapter we discuss briefly the global cycles of carbon, nitrogen, phosphorus, and sulfur, both as general background information and to illustrate the various transformations—the chemical reactions—that are the subjects of the following chapters.

So easily intuitive seems the idea of mass conservation that its application into workable mathematical expressions can be ironically frustrating. In the study of aquatic transport phenomena, the difficulty lies in writing—and then solving—the differential equations expressing the conservation of solvent and solutes that are subjected to displacement and volume changes. In the remainder of this chapter, we study how conservation principles are applied to aqueous chemical systems in which multiple reactions, but no bulk transport, are taking place. The underlying mathematics are more in the field of linear algebra than that of differential analysis. The key concept to be mastered is that of components, the fundamental entities to which the conservation principle is applied in the form of mole balance equations.

1. GLOBAL CYCLES

As our introductory platonic example has already suggested, the global cycle of an element has two fundamental features: the distribution of the element among the various reservoirs and the flux, or transfer, of the element between reservoirs. Both of these aspects are important because we are most often interested in the amount of an element in a relatively "minor" compartment— minor, that is, in terms of the total mass of the element. For example, the amount of carbon in the atmosphere is of concern because atmospheric CO_2 traps energy radiated from the earth's surface, the so-called greenhouse effect. The concentration of CO_2 in the atmosphere is determined by the balance of processes that consume or produce CO_2, predominantly photosynthesis, respiration, and air–water exchange. Human activities such as fossil fuel combustion and deforestation have resulted in a steady increase in the atmospheric CO_2 concentration. The fluxes of carbon into and out of the atmospheric reservoir, including anthropogenic fluxes, are vitally important in determining the extent of global warming. The atmospheric reservoir and the fluxes through this reservoir are insignificant, however, in the global distribution of carbon, of which 99.94% is contained in rocks and sediments and only 0.001% in the atmosphere. Indeed, the lithosphere is the predominant reservoir for carbon, nitrogen, phosphorus, and sulfur.

Before we examine the global elemental cycles in more detail, we should consider the energy and hydrologic cycles, which drive all the rest. Solar radiation is the primary energy input to the earth's surface; a small fraction, 0.02%, is contributed from heat flow from the interior and tidal energy. Of the incoming solar radiation, approximately 30% is reflected back into space and the remaining 70% is absorbed. The energy absorbed must be balanced by an equivalent loss since the average temperature of the earth remains constant (over geologically short times of a few thousand years). The marked latitudinal variations in the earth's heating cycle produce convection in both the atmosphere and oceans, govern the distribution of the planet's freshwater resources (of which

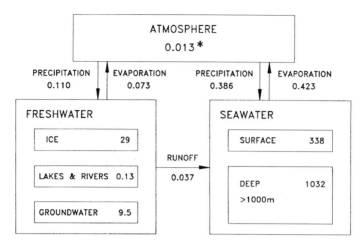

Figure 1.1   The hydrologic cycle: distribution of water among atmospheric, freshwater, and seawater reservoirs and fluxes between reservoirs. Units: fluxes in $10^6 \, km^3/year$; inventories in $10^6 \, km^3$. The atmospheric inventory (starred value) is reported as the liquid equivalent of water vapor.

78% are stored in polar ice), and drive the hydrologic cycles of evaporation, precipitation, and surface and underground water transport (Figure 1.1)[1].

Together, the energy and water cycles sustain the global elemental cycles. The conversion of solar energy to chemical energy by plants and photosynthetic bacteria supports element fixation, the incorporation of inorganic nutrients into biomass. Remineralization accompanies the decomposition of organic matter. The hydrologic cycle results in the weathering of continental rocks and the transport of continental material to the world's oceans.

Here we begin to see the types of processes that drive the elemental fluxes between global reservoirs even though we have not yet clearly defined those reservoirs. This is not as trivial an exercise as it might seem at first. We must decide, for example, whether the surface waters of the ocean, which are in contact with the atmosphere, should be considered separately from the deep waters, which are isolated from the atmosphere. The global carbon cycle in Figure 1.2 illustrates a common choice of global reservoirs: the atmosphere, the hydrosphere, the terrestrial biosphere and pedosphere, and the lithosphere. All of these contain smaller compartments, such as the ocean deep water compartment within the hydrosphere.

Several aspects of the global carbon cycle should be noted. Carbon inventories in the hydrosphere are divided into inorganic and organic carbon with inorganic carbon (i.e., carbonate) being vastly more abundant. The system is not at steady state. As a result of fossil fuel combustion (which is inordinately rapid compared to its production), the inventories of carbon in the atmosphere and hydrosphere are increasing and that of the lithosphere, presumably, decreasing. Perhaps most importantly, the largest fluxes involve transfer to and from one of the smallest

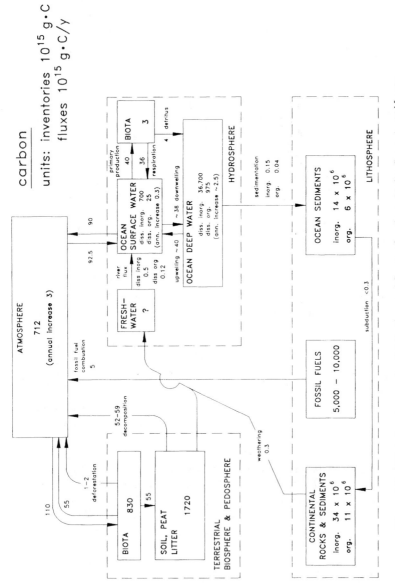

Figure 1.2 Global carbon cycle: distribution and fluxes of carbon. Units: fluxes in 10^{15} g C/year, inventories in 10^{15} g C. Oceanic and sedimentary carbon inventories are characterized as inorganic or organic. Inorganic carbon includes dissolved carbonate species and carbonate minerals. Organic carbon consists of, or is derived from, products of biological activity. Data from Bolin and Cook (1983).[2]

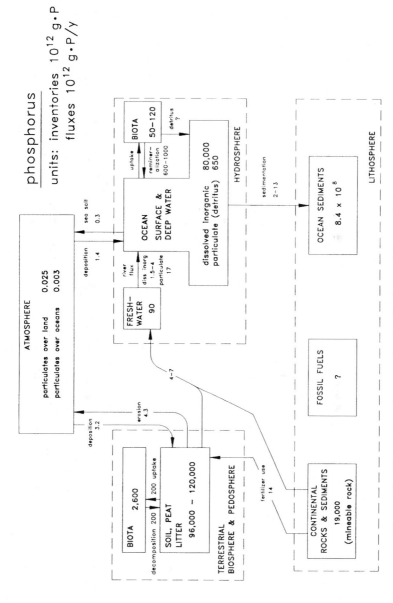

Figure 1.3 Global phosphorus cycle: distribution and fluxes of phosphorus. Units: fluxes in 10^{12} g P/year, inventories in 10^{12} g P. Data from Bolin and Cook (1983).[2]

reservoirs, the atmosphere. The residence time of carbon in the atmosphere, defined as inventory divided by the flux, is correspondingly one of the shortest. In the global cycles of carbon, nitrogen, and sulfur, the atmosphere serves as an important conduit between reservoirs with much larger inventories.

This is not the case for phosphorus whose atmospheric inventory consists solely of particulate phosphorus. This cycle (Figure 1.3) is also an exception in that it is not affected by fossil fuel combustion. This does not mean that the phosphorus cycle is unperturbed by human activity. The significant flux from continental rocks and sediments to soils is due to the application of fertilizers in agriculture.

The nitrogen cycle (Figure 1.4) illustrates that quantifying the reservoirs and fluxes of the total mass of an element is insufficient to comprehend the processes governing its cycles. Several forms of inorganic nitrogen, such as N_2, NO_3^-, and NH_3, must be considered separately because of their very different biological availabilities, chemical reactivities, and volatilities. Note that the cycling (or interconversions) of these different chemical species within a given global reservoir is not included. In this case, a great deal of information would be lost if only the conservation of elemental nitrogen were considered.

The sulfur cycle (Figure 1.5) has been profoundly influenced by human activities. Anthropogenic inputs of sulfur to the atmosphere have resulted in increased deposition of sulfate to soils and surface waters. This acid deposition has caused acidification of surface waters in the northern United States, Canada, and Scandinavia and has been implicated in forest decline in central Europe. Sulfur, like nitrogen, exists in several chemical forms. In the hydrosphere, sulfur is present in an oxidized form as sulfate, $S(VI)O_4^{2-}$. In the lithosphere, oxidized sulfur occurs in sulfate-containing minerals such as gypsum, $CaSO_4 \cdot 2H_2O$, and reduced sulfur in sulfide minerals such as chalcopyrite, $CuFeS(-II)_2$. (NB: the roman numerals indicate the oxidation state of sulfur as discussed in Chapter 7.)

Although we have discussed each of the elemental cycles individually, they are strongly interdependent. This is obvious when we consider that carbon fixation by primary producers (i.e., photosynthetic organisms) is accompanied by fixation of other inorganic nutrient elements in the approximate ratio $C_{106}H_{263}O_{110}N_{16}P_1$. This ratio, which describes the average proportions of the major elements in biomass (excluding wood), is termed the Redfield ratio. Thus, particularly for the parts of the global cycles that directly involve the biota or biological detritus, the interdependence of the elemental cycles must also be considered.

Obviously, we are dealing with a system of enormous complexity. This is true even if we limit our attention to the hydrosphere and its direct interactions with other global reservoirs, as is done in this text. We have seen that the principle of elemental conservation, although useful, is not sufficient, for we must be able to treat the transformation, or lack of transformations, between different chemical forms of an element.

In this text we consider reactions occurring in water or at the interfaces between water and the biota, the atmosphere, or sediments in terms of the

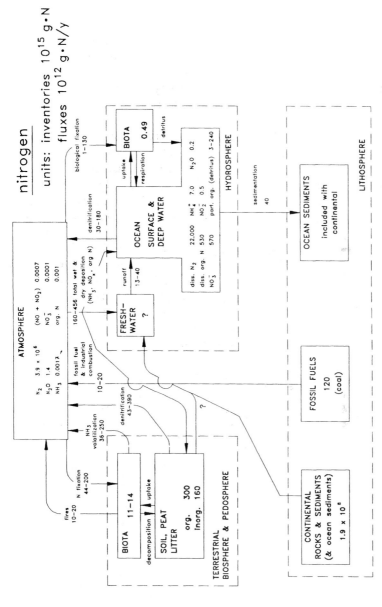

Figure 1.4 Global nitrogen cycle: distribution and fluxes of nitrogen. Units: fluxes in 10^{12} g N/year, inventories in 10^{15} g N. Contributions of various inorganic nitrogen species, for example nitrate (NO_3^-) or ammonia (NH_3), are noted individually. Data from Bolin and Cook (1983).[2]

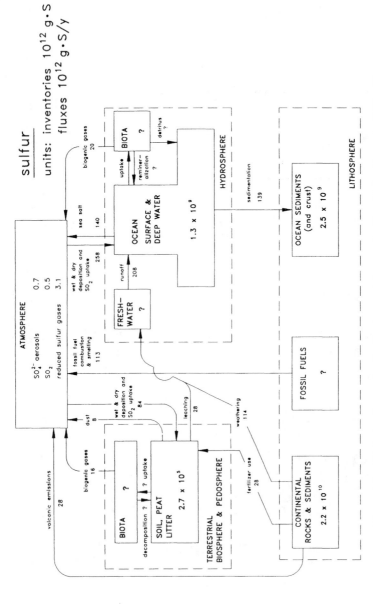

Figure 1.5 Global sulfur cycle: distribution and fluxes of sulfur. Units: fluxes in 10^{12} g S/year, inventories in 10^{12} g S. Data from Bolin and Cook (1983).[2]

effects of such reactions on the composition of the aqueous phase. The reactions we examine include those that result in the partitioning of chemical species between water and other compartments or in the transformation of dissolved chemical species. A concise and consistent description of interacting species and their reactions should lend greater insight into the global cycles outlined in this chapter.

2. MOLE BALANCE EQUATIONS

2.1 Preliminary Notion

Having considered the global transfers and transformations of some elements, let us now focus on 1 L of a natural water conceptually isolated to examine its chemistry. As a first approximation of the real thing we imagine that 100 mg of $CaCO_3$ and 48.4 mg of CO_2 have been dissolved in this model liter of water. To discuss the application of the conservation principle to this system we need to know what changes (i.e., what chemical reactions) may be occurring and what chemical species are present. In this example we assume that the dissociation of water ($H_2O = H^+ + OH^-$) and the formation of bicarbonate from CO_2 ($CO_2 + OH^- = HCO_3^-$) are the only reactions occurring and that H^+, OH^-, CO_2, HCO_3^-, and Ca^{2+} are the only species in solution. The dissolution of $CaCO_3$ in CO_2-bearing water ($CaCO_3 + CO_2 + H_2O \rightarrow Ca^{2+} + 2HCO_3^-$) is deemed to have gone to completion and is not "occurring" in the water. This reaction is implicitly included in the description of the system by considering the species Ca^{2+} and HCO_3^- to exist in the system but not $CaCO_3$.

A complete chemical definition of this model system thus includes a recipe, a list of reactions, and a list of species.

Recipe $\qquad [H_2O]_T = 55.4\,M$ (molarity of water in water at 25°C:

$$997\,g\,L^{-1}/18.01\,g\,mol^{-1} = 55.4\,M)$$

$$[CaCO_3]_T = 100\,mg\,L^{-1}/100\,g\,mol^{-1} = 1\,mM$$

$$[CO_2]_T = 48.4\,mg\,L^{-1}/44\,g\,mol^{-1} = 1.1\,mM$$

Reactions $\qquad\qquad H_2O = H^+ + OH^- \qquad\qquad (1)$

$$CO_2 + OH^- = HCO_3^- \qquad\qquad (2)$$

Species $\qquad H_2O, H^+, OH^-, CO_2, HCO_3^-, Ca^{2+}$

(The molar concentrations of the species are symbolized by brackets: $[H_2O]$, $[H^+]$, etc.)

A mathematical description of mass conservation in this system corresponds to a statement that transmutation does not happen. Thus the concentration of each element present in the various species in the system is equal to the concentration of the element present in the compounds included in the recipe of the system:

$$TOTH = 2[H_2O] + [H^+] + [OH^-] + [HCO_3^-]$$
$$= 2[H_2O]_T \simeq 110.8 \, M \tag{3}$$

$$TOTO = [H_2O] + [OH^-] + 2[CO_2] + 3[HCO_3^-]$$
$$= [H_2O]_T + 3[CaCO_3]_T + 2[CO_2]_T \simeq 55.4 \, M \tag{4}$$

$$TOTC = [CO_2] + [HCO_3^-]$$
$$= [CaCO_3]_T + [CO_2]_T = 2.1 \times 10^{-3} \, M \tag{5}$$

$$TOTCa = [Ca^{2+}]$$
$$= [CaCO_3]_T = 10^{-3} \, M \tag{6}$$

The mole balance Equations 3, 4, 5, and 6, based on the principle of elemental conservation, provide a perfectly good description of mass conservation.

Suppose, however, that we now add $10^{-4} \, M$ of dissolved oxygen to our basic recipe and that no further reaction occurs. The recipe is modified by adding $[O_2]_T = 10^{-4} \, M$ and the species list by adding $[O_2 \cdot aq]$. The conservation equations for H, C, and Ca (Equations 3, 5, and 6) are obviously unchanged and that for O becomes

$$TOTO = [H_2O] + [OH^-] + 2[CO_2] + 3[HCO_3^-] + 2[O_2 \cdot aq]$$
$$= [H_2O]_T + 3[CaCO_3]_T + 2[CO_2]_T + 2[O_2]_T$$
$$\simeq 55.4 \, M \tag{7}$$

Although Equations 3, 5, 6, and 7 are still perfectly valid, they no longer provide an adequate expression of conservation in this system. Since the oxygen that has been added has not reacted, it must be conserved and the equation

$$[O_2 \cdot aq] = [O_2]_T = 10^{-4} \, M \tag{8}$$

must also be applicable. The conservation equation (8) is independent of the four equations of elemental conservation (3, 5, 6, 7) which are thus insufficient to describe mass conservation in the new system.

As seen in this simple example, conservation of elements does provide a first intuitive notion for conservation equations in chemical systems: mole balance equations are obtained by writing that what is present in the system (species concentrations) is somehow equal to what is put into the system (recipe). However, in some cases the equations resulting from elemental conservation are insufficient to describe properly the conservation principle, and in others they are in fact redundant. There is thus more generality and subtlety to mole balance equations than mere elemental conservation.

We are about to embark on an introduction to the *Tableau* method (adopting the French word and its plural, *tableaux*, for this specific purpose). This formalism may initially seem both pedestrian and counterintuitive to some. Our eventual reward will be the ease with which we shall solve complex equilibrium problems. There is an elegance and simplicity to the *Tableau* method that belies its initial awkwardness.

2.2 Definition of Components

Because "balancing" chemical reactions is prima facie an expression of the conservation of elemental mass, chemical elements are seen intuitively as the fundamental conservative entities. As chemical accountants, we take grams or moles of hydrogen, carbon, oxygen and calcium as our basic currencies. To dispel this misconception requires almost an unlearning of chemistry, viewing it (as did J. W. Gibbs) without a complete knowledge of elements or elemental composition. For example, when we write the chemical reaction

$$CO_2 + OH^- = HCO_3^- \tag{9}$$

we demonstrate more knowledge than the reaction itself is supposed to formalize. If asked, at this point, for the "formula" of bicarbonate, we would undoubtedly write:

$$bicarbonate = 1 \text{ hydrogen atom} + 1 \text{ carbon atom}$$
$$+ 3 \text{ oxygen atoms (and 1 negative charge)}$$

This is a direct translation of our symbolic chemical formalism, a piece of information truly extraneous to the writing of the chemical reaction itself which expresses only that 1 carbon dioxide molecule and 1 hydroxide ion combine to form 1 bicarbonate ion. The corresponding formula is

$$bicarbonate = carbon \text{ dioxide} + hydroxide \text{ ion}$$

that is,

$$HCO_3^- = (CO_2)_1 (OH^-)_1$$

The correct stoichiometry for the reaction expresses the conservation of CO_2 and OH^-, not that of hydrogen, carbon, and oxygen. We need not even know that these elements exist.

In the same way, Reaction 1 expresses the conservation of H^+ and OH^- and defines the stoichiometric formula of water as

$$water = (H^+)_1 (OH^-)_1$$

not as $(H)_2(O)_1$. Thus all the species in our model system can be expressed as stoichiometric formulae of the components H^+, OH^-, CO_2, and Ca^{2+}, rather than the chemical elements,

$$H_2O = (H^+)_1 (OH^-)_1$$
$$H^+ = (H^+)_1$$
$$OH^- = (OH^-)_1$$
$$HCO_3^- = (CO_2)_1 (OH^-)_1$$
$$Ca^{2+} = (Ca^{2+})_1$$
$$CO_2 = (CO_2)_1$$

and so can the chemicals used in the recipe of the system:

$$H_2O = (H^+)_1 (OH^-)_1$$
$$CaCO_3 = (Ca^{2+})_1 (CO_2)_1 (OH^-)_1 (H^+)_{-1}$$
$$CO_2 = (CO_2)_1$$

The corresponding mole balance equations expressing conservation in the system are then:

$$TOTH^+ = [H_2O] + [H^+] = -[CaCO_3]_T + [H_2O]_T \approx 55.4M \tag{10}$$

$$TOTOH^- = [H_2O] + [OH^-] + [HCO_3^-] = [CaCO_3]_T + [H_2O]_T \simeq 55.4 \ M \tag{11}$$

$$TOTCO_2 = [CO_2] + [HCO_3^-] = [CaCO_3]_T + [CO_2]_T = 2.1 \times 10^{-3} \ M \tag{12}$$

$$TOTCa^{2+} = [Ca^{2+}] = [CaCO_3]_T = 10^{-3} M \tag{13}$$

where each of the quantities above represents a concentration in moles per liter. A compact and convenient way to represent the relationship between components, species, formulae, and mole balance equations is to organize the stoichiometric data in the form of a "tableau."

As illustrated in Tableau 1.1, a tableau is a matrix of stoichiometric coefficients in which the components define the columns and the species present in the system (top part) and those used in its recipe (bottom part) define the lines. The stoichiometric formulae of the species as a function of the components are written horizontally (lines); the coefficients of the mole balance equations are then given vertically (columns), as can be seen by comparing the first column with Equation 10, the second with Equation 11, the third with Equation 12, and the fourth with Equation 13.

With this example in mind, we now define components as *a set of chemical entities that permits a complete description of the stoichiometry of a system.* Two

TABLEAU 1.1

		Components			
		H^+	OH^-	CO_2	Ca^{2+}
Species	H_2O	1	1		
	H^+	1			
	OH^-		1		
	CO_2			1	
	HCO_3^-		1	1	
	Ca^{2+}				1
Recipe	$CaCO_3$	-1	1	1	1
	CO_2			1	
	H_2O	1	1		

equivalent concepts are meant by this definition:

1. The mole balances corresponding to the components must provide a complete expression of conservation for the chemical system.
2. Components must provide a complete and unique stoichiometric formula for each chemical species in the system.

For example, addition of $10^{-4} M$ O_2 in the recipe of our model system adds the species O_2, which cannot be expressed as a formula of the chosen components. Thus another component, such as O_2 itself, must be added to the chosen set and the corresponding mole balance equation added to the list.

In any closed chemical system (defined as a set of chemical species and a set of chemical reactions among these species) at fixed temperature and pressure, the only possible changes are changes in the concentration of species due to the progress of one or several reactions. The conservative quantities in such a system are thus defined by the set of reactions considered. The chemical components that we can choose as basic accounting currencies, to insure that mass is conserved, are then strictly subordinate to the possible reactions among the species, not to the elemental composition of the species. Only if the formation or dissociation of the species from or into their elements were considered as possible reactions in the system, would the elements themselves automatically become possible choices for components. All the stoichiometric information relevant to mass conservation in a chemical system is contained in the stoichiometry of the possible reactions—no more, no less.

To emphasize this point consider a system where the reactions of interest include the radioactive decay of thorium (228) into lead (208) (through short-lived radon and polonium intermediates). In this situation neither of the elements lead or thorium are conserved, but their sum is. We cannot write expressions of conservation for these elements, but we can write one for the *component*

Pb with which we define the stoichiometric formulae, $Pb = (Pb)_1$ and $Th = (Pb)_1$, according to the radioactive decay reaction.

We also note in passing that, with respect to mass conservation, chemical reactions are strictly stoichiometric relationships: so many moles of each reactant yield so many moles of each product. The actual reaction process, the path of the reaction, might be very different from that implied by its symbolic representation. For example, one or more intermediate compounds not explicitly shown may be formed during the reaction. As a consequence, chemical reactions may be added together, or subtracted from each other, to yield new reactions equally valid for the chemical system under consideration—validity being determined from the point of view of mass conservation and energetic principles.

3. PROPERTIES OF COMPONENTS

3.1   Independence of Components

If in the previous example we had replaced the second reaction by an equivalent one showing the formation of bicarbonate by the hydration of CO_2 and subsequent proton (hydrogen ion) release,

$$CO_2 + H_2O = HCO_3^- + H^+ \tag{14}$$

we might have written the formula for HCO_3^- as

$$HCO_3^- = (CO_2)_1 (H_2O)_1 (H^+)_{-1}$$

and decided to include H_2O in our list of components: H^+, OH^-, CO_2, H_2O, and Ca^{2+}.

We would then have been faced with the dilemma of expressing H_2O either as $(H_2O)_1$ or as $(H^+)_1 (OH^-)_1$, and HCO_3^- either as $(CO_2)_1 (H_2O)_1 (H^+)_{-1}$ or as $(CO_2)_1 (OH^-)_1$. The stoichiometric representation of the system as a function of our components would not have been unique and we would not have had a proper accounting system to express conservation of mass.

Although this necessity of "independence" of components does not always take such an obvious form, we can make it a general rule: *no component should be expressible as a formula of other components; or equivalently: a set of components should yield a unique formula for each species in the system.**

Note also that if we had considered all three possible reactions simultaneously,

$$H^+ + OH^- = H_2O$$
$$CO_2 + OH^- = HCO_3^-$$
$$CO_2 + H_2O = HCO_3^- + H^+$$

---

*For mathematically inclined readers it becomes clear at this point that we are defining a set of components as a basis in the vector space of chemical species. This linear algebraic nature of chemical systems will be transparent throughout this chapter and the next.

our choice of components and the corresponding mole balance equations in Tableau 1.1 would still be valid. Indeed the third reaction provides no independent stoichiometric information; it is merely obtained by subtracting the first reaction from the second.

3.2 Alternative Choices of Components

In the previous example, we cannot choose H^+, OH^-, and H_2O together as components because they are not independent of each other. We can, however, choose any two of these three; thus H^+, H_2O, CO_2 and Ca^{2+} is an acceptable alternative set of components. According to Tableau 1.2, the corresponding mole balance equations are written as follows:

$$TOTH^+ = [H^+] - [OH^-] - [HCO_3^-] = -2[CaCO_3]_T = -2 \times 10^{-3} M \tag{15}$$

$$TOTH_2O = [H_2O] + [OH^-] + [HCO_3^-] = [CaCO_3]_T + [H_2O]_T \simeq 55.4 M \tag{16}$$

$$TOTCO_2 = [CO_2] + [HCO_3^-] = [CaCO_3]_T + [CO_2]_T = 2.1 \times 10^{-3} M \tag{17}$$

$$TOTCa^{2+} = [Ca^{2+}] = [CaCO_3]_T = 10^{-3} M \tag{18}$$

In the same way, OH^-, H_2O, CO_2, and Ca^{2+} is a proper component set which leads to Tableau 1.3 and the following mole balances:

$$TOTOH^- = -[H^+] + [OH^-] + [HCO_3^-] = 2[CaCO_3]_T = 2 \times 10^{-3} M \tag{19}$$

$$TOTH_2O = [H_2O] + [H^+] = -[CaCO_3]_T + [H_2O]_T \simeq 55.4 M \tag{20}$$

$$TOTCO_2 = [CO_2] + [HCO_3^-] = [CaCO_3]_T + [CO_2]_T = 2.1 \times 10^{-3} M \tag{21}$$

$$TOTCa^{2+} = [Ca^{2+}] = [CaCO_3]_T = 10^{-3} M \tag{22}$$

Since they are alternative expressions of conservation for the same chemical system, the mole balance equations corresponding to Tableaux 1.1 (Equations 10–13), 1.2 (Equations 15–18), and 1.3 (Equations 19–22) must be mathematically equivalent to one another. We can verify that they are indeed linear combinations of each other. For example, Equation 15 is obtained by subtracting Equation 11 from Equation 10, and Equation 19 by changing the sign of Equation 15.

If we consider a given component, say H^+ in our current example, the corresponding mole balance equation is strictly dependent on the choice of other components. By inspection of the first column of Tableaux 1.1 and 1.2, we can see that the $TOTH$ equations are quite different from each other, the only coefficient in common being that of H^+ itself.

Vice versa, two different sets of components can yield an identical mole balance equation. For example, the mole balances for H^+ in Tableaux 1.2 and for OH^- in Tableau 1.3 are identical, after adjustment of the sign.

TABLEAU 1.2

		Components			
		H^+	H_2O	CO_2	Ca^{2+}
Species	H_2O		1		
	H^+	1			
	OH^-	−1	1		
	CO_2			1	
	HCO_3^-	−1	1	1	
	Ca^{2+}				1
Recipe	$CaCO_3$	−2	1	1	1
	CO_2			1	
	H_2O		1		

TABLEAU 1.3

		Components			
		OH^-	H_2O	CO_2	Ca^{2+}
Species	H_2O		1		
	H^+	−1	1		
	OH^-	1			
	CO_2			1	
	HCO_3^-	1		1	
	Ca^{2+}				1
Recipe	$CaCO_3$	2	−1	1	1
	CO_2			1	
	H_2O		1		

Components and their corresponding mole balance equations are thus defined as a set, not as individuals, and any individual species can always be made a component.

3.3 Components Need Not Be Species

So far, to stay within what is chemically acceptable, we have chosen components that were also existing species in the system. Such a limitation is not in fact necessary and a chemical entity such as CO_3^{2-}, which we have not considered as an existing species in the system, can quite properly be chosen as a component, as shown in Tableau 1.4.

To go further, there is in fact no need for components to have any sort of chemical significance as exemplified in Tableau 1.5.

TABLEAU 1.4

		Components			
		H^+	H_2O	CO_3^{2-}	Ca^{2+}
Species	H_2O		1		
	H^+	1			
	OH^-	-1	1		
	CO_2	2	-1	1	
	HCO_3^-	1		1	
	Ca^{2+}				1
Recipe	$CaCO_3$			1	1
	CO_2	2	-1	1	
	H_2O		1		

TABLEAU 1.5

		Components			
		H_5O^{3+}	CaO_2^{2-}	CH_6^{10+}	CaH_2^{4+}
Species	H_2O	1/2	1/4		$-1/4$
	H^+	1/6	$-1/12$		1/12
	OH^-	1/3	1/3		$-1/3$
	CO_2	$-2/3$	4/3	1	$-4/3$
	HCO_3^-	$-1/3$	5/3	1	$-5/3$
	Ca^{2+}	$-1/3$	1/6		5/6
Recipe	$CaCO_3$	$-5/6$	23/12	1	$-11/12$
	CO_2	$-2/3$	4/3	1	$-4/3$
	H_2O	1/2	1/4		$-1/4$

The components $H_5O^{3+}, CaO_2^{2-}, CH_6^{10+}$, and CaH_2^{4+} correspond to no chemical reality whatsoever, yet they define a perfectly acceptable accounting currency for the species in our system. Because an accounting method only requires information relating to the number of moles, not to actual chemical processes, the components and the stoichiometric formulation of the species need not have any structural or chemical meaning. Note that the six "reactions" corresponding to each line of the tableau have no more chemical significance than the components that they contain. However, by linear combination of these six "reactions," the original reactions may be readily obtained.

3.4 Number of Components

It is apparent at this point that the number of components (N_C) for a given chemical system is independent of the particular set of components that is chosen, and that it is in fact equal to the number of species (N_S) minus the number of (independent) chemical reactions (N_R) among these species. Since each reaction defines a stoichiometric relationship among species and allows the expression of one of them as a formula of the others, the minimum number of species necessary to formulate all others is precisely $N_C = N_S - N_R$. There is one caveat to this rule; the reactions must be stoichiometrically independent; that is, no reaction must be obtained as a linear combination of others. Even if they correspond to separate chemical processes taking place simultaneously in a system, reactions or sets of reactions that contain the same stoichiometric information are redundant from the point of view of mass conservation. For example, of the seven reactions

$$CO_2(g) = CO_2(aq) \tag{23}$$

$$CO_2(aq) + H_2O = H_2CO_3 \tag{24}$$

$$CO_2(aq) + OH^- = HCO_3^- \tag{25}$$

$$CO_2(aq) + OH^- = CO_3^{2-} + H^+ \tag{26}$$

$$H_2CO_3 = HCO_3^- + H^+ \tag{27}$$

$$HCO_3^- = CO_3^{2-} + H^+ \tag{28}$$

$$H_2O = H^+ + OH^- \tag{29}$$

only five are independent, yielding three components for this system (eight species). In this example, if the reaction between the gas and the aqueous phase did not take place $[CO_2(g) = CO_2(aq)]$, say for reasons of slow kinetics, the system would include four components, not three. The gaseous species $CO_2(g)$ would not be related to any of the aqueous species and two separate mole balance equations would have to be written for carbon, one in the gas and one in the aqueous phase.

3.5 Rules for Choosing Components

Consider the mole balance equations for H^+ and OH^- corresponding to Tableau 1.1:

$$TOTH^+ = [H_2O] + [H^+] \simeq 55.4\,M \tag{30}$$

$$TOTOH^- = [H_2O] + [OH^-] + [HCO_3^-] \simeq 55.4\,M \tag{31}$$

Reasoning that HCO_3^- cannot exceed 0.0011 M (the total carbon concentration in the system), and arguing that the hydrogen and hydroxide ion concentrations,

$[H^+]$ and $[OH^-]$, are always much smaller than that of water, we can neglect the small concentrations and simplify both equations to

$$[H_2O] \simeq 55.4\,M$$

This is hardly a surprising result for a dilute aqueous solution. However, in the process of making these approximations, which are perfectly appropriate and often occur by simple round-off errors, critical information has been lost regarding the conservation of hydrogen and hydroxide ions in the system. The reason is seen by rearranging $TOTOH^-$ Equation 31:

$$[OH^-] + [HCO_3^-] = 55.4 - [H_2O]$$

This is an expression of a small number as a difference between two large numbers, a situation typically leading to poor numerical behavior. One way out of this difficulty is to subtract the two equations:

$$TOTH^+ - TOTOH^- = [H^+] - [OH^-] - [HCO_3^-] = -2[CaCO_3]_T$$
$$= -2 \times 10^{-3}\,M \tag{32}$$

thus obtaining a numerically convenient equation.

To eliminate the possibility of errors, it would be helpful to obtain this equation directly without having to examine the mole balance equations in detail and to manipulate them. Since we want to eliminate the species $[H_2O]$—whose high concentration swamps all others—from all mole balance equations but one, a simple solution is to choose H_2O itself as one of the components. The formula of H_2O is then simply $H_2O = (H_2O)_1$ and, in the corresponding tableau, the species H_2O has a zero coefficient in all columns except that of the component H_2O. Herein lies the true elegance of components: any component set provides a simple and systematic way of writing a complete set of mole balance equations; in addition, a judicious choice of components leads directly to a convenient set of mole balance equations. For example, in Tableau 1.2 which included H_2O as a component, we obtained directly the $TOTH^+$ equation in the desired form (15 and 32). Modification of the set of components results in new mole balance expressions that are more convenient linear combinations of the old ones.

Because we are dealing with dilute solutions, the reasoning carried out for this example is always applicable and H_2O is always chosen as a component. Further, the resulting mole balance equation has a rather trivial solution ($[H_2O] = 55.4\,M$), and since the degree of hydration of the various species (i.e., their stoichiometric coefficient for H_2O) need not be considered for calculating the composition of the system, H_2O is omitted from the tableaux. (It is good practice, however, to include it explicitly until one is familiar with the use of components.) In other words, an arbitrary number of H_2O's is implicitly included in the formulae of the various species. For example, we write

$$OH^- = (H^+)_{-1}$$

instead of

$$OH^- = (H^+)_{-1}(H_2O)_1$$

Together with this choice of H_2O as an "understood" component, H^+ is also chosen as a component systematically, explicitly, and strictly by convention. The corresponding proton conservation equation, $TOTH^+$, will be the focus of much of our attention.

We can now summarize the various rules we have enunciated for choosing components for a particular chemical system.

In a proper component set:

1. *All species can be expressed stoichiometrically as a function of the components, the stoichiometry being defined by the chemical reactions.*
2. *Each species has a unique stoichiometric expression as a function of the components.*

A necessary, but not sufficient, condition to fulfill these requirements is that the number of components be equal to the number of species minus the number of independent reactions considered to take place in the system.

In addition, to obtain a convenient set of mole balance equations:

3. H_2O *should always be chosen as a component.* Because the corresponding mole balance equation, and the corresponding column in the tableau yield no useful information, in this text H_2O is usually omitted from the tableau. It is however always included implicitly in the component set and it may be good practice for beginners to include it explicitly in the tableau.
4. H^+ *should always be chosen as a component.* This is arbitrary but convenient, as expounded on in Chapters 2 and 4.

For the other components, unless there are reasons to do otherwise, the simplest "building blocks" (basic components) can be chosen. In addition, if we wish for some reason to isolate a particular species in one equation, it is sufficient to choose the corresponding species as a component. Other practical rules are derived in Chapter 2.

3.6   The Electroneutrality Condition

Since our model system (like any real chemical system) has been made up of neutral chemical constituents and no electrical charge can be created or lost in any of the reactions, the electroneutrality equation expressing the balance of

TABLEAU 1.6

		H^+	CO_2	$Ca(OH)_2$	(H_2O)
			Components		
Species	H_2O				(1)
	H^+	1			
	OH^-	-1			(1)
	CO_2		1		
	HCO_3^-	-1	1		(1)
	Ca^{2+}	2		1	(-2)
Recipe	$CaCO_3$		1	1	(-1)
	CO_2		1		
	H_2O				(1)

positive and negative charges in the system must be verified:

$$[H^+] + 2[Ca^{2+}] = [OH^-] + [HCO_3^-] \tag{33}$$

If our set of mole balance Equations 10–13 is truly complete, it must somehow contain Equation 33. This is verified by subtracting Equation 11 from Equation 10 and adding twice Equation 13.

We should be able to make a choice of components that would directly yield the electroneutrality expression as one of the mole balance equations. Keeping with our convention to use H^+ as a component, we can insure that each species' coefficient in the H^+ column of the tableau equals its electrical charge by choosing all other components (besides H^+) to be neutral. Under these conditions, the $TOTH^+$ equation should then logically be the electroneutrality expression. Let us choose, for example, the components H^+, CO_2, $Ca(OH)_2$, (and H_2O!) as shown in Tableau 1.6.

The corresponding $TOTH^+$ expression is indeed the sought-for electroneutrality condition: $TOTH^+ = [H^+] - [OH^-] - [HCO_3^-] + 2[Ca^{2+}] = 0$.

Note again that in this tableau the chemical formulae of the various species as given by the lines do not strictly correspond to chemical reactions since the components are not species actually present in the system.

4. AN EXAMPLE OF CHEMICAL ACCOUNTING

To illustrate how to apply our rules for choosing components, set up a tableau, and write a complete set of mole balance equations, we now consider a somewhat more realistic model of a natural water. To our original recipe we add $10\,mM$ of a background salt, NaCl, and $0.4\,mM$ of an organic acid of undefined composition, denoted HA. To the list of species we add Na^+, Cl^-, HA, and A^-, and also two carbonate species we have so far ignored, carbonic acid H_2CO_3 (hydrated CO_2) and the carbonate ion CO_3^{2-}. The complete recipe, list of chemical species, and set of chemical reactions are given as follows:

Recipe $[CaCO_3]_T = 10^{-3} M$; $[CO_2]_T = 1.1 \times 10^{-3} M$; $[NaCl]_T = 10^{-2} M$;
$[HA]_T = 4 \times 10^{-4} M$ (and implicitly $[H_2O]_T = 55.4 M$)

Species H^+, OH^-, Na^+, Cl^-, Ca^{2+}, HA, A^-, CO_2, H_2CO_3, HCO_3^-, CO_3^{2-},
(and implicitly H_2O)

Reactions

$$H_2O = H^+ + OH^- \tag{34}$$

$$HA = H^+ + A^- \tag{35}$$

$$CO_2 + H_2O = H_2CO_3 \tag{36}$$

$$H_2CO_3 = H^+ + HCO_3^- \tag{37}$$

$$CO_2 + OH^- = HCO_3^- \tag{38}$$

$$HCO_3^- = H^+ + CO_3^{2-} \tag{39}$$

$$H_2CO_3 = 2H^+ + CO_3^{2-} \tag{40}$$

Before proceeding we inspect the list of reactions to make sure they are independent. They are not. For example, Equation 40 is merely the addition of Equations 37 and 39. After some thoughtful elimination we are left with five *independent reactions*:

$$H_2O = H^+ + OH^- \tag{34}$$

$$HA = H^+ + A^- \tag{35}$$

$$CO_2 + H_2O = H_2CO_3 \tag{36}$$

$$H_2CO_3 = H^+ + HCO_3^- \tag{37}$$

$$HCO_3^- = H^+ + CO_3^{2-} \tag{39}$$

Given a list of 11 species and 5 independent reactions, we are seeking a set of 6 components (not counting H_2O as either a species or a component) from which all species can be expressed in a unique way.

Besides H^+, which is always taken as a component, a rather obvious choice is given by the *basic* components Na^+, Cl^-, Ca^{2+}, A^-, and CO_3^{2-}. The resulting Tableau 1.7 yields the mole balances Equations 41–46 corresponding to each

TABLEAU 1.7

	H^+	Na^+	Cl^-	Ca^{2+}	A^-	CO_3^{2-}	(H_2O)
H^+	1						
OH^-	-1						(1)
Na^+		1					
Cl^-			1				
Ca^{2+}				1			
A^-					1		
HA	1				1		
CO_3^{2-}						1	
HCO_3^-	1					1	
H_2CO_3	2					1	
CO_2	2					1	(-1)
(H_2O)							(1)
$CaCO_3$				1		1	
CO_2	2					1	(-1)
NaCl		1	1				
HA	1				1		

column. These mole balance equations provide a complete expression of mass conservation in this system:

$$TOTH = [H^+] - [OH^-] + [HA] + [HCO_3^-] + 2[H_2CO_3] + 2[CO_2]$$
$$= 2[CO_2]_T + [HA]_T = 2.6 \times 10^{-3} \qquad (41)$$

$$TOTNa = [Na^+] = [NaCl]_T = 10^{-2} \qquad (42)$$

$$TOTCl = [Cl^-] = [NaCl]_T = 10^{-2} \qquad (43)$$

$$TOTCa = [Ca^{2+}] = [CaCO_3]_T = 10^{-3} \qquad (44)$$

$$TOTA = [A^-] + [HA] = [HA]_T = 4 \times 10^{-4} \qquad (45)$$

$$TOTCO_3 = [CO_3^{2-}] + [HCO_3^-] + [H_2CO_3] + [CO_2]$$
$$= [CaCO_3]_T + [CO_2]_T = 2.1 \times 10^{-3} \qquad (46)$$

For simplicity, the charges are now omitted in the designation of the mole balance equations ($TOTH$, $TOTNa$, etc.) and the molar units from the numerical values of the concentrations (10^{-2} for $10^{-2}\,M$). It is worth pointing out once more how different these mole balances are from expressions of elemental conservation. For example, $TOTA$ expresses the conservation of an organic anion which may contain only carbon, oxygen, and hydrogen (e.g., CH_3COO^-).

TABLEAU 1.8

	H^+	Na^+	Cl^-	Ca^{2+}	A^-	HCO_3^-
H^+	1					
OH^-	-1					
Na^+		1				
Cl^-			1			
Ca^{2+}				1		
A^-					1	
HA	1				1	
CO_3^{2-}	-1					1
HCO_3^-						1
H_2CO_3	1					1
CO_2	1					1
$CaCO_3$	-1			1		1
CO_2	1					1
NaCl		1	1			
HA	1				1	

Thus there are seven components (including H_2O) and only six elements (C, H, O, Na, Cl, Ca). Note also that in the $TOTH$ expression, the coefficient of $[H_2CO_3]$ is 2 while that for $[OH^-]$ is -1, illustrating again that the coefficients of the mole balance equations are not the stoichiometric coefficients of the elements but those of the components.

Among the possible choices of components the most obvious and convenient consist of species present in the system. Here, besides our present choice, convenient sets of components are those in which we replace A^- by HA, or CO_3^{2-} by HCO_3^- or H_2CO_3 (or equivalently CO_2). As an example, consider the set of components: H^+, Na^+, Cl^-, Ca^{2+}, A^-, and HCO_3^-. The corresponding Tableau 1.8 is identical to the previous one (1.7) except for the coefficients in the H^+ column. Thus only the $TOTH$ equation is modified. The corresponding sets of mole balance Equations 47–52 are equivalent mathematically, but not numerically to Equations 41–46. One system may yield more readily an approximate hand solution than the other or be more efficiently solved by a computer algorithm. As discussed in Chapter 2, choosing a set of components that yields a numerically convenient $TOTH$ equation is most of the art in calculating the equilibrium composition of a system.

Mole balance equations:

$$TOTH = [H^+] - [OH^-] + [HA] - [CO_3^{2-}] + [H_2CO_3] + [CO_2]$$

$$= -[CaCO_3]_T + [CO_2]_T + [HA]_T = 5 \times 10^{-4} \tag{47}$$

$$TOTNa = [Na^+] = [NaCl]_T = 10^{-2} \tag{48}$$

$$TOTCl = [Cl^-] = [NaCl]_T = 10^{-2} \tag{49}$$

$$TOTCa = [Ca^{2+}] = [CaCO_3]_T = 10^{-3} \tag{50}$$

$$TOTA = [A^-] + [HA] = [HA]_T = 4 \times 10^{-4} \tag{51}$$

$$TOTHCO_3 = [CO_3^{2-}] + [HCO_3^-] + [H_2CO_3] + [CO_2]$$
$$= [CaCO_3]_T + [CO_2]_T = 2.1 \times 10^{-3} \tag{52}$$

5. NOTATION, SYMBOLS, UNITS, AND TERMINOLOGY

Throughout Sections 2–4 we used particular units and introduced a few symbols. We also employed the word "component" with a special meaning. Let us compile here these various conventions while extending and explaining them a little more.

5.1 Equal Signs

Equal signs are used here for three different purposes:

1. To equate algebraic or arithmetic quantities
2. To define the stoichiometric formulae of compounds
3. As a shorthand notation for chemical reactions (replacing $\rightleftharpoons$)

For all three types of expressions written with an equal sign, the numerical or chemical entities on each side of the corresponding "equation" can be freely subjected to all the usual linear operations, such as multiplication by a number and addition of equations. See, for example, Table 1.1.

5.2 Types of Concentration

Consider a system made up of a 3-mM concentration of NaCl, a 1-mM concentration of NaOH, and a 2-mM concentration of HCl, yielding the species H^+,

TABLE 1.1

	Chemical Reactions	Equivalent Formulae
$1 \times$	$CO_2 + H_2O = H_2CO_3$	$H_2CO_3 = (CO_2)_1(H_2O)_1$
$-1 \times$	$CO_3^{2-} + 2H^+ = H_2CO_3$	$H_2CO_3 = (CO_3^{2-})_1(H^+)_2$
$2 \times$	$H^+ + OH^- = H_2O$	$H_2O = (H^+)_1(OH^-)_1$
	$CO_2 + 2OH^- = CO_3^{2-} + H_2O$	$CO_3^{2-} = (CO_2)_1(OH^-)_2(H_2O)_{-1}$

OH^-, Na^+, Cl^- (and H_2O). Four different types of concentrations involving sodium may be defined for such a system:

1. $[NaCl]_T = 3 \times 10^{-3} M$ and $[NaOH]_T = 10^{-3} M$. These are simply the molar concentrations of chemical compounds used in the recipe of the system. In this case, neither NaCl nor NaOH is present as a species in the system.

2. $[Na^+]$. This is the molar concentration of the species Na^+ (sodium ion) and its numerical value is not known a priori. Although this is not the case here, other species involving sodium could in principle be present in the system (e.g., $NaCO_3^-$, $NaSO_4^-$), each at its own individual concentration (e.g., $[NaCO_3^-]$, $[NaSO_4^-]$).

3. TOTNa. This is the value of the mole balance equation for Na^+ if Na^+ has been chosen as a component. As before, the symbol for the charge is omitted for simplicity. TOTNa is strictly dependent upon the particular choice of the other components and may be negative or null, although it has dimensions of moles per liter. For example, with the choices of components H^+, Na^+, Cl^- and H^+, Na^+, NaCl, two widely different values are obtained for TOTNa as shown in Tableau 1.9.

$$[TOTNa]_a = [Na^+] = [NaCl]_T + [NaOH]_T$$
$$= 3 \times 10^{-3} + 10^{-3} = 4 \times 10^{-3} \tag{53}$$

$$[TOTNa]_b = [Na^+] - [Cl^-] = [NaOH]_T - [HCl]_T$$
$$= 10^{-3} - 2 \times 10^{-3} = -10^{-3} \tag{54}$$

Note that the mole balance equations consist of two equal quantities: the left-hand side, which is a sum of concentrations of species present in the system (the value of each of these is usually unknown), and the right-hand side, which is a sum of concentrations of chemical compounds used in the recipe of the system.

TABLEAU 1.9

	(a)			(b)		
	H^+	Na^+	Cl^-	H^+	Na^+	NaCl
H^+	1	0	0	1	0	0
OH^-	-1	0	0	-1	0	0
Na^+	0	1	0	0	1	0
Cl^-	0	0	1	0	-1	1
NaCl	0	1	1	0	0	1
NaOH	-1	1	0	-1	1	0
HCl	1	0	1	1	-1	1

4. Na_T. This is the analytical concentration of sodium, not previously introduced in this chapter. The charge is again omitted for simplicity; brackets may be used for clarity whenever the chemical formula is complicated such as $[CO_3]_T$. This concentration, sometimes called the total concentration, is defined as that which would be measured if we analyzed the system for sodium (or carbonate). It is equal to the sum of the concentrations of all species containing sodium—in our case only Na^+—and to the sum of the concentrations of the chemicals containing sodium used in the recipe of the system, if the system is so defined:

$$Na_T = [Na^+] = [NaCl]_T + [NaOH]_T = 4 \times 10^{-3} M \qquad (55)$$

From a comparison of Equations 53 and 55 it is clear that some of the more common choices for components yield the equality

$$TOTNa = Na_T$$

This is true when none of the other components contain the species or element of interest. For example, in Chapter 4 we use chiefly three alternative sets of components to study the carbonate system:

(a) H^+, CO_3^{2-}, + other components containing no carbonate
(b) H^+, HCO_3^-, + other components containing no carbonate
(c) H^+, H_2CO_3, + other components containing no carbonate

Then, for any given system,

$$TOTCO_3 = TOTHCO_3 = TOTH_2CO_3 = [CO_3]_T$$

Note that the concentrations defined in (1) and in (4) are conceptually related and the corresponding notations are similar. The numerical values of the other types of concentrations may sometimes be the same, as illustrated by Equations 53 and 55; however, they are based on different concepts and are designated by clearly different notations.

5.3 Concentration Units

So far, we have expressed all concentrations on the molar scale; we shall continue to do so. However, it is necessary to be aware of other commonly used concentration scales and to understand their interrelationships.

WEIGHT FRACTION

Symbols   % (percent); ‰ (per mil = parts per thousand); ppm (parts per million); ppb (parts per billion).
Dimensions None

Definition   The weight of the species or element of interest per total weight of the system. This is a commonly used analytical scale, particularly useful for solids (dry weight). In water its relation to the molar scale is a function of the molecular weight of the species in question and the density of the solution. In using this scale it is critical to specify the particular species being considered. Contemplate, for example, the not uncommon and truly ambiguous situation in which, say, a 3-ppm concentration of phosphate is reported. As P? As PO_4? As H_3PO_4? Such a unit, which does not provide direct information on the number of chemical entities (atoms, molecules, ions), is inherently less chemically relevant than a unit based on the number of moles.

VOLUME FRACTION

Symbols   %, etc.
Dimensions   None
Definition   The volume of the species of interest per total volume of the system. Used chiefly for liquid mixtures such as concentrated acid solutions, this concentration scale is of little interest in aquatic chemistry.

MOLAL CONCENTRATIONS

Symbols   m, mm, μm, etc.
Dimensions $mol\,kg^{-1}$.
Definition   The number of moles of the species of interest per kilogram of solvent.

MOLAR CONCENTRATION

Symbols M, mM, μM, etc.
Dimensions $mol\,liter^{-1}$.
Definition   The number of moles of the species of interest per liter of solution. The molal scale is in principle thermodynamically preferable to the molar scale because it is independent of the effects of temperature and solutes on the density of the solution or its molar volume. However, the molar scale is analytically more convenient and in aquatic chemistry, which deals mostly with dilute solutions over a small range of temperatures, the two scales are almost equivalent: at 20°C, with a total salt content of less than 3%, molar and molal concentrations differ by less than 1%. For convenience, in this text the symbol M is often omitted, particularly in the derivation of equations. The notation pX is used to indicate the negative logarithm of the molar concentration of X:

$$pX = -\log_{10}[X]$$

ATOM CONCENTRATIONS

Symbols $g\text{-at}\,L^{-1}, mg\text{-at}\,L^{-1}, \mu g\text{-at}\,L^{-1}$, etc.

Dimensions gram-atom liter^{-1}.

Definition The number of gram-atoms of an element per liter of solution. This scale is equivalent to the molar scale, but normalizes the concentration to the number of atoms of an element of interest rather than to the number of moles of species. For example, 1 M urea $(CO(NH_2)_2) = 2\,g\text{-at}\,L^{-1}\,N$. This scale is widely used by biologists interested in elemental ratios.

EQUIVALENT CHARGE CONCENTRATION

Symbols $eq\,L^{-1}, meq\,L^{-1}, \mu eq\,L^{-1}$, etc.

Dimensions Equivalent liter^{-1} (= 96,500 coulombs liter^{-1}).

Definition The number of equivalent charges of a given ion per liter of solution. Again, this scale is similar to the molar scale, but here the molar concentration of a given ion is multiplied by the absolute charge number of the ion. For example,

$$1\,M\,H^+, OH^-, HCO_3^-, \text{ or } NO_3^- = 1\,eq\,L^{-1}$$
$$1\,M\,SO_4^{2-}, Ca^{2+}, \text{ or } CO_3^{2-} = 2\,eq\,L^{-1}$$

MOLE FRACTION

Symbols $\%$, etc.

Dimensions None.

Definition The number of moles of the species of interest per total number of moles in the system. This is the thermodynamic scale par excellence. Its occasional use is restricted to thermodynamic developments.

Other concentration units used for historical reasons or for particular applications are not widely encountered in aquatic chemistry.

5.4 A Matter of Terminology and History

There is some confusion in the literature regarding the meaning of the word "component." It is fairly common practice to limit the use of the word to designate uncharged species, those that can exist as individual salts or pure substances. In this tradition, components are considered to be the minimum set of chemicals that have to be taken off the shelf to duplicate the system under consideration. We shall not follow this practice, which appears to originate from a confusion regarding the meaning that J. W. Gibbs intended for the word "component."

TABLEAU 1.10

	H^+	(H_2O)
H_2O	0	(1)
H^+	1	(0)
OH^-	-1	(1)

To illustrate the differences between these two approaches, consider the simplest of all aqueous solutions, pure water. Three species, H_2O, H^+, and OH^-, and one reaction have to be considered:

$$H_2O = H^+ + OH^-$$

According to our definition, two components must be chosen; according to our newly set rules, these are H_2O and H^+, as shown in Tableau 1.10.

In our approach, two independent mole balance equations can be written, the second one being of little interest:

$$TOTH = [H^+] - [OH^-] = 0 \tag{56}$$

$$TOTH_2O = [H_2O] + [OH^-] \simeq 55.4\,M \tag{57}$$

If we limited the choice of components to chemicals necessary to make up the system, only one component would and could be chosen: H_2O. To obtain Equation 56 we would then impose the electroneutrality condition as an additional constraint on the system, independent of the conservation constraints. This appears somewhat illogical since electroneutrality of the system results simply from the fact that it is originally made up of an electrically neutral compound: H_2O. There are chemical systems or subsystems that are not electrically neutral. Some conservation equation for electrical charge must then obtain in such systems but electroneutrality does not. Conservation of charge (i.e., electroneutrality in a neutral system) is included implicitly in the mole balance equations when the component set is complete and accounts for the electrical charges of the ions.

The historical justification for our use of the word "component" rests with the writings of the man who originally coined its chemical meaning.

In his classical treatise *On the Equilibrium of Heterogeneous Substances*, J. W. Gibbs[3] introduced the notion of component in a way that is very similar to the differential approach of the next section. In the following quote, which may be difficult to fully understand out of context, he defines components as independent entities that can account for all possible variations in the system:

The substances $S_1, S_2 \ldots S_n$ of which we consider the mass composed, must of course be such that the values of the differentials $dm_1, dm_2 \ldots dm_n$ shall be independent, and shall express every possible variation in the composition of the homogeneous mass considered, including those produced by the absorption of substances different from any initially present. It may therefore be necessary to have terms in the equation relating to *component substances* which do not initially occur in the homogeneous mass considered...

Gibbs's arguments are fundamentally mathematical and he is quite clear that the notion of component is not a chemical one:

...the choice of the substances which we are to regard as the components of the mass considered, may be determined entirely by convenience, and independently of any theory in regard to the internal constitution of the mass. The number of components will sometimes be greater, and sometimes less, than the number of chemical elements present.

Writing at a time when the ionic composition of solutions was just beginning to be discovered, he specifically allowed the possibility of choosing ions as components.

...It will be observed that the choice of the substances which we regard as the components of the fluid is to some extent arbitrary.... For our purpose, which has nothing to do with any theories of molecular constitution, we may choose such a set of components as may be convenient, and call those ions...without farther limitation.

6. A DIFFERENTIAL APPROACH TO THE PROBLEM

As mentioned before, it is possible to develop rigorously the concept of chemical components in a mathematical framework. Although we do not wish to be so theoretical, it is instructive to examine the essential aspects of such a development.

In mathematical terms, a conservative quantity is a parameter that remains constant while all possible variations are considered in the system. In other words, it is a quantity whose total differential is identically null. The only possible variations in a closed chemical system at constant pressure and temperature are changes in the concentrations of the species due to the advancement of one or several reactions. Our purpose is then simply to formulate chemical quantities that are invariant with respect to the advancement of the reactions in a given system.

Consider our original example consisting of the six species H_2O, H^+, OH^-, CO_2, HCO_3^-, and Ca^{2+} and the two reactions

$$H_2O = H^+ + OH^- \tag{58a}$$

$$CO_2 + OH^- = HCO_3^- \tag{58b}$$

The conservation principle can be expressed by differential equations relating the changes in the number of moles of each of the species to the advancement of each of the chemical reactions 58a and 58b. This is most simply achieved by defining parameters ξ_1 and ξ_2 corresponding to the advancement, from the left to the right, of Reactions 58a and 58b respectively:

$$\frac{\partial[H_2O]}{\partial\xi_1} = -1; \qquad \frac{\partial[H^+]}{\partial\xi_1} = +1; \qquad \frac{\partial[OH^-]}{\partial\xi_1} = +1; \qquad \text{all other } \frac{\partial[\;]}{\partial\xi_1} = 0$$

$$\tag{59}$$

$$\frac{\partial[CO_2]}{\partial\xi_2} = -1; \qquad \frac{\partial[OH^-]}{\partial\xi_2} = -1; \qquad \frac{\partial[HCO_3^-]}{\partial\xi_2} = +1; \qquad \text{all other } \frac{\partial[\;]}{\partial\xi_2} = 0$$

The parameters ξ_1 and ξ_2 are actually defined here by the above equations themselves and have units of moles per liter. Since Reactions 58a and 58b are the only ones considered to take place, the total differentials are written

$$d[H_2O] = -d\xi_1 \tag{60a}$$

$$d[H^+] = +d\xi_1 \tag{60b}$$

$$d[OH^-] = d\xi_1 - d\xi_2 \tag{60c}$$

$$d[CO_2] = -d\xi_2 \tag{60d}$$

$$d[HCO_3^-] = +d\xi_2 \tag{60e}$$

$$d[Ca^{2+}] = 0 \tag{60f}$$

We can combine these equations to eliminate the variables $d\xi_1$ and $d\xi_2$ as in the following:

$$d[H_2O] + d[H^+] = 0$$
$$d[CO_2] + d[HCO_3^-] = 0$$
$$d[H_2O] + d[OH^-] + d[HCO_3^-] = 0$$
$$d[Ca^{2+}] = 0$$

These exact differentials can now be integrated to yield the desired mole balance equations:

$$[H_2O] + [H^+] = \text{constant } [= TOTH] \tag{61a}$$

$$[CO_2] + [HCO_3^-] = \text{constant } [= TOTCO_2] \tag{61b}$$

$$[H_2O] + [OH^-] + [HCO_3^-] = \text{constant} \ [= TOTOH] \qquad (61c)$$

$$[Ca^{2+}] = \text{constant} \ [= TOTCa] \qquad (61d)$$

Other combinations of Equations 60 to eliminate the variables $d\xi_1$ and $d\xi_2$ would result in different sets of mole balance equations that are linear combinations of Equations 61.

The general problem involving a chemical system with N_S species, each with a concentration (S_i), and N_R reactions, each characterized by a degree of advancement ξ_j, would be obviously cumbersome to approach in this manner. A more elegant solution can be written with matrix algebra, but this is beyond the scope of this chapter. Here it is sufficient to note that we can always find $N_C = N_S - N_R$, and no more than N_C, independent ways to eliminate the $d\xi_j$'s from the N_S total differentials of species concentrations:

$$d[S_i] = \sum \frac{\partial [S_i]}{\partial \xi_j} d\xi_j$$

The coefficients of these N_C linear combinations to eliminate $d\xi_j$'s define the stoichiometry of the compounds. By integration, N_C mole balance equations are obtained that express fully the notion of mass conservation in the system.

Such a differential approach provides a fundamental mathematical definition of components and mole balance equations. However, we have seen that components of a chemical system can usually be defined intuitively and the stoichiometric relations between species, components, and mole balances are readily expressed in the form of a tableau. In effect, this intuitive approach is no more than an efficient utilization of our chemical symbolism, which expresses inherently the notion of mass conservation in the elemental stoichiometry of chemical species.

7. USE OF MOLE BALANCES IN TRANSPORT EQUATIONS

In natural waters, chemical species are subjected to transport by advective and diffusive processes at the same time that they are reacting chemically. In many instances, the concept of components can be used to simplify the problem of simultaneous transport and reactions of chemical species.[4,5] Let us first show the general approach and then illustrate it with a specific example.

NOTATION

t	time
x	one-dimensional coordinate
U	advective velocity in the x direction
$[S_k]$	molar concentration of the kth species; $k = 1, 2, \ldots N_S$
D_k	diffusion coefficient of S_k (turbulent diffusion coefficient if applicable)

TABLEAU 1.11

	C_1	C_2	$\cdots$	C_i	$\cdots$
S_1	α_{11}	α_{12}	$\cdots$	α_{1i}	$\cdots$
S_2	α_{21}	α_{22}	$\cdots$	α_{2i}	$\cdots$
$\vdots$	$\vdots$	$\vdots$		$\vdots$	
S_k	α_{k1}	α_{k2}		α_{ki}	
$\vdots$	$\vdots$	$\vdots$		$\vdots$	

R_{kj} rate of production of S_k in the jth reaction; $j = 1, 2, \ldots N_R$

$d\xi_j$ degree of advancement of the jth reaction

v_{jk} stoichiometric coefficient of S_k in the jth reaction ($v > 0$ for products; $v < 0$ for reactants)

$TOTC_i$ mole balance equation for the component C_i; $i = 1, 2, \ldots N_C$

α_{ki} stoichiometric coefficient of S_k as a function of the component C_i

Tableau 1.11 is implied by this notation. The mole balance equation for the component C_i is thus written

$$TOTC_i = \sum_k \alpha_{ki}[S_k] \tag{62}$$

The condition of conservation for the component C_i in the jth chemical reaction symbolized by

$$0 = v_{j1}S_1 + v_{j2}S_2 + \cdots + v_{jN_S}S_{N_S}$$

results in the identity

$$\sum_k v_{jk}\alpha_{ki} = 0 \tag{63}$$

To keep the notation elementary, let us consider the one-dimensional transport problem and write the conservation equation

$$\frac{\partial[S_k]}{\partial t} = -U\frac{\partial[S_k]}{\partial x} + D_k\frac{\partial^2[S_k]}{\partial x^2} + \sum_j R_{kj} \tag{64}$$

The reaction rates (R_{kj}'s) can have complicated and often poorly known functionalities, so the solution to these N_S coupled differential equations can be very difficult even if the advective velocity field is obtained independently by neglecting the effects of the solutes on the solvent motion—a reasonable approximation in almost all cases.

From our definition of the degree of advancement, ξ_j, of the jth reaction, R_{kj} can be written more explicitly as

$$R_{kj} = \frac{\partial[S_k]}{\partial \xi_j} \frac{d\xi_j}{dt} \tag{65}$$

which results in

$$R_{kj} = v_{jk} \frac{d\xi_j}{dt} \tag{66}$$

Introducing Equation 66 in Equation 64, multiplying by α_{ki}, and summing over k:

$$\sum_k \alpha_{ki} \frac{\partial[S_k]}{\partial t} = -U \sum_k \alpha_{ki} \frac{\partial[S_k]}{\partial x} + \sum_k \alpha_{ki} D_k \frac{\partial^2[S_k]}{\partial x^2} + \sum_k \alpha_{ki} \sum_j v_{jk} \frac{d\xi_j}{dt} \tag{67}$$

Using Equations 62 and 63 and rearranging:

$$\frac{\partial TOTC_i}{\partial t} = -U \frac{\partial TOTC_i}{\partial x} + \sum_k D_k \frac{\partial^2 \alpha_{ki}[S_k]}{\partial x^2} \tag{68}$$

The N_S equations describing the rate of change of the species concentrations have now been replaced by N_C ($< N_S$) equations describing the transport of the component concentrations. The chemical reaction rates have been effectively eliminated from the equations. By judicious choice of components and reasonable approximations, Equation 68 can be made to have particularly simple forms and convenient boundary conditions.

It is, for example, possible to choose components so as to insure that all species in a given mole balance equation are in solution. It is then sometimes a good approximation to take all diffusion coefficients as equal:

$$D_k = D \tag{69}$$

$$\frac{\partial TOTC_i}{\partial t} = -U \frac{\partial TOTC_i}{\partial x} + D \frac{\partial^2 TOTC_i}{\partial x^2} \tag{70}$$

Equation 70 (which could be obtained in three dimensions as well as in one) is surprisingly simple. It expresses the equation of transport of the component C_i as a conservative entity. The only condition for the applicability of such an equation is that the diffusion coefficients be approximately equal for all species in the corresponding mole balance equation.

Let us take as an example the dissociation of mercuric sulfide in a high chloride medium (say, $0.5\,M$ NaCl). The reactions to be considered are:

$$HgS(s) = Hg^{2+} + S^{2-}$$
$$S^{2-} + H^+ = HS^-$$
$$HS^- + H^+ = H_2S(aq)$$
$$H_2S(aq) = H_2S(g)$$
$$H_2O = H^+ + OH^-$$
$$Hg^{2+} + 2Cl^- = HgCl_2$$
$$HgCl_2 + Cl^- = HgCl_3^-$$
$$HgCl_3^- + Cl^- = HgCl_4^{2-}$$

This is a complicated situation with dissolution of a solid phase, loss of a gas phase, and a complex set of coordination reactions in the aqueous phase. The concentration of each species at a given location is a function of the rate of the various reactions and of advective and diffusive processes. It would be described by the differential equation 67 replacing the coefficients by the applicable parameters (if known). Consider, however, the choice of components $H^+, Na^+,$ $Cl^-, H_2S,$ and HgS (and H_2O) shown in Tableau 1.12. The corresponding $TOTH$ equation is given as

$$TOTH = [H^+] - [OH^-] - [HS^-] - 2[S^{2-}] + 2[Hg^{2+}]$$
$$+ 2[HgCl_2] + 2[HgCl_3^-] + 2[HgCl_4^{2-}] = 0 \qquad (71)$$

TABLEAU 1.12

	H^+	Na^+	Cl^-	H_2S	HgS
Na^+		1			
Cl^-			1		
H_2S				1	
$H_2S(g)$				1	
$HgS(s)$					1
H^+	1				
OH^-	-1				
HS^-	-1			1	
S^{2-}	-2			1	
Hg^{2+}	2			-1	1
$HgCl_2$	2		2	-1	1
$HgCl_3^-$	2		3	-1	1
$HgCl_4^{2-}$	2		4	-1	1
$NaCl$		1	1		
HgS					1

If the diffusion coefficients of the individual species included in this equation can be considered as approximately equal, Equation 70 is applicable to $TOTH$. This is a good approximation since all the species of interest are small solutes. Thus in the absence of external sources of acids or bases, $TOTH$ will be conservative in this system and remain equal to zero if it is initially null. It is quite remarkable that in this likely situation, this particular proton balance will remain null at all times and in all places regardless of the rate of dissolution of HgS(s), of the rate of volatilization of $H_2S(g)$, or of the rate of formation of the various mercuric chloride species, and regardless of advective and diffusive processes. With some assumptions on the controlling reactions (i.e., the slow ones) and reasonably simple transport conditions, this equation allows us to calculate all species concentrations as a function of time and place.

REFERENCES

1. E. K. Berner and R. A. Berner, *The Global Water Cycle*, Prentice-Hall, Englewood Cliffs, NJ, 1987, Chapter 2.
2. B. Bolin and R. B. Cook, Eds., *The Major Biogeochemical Cycles and their Interactions*, SCOPE Report 21, Wiley, Chichester, 1983, Chapter 2.
3. J. W. Gibbs, *On the Equilibrium of Heterogeneous Substances: The Collected Works*, Vol. 1, Yale University Press, New Haven, CT, 1906, pp. 63, 93, 332.
4. J. C. Westall, F. M. M. Morel, and D. N. Hume, *Anal. Chem.*, **51**, 1792 (1979).
5. D. M. DiToro, in *Modeling Biochemical Processes in Aquatic Systems*, R. C. Canale, Ed., Ann Arbor Science, Ann Arbor, MI, 1976.

PROBLEMS

1.1   Write the "formulae" for the species indicated in italics in the following chemical reactions using (H_2O), (H^+), and whatever other species involved in each reaction are necessary:

a. $NH_3 + H_2O = NH_4^+ + OH^-$

b. $H_2S = S^{2-} + 2H^+$

c. $Cl_2(g) + H_2O = HOCl + H^+ + Cl^-$

d. $Ca(OH)_2(s) = Ca^{2+} + 2OH^-$

e. $CaCO_3(s) + H_2CO_3 = Ca^{2+} + 2HCO_3^-$

f. $FeCO_3(s) + 2H^+ = Fe^{2+} + CO_2(g) + H_2O$

g. $Zn^{2+} + 3OH^- = Zn(OH)_3^-$

h. $NaAlSi_3O_8(s) + CO_2(g) + \frac{11}{2}H_2O = Na^+ + HCO_3^- + 2H_4SiO_4 + \frac{1}{2}Al_2Si_2O_5(OH)_4(s)$

i. $Fe^{3+} + 3OH^- = Fe(OH)_3(s)$

j. $Fe(OH)_3(s) = FeOH^{2+} + 2OH^-$

k. $Fe(OH)_3(s) = Fe(OH)_2^+ + OH^-$

l. $2Cu^{2+} + 2OH^- = Cu_2(OH)_2^{2+}$

m. $Cu_2(OH)_2CO_3(s) + 4H^+ = 2Cu^{2+} + 3H_2O + CO_2(g)$

n. $Pb^{2+} + 2OH^- = Pb(OH)_2(s)$

1.2 A system is made up of a solution of Na_2SO_4 and H_2S.

Species $\quad H_2S, HS^-, S^{2-}, Na^+, SO_4^{2-}, H^+, OH^-$

Reactions $H_2S = HS^- + H^+$

$\qquad\qquad HS^- = S^{2-} + H^+$

Which of the following component choices are acceptable, even if not very practical?

a. $SO_4^{2-}, H_2S, Na^+, OH^-, (H_2O)$

b. $H_2SO_4, NaHS, NaOH, H^+, (H_2O)$

c. $Na_2SO_4, S^{2-}, Na_2S, NaOH, (H_2O)$

d. $Na_2SO_4, H_2SO_4, Na_2S, NaOH, (H_2O)$

e. $Na_2SO_4, H_2SO_4, Na_2S, Na^+, (H_2O)$

f. $SO_4^{2-}, H_2S, S^{2-}, Na^+, (H_2O)$

g. $Na_2SO_4, SO_4^{2-}, S^{2-}, Na^+, (H_2O)$

h. $S_8^0, O_2^0, Na^0, H^+, (H_2O)$

i. $S^{2-}, O_2^0, Na^+, H^+, (H_2O)$

j. $Na_2HSO_4^+, H_2SO_4, Na_3S^+, HS^-, (H_2O)$

1.3 **a.** Given the following definition of aqueous chemical systems, write complete sets of mole balance equations for each system; ($[H_2O]_T = 55.4\ M$ in all recipes, and H_2O is always a species).

1 Recipe $\quad [CH_2O]_T = [CO_2]_T = 10^{-4}\ M$

Species $\quad CH_2O, O_2(g), CO_2(g)$

Reactions $\quad CH_2O + O_2(g) = CO_2(g) + H_2O$

2 Recipe $\quad [HNO_3]_T = 10^{-3}\ M; [NH_4Cl]_T = 10^{-4}\ M$

Species $\quad NH_4^+, NH_3, NO_3^-, Cl^-, H^+, OH^-$

Reactions $\quad NH_4^+ = NH_3 + H^+$

$\qquad\qquad NH_4^+ + OH^- = NH_3 + H_2O$

$\qquad\qquad H_2O = H^+ + OH^-$

3 Recipe $\quad [CO_2]_T = 10^{-3}\ M; [NaHCO_3]_T = 10^{-4}\ M$

Species $\quad H_2CO_3, HCO_3^-, CO_3^{2-}, Na^+, H^+, OH^-$

Reactions $\quad H_2CO_3 = HCO_3^- + H^+$

$\qquad\qquad HCO_3^- = CO_3^{2-} + H^+$

$\qquad\qquad H_2CO_3 = CO_3^{2-} + 2H^+$

$\qquad\qquad HCO_3^- + OH^- = CO_3^{2-} + H_2O$

$\qquad\qquad H_2O = H^+ + OH^-$

4 Recipe $\quad [NaHCO_3]_T = [Na_2CO_3]_T = [NaOH]_T$

$\qquad\qquad\qquad = [CH_3COONa]_T = 10^{-3}\ M$

Species $\quad$ Same as system $3 + CH_3COOH, CH_3COO^-$

Reactions $\quad$ Same as system $3 + CH_3COOH = CH_3COO^- + H^+$

b. Show that electroneutrality is satisfied by each system of mole balance equations in part a.

c. Suppose that half of the carbon is eliminated from systems 3 and 4 by bubbling some inert gas:

$$H_2CO_3 \rightarrow H_2O + CO_2(g)\uparrow$$

Write new mole balance equations for 3 and 4 (for the aqueous phase only).

1.4 Given the system defined by

Recipe $[NaHCO_3]_T = 10^{-3} M$
  $[CO_2]_T = 2 \times 10^{-3} M$

Species $(H_2O), H^+, OH^-, Na^+, CO_2, H_2CO_3, HCO_3^-, CO_3^{2-}$

Reactions $H_2O = H^+ + OH^-$
  $CO_2 + H_2O = H_2CO_3$
  $H_2CO_3 = HCO_3^- + H^+$
  $HCO_3^- = CO_3^{2-} + H^+$

Choose components and write tableaux and mole balance equations, first isolating HCO_3^- in one mole balance equation and second isolating H_2CO_3 in one equation. Verify that the two sets of equations are equivalent and that the electroneutrality condition is satisfied.

1.5 In systems containing solid and gas phases, it is usually convenient to isolate each solid or gaseous species in one equation. Do this for a chemical system similar to Problem 1.4, but add

Recipe $[CaSO_4]_T = 10^{-1} M$
  $[NaOH]_T = 10^{-1} M$

Species $Ca^{2+}, SO_4^{2-}, CaCO_3(s), Ca(OH)_2(s), CaSO_4(s)$

Reactions $Ca^{2+} + CO_3^{2-} = CaCO_3(s)$
  $Ca^{2+} + 2OH^- = Ca(OH)_2(s)$
  $Ca^{2+} + SO_4^{2-} = CaSO_4(s)$

1.6 Consider a chemical system containing only three species, X, Y, and Z. Only one reaction takes place:

$$xX + yY = zZ$$

(x, y, and z are stoichiometric coefficients). Taking X and Y as components, write the mole balance equation for the system, first using molar concentration units and then mole fractions. Apply your result to the H_2O, H^+, OH^- system.

CHAPTER 2

CHEMICAL EQUILIBRIUM AND ENERGETICS

Our ultimate purpose in aquatic chemistry is to understand the chemical behavior of natural waters. For a particular body of water, perfect intelligence would be to know the concentration of all chemical species at all times and in all places. Given the necessary analytical information we have two basic chemical theories to help us attempt such perfect insight: equilibrium and kinetics. Leaving the more complicated topic of kinetics for Chapter 3, we focus here on chemical equilibrium, the proper subject of classical chemical thermodynamics. In effect, rather than considering what the detailed time history of the system might be, we now ask what its final composition should be—supposing that we wait long enough for the reactions to reach equilibrium.

The fundamental reason why equilibrium is easier to describe than kinetics rests with our greater ability to study the differences in energy between different states of a system rather than the dynamics of the change from one state to another. Think, for example, of the difficulty in predicting the flight of a feather, while its equilibrium position on the floor is a foregone conclusion. The decrease in the feather's gravitational energy and its minimum on the floor are easily calculated; the elegance of its downward glide defies simple mathematics. To be sure, if the time course of the process is precisely what matters, then knowledge of the final equilibrium condition is of little help. However, chemical equilibrium often provides a good approximation of the composition of natural waters. Further, chemical thermodynamics is more than a study of ultimate composition; it also provides information on the direction of spontaneous change and on the energy available from or required for a particular reaction. In some cases, energetics may determine the kinetics of a chemical reaction.

Many chemical reactions taking place in natural waters are quite fast and can be considered to be at equilibrium. For example, the acid–base reactions of the carbonate species are complete within seconds or minutes, well within the time frame of most of our observations. By ignoring slow reactions such as gas exchange or mineral dissolution, one obtains a *partial equilibrium model* of the real system, which includes some kinetic information just by the choice of the reactions considered to take place. This notion of partial equilibrium is a chemical, not a thermodynamic, one. Once the system is defined with *all its possible variations* (i.e., all its chemical reactions), its equilibrium state is uniquely defined. If additional independent variations are considered, the system is thermodynamically different and has a different equilibrium state. In our feather experiment we may wish to consider the *possibility* of the feather floating through the window. Its equilibrium state is then the street, not the floor. The equilibrium pH of a water sample is usually not the same when kept in a closed container as in an open one able to exchange CO_2 (a weak acid) with the atmosphere.

A kinetic description of chemical reactions that are neither very fast nor very slow (compared to some time scale of interest, such as the residence time of the water) can, under certain conditions, be superimposed on an equilibrium model for the fast reactions, then called a *pseudoequilibrium* model. In this way we can describe the variations in the carbonate chemistry and pH of a lake by calculating the atmospheric exchange of CO_2 with time and maintaining equilibrium among the dissolved species.

Even when equilibrium is not reached, chemical thermodynamics tells us the direction of spontaneous change. From energetic considerations we can decide, for example, whether or not particular redox species, say ferric and ferrous iron, are at equilibrium. If they are not at equilibrium, we can estimate how much energy can be obtained from the existing disequilibrium state, or may be required to maintain it. We can then investigate the nature of this disequilibrium, which may be caused by kinetic hindrance or an energy-consuming process mediated by light or by the biota. Since microbial communities seem to have evolved to exploit the most energetically favorable reactions first, such energetic analysis allows us to predict, or at least rationalize, the sequence of biologically mediated chemical events in many aquatic systems.

As we shall see in Chapter 3, energetic principles are the cornerstone of the principal modern theory of chemical kinetics, the *transition state theory*. Understanding thermodynamics is thus a prerequisite to understanding kinetics. In addition, there are many cases of correlation between the kinetic and the thermodynamic parameters describing particular families of chemical reactions. In the absence of precise kinetic data, equilibrium information can then serve to guide us in describing the kinetics of some chemical processes.

In this chapter we do not attempt to provide a compact classical presentation of chemical thermodynamics.[1-3] Although a good grasp of this wide subject is indispensable to an advanced study of the physical chemistry of solutions, our purpose here is only to provide an intuitive basis for the essential thermodynamic concepts used in aquatic chemistry. This particular point of view biases considerably the relative importance of the various aspects of the subject. For example, a familiarity with the relationship between free energies and equilibrium constants is more important to us than a discussion of the philosophical and scientific underpinning of the Second Law. (We shall simply decree that the free energy of a system spontaneously decreases and achieves a minimum at its equilibrium state rather than discuss entropy increases.) As a result of this rather pragmatic approach, our presentation is more intuitive than rigorous and is not intended as a substitute for the many excellent texts on chemical thermodynamics. Following a brief discussion of the thermodynamics of ideal chemical systems at fixed pressure and temperature, we examine the practical issue of equilibrium calculations. The complications introduced by nonideal effects and by variations in pressure and temperature are discussed in the third and final section.

1. THERMODYNAMICS OF CHEMICAL SYSTEMS

To make our discussion concrete, let us consider our original example of Chapter 1, consisting of $1 \, mM$ $CaCO_3$ and $1.1 \, mM$ CO_2 totally dissolved in water and resulting in the dissolved species H^+, OH^-, CO_2, HCO_3^-, and Ca^{2+} which are involved in two reactions:

$$H_2O = H^+ + OH^- \tag{1}$$

$$CO_2 + OH^- = HCO_3^- \tag{2}$$

We examined in Chapter 1 the question of mass conservation for this simple model of a natural water; we now explore the issue of chemical equilibrium and energetics.

As a first and major simplification we consider that our solution is under atmospheric pressure (1 atm) and has a fixed temperature of 25°C. Our short exploration into thermodynamics is thus limited to examining the energetic consequences of changes in the composition of a *closed chemical system*, a system that does not exchange matter with its surroundings, at *fixed pressure and temperature* (P and T). These changes in composition are brought about by the progress of the two chemical reactions 1 and 2. Our goals are to be able to calculate the equilibrium composition of the system and to estimate the energy associated with a particular change in composition.

Taking a simple mathematical viewpoint of the equilibrium problem we find that we have five unknowns, the equilibrium concentrations of our five dissolved species (ignoring H_2O). In Chapter 1 we derived three mole balance equations (not including $TOTH_2O$); we thus need two additional independent equations to obtain a mathematical solution for our equilibrium problem. These two equations are, of course, the two mass action laws corresponding to Reactions 1 and 2:

$$[H^+][OH^-] = K_1 (= 10^{-14.0}) \tag{3}$$

$$\frac{[HCO_3^-]}{[CO_2][OH^-]} = K_2 (= 10^{7.7}) \tag{4}$$

In this section we rederive these mass law equations to establish the fundamentals of chemical energetics and to make clear what assumptions are involved in this mathematical formulation of chemical equilibrium.

1.1 The Free Energy of a System

For the purpose of our discussion we define the free energy of a system as the amount of work necessary to reproduce the system from an arbitrary reference state (supposing 100% efficiency). For example, when we talk of the gravitational energy of an object of mass m at some elevation z above ground as being mgz, we have implicitly defined the system to be the object plus the earth's gravitation (acceleration g) and the reference state (= 0 energy) to be the object resting on the ground ($z = 0$). Following international conventions, we count positively the work done to or stored by the system of interest and negatively the work done by or extracted from the system. We further consider that the free energy of a system is the sum of the free energies of its chemical species. Thus in our example the total free energy G is given by

$$G = n_{H_2O} \mu_{H_2O} + n_{H^+} \mu_{H^+} + n_{OH^-} \mu_{OH^-} + n_{CO_2} \mu_{CO_2}$$
$$+ n_{HCO_3^-} \mu_{HCO_3^-} + n_{Ca^{2+}} \mu_{Ca^{2+}} \tag{5}$$

where n indicates the number of moles of each species and μ its molar free energy. More generally,

$$G = \sum_i n_i \mu_i \tag{6}$$

Note that according to this equation the work necessary to assemble the different molecules of a system must, in some way or another, be reflected in the free energies of the individual molecules.

The free energy of a system is a *state property*; in other words, the work required to reproduce the system from the reference state is independent of the way we do it (independent of the path), as long as we are perfectly efficient of course. This property allows us to consider the free energy of a system as a sum of energies of different types. For example, one way to reproduce the system may be to elevate and pressurize the reference state system sequentially and bring it into an electrical field. Thus the gravitational, mechanical, and electrical energies are regarded as additive components of the total free energy of the system.

We now define the equilibrium of a system (i.e., its stable resting state at some P and T) as given by its lowest possible free energy state. The adjective "possible" reminds us that the proper definition of a system includes a definition of the *possible variations* in the system, foremost among those, the advancements of the chemical reactions. We further postulate that, in the absence of an external energy source, a system proceeds spontaneously to a lower free energy state. For example, chemical reactions must take place in such a way as to decrease the free energy of a system.

Our simple definition of G is intended to bypass the more general definitions of thermodynamic quantities and our "definition" of equilibrium is a statement of the Second Law. The ultimate justification of our "definitions" rests on intuitive correctness and experimental verifications of their consequences. We expect the feather to float down, not up, and by analogy, to witness a spontaneous decrease, not an increase, in the free energy of any system. We also consistently fail to build perpetual motion machines or to cool water by placing it on the heater. Pragmatically, the conceptual and mathematical framework with which the concepts and formulae of thermodynamics allow us to structure our observations simply enhances the success of our scientific predictions.

1.2   Molar Free Energies in Ideal Systems

The free energy contributed by each mole of a given species, say Ca^{2+}, to the overall free energy of a system can be divided into three parts: (1) a free energy corresponding to the chemical nature of the species and representing the energy of formation of the species from some reference state (e.g., the formation of calcium ion in solution from metallic calcium); (2) a free energy corresponding to the concentration of the species in the system and representing the energy necessary to obtain the desired concentration (e.g. $[Ca^{2+}] = 1 \text{ m}M$); (3) a free energy corresponding to the interactions among the species in the system and representing the fraction of the energy necessary to put the mixture together that we attribute to the species of interest.

The third term is obviously troublesome; we shall ignore it for now and postpone studies of species interactions to the third section of this chapter. Following a long scientific tradition we define an *ideal system* as one in which the free energy of a species is independent of the nature and concentration of other species. We thus write

$$\mu_{Ca^{2+}} = \mu^o_{Ca^{2+}} + RT \ln X_{Ca^{2+}} \tag{7}$$

where $\mu^o_{Ca^{2+}}$ is the standard free energy of the calcium ion and $RT \ln X_{Ca^{2+}}$ represents the free energy corresponding to the concentration of Ca^{2+} defined by the mole fraction

$$X_{Ca^{2+}} = \frac{n_{Ca^{2+}}}{n_T} \tag{8}$$

where n_T is the total number of moles in the system:

$$n_T = n_{H_2O} + n_{H^+} + n_{OH^-} + n_{CO_2} + n_{HCO_3^-} + n_{Ca^{2+}} \tag{9}$$

R is the universal gas constant ($8.3143 \, J \, deg^{-1} \, mol^{-1}$) and T is the absolute temperature; ln designates the Naperian (natural) logarithm. ($2.3 \, RT \simeq 5.7 \, kJ \, mol^{-1} \simeq 1.36 \, kcal \, mol^{-1}$ at 25°C.)

We now examine each of the two terms of the molar free energies in ideal systems and leave considerations of nonideal effects to the third section of this chapter. The general expressions for Equations 7–9 are

$$\mu_i = \mu_i^o + RT \ln X_i \tag{10}$$

$$X_i = n_i/n_T \tag{11}$$

$$n_T = \sum_i n_i \tag{12}$$

Practical expressions of molar free energies using the molar concentration scale rather than the mole fraction scale are given in Table 2.1.

Standard Molar Free Energy   Chemical species are assemblages of atoms of various elements and thus possess a free energy different from that of the pure elements. By defining as the reference state for the pure elements a chemical standard state corresponding to their most stable forms (solid, liquid, or gas) under standard pressure (1 atm = 1.0133×10^5 Pascals) and temperature (25°C = 298.15 K), we can define the chemical free energy of all chemical species under any condition. For example, the molar free energy of liquid water under arbitrary conditions of Pressure (P) and temperature (T) is given by the work necessary (= minus the energy released) to carry out the overall reaction:

$$P = 1 \, atm, \ T = 25°C; \ H_2(g) + \tfrac{1}{2}O_2(g) \ \rightarrow \ H_2O(l); \ P, T$$

This work includes the energy of the gas phase reaction at standard pressure and temperature, the energy of condensation of water vapor, and the energy necessary to heat (or cool) and pressurize (or depressurize) one mole of water from standard conditions to the temperature and pressure considered. The standard molar free energy thus includes bond breaking and bond making energy (enthalpy) as well as energy corresponding to the ordering or disordering

TABLE 2.1   Practical Expressions of Molar Free Energies

Recognizing our predilection for moles per liter (M) rather than mole fractions as concentration units, we establish here some convenient formulae for molar free energies, starting with the ideal formula, Equation 10.

Solids

For a species constituting a pure solid phase, the mole fraction is identically unity, and its logarithm is thus identically null:

$$X = 1$$

therefore,

$$\mu_i = \mu_i^o$$

Solutes

For a species S_i in the solution phase, the mole fraction is given by

$$X_i = \frac{n_i}{n_w}$$

therefore

$$\mu_i = \mu_i^o + RT \ln \frac{n_i}{n_w}$$

where n_w is the total number of moles in the solution. We can introduce molar concentrations in this equation by dividing by the volume V:

$$[S_i] = \frac{n_i}{V}$$

$$\mu_i = \mu_i^o - RT \ln \frac{n_w}{V} + RT \ln \frac{n_i}{V}$$

For dilute solutions, n_w/V, the total number of moles per liter of solution is approximately constant and equal to the number of moles of water (55.4 M). The second term of the equation can be thus incorporated in the standard free energy value μ_i^o:

$$\mu_i = \mu_i^o + RT \ln [S_i]$$

This is the equation that we use in practice for solutes, and it should be remembered that the corresponding standard free energy μ_i^o is only valid for the chosen molar concentration scale.

Water

The mole fraction of water in dilute aqueous solutions is approximately unity, and we

TABLE 2.1 (*Continued*)

shall consider the molar free energy of water to be constant:

$$\mu_{\text{water}} \simeq \mu^o_{\text{water}}$$

This approximation, which is satisfactory for the study of chemical phenomena in dilute solutions, does not permit the study of physical phenomena such as the depression in freezing point, osmotic pressure, or other colligative properties.[a]

Gases

For gases, the mole fraction is equal to the ratio of the partial pressure of the gas to the total pressure of the gas phase:

$$X_i = \frac{n_i}{n_G} = \frac{P_i}{P}$$

$$\mu_i = \mu^o_i + RT \ln \frac{P_i}{P}$$

where n_G is the total number of moles in the gas phase. If the total pressure is taken as constant (e.g., $P = 1$ atm) and $-RT \ln P$ is included in the standard free energy value μ^o_i, the practical expression becomes

$$\mu_i = \mu^o_i + RT \ln P_i$$

[a]The colligative properties of a solvent are those that depend on the number of particles, such as molecules or ions, in the solvent.

of the system (entropy). The nature of the solvent is important in determining the free energy of a solute since the energy of solvation involves intermolecular forces (enthalpy) and an ordering of the solvent (entropy). In the case of ions (which have a high negative entropy of solvation in water) it is necessary to define arbitrarily one additional reference state to obtain expressions for the free energy of individual ions. By convention, the free energy of formation of the hydrogen ion H^+ (= hydrated proton) is taken to be zero. The free energy of an ion such as chloride can then be obtained from the energy required for the formation of a completely ionized hydrochloric acid solution:

$$\tfrac{1}{2}H_2(g) + \tfrac{1}{2}Cl_2(g)(+ \text{water}) = H^+(aq) + Cl^-(aq)$$

The free energies of other ions can consequently be obtained by considering the work necessary for the formation of other electrolyte solutions; for example,

$$Ca(s) + Cl_2(g)(+ \text{water}) = Ca^{2+}(aq) + 2Cl^-(aq)$$

The standard molar free energy $\mu^o_{Ca^{2+}}$, obviously depends on the *reference state* of the calcium ion, including the pressure, temperature, and composition of the solution in which we are "forming" Ca^{2+} from elemental calcium in its *standard state*. For us the reference pressure and temperature are obviously those of the system (say $P = 1$ atm and $T = 25°C$) and the composition of the solution is so chosen as to make our assumption of "ideal" behavior as good as possible. The free energy of a solute may be considered independent of the nature and concentration of all others in either of the two following cases:

1. The system is very dilute, and all individual solute molecules are far apart and effectively "ignorant" of each other (i.e., they have no energetic interactions, and their individual free energies are unaffected by each other's presence). This is the "infinite dilution" reference state.

2. The major solutes (those accounting for the bulk of the dissolved species) are considered to be at a fixed concentration and whatever effects they have on the free energy of another species are accounted for in the standard value μ^o_i of the chemical free energy of that species. This is the "fixed composition" reference state.

Both types of reference state—infinite dilution or fixed composition of background solutes—are used in practice to express the various thermodynamic parameters. In the first case the standard molar free energy $\mu^o_{Ca^{2+}}$ is the work necessary to produce an infinitesimal concentration of Ca^{2+} from elemental calcium (and H^+), all solutes being present at vanishingly low concentrations. In the second, $\mu^o_{Ca^{2+}}$ is the work necessary to produce an infinitesimal concentration of Ca^{2+} in the presence of a given concentration of other solutes.

Because of the logarithmic expression $\ln X_{Ca^{2+}}$ in Equation 7, the standard free energy $\mu^o_{Ca^{2+}}$ is the value of $\mu_{Ca^{2+}}$ when the concentration of Ca^{2+} is unity ($X_{Ca^{2+}} = 1$). This situation seems somewhat paradoxical since a unit concentration can hardly be considered infinitesimal. This is merely a mathematical contortion: in effect, to obtain the free energy at a given concentration, X_i, from that obtained (theoretically) at an infinitesimal concentration, we first extrapolate up to unit concentration and then back down to $X_{Ca^{2+}}$. As seen later, the only problem resulting from such a convention is that the standard free energies are dependent not only on the chosen reference state but also on the concentration scale. The standard state is defined by $\mu_i = \mu^o_i$, thus by a unit concentration of the species on whatever scale.

Concentration Term (Definition of Phases)   We know that even in the absence of any mechanical mixing, gases and solutes spontaneously diffuse to achieve a homogeneous concentration distribution. We have an intuitive feeling for the diffusion of sugar in our coffee as we do for falling cups and feathers: both processes occur spontaneously and must thus be energetically favorable.

To obtain a mathematical expression for the concentration term of the molar free energy we need to define the possible variations so that we can decide what geometrical limits there are to the diffusion process. This is done through the familiar concept of *phases*, which distinguishes typically gas, solution, and solid phases, each referring to the particular physical state of the constitutive matter. Normally in aquatic chemistry we consider only one gas phase (air) and one solution phase (water); nonetheless, it is sometimes useful to consider others, such as the intracellular solutions in aquatic microorganisms. Gases are free to diffuse throughout the air and solutes throughout the water phase. For simplicity, we may consider at this point that all solid phases are pure phases (i.e., they consist of only one chemical species) and that no diffusion takes place in solid phases. This restriction leads to the obvious result that there is *no* concentration term for the free energy of solids.

For gases and solutes, diffusion clearly corresponds to an increase in the entropy of the system and is a reflection of the second law of thermodynamics. We can thus recognize the concentration term of the molar free energy expression as an explicit entropic contribution of the species to the total free energy of the system. (In contrast, as discussed later, the standard free energy terms μ_i^o represents *both* enthalpic and entropic contributions.)

A justification for the logarithmic formulation of the concentration term (e.g., $RT \ln X_{Ca^{2+}}$) is most easily seen by first considering gaseous species. The minimum work necessary to bring an ideal gas from pressure P_1 to P_2 is given by the expression

$$w = RT \ln \frac{P_2}{P_1} \tag{13}$$

This expression can be derived by calculating the mechanical pressure–volume work on the basis of Boyle–Mariotte's law. The result can then be extended to partial pressures of gases in mixtures and then to mole fractions of species in solution by application of Raoult's and Henry's laws for the equilibrium between concentrations of solvent and solutes and their vapor pressures. Thus the minimum work necessary to bring a solute from a concentration X_1 to a concentration X_2 is

$$w = RT \ln \frac{X_2}{X_1} \tag{14}$$

Many other experimental justifications of Equation 14 are provided by verification of its consequences, for example, the mass law expression for reactions among solutes and the expression of solubility of solids. Although it draws on material to be studied in another chapter, the most compelling justification for Equation 14 is perhaps given by the Nernst equation, which demonstrates directly the logarithmic relationship between concentration and electrical energy. As an alternative to empirical justification, Equation 14 can be obtained from the theory of statistical mechanics. The Maxwell–Boltzmann distribution law, which relates concentrations of particles to the exponential of their energy, is clearly kindred to Equation 14. For our purposes this equation is simply considered given; it leads directly to the formulation of the molar free energy given in Equation 10 by including the reference concentration term $(-RT \ln X_1)$ in μ^o.

1.3 Energetics of Chemical Reactions

On the basis of the formulae for the total free energy of a system (Equations 5 and 6) and of the molar free energy of individual species (Equations 7 and 10) we can study the energetic consequences of the advancement of a chemical reaction. What is the differential change dG in the free energy G of the system when Reaction 1 proceeds by $d\xi$ number of moles where $d\xi$ is equal to $-dn_{H_2O}$, the additional number of moles of H_2O dissociated?

The differential of the general equation (6) gives

$$dG = \sum_i \mu_i \frac{\partial n_i}{\partial \xi} d\xi + \sum_i n_i \frac{\partial \mu_i}{\partial \xi} d\xi \tag{15}$$

The second term is identically null (see Sidebar 2.1 on the Gibbs–Duhem relation) and, as seen in Chapter 1, the partial differentials $\partial n_i/\partial \xi$ are simply the stoichiometric coefficients of the species in Reaction 1:

$$\frac{\partial n_{CO_2}}{\partial \xi} = \frac{\partial n_{HCO_3^-}}{\partial \xi} = \frac{\partial n_{Ca^{2+}}}{\partial \xi} = 0$$

and

$$\frac{\partial n_{H_2O}}{\partial \xi} = -1; \quad \frac{\partial n_{H^+}}{\partial \xi} = +1; \quad \frac{\partial n_{OH^-}}{\partial \xi} = +1$$

Thus

$$\frac{dG}{d\xi} = -\mu_{H_2O} + \mu_{H^+} + \mu_{OH^-} \tag{16}$$

The differential increase in the free energy due to the advancement of the reaction—called the *free energy change* of the reaction and denoted ΔG—is simply the difference between the free energy of the products and that of the reactants. In general,

$$\Delta G = \frac{dG}{d\xi} = \sum_i v_i \mu_i \tag{17}$$

SIDEBAR 2.1

Gibbs–Duhem Relation

Based on the given expression for molar free energies

$$\mu_i = \mu_i^o + RT \ln X_i$$

it is easy to show that the Gibbs–Duhem relation

$$\sum n_i \, d\mu_i = 0$$

holds.

From the definition of $X_i = n_i/n_T$ we obtain the two relations

$$n_1/X_1 = n_2/X_2 = \cdots n_i/X_i = n_T$$

and

$$\sum X_i = 1$$

Differentiation of the molar free energy yields

$$d\mu_i = 0 + RT \frac{dX_i}{X_i}$$

Thus,

$$\sum n_i d\mu_i = RT \sum \frac{n_i}{X_i} dX_i$$
$$= RT \, n_T \sum dX_i$$
$$= RT \, n_T \, d(\sum X_i)$$
$$= 0$$

In this approach the Gibbs–Duhem relation is thus taken as a mere mathematical consequence of the functionality of the concentration term in the expression of the molar free energy. The property of μ_i to be a partial differential of $G(\mu_i = \partial G/\partial n_i$; hence the justification of its appellation as a *partial* molar free energy) follows from the Gibbs–Duhem relation and

(continued)

from the fundamental expression of additivity of free energies $G = \sum n_i \mu_i$. In classical thermodynamic texts, μ_i is usually defined a priori as a partial differential, $\mu_i = \partial G / \partial n_i$, and G as an extensive property of the system (by which it is meant to be a homogeneous function of degree 1 in n_i). Use of Euler's theorem for homogeneous functions then yields $G = \sum n_i \mu_i$ and, by differentiation, the Gibbs–Duhem relation, $\sum n_i d\mu_i = 0$. The approach taken here appears no less intuitive, and it stresses the need for the molar free energies to verify the Gibbs–Duhem relation in order for the fundamental thermodynamic equations to be strictly applicable. For example, neither the expression of the free energy change of a reaction nor the mass law equation is strictly correct when the molar scale rather than the mole fraction scale is used.

where the v_i's are the stoichiometric coefficients of the reaction whose advancement is measured by $d\xi$.

Introducing the expression of the molar free energies (Equation 10) in Equation 16 gives:

$$\underbrace{\frac{dG}{d\xi}}_{\Delta G_1 =} = \underbrace{(-\mu^o_{H_2O} + \mu^o_{H^+} + \mu^o_{OH^-})}_{\Delta G^o_1} + \underbrace{RT \ln \frac{X_{H^+} \cdot X_{OH^-}}{X_{H_2O}}}_{+ RT \ln \quad Q_1} \quad (18)$$

A similar expression is obtained for the other reaction in the system:

$$\Delta G_2 = -\mu^o_{CO_2} - \mu^o_{OH^-} + \mu^o_{HCO_3^-} + RT \ln \frac{X_{HCO_3^-}}{X_{CO_2} \cdot X_{OH^-}} \quad (19)$$

The free energy change of a reaction is thus made up of two terms: (1) a constant, ΔG^o, called the *standard free energy change* of the reaction, which is given by the stoichiometry of the reaction and the standard free energies of the species involved, and (2) a logarithmic term that depends on the composition of the system through the product Q, called the *reaction quotient*.

Reactions that have very large standard free energies, positive or negative, (say hundreds of kilojoules per mole) tend to proceed to completion in one direction or the other. Such reactions are very far from equilibrium when products and reactants are present in comparable concentrations, and in this case the ΔG^o term is normally much larger than the $RT \ln Q$ term. The change in the free energy of the system, as the reaction proceeds, is then approximately constant and equal to the standard free energy change of the reaction (per mole reacted):

$$\Delta G \simeq \Delta G^o \simeq \sum_i v_i \mu^o_i$$

This relationship is particularly useful for examining the energetics of redox reactions carried out by organisms: photosynthesis, respiration, nitrification, nitrogen fixation, sulfate reduction, and others. Note that it is only an approximate relationship, however, and in many cases the concentration of some reactants such as H^+ can have a large effect on the free energy change of the reaction (see Chapter 7).

A reaction whose standard free energy is not too large (say a few kilojoules per mole) will proceed up to the point where the reaction quotient term balances the standard free energy term. This is the equilibrium condition discussed in the next section.

1.4   Chemical Equilibrium: Mass Law Equations

Consider a set of independent chemical reactions describing all the possible independent variations in the composition of a closed system. The minimum of the free energy of the system and hence its equilibrium state is obtained mathematically by equating to zero all the differentials of G with respect to the degree of advancement ξ of each reaction, subject to constraints of mass conservation (i.e., mole balances) and positivity of concentrations:

$$\Delta G = \frac{\partial G}{\partial \xi} = 0 \tag{20}$$

This is the expression of equilibrium for the reaction with advancement ξ: there is a zero gain or loss of free energy in the system as the reaction proceeds infinitesimally in either direction. The system is thus at equilibrium when all the possible reactions are at equilibrium, and vice versa. In our example,

$$\Delta G_1 = \Delta G_1^\circ + RT \ln Q_1 = 0 \tag{21}$$

$$\Delta G_2 = \Delta G_2^\circ + RT \ln Q_2 = 0 \tag{22}$$

These equations are usually written in their exponential form, known as the mass law equation:

$$Q = \exp\left[-\frac{\Delta G^\circ}{RT} \right] \tag{23}$$

that is,

$$\frac{X_{H^+} \cdot X_{OH^-}}{X_{H_2O}} = K_1 = \exp\left[-\frac{\Delta G_1^\circ}{RT} \right] \tag{24}$$

$$\frac{X_{HCO_3^-}}{X_{CO_2} \cdot X_{OH^-}} = K_2 = \exp\left[\frac{-\Delta G_2^\circ}{RT} \right] \tag{25}$$

K_1 and K_2 are called the *equilibrium constants*, and are directly related to the standard free energies of the reactions (ΔG°) by their exponentials in Equations 24 and 25. The equivalent logarithmic expression is

$$-RT \ln K = \Delta G^\circ = \sum v_i \mu_i^\circ \tag{26}$$

For convenience, the mass law equation is usually written with molar concentrations and partial pressures rather than mole fractions. This is simply achieved by multiplying each side of Equations 24 and 25 by the factor (n_T/volume) at the appropriate power. Recalling that n_T/volume $\simeq n_{H_2O}$/volume $\simeq 55.4\,M$ (and thus that $X_{H_2O} \simeq 1$), we obtain the familiar form of the mass laws:

$$(24) \times (n_T/\text{volume})^2 \quad \Rightarrow \quad [H^+][OH^-] = K_1'$$

$$(25) \times (n_T/\text{volume})^{-1} \quad \Rightarrow \quad \frac{[HCO_3^-]}{[CO_2][OH^-]} = K_2'$$

The molar formulations of mass law equations involving various types of chemical species are provided in Table 2.2. These equations are obtained directly by equating to zero the free energy changes of the reactions when the molar free energies are given by their practical expressions as in Table 2.1.

1.5   Uniformity of Molar Free Energies at Equilibrium

Consider the species Ca^{2+} in two different parts A and B of our model system at equilibrium. In general A and B may be two different phases; in our case they are simply different locations in the aqueous phase. Since Ca^{2+} is free to transfer from A to B, we can represent that possible variation in the system by the advancement of the reaction

$$Ca_A^{2+} = Ca_B^{2+} \tag{27}$$

The equilibrium condition is thus

$$\frac{\partial G}{\partial \xi} = \Delta G = \mu_{Ca^{2+}}^B - \mu_{Ca^{2+}}^A = 0$$

or

$$\mu_{Ca^{2+}}^B = \mu_{Ca^{2+}}^A \tag{28}$$

This result can of course be applied to any part of the system. We have thus demonstrated that at equilibrium the molar free energy of any species must be uniform throughout the system (if there is no absolute barrier to transfer). This necessary condition is useful in studying equilibrium among phases. Applied to a solution phase, it results directly in a condition of uniformity of concentrations

TABLE 2.2 Practical Forms of Mass Law Equations

Solid–solutes

$$CaCO_3(s) = Ca^{2+} + CO_3^{2-}$$

$$[Ca^{2+}][CO_3^{2-}] = K_{CaCO_3} \text{ (solubility constant)}$$

Gas–solutes

$$CO_2(g) = CO_2(aq)$$

$$\frac{[CO_2 \cdot aq]}{P_{CO_2}} = K_H \text{ (Henry's law constant)}$$

Solutes–solvent

$$CO_2(aq) + H_2O = H_2CO_3$$

$$\frac{[H_2CO_3]}{[CO_2 \cdot aq]} = K_{H_2O} \text{ (hydration constant)}$$

or

$$H_2O = H^+ + OH^-$$

$$[H^+][OH^-] = K_w \text{ (ion product of water)}$$

Solutes

$$HAc = H^+ + Ac^-$$

$$\frac{[H^+][Ac^-]}{[HAc]} = K_a \text{ (acidity constant)}$$

$$Hg^{2+} + 3Cl^- = HgCl_3^-$$

$$\frac{[HgCl_3^-]}{[Hg^{2+}][Cl^-]^3} = K \text{ (complexation constant)}$$

throughout the solution:

$$\mu_{Ca^{2+}}^o + RT \ln [Ca^{2+}]_A = \mu_{Ca^{2+}}^o + RT \ln [Ca^{2+}]_B$$

or

$$[Ca^{2+}]_A = [Ca^{2+}]_B \tag{29}$$

We have thus rediscovered the wheel and the thermodynamic necessity of diffusion of sugar (and calcium) in our coffee cup.

2. SOLUTION OF CHEMICAL EQUILIBRIUM PROBLEMS

The equilibrium state of a closed chemical system is defined by its minimum free energy within the constraints of mass conservation. In Chapter 1 we saw how a proper choice of components leads to a set of mole balance equations that expresses fully the mass conservation condition. In the first section of this chapter we showed how the set of mass law equations corresponding to all the independent chemical reactions is equivalent to the condition of free energy minimum of a system. These two sets of equations, mole balances and mass laws, define a well-posed mathematical problem that can be shown to have one, and only one, solution: the concentrations of the species at equilibrium. However, the solution to that chemical equilibrium problem may not be straightforward, particularly when it involves a large number of reacting species, as it normally does in aquatic chemistry. In this section we focus our attention on such a chemical equilibrium problem, examining how to best organize and most simply solve it.

Posing a chemical equilibrium problem in proper mathematical terms is a matter of insuring that all the necessary analytic and thermodynamic information is translated into a complete set of algebraic equations. This is best achieved by using the "tableau" methodology already introduced in Chapter 1 as a convenient way to express the stoichiometric relations between species and components. Once the mathematical problem is properly posed, its solution would appear to require only numerical manipulations of little scientific interest and best left to the mindless patience of computers. Useful though they are, computer solutions do not provide much insight and have limited didactic value. The mathematics of chemical equilibrium present some remarkable particularities that make them amenable to convenient hand methodologies such as drastic approximations, graphical solutions, and simple iterative schemes. Because the mathematical problem is a representation of the chemical reality, its numerical particularities correspond to actual chemical behavior. There is thus some chemistry to be learned by solving chemical equilibrium problems, and vice versa, chemical intuition can greatly simplify one's way through the numerical maze of mole balance and mass law equations.

2.1 Organization: Tableaux

Using again the first example of Chapter 1, we consider an equilibrium problem specified in canonical form by:

(1) A recipe for how the system is made up (Recipe 1):

$$[CaCO_3]_T = 10^{-3} M$$

$$[CO_2]_T = 1.1 \times 10^{-3} M$$

TABLEAU 2.1

	H^+	CO_2	Ca^{2+}	$\log K$
H^+	1			
OH^-	-1			-14.0
HCO_3^-	-1	1		-6.3
CO_2		1		
Ca^{2+}			1	
$CaCO_3$	-2	1	1	
CO_2		1		

(2) A list of species in the system:

$$(H_2O), H^+, OH^-, CO_2, HCO_3^-, \text{ and } Ca^{2+}$$

(3) A list of independent reactions with their equilibrium constants:

$$H_2O = H^+ + OH^-; \qquad K_1 = 10^{-14.0} \tag{30}$$

$$CO_2 + OH^- = HCO_3^-; \qquad K_2 = 10^{7.7} \tag{31}$$

In order to write a complete set of equations—mole balances and mass laws—for the equilibrium composition of this system, it is convenient to organize the information from (1)–(3) in a tableau as we did in Chapter 1. Taking H^+, CO_2, Ca^{2+}, and H_2O implicitly as components yields Tableau 2.1.

In this tableau a column of equilibrium constants has been added. The constants are written for the mass laws corresponding to the formation of the species from the components. They are readily obtained from those given in (3) by considering the reactions corresponding to each line of the tableau. The constant for the hydroxide ion, OH^-, is straightforward: the mass law equation $[OH^-] = K[H^+]^{-1}$ corresponds to the reaction $H_2O = H^+ + OH^-$, thus $\log K = \log K_w = -14.0$. Calculation of the bicarbonate constant requires some manipulation: the mass law equation $[HCO_3^-] = K[CO_2][H^+]^{-1}$ corresponds to the reaction $CO_2 + H_2O = H^+ + HCO_3^-$ which is obtained by adding Reactions 30 and 31:

$$
\begin{array}{ll}
H_2O = H^+ + OH^- & \log K_w = -14.0 \\
\underline{CO_2 + OH^- = HCO_3^-} & \underline{\log K = 7.7} \\
CO_2 + H_2O = H^+ + HCO_3^- & \log K = -6.3
\end{array}
$$

Note that the logarithms of the equilibrium constants are subjected to the same linear operations as the corresponding reactions. The exponents in the mass law expressions are precisely the coefficients of the corresponding lines of the tableau. A constant of unity is implied for the expressions of the species that are also components.

Tableau 2.1 in this form contains all the necessary information for the equilibrium problem and a complete set of equations is readily written: Mole balances (from each column)

$$TOTH = [H^+] - [OH^-] - [HCO_3^-] = -2[CaCO_3]_T = -2 \times 10^{-3} M \tag{32}$$

$$TOTCO_2 = [HCO_3^-] + [CO_2] = [CaCO_3]_T + [CO_2]_T = 2.1 \times 10^{-3} M \tag{33}$$

$$TOTCa = [Ca^{2+}] = [CaCO_3]_T = 10^{-3} M \tag{34}$$

Mass laws (from each line)

$$[OH^-] = 10^{-14} [H^+]^{-1} \tag{35}$$

$$[HCO_3^-] = 10^{-6.3} [H^+]^{-1} [CO_2] \tag{36}$$

There are five unknowns in this equilibrium problem (the concentrations of each of the five species) and five equations (three mole balances and two mass laws). Were we to consider $[H_2O]$ as an additional unknown we would simply have to consider the corresponding column and mole balance equation; the result would be, approximately, $[H_2O] = 55.4 M$.

2.2 The Mathematical Problem

Since the species' concentrations are given explicitly as a function of the components by the mass law equations, it is always possible to reduce the number of equations and the number of unknowns by substituting the mass laws into the mole balances. One is then left with a system of nonlinear equations equal to the number of components and with the concentrations of the components themselves as the only unknowns. Although such substitution and reduction in the number of equations is tempting and is in fact practical for computer algorithms, it should be firmly resisted in hand calculations. Instead we want to show how one can solve the problem without manipulating equations or solving high degree polynomials.

The lazy method is trial and error; one's constant hope (hypothesis) is that the solution of the problem is trivial. If we consider the four mole balance equations, one is indeed trivial and needs no further consideration:

$$[Ca^{2+}] = 10^{-3} M$$

Since the calcium ions are not reacting with any other species, they have no influence on the equilibrium composition of the system and they can simply be ignored in the solution of the problem.

For the other mole balance equations, 32 and 33, it would be very convenient if all the terms but one could be neglected from the sum; that is, if the concentration of one species were much larger than that of the others and effectively equal to the total concentration of the corresponding component. This is usually the case. If $TOTX = 0$, however, then two terms of opposite signs have to be equal.

Suppose that we choose arbitrarily one species and hypothesize that its concentration is much larger than that of the other species appearing in the same mole balance. There are three possible outcomes to such an arbitrary choice: (A) we are right, (B) we are wrong because some other concentration is in fact much larger, (C) we are wrong because some other concentration(s) is (are) of comparable magnitude. We wish to show that in Case A the problem is readily solved; in Case B the inappropriateness of the choice is immediately apparent, leading to a better alternative choice, and in Case C the problem is solved by a simple iteration procedure. Once we improve our chemical intuition, the choice of the dominant species is not totally arbitrary, so that Case A is the one most often encountered.

Case A

Giving ourselves the benefit of a good chemical or mathematical intuition, let us assume: $[HCO_3^-] \gg$ all other concentrations. In this case the $TOTH$ Equation 32 yields:

$$[HCO_3^-] = 2 \times 10^{-3} M$$

The concentration of CO_2 is then obtained by difference from Equation 33

$$[CO_2] = 10^{-4} M$$

The values of $[HCO_3^-]$ and $[CO_2]$ can be introduced into the mass laws to yield

$$[H^+] = 10^{-7.6}; \quad [OH^-] = 10^{-6.4}$$

We can verify that the assumption ($[HCO_3^-] \gg$ all other concentrations) is valid by substituting the calculated concentrations back into the original $TOTH$ equation. The error is about 0.01%, which is acceptable.

Had we known a priori that HCO_3^- is a dominant species, we should have taken it as a component (for the same reason that we found it helpful to choose H_2O as a component). The mole balance equations corresponding to the components H^+, HCO_3^-, and Ca^{2+} are

$$TOTH = [H^+] - [OH^-] + [CO_2] = 10^{-4} \tag{37}$$

$$TOTHCO_3 = [HCO_3^-] + [CO_2] = 2.1 \times 10^{-3} \tag{38}$$

$$TOTCa = [Ca^{2+}] = 10^{-3} \tag{39}$$

This choice of components, which isolates the dominant species in one equation, would have lead us readily to the solution by assuming $[CO_2]$ to be the dominant term in the $TOTH$ Equation 37.

Case B

Using our original set of Equations 32–36 we now consider less felicitous choices for our initial guess of the dominant term in the $TOTH$ equation:

(1) $[H^+] \gg$ all other concentrations

This is clearly impossible since $[H^+]$ has a positive sign in the $TOTH$ equation and a concentration cannot be negative.

(2) $[OH^-] \gg$ all other concentrations

$$[OH^-] = 2 \times 10^{-3}$$

Substituting into the mass law equations yields

$$[H^+] = 10^{-11.3}$$

$$[HCO_3^-] = 10^{5.0} [CO_2]$$

From which we conclude

$$[HCO_3^-] \gg [CO_2]$$

The $TOTCO_2$ equation then simplifies to

$$[HCO_3^-] = 2.1 \times 10^{-3}$$

which contradicts our hypothesis $[OH^-] \gg [HCO_3^-]$. However, it is now apparent that $[HCO_3^-]$ is likely the dominant term in the $TOTH$ equation and the trial and error procedure can be restarted with the appropriate initial guess.

A more chemically insightful approach to this problem is to consider that the base CO_3^{2-} and the acid CO_2 neutralize each other, $CO_3^{2-} + CO_2 + H_2O \rightarrow 2HCO_3^-$, up to exhaustion of the base. The result is the formation of $2\,mM\ HCO_3^-$ with $10^{-4}\,M\ CO_2$ remaining unreacted. Such a chemically intuitive approach provides at least a good initial guess to solve the chemical equilibrium problem.

Case C

To illustrate this case let us revisit the second example of Chapter 1, which includes additional background salt and weak acid and considers the formation of carbonic acid $[H_2CO_3]$ and of the carbonate anion $[CO_3^{2-}]$ in addition to CO_2 and HCO_3^-.

Recipe $[CaCO_3]_T = 10^{-3} M; [CO_2]_T = 1.1 \times 10^{-3} M; [NaCl]_T = 10^{-2} M;$

$[HA]_T = 4 \times 10^{-4} M$

Species $\quad$ $H^+, OH^-, Na^+, Cl^-, Ca^{2+}, HA, A^-, H_2CO_3^*, HCO_3^-, CO_3^{2-}$

Reactions $\quad$ $H_2O = H^+ + OH^-$; $\qquad \log K_w = -14.0$ $\qquad$ (40)

$$HA = H^+ + A^-; \qquad \log K_a = -4.3 \qquad (41)$$

$$H_2CO_3^* = H^+ + HCO_3^-; \qquad \log K_{a1} = -6.3 \qquad (42)$$

$$HCO_3^- = H^+ + CO_3^{2-}; \qquad \log K_{a2} = -10.3 \qquad (43)$$

Note. For simplicity the sum of the concentrations of the dissolved CO_2 and carbonic acid, H_2CO_3, is denoted by the species $H_2CO_3^*$. In fact, $H_2CO_3^*$ is 99.85% CO_2 and only 0.15% H_2CO_3.

Choice of Components $\quad$ Aside from nonideal effects, which are discussed in the next section, the addition of background electrolyte (Na^+, Cl^-) should have no effect on the chemistry of this system. Here H^+, Na^+, Cl^-, Ca^{2+} are obvious choices for components. The addition of 0.4 mM organic acid should not alter dramatically the pH of a system containing an excess (ca. 2 mM) of bicarbonate. We thus use the solution of the previous problem to guess that HCO_3^- is the dominant carbonate species. Then HCO_3^- is taken as a component; so is A^-, since a relatively strong acid is expected to be mostly dissociated in a nonacidic solution ($[H^+] < K_a$). Thus, $H^+, Na^+, Cl^-, Ca^{2+}, HCO_3^-$, and A^- are taken as components in Tableau 2.2.

Mole Balance Equations

$$TOTH = [H^+] - [OH^-] + [H_2CO_3^*] - [CO_3^{2-}] + [HA]$$
$$= -[CaCO_3]_T + [CO_2]_T + [HA]_T = 5 \times 10^{-4} M \qquad (44)$$

$$TOTNa = [Na^+] = [NaCl]_T = 10^{-2} M \qquad (45)$$

$$TOTCl = [Cl^-] = [NaCl]_T = 10^{-2} M \qquad (46)$$

$$TOTCa = [Ca^{2+}] = [CaCO_3]_T = 10^{-3} M \qquad (47)$$

$$TOTHCO_3 = [H_2CO_3^*] + [HCO_3^-] + [CO_3^{2-}]$$
$$= [CaCO_3]_T + [CO_2]_T = 2.1 \times 10^{-3} M \qquad (48)$$

$$TOTA = [HA] + [A^-] = [HA]_T = 4 \times 10^{-4} M \qquad (49)$$

TABLEAU 2.2

	H^+	Na^+	Cl^-	Ca^{2+}	HCO_3^-	A^-	$\log K$
H^+	1						
OH^-	-1						-14.0
Na^+		1					
Cl^-			1				
Ca^{2+}				1			
$H_2CO_3^*$	1				1		$+6.3$
HCO_3^-					1		
CO_3^{2-}	-1				1		-10.3
HA	1					1	$+4.3$
A^-						1	
$CaCO_3$	-1			1	1		
CO_2	1				1		
HA	1					1	
$NaCl$		1	1				

Mass Laws

$$[OH^-] = 10^{-14}[H^+]^{-1} \tag{50}$$

$$[H_2CO_3^*] = 10^{6.3}[H^+][HCO_3^-] \tag{51}$$

$$[CO_3^{2-}] = 10^{-10.3}[H^+]^{-1}[HCO_3^-] \tag{52}$$

$$[HA] = 10^{4.3}[H^+][A^-] \tag{53}$$

As expected, the mole balance equations (45–47) are trivial and give

$$[Na^+] = [Cl^-] = 10^{-2}$$
$$[Ca^{2+}] = 10^{-3}$$

In accordance with our choice of principal components, HCO_3^- and A^-, we start with the initial guess that their concentrations are dominant in the corresponding mole balance Equations 48 and 49:

$$[HCO_3^-] = 2.1 \times 10^{-3} = 10^{-2.68} \gg [H_2CO_3^*], [CO_3^{2-}]$$
$$[A^-] = 4 \times 10^{-4} = 10^{-3.4} \gg [HA]$$

As is usually the case, we now need to find the dominant terms in the $TOTH$ Equation 44. The negative terms $[OH^-]$ and $[CO_3^{2-}]$ can be safely eliminated, leaving $[H^+]$, $[H_2CO_3^*]$, and $[HA]$ as possible candidates.

(1) $[H^+] = 5 \times 10^{-4}$ leads to $[H_2CO_3^*] \gg [HCO_3^-]$ according to Equation 51, thus contradicting our initial guess.

(2) $[H_2CO_3^*] = 5 \times 10^{-4}$ leads to no major contradiction. The initial guess for $[HCO_3^-]$ needs to be modified, however, as discussed below.

(3) $[HA] = 5 \times 10^{-4}$ is impossible according to Equation 49 and contradicts our initial guess that $[A^-] \gg [HA]$.

Thus (2) is the only reasonable possibility. The solution to the mole balance Equation 48 must be modified accordingly, leading to

$$[HCO_3^-] = 1.6 \times 10^{-3} = 10^{-2.8}$$

Substituting $10^{-3.3}$ ($= 5 \times 10^{-4}$) for $[H_2CO_3^*]$ and $10^{-2.8}$ for $[HCO_3^-]$ in mass law 51 leads to

$$[H^+] = 10^{-6.8}$$

Further substitution in the other mass laws leads to a complete solution:

$$[OH^-] = 10^{-7.2}$$
$$[CO_3^{2-}] = 10^{-6.3}$$
$$[HA] = 10^{-5.9}$$

By substituting these concentrations into the original mole balance equations 44–49, we can verify that all our approximations are reasonable. Equation 44 is verified within 0.2%, Equation 48 within 1%, and Equation 49 within 0.3%. Much of the remaining error is due to the two decimal digit round-off we used in the logarithmic transformations.

Note how little algebra has been performed to obtain this answer. It is typical of the mathematics of equilibrium problems that concentrations of species are orders of magnitude different from each other as a result of the multiplicative nature of mass laws and that only a few terms are actually important in each mole balance equation. Even when initial approximations are not strictly valid, it is usually easier to iterate to a more precise solution than to try solving a polynomial of high degree. Intelligent reasoning and guessing make iterative solutions relatively simple, rarely requiring more than a couple of attempts before success is achieved.

3. THE GRAPHICAL METHOD

Another way to take advantage of the peculiar mathematics involved in chemical equilibrium problems is to utilize logarithmic graphs, typically a log-concentration vs. pH diagram. Consider the equations that include the concentrations of carbonate species in our examples:

Mole balance equation:

$$TOTHCO_3 = C_T = [H_2CO_3^*] + [HCO_3^-] + [CO_3^{2-}] = 2.1 \times 10^{-3} = 10^{-2.68} \tag{54}$$

Mass law equations:

$$[H_2CO_3^*] = 10^{6.3}[HCO_3^-][H^+] \tag{55}$$

$$[HCO_3^-] = 10^{10.3}[CO_3^{2-}][H^+] \tag{56}$$

By themselves these three equations completely define the concentrations of the carbonate species as a function of $[H^+]$ and are always applicable, no matter what concentrations of other acids and bases are also in the system. For a given total concentration of the dissolved carbonate species C_T, a unique graph of carbonate species concentrations vs. hydrogen ion concentration can thus be developed. This graph has a particularly convenient shape on a log vs. log scale, that is a log-concentration vs. pH diagram. For the time being, we define pH as the negative logarithm (base 10) of the hydrogen ion concentration: $pH = -\log[H^+]$.

3.1 Log-Concentration versus pH Diagram for Carbonate

To develop the sought-for graph of species concentrations as a function of pH, shown in Figure 2.1, we use the logarithmic forms of the governing equations and make all appropriate approximations. First the concentration plots for $[H^+]$ and $[OH^-]$ are obtained directly from the definition of pH and the ion

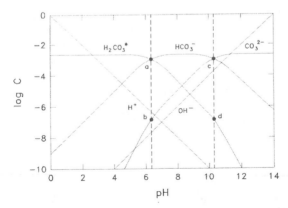

Figure 2.1 Log C–pH diagram for the carbonate system ($C_T = 10^{-2.68}\ M$).

product of water, Equation 50:

$$\log[H^+] = -pH; \quad \text{a straight line of slope } -1.$$
$$\log[OH^-] = -14 + pH; \quad \text{a straight line of slope } +1.$$

The carbonate species are slightly more difficult to graph because their concentrations change dramatically near the relevant pK_a's (negative logs of acidity constants). For example, consider the mass laws for the various carbonate species as the pH is varied from low to high values:

Case A: pH < 6.3

In view of mass laws 55 and 56 and $[H^+] \gg 10^{-6.3}$ we may conclude that $H_2CO_3^*$ is the dominant species at this pH:

$$[H_2CO_3^*] = 10^{6.3}[H^+][HCO_3^-]; \quad [H_2CO_3^*] \gg [HCO_3^-]$$
$$[HCO_3^-] = 10^{10.3}[H^+][CO_3^{2-}]; \quad [HCO_3^-] \gg [CO_3^{2-}]$$

On the basis of mole balance 54, the concentration of $H_2CO_3^*$ is thus practically equal to the total carbonate concentration: $\log[H_2CO_3^*] = \log C_T = -2.68$; a straight line of slope 0 (horizontal). Substituting this result into the logarithmic forms of the mass laws, we then obtain the graphs for HCO_3^- and CO_3^{2-} in the pH range below 6.3:

$$\log[HCO_3^-] = \log[H_2CO_3^*] - 6.3 - \log[H^+]$$
$$= -8.98 + pH; \quad \text{a straight line of slope } +1$$

$$\log[CO_3^{2-}] = \log[HCO_3^-] - 10.3 - \log[H^+]$$
$$= -19.28 + 2pH; \quad \text{a straight line of slope } +2$$

Case B: pH = 6.3

Substituting $[H^+] = 10^{-6.3}$ into the mass law equations 55 and 56 gives:

$$[H_2CO_3^*] = [HCO_3^-]$$
$$[CO_3^{2-}] = 10^{-4.0}[HCO_3^-] \ll [HCO_3^-]$$

Neglecting CO_3^{2-} in the carbonate mole balance Equation 54 yields the following:

$$[H_2CO_3^*] = [HCO_3^-] = 0.5 \times 10^{-2.68} = 10^{-2.98}$$
$$\log[H_2CO_3^*] = \log[HCO_3^-] = -2.98; \quad \text{point a}$$
$$\log[CO_3^{2-}] = -4.0 - 2.98 = -6.98; \quad \text{point b}$$

Case C: $6.3 < pH < 10.3$

Given $10^{-6.3} \gg [H^+] \gg 10^{-10.3}$, and following the same reasoning as in Case A, we find that $[HCO_3^-]$ is the dominant carbonate species and thus:

$$\log[HCO_3^-] = \log C_T$$
$$= -2.68; \text{ horizontal line}$$
$$\log[H_2CO_3^*] = 6.3 + \log[H^+] + \log[HCO_3^-]$$
$$= 3.62 - pH; \text{ straight line of slope } -1$$
$$\log[CO_3^{2-}] = -10.3 + \log[HCO_3^-] - \log[H^+]$$
$$= -12.98 + pH; \text{ straight line of slope } +1$$

Case D: $pH = 10.3$

Reasoning as in Case B, we neglect $H_2CO_3^*$ in the carbonate mole balance to obtain:

$$[HCO_3^-] = [CO_3^{2-}] = 0.5 \times 10^{-2.68} = 10^{-2.98}$$
$$[H_2CO_3^*] = 10^{-4}[HCO_3^-] \ll [HCO_3^-]$$
$$\log[HCO_3^-] = \log[CO_3^{2-}] = -2.98; \text{ point c}$$
$$\log[H_2CO_3^*] = -6.98; \text{ point d}$$

Case E: $pH > 10.3$

Here, according to the mass laws, CO_3^{2-} is the dominant carbonate species:

$$\log[CO_3^{2-}] = \log C_T$$
$$= -2.68; \text{ horizontal line}$$
$$\log[HCO_3^-] = 10.3 + \log[H^+] + \log[CO_3^{2-}]$$
$$= 7.62 - pH; \text{ straight line of slope } -1$$
$$\log[H_2CO_3^*] = 6.30 + \log[H^+] + \log[HCO_3^-]$$
$$= 13.92 - 2pH; \text{ straight line of slope } -2$$

Interpolation of the various lines in the neighborhood of points a, b, c, and d yields the diagram of Figure 2.1. The characteristics of such a diagram are universal among all log-concentration vs. pH graphs for weak acids and bases: (1) various species are dominant in pH ranges delineated by the pK_a's and their concentrations are constant (horizontal lines) and equal to the total weak acid–weak base concentration within that pH range; (2) the graphs of the nondominant species are given by straight lines of integer slopes from $+4$ to

−4, depending on how many protons they must gain or lose to yield the dominant species in that pH range; (3) at the various pK_a's the concentrations of the species that dominate on each side of the pK_a's are both equal to one-half the total weak acid–weak base concentration.

Note that the graphs for $[H^+]$ and $[OH^-]$ are independent of the particular system considered. Note also that the graphs for the weak acid–weak base species ($H_2CO_3^*$, HCO_3^-, CO_3^{2-}) are only a function of the total concentration (given fixed equilibrium constants) and that they translate up and down as a block according to the particular total concentration in the system (C_T in our example).

3.2 The Graphical Method for Problem 1

Let us go back to our first example (see Tableau 2.1) in which we ignored the species CO_3^{2-} and H_2CO_3. The corresponding log C–pH diagram is thus shown in Figure 2.2. The convenient $TOTH$ equation is the one obtained using HCO_3^- as a component (along with (H_2O), H^+, and Ca^{2+}):

$$TOTH = [H^+] - [OH^-] + [CO_2] = 10^{-4} \qquad (57)$$

According to our general hypothesis of great differences among concentrations, the solution to this equation is given by one of the following conditions:

(i) $[H^+] = 10^{-4} \gg [OH^-], [CO_2]$

(ii) $[CO_2] = 10^{-4} \gg [H^+], [OH^-]$

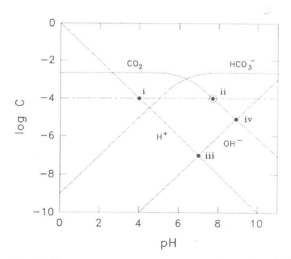

Figure 2.2 Log C–pH diagram for the carbonate system ($C_T = 10^{-2.68}\,M$); graphical solution for Problem 1 (point ii).

(iii) $[H^+] = [OH^-] \gg [CO_2], \, 10^{-4}$

(iv) $[OH^-] = [CO_2] \gg [H^+], \, 10^{-4}$

The points corresponding to the equalities (i)–(iv) are marked on the graph (see Figure 2.2). The only one of these points for which the inequalities are verified is (ii) and a pH of 7.6 is read on the graph.

Note 1. Even if the graph is so rough that it permits only a very vague estimation of pH (say ± 1 unit), it still allows an immediate realization that the only possible solution is provided by (ii), the intersection of the $[CO_2]$ and (-4) lines. From that realization the problem is readily solved analytically.

Note 2. To obtain a graphical solution it is critical that the proper $TOTH$ equation be utilized, that corresponding to the major species as components (in this case HCO_3^-). For example, the $TOTH$ equation corresponding to CO_2 as a component

$$TOTH = [H^+] - [OH^-] - [HCO_3^-] = -2[CaCO_3]_T = -2 \times 10^{-3} \quad (58)$$

would yield four possibilities, according to the assumption of great differences among concentrations:

(i) $[OH^-] = 2 \times 10^{-3} = 10^{-2.7} \gg [H^+], [HCO_3^-]$

(ii) $[HCO_3^-] = 10^{-2.7} \gg [H^+], [OH^-]$

(iii) $[H^+] = [OH^-] \gg [HCO_3^-], \, 10^{-2.7}$

(iv) $[H^+] = [HCO_3^-] \gg [OH^-], \, 10^{-2.7}$

Upon inspection of Figure 2.2, we see that the correct hypothesis is (ii). However, the intersection of the $[HCO_3^-]$ and the (-2.7) graphs is poorly defined on our log–log diagram and does not permit a graphical evaluation of the pH.

3.3 The Graphical Method for Example 2

Consider now our second example which takes into account all the carbonate species and adds a background electrolyte and some weak acid (see Tableau 2.2). The $TOTH$ equation with $[HCO_3^-]$ as a component is written

$$TOTH = [H^+] - [OH^-] + [H_2CO_3^*] - [CO_3^{2-}] + [HA]$$

$$= -[CaCO_3]_T + [CO_2]_T + [HA]_T = 10^{-3.3} \quad (59)$$

To obtain a graphical solution, we add to the diagram of Figure 2.1, the organic acid species as shown on Figure 2.3. Of the many possible simple solutions to

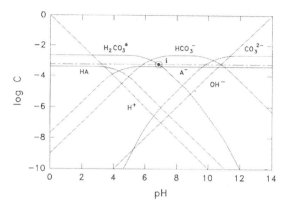

Figure 2.3   Log C–pH diagram for carbonate ($C_T = 10^{-2.68}$ M) and a weak organic acid ($HA_T = 10^{-3.4}$ M; $pK_a = 4.3$); graphical solution for Problem 2 (point i).

Equation 59, the only one consistent with Figure 2.3 is given by

$$[H_2CO_3^*] = 10^{-3.3} \gg \text{all other concentrations}$$

corresponding to point (i) on the graph. The pH of approximately 6.9 can then be read on the graph, as can the corresponding concentrations of the other species.

Note 3. As is now amply clear from the graph of Figure 2.3, the pH is a critical variable in all problems considering weak acid–base systems; that is, all problems of interest in natural waters. For example, some approximate knowledge of the equilibrium pH value permits an a priori choice of the most abundant species as components and leads directly to many simplifications in the mole balance equations. In general, it is in fact practical to consider pH (or $[H^+]$) as the principal unknown and to start the calculation with a rough guess of its value.

Note 4. In this example, the organic acid HA is completely dissociated and behaves like a strong acid. The inclusion of the species HA in the problem and in the graph of Figure 2.3 is a realistic but unnecessary complication.

4. SOME PRACTICAL CONSIDERATIONS

In various places in this section we have pointed out some particularly convenient ways to pose and solve chemical equilibrium problems. Let us summarize here some of these considerations which provide a practical methodology. Some of the "rules" set out here for solids and gases are made in anticipation of future chapters.

A. As much as is possible, depending on one's intuition for chemistry or mathematics or on one's knowledge of an actual system being modelled, an attempt should be made to obtain some range of values for the critical parameters; foremost among these is pH.

B. Sketch out useful graphs (mostly log C–pH diagrams).

C. Choose the components of the system in the following order:

(1) H_2O implicitly

(2) H^+

(3) species with fixed "activities" (see below), such as solids or gases at fixed partial pressure

(4) "major" (i.e., most abundant) dissolved species to round out the set

Such a set of components is called the *principal components*. It is not always possible to choose the principal components a priori, but as the trial and error procedure for solving an equilibrium problem progresses, so does the choice of principal components.

D. Write the corresponding tableau.

E. Write all the equations and solve them by trial and error. To do this, make extensive use of log C–pH diagrams whenever it is practical and always start with the general hypothesis that concentrations are widely different from each other. If the system of equations appears numerically ill behaved, it is probable that the set of components should be modified.

F. For more complicated problems use general computer programs for chemical equilibrium calculations. The major family of such programs— MINEQL and its derivatives, GEOCHEM, MINTEQ, MINEQL+[3a]—is based on the tableau method.

5. COMPLICATIONS: EFFECTS OF IONIC STRENGTH, PRESSURE, AND TEMPERATURE

In our definition of an *ideal* thermodynamic system, we postulated in Equation 10 that the molar free energy μ_i of a species S_i depends exclusively on the mole fraction of that species. In effect, we implied that the composition of the rest of the system has no effect on μ_i whose value is then dependent only on the nature and concentration of S_i. To account for *nonideal effects*— the effects of other solutes on the free energy of the calcium ion for example—we introduce a correction term in the expression of the molar free energy (using the molar scale for simplicity):

$$\mu_{Ca^{2+}} = \mu^o_{Ca^{2+}} + RT \ln [Ca^{2+}] + RT \ln \gamma_{Ca^{2+}} \tag{60}$$

The nonideality correction term is given as a logarithmic expression for convenience and the parameter $\gamma_{Ca^{2+}}$ is called the *activity coefficient* of the calcium ion in solution. Equation 60 can be rewritten

$$\mu_{Ca^{2+}} = \mu^o_{Ca^{2+}} + RT \ln \{Ca^{2+}\} \tag{61}$$

where $\{Ca^{2+}\} = [Ca^{2+}] \gamma_{Ca^{2+}}$ is the *activity* of the calcium ion. The activity of Ca^{2+} is a measure of its "reactivity" which can be decreased by interaction with other ions (Ca^{2+} is stabilized; its free energy is decreased) or increased.

5.1 Interactions among Solutes

Chemical species in solution can interact in a variety of ways ranging from co-valent bonding, to London–van der Waals (i.e., dipole) interactions, to long-range coulombic repulsion and attraction, or even simply to volume exclusion effects when solutes are sufficiently concentrated that they crowd each other. All these interactions are of course superimposed on the interaction between solute and solvent molecules, the solvation effect, which in aqueous solutions is primarily an electrostatic interaction between ions or polar molecules and the polar water molecules. The intensity of these various interactions among molecules in solution depends on the distances between molecules and thus on their concentrations.

The average distance between the centers of molecules at a molar concentration C is approximately $d = 1.2 C^{-1/3}$ nm [$= (\mathcal{N} C)^{-1/3}$]. In a NaCl solution, the sodium and chloride ions whose hydrated radii are 0.36 and 0.33 nm respectively must "touch" each other when the concentration increases above 2.5 M ($= [Na^+] = [Cl^-]$). We can thus expect the activity of all species in solution to be *increased* by volume exclusion at concentrations about a tenth of this value and above.

Interactions involving fixed charges with dipoles, and dipoles with dipoles are a function of the nature of the chemical species whose electronic structure defines the dipole moments. Such specific interactions decrease very rapidly with increasing distance and become negligible when the distance between ions or molecules exceeds 1 nm. The net result is that these interactions are important in solutions near 0.1 M and above. All electrostatic interactions involving dipoles are attractive and thus *decrease* the activities of solutes.

The electrostatic interactions which extend the farthest are those involving fixed ionic charges. These *coulombic* interactions (given by Coulomb's law) depend of course on the charge of the ions and on the dielectric constant of water. The resulting electrostatic energy of interaction reaches values on the order of 0.1 kJ mol^{-1} ($\sim 0.04\, RT$) for singly charged ions at distances of about 10 nm. The net result is that coulombic interactions become important in ionic solutions at concentrations of $10^{-3} M$ and above. Coulombic interactions—attractions of ions of opposite charge and repulsion of ions of like charge—are thus the principal causes of nonideal effects in natural waters. As a result of coulombic forces, the distribution of ions in solution is not uniform. This separation of charges at the molecular scale leads to local variations in the electrical potential of the solution which effectively decrease the total free energy of the system.

5.2 The Debye–Hückel Theory

This concept is the basis of the theory of nonideal effects in dilute solutions—the Debye-Hückel theory—in which it is assumed that the only interactions among solutes not accounted for by chemical reactions are due to their electrical charge (considered as point charges), not their chemical nature. The development of the theory consists in evaluating the electrostatic energy necessary to charge an ion (which is related to the surface potential) and including it in the expression of the free energy of that ion.

Like all theories whose purpose is to calculate a coulombic contribution to free energies (other such theories for polyelectrolytes and for solid surfaces are discussed in Chapters 6 and 8), the Debye–Hückel theory consists primarily of solving the appropriate Poisson–Boltzmann equation, an ad hoc blend of electrostatics and thermodynamics:

1. The Poisson equation of electrostatics relates the Laplacian of the electrical potential, Ψ, to the charge density in the medium, ρ (ε is the dielectric constant and ε_0 is the vacuum permittivity):

$$\nabla^2 \Psi = \frac{-\rho}{\varepsilon\varepsilon_0} \tag{62}$$

where $\nabla^2 \Psi$ is the Laplacian of Ψ; in rectangular coordinates

$$\nabla^2 \Psi = \frac{\partial^2 \Psi}{\partial x^2} + \frac{\partial^2 \Psi}{\partial y^2} + \frac{\partial^2 \Psi}{\partial z^2}$$

2. The Boltzmann distribution relates the concentration of S_i to the exponential of the potential (Z_i is the charge of S_i and F, the Faraday constant):

$$[S_i] = [S_i]_0 \, e^{-(Z_i F/RT)\Psi} \tag{63}$$

It can be derived by expressing the uniformity of the molar free energy of S_i throughout the system (using the extended formula, see Section 5.7) and taking $[S_i]_0$ as the average bulk concentration and $\Psi = 0$ as the corresponding reference potential:

$$\mu_i = \mu_i^\circ + RT \ln [S_i] + Z_i F \Psi = \mu_i^\circ + RT \ln [S_i]_0 \tag{64}$$

Since the local charge density is simply due to the excess of cations over anions or vice versa ($\rho = \sum Z_i F[S_i]$), Poisson and Boltzmann together give us the Poisson–Boltzmann equation:

$$\nabla^2 \Psi = -\frac{1}{\varepsilon\varepsilon_0} \sum Z_i F[S_i]_0\, e^{-(Z_i F/RT)\Psi} \tag{65}$$

whose solution derived by Debye and Hückel for the case of dilute solutions is given in Sidebar 2.2.

The three important results of the Debye–Hückel theory one should remember are:

1. The definition of the *ionic strength* to characterize the role of ionic solutes in modulating coulombic interactions:

$$I = \tfrac{1}{2}\sum Z_i^2\, [S_i]_0 \tag{66}$$

2. The square root dependency of the logarithm of the activity coefficient on I for small values of I:

$$\log \gamma_i \simeq -0.5\, Z_i^2\, I^{1/2} \tag{67}$$

3. The magnitude of the ionic strength effect in dilute solutions which yields approximately $\log \gamma = -0.1$ at $0.1\,M$ ionic strength for a monovalent ion.

SIDEBAR 2.2

Solutions to the Poisson–Boltzmann Equation:
I. DEBYE–HÜCKEL THEORY (1923)

Our first objective is to find the electrostatic potential at the surface of an ion in solution as a function of its charge. In the Debye–Hückel model, the ion is pictured as a sphere of radius R_0 and surface charge density σ. The ionic medium surrounding this central ion is assumed to be a region of continuously varying charge density ρ. For a spherically symmetrical system, then, the Poisson–Boltzmann equation becomes

$$\frac{d^2\Psi}{dr^2} + \frac{2}{r}\frac{d\Psi}{dr} = \frac{-1}{\varepsilon\varepsilon_0}\sum Z_i F[S_i]_0\, e^{-Z_i F\Psi/RT}$$

*This and other sidebars on the Poisson–Boltzmann Equation were written with the assistance of B. M. Bartschat.

(*continued*)

Since solving nonlinear differential equations was not so easy before the invention of digital computers, Debye and Hückel linearized this equation with the approximation $e^x = 1 + x$. This approximation is valid when potentials are small ($\Psi < (RT/F \simeq 25\,\text{mV})$) and leads to

$$\frac{d^2\Psi}{dr^2} + \frac{2}{r}\frac{d\Psi}{dr} = \frac{-F}{\varepsilon\varepsilon_0}\left[\sum Z_i[S_i]_0 - \sum Z_i^2[S_i]_0\frac{F\Psi}{RT}\right]$$

Because of the electroneutrality condition for the bulk solution, the first summation term is equal to zero. In the second term, only Ψ is a function of r and the equation can be written as

$$\frac{d^2\Psi}{dr^2} + \frac{2}{r}\frac{d\Psi}{dr} = \kappa^2\Psi$$

where κ, the inverse of the *Debye length* is given by

$$\kappa = \left[\frac{F^2}{\varepsilon\varepsilon_0 RT}\sum Z_i^2[S_i]_0\right]^{1/2}$$

The general solution is

$$\Psi(r) = C_1\frac{e^{-\kappa r}}{r} + C_2\frac{e^{\kappa r}}{r}$$

Since we expect the local concentration of co- and counter ions to approach the bulk concentration as we move away from the influence of the central ion, we can write the first boundary condition as

$$\lim_{r\to\infty}\Psi(r) = 0$$

The second boundary condition is a form of Gauss's law and relates the slope of Ψ versus r at the surface of the ion to the surface charge density:

$$\left[\frac{d\Psi}{dr}\right]_{R_0} = \frac{-\sigma}{\varepsilon\varepsilon_0} = \frac{-q}{4\pi\varepsilon\varepsilon_0 R_0^2}$$

where q is the total charge of the sphere. By inserting the boundary conditions we find the values of the constants C_1 and C_2:

$$\Psi(r) = \frac{q}{4\pi\varepsilon\varepsilon_0}\left[\frac{r^{\kappa R_0}}{1 + \kappa R_0}\right]\frac{e^{-\kappa R}}{r}$$

(continued)

We now wish to obtain the electrostatic correction to the chemical potential of a species by calculating the work required to charge the ionic sphere while allowing the ionic atmosphere around the ion to rearrange itself after each infinitesimal change in charge. Since work is given by electrostatic potential multiplied by charge, this work function for an ion of charge number Z and radius R is given by

$$W_{el} = \int_0^{Z_e} \Psi_{R_0}(q)dq = \frac{Z^2 e^2}{8\pi\varepsilon\varepsilon_0 R_0}\left[\frac{1}{\kappa R_0 + 1}\right]$$

where Ψ_{R_0} is the potential at the surface. According to these results, our expression for the activity coefficient of an ion of charge number Z, or γ_z, should then be

$$\ln \gamma_z = \frac{\mathcal{N} W_{el}}{RT} = Z^2 \left[\frac{\mathcal{N} e^2}{8\pi\varepsilon\varepsilon_0 RT}\right]\left[\frac{1}{R_0}\right]\left[\frac{1}{\kappa R_0 + 1}\right]$$

However, this result is not consistent with our choice of reference state. The solution should be ideal (and $\ln \gamma_z$ equal to zero) at infinite dilution of ions (when κ is equal to zero). Thus we subtract out the contribution to W_{el} when κ is equal to zero and our expression becomes

$$\ln \gamma_z = -Z^2 \left[\frac{\mathcal{N} e^2}{8\pi\varepsilon\varepsilon_0 RT}\right]\left[\frac{\kappa}{\kappa R_0 + 1}\right]$$

Note that $\ln \gamma_z$ is now negative; it takes less work to charge the ion in the presence of an ionic atmosphere than in its absence. Ions are stabilized by other ions in a solution. The sum of concentrations of ions in the bulk phase weighted by the square of their charge is defined as the ionic strength I of the solution:

$$I = \tfrac{1}{2}\sum_i Z_i^2 [S_i]_0$$

After substituting the physical constants we finally obtain a familiar equation for Debye–Hückel activity coefficients:

$$\ln \gamma_z = -AZ^2 \frac{I^{1/2}}{1 + BR_0 I^{1/2}}$$

The constants A and B depend on the dielectric constant ($\varepsilon\varepsilon_0$) and absolute temperature T of the system. In water at 25°C, their values are

$$A = 1.17 \, \text{mol}^{-1/2} \, \text{L}^{1/2}; \qquad B = 0.329 \, \text{Å}^{-1} \, \text{mol}^{-1/2} \, \text{L}^{1/2}$$

(*continued*)

The ionic radius R_0 (a in Fig. 2.4) is a fitting parameter whose value varies between 3 and 10 Å. At low ionic strengths, or for point charges ($R_0 = 0$), the second term in the denominator can be neglected. Then $\ln \gamma_z$ is proportional to the square root of ionic strength. This relationship is sometimes called the Debye–Hückel limiting law and has been verified experimentally.

5.3 "Nonideal" Nonideal Effects

For all its approximations, the Debye–Hückel theory provides a remarkably accurate prediction of nonideal effects in dilute solutions. As discussed above, however, when a solution becomes sufficiently concentrated, molecules interact through means other than long-range Coulombic forces; interactions involving the specific dipole properties of molecules become important. Further, the point charge approximation becomes invalid and molecules begin crowding each other. The activity coefficients of ions then do not exactly follow the "ideal" nonideal formula of Debye and Hückel, and the activity coefficients of neutral molecules are observed to deviate from unity. The complexity of the interactions among molecules in (moderately) concentrated solutions is such that theories are of little help. The approach then consists of introducing empirical modifications to the Debye–Hückel formula.

Empirical Formulae for Activity Coefficients at High Ionic Strength Although activity coefficients of individual ions cannot be measured, the mean activity coefficients of binary electrolytes can be obtained experimentally by various methods such as solubility measurements or electrochemical determinations of free energies. The mean activity coefficient is related to the single ion activity coefficients γ_+ and γ_- by the equation

$$(Z_+ + Z_-)\log \gamma_\pm = Z_- \log \gamma_+ + Z_+ \log \gamma_- \tag{68}$$

where Z_+ and Z_- are the charge numbers of the cation and the anion, respectively. As can be seen in Figure 2.4, the Debye–Hückel formula does not accurately predict the activity coefficients of simple electrolytes at high ionic strength; even at an ionic strength of 0.1 M, the activity coefficient of such an important salt as NaCl is measurably underpredicted by the formula. The chemical nature of the species, not just their electrical charge, then plays a role in their mutual interactions.

To extend the applicability of the Debye–Hückel formula to higher ionic strength systems, a variety of empirical and semiempirical expressions have been proposed. One of the simplest and most widely used expressions is that of Davies, who eliminated the size parameter (R_0) and added a linear term to the Debye–Hückel formula given in the Sidebar 2.2:

$$\ln \gamma_i = - AZ_i^2 \left[\frac{I^{1/2}}{1 + I^{1/2}} - bI \right] \tag{69}$$

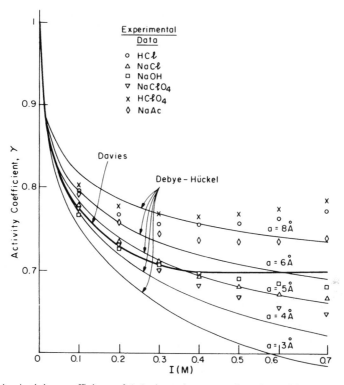

Figure 2.4 Activity coefficient of 1:1 electrolytes as a function of ionic strength. The experimental data represent the *mean* activity coefficients of the respective electrolytes (taken from Harned and Owen, 1958[2]). The lines correspond to the predictions of the Debye–Hückel and the Davies formulae for singly charged ions. The original value $b = 0.2$ has been used for the empirical parameter of the Davies formula; Davies himself later suggested a slightly higher value: $b = 0.3$.

By optimizing for measured activity coefficients below $I = 0.1\,M$, the empirical parameter b was estimated by Davies to be in the range 0.2 to 0.3. The principal advantage of Davies formula is to provide, as exhibited by the data, a quasi-constant value of the activity coefficients in the range $I = 0.3$ to $0.7\,M$.

In aquatic systems the ionic strength rarely exceeds $0.7\,M$, and the inaccuracies introduced by using the Davies equation are usually smaller than other sources of errors and uncertainties. Numerical values of activity coefficients based on the Davies equation are given in Table 2.3. Throughout this text we use these values consistently to account for nonideal effects.

When the objective is to analyze accurately the interactions of seawater ions or to study the chemistry of concentrated brines, an approach more sophisticated than Davies semiempirical formula becomes necessary. The activity coefficients must account for specific as well as nonspecific ion interactions and exhibit a large increase at high ionic strengths.

The most commonly used solution to this problem is the Brønsted–

TABLE 2.3   Activity Coefficients for Dissolved Ionic Species

I	$-\log \gamma_z$			
(mol L^{-1})	$Z = 1$	$Z = 2$	$Z = 3$	$Z = 4$
0.0001	0.005	0.02	0.05	0.08
0.0005	0.01	0.04	0.10	0.18
0.001	0.02	0.06	0.14	0.25
0.005	0.03	0.13	0.30	0.53
0.01	0.05	0.18	0.40	0.72
0.05	0.09	0.35	0.78	1.39
0.1	0.11	0.44	0.99	1.76
0.3	0.13	0.52	1.17	2.08
0.5	0.15	0.60	1.35	2.40
1.0	0.14	0.56	1.26	2.24
2.0	0.11	0.44	0.99	1.76
3.0	0.07	0.28	0.63	1.12
4.0	0.03	0.12	0.27	0.48

Based on Davies Eq. with A $= 1.17\ M^{-1/2}$ and B $= 0.3\ M^{-1/2}$.

Guggenheim[4,5] specific ion interaction model and its extensions by Scatchard, Mayer,[6-8] and others. The general formula for this model consists of a *virial* expansion of the form

$$\ln \gamma_i = \ln \gamma_{DH} + \sum_j B_{ij}[S_j] + \sum_j \sum_k C_{ijk}[S_j][S_k] + \cdots \qquad (70)$$

The first term is simply the Debye–Hückel activity coefficient. The *second virial coefficients* (B_{ij}) account for specific interactions among pairs of ions, the *third virial coefficients* (C_{ijk}) for specific interactions among three ions, and so on. (Note that in this formula [S_j] represents the *molar* concentrations of S_j.) Like γ_{DH}, the higher-order virial coefficients are functions of the ionic strength, and successful empirical formulations for these coefficients are at the core of the thermodynamic description of concentrated electrolytes.[9] Up to ionic strengths of about $4\ M$, good agreement with experimental data can be obtained using an expansion that stops with the second virial coefficients. The higher-order terms become important only in the most concentrated systems.

Activity Coefficients of Neutral Molecules   Neutral molecules also are found to behave nonideally in solution. This is most often observed as a *salting out* effect such as the decrease in the solubility of sucrose in the presence of sodium chloride. This effective *increase* in the activity coefficient of sucrose ($\gamma_{sucrose} > 1$) is mostly due to the interaction between sucrose and the sodium and chloride ions which can be interpreted as a modification of the dielectric constant of water. The added sucrose lowers the dielectric constant of water, thus increases the free energy of the charged ions and, in effect, contributes to the total free energy of the system beyond its "own" free energy: $dG > \mu_{sucrose}\ dn_{sucrose}$. This effect is taken into account by assigning an activity coefficient greater than 1 to sucrose. In practice, a useful approximate empirical formula for the activity coefficient of neutral molecules is given by

$$\log \gamma \simeq 0.1 \, I$$

Thus, the activity coefficients of neutral molecules are negligibly different from 1 at ionic strengths less than 0.1 M and we usually neglect such nonideality effects.*

Note also that our whole discussion of nonideal behavior has been limited to the aqueous phase. At high pressures—a condition of little interest to us—gases also deviate from ideal behavior; they no longer verify Boyle–Mariotte's law. This is accounted for by defining *fugacity coefficients* and *fugacities*, which are to partial pressures what activity coefficients and activities are to concentrations.

5.4   Nonideal Effects on Equilibrium

The problem of introducing the nonideal term $RT \ln \gamma$ into the mass law equation can be treated in one of two ways:

1. As shown earlier, the last two terms of Equation 60 can be combined to define the *activity* $\{S_i\}$ of the species S_i. The mass law is thus applied to the activities of the reactants and products of a reaction rather than to their concentrations. For example, Equations 3 and 4 are rewritten

$$\{H^+\}\{OH^-\} = K_1 = 10^{-14.0} \tag{71}$$

$$\frac{\{HCO_3^-\}}{\{CO_2\}\{OH^-\}} = K_2 = 10^{7.7} \tag{72}$$

2. Alternatively, the first and the last term of Equation 60 can be combined to define the *concentration equilibrium constant* ^cK:

$$\ln {}^cK = -\frac{1}{RT}\sum_i v_i \mu_i^\circ - \sum_i v_i \ln \gamma_i \tag{73}$$

In this case the mass law equation is applied to the concentrations of reactants and products in a reaction but the equilibrium constant is modified to account

---

*One should not confuse the mean activity coefficient of electrolytes with the activity coefficients of neutral molecules. The former represents a mean thermodynamic quantity for both the cation and the anion in a fully dissociated electrolyte. The latter describes the nonideal behavior of the single undissociated species in solution.

for nonideal effects:

$$[H^+][OH^-] = K_1[\gamma_{H^+} \cdot \gamma_{OH^-}]^{-1} = {}^c K_1 \qquad (74)$$

$$\frac{[HCO_3^-]}{[CO_2][OH^-]} = K_2 \frac{\gamma_{CO_2} \gamma_{OH^-}}{\gamma_{HCO_3^-}} = {}^c K_2 \qquad (75)$$

In both cases the convenient form of the mass law equation is retained. In the first case this is achieved by replacing each species concentration by an idealized quantity, the species' activity. This activity is smaller than the concentration, and its difference from the concentration is a measure of the decreased reactivity of the ion due to the stabilizing effect of its electrostatic interactions with other ions. In the second case, the standard free energy change of each reaction is decreased, and the ions themselves are considered to behave ideally. This is strictly identical to considering a reference state that includes the effects of the major solutes (those contributing most of the ionic strength of the system) on the free energy of the ions: $RT \ln \gamma_i$ is simply included in μ_i^o. This second approach is a great deal more convenient from a numerical point of view, and we use it preferentially.

The activity, rather than the concentration, of the hydrogen ion is usually measured with pH electrodes. We thus now define the pH as the negative logarithm of the activity of the hydrogen ion:

$$pH = -\log\{H^+\} \qquad (76)$$

As a result, *mixed acidity constants*, which involve the activity of H^+ and the concentrations of the other reacting solutes, are often used for acid–base reactions. For example, the mixed acidity constant for the dissociation reaction

$$HA = H^+ + A^-$$

is defined as

$$K_a' = \frac{\{H^+\}[A^-]}{[HA]} = K_a \frac{\gamma_{HA}}{\gamma_{A^-}} \qquad (77)$$

Tabulations of equilibrium constants (e.g., Sillen and Martell,[10] Martell and Smith, and Smith and Martell[11]) normally contain information on the pertinent ionic strength condition as well as temperature and pressure. Often the values of the constants are extrapolated to zero ionic strength, and these "infinite dilution constants" are those given in the appropriate tables at the beginning of the following chapters.

Let us now examine the practical consequences of nonideal effects on equilibrium. To do so we go back to our first example, which we solved in Section 2.2 without taking into account any ionic strength effect.

As calculated on the basis of our previous solution, the ionic strength of this system is $3\,\mathrm{m}M$:

$$I \simeq \tfrac{1}{2}(4[\mathrm{Ca}^{2+}] + [\mathrm{HCO_3^-}] + \text{negligible contributions from } \mathrm{H^+} \text{ and } \mathrm{OH^-})$$
$$= 3 \times 10^{-3}\,M$$

The concentration equilibrium constants can thus be calculated from Equations 74 and 75 by taking the appropriate activity coefficients from Table 2.3:

$$\log {}^c\!K_1 = \log K_1 - \log \gamma_{\mathrm{H^+}} - \log \gamma_{\mathrm{OH^-}}$$
$$\log {}^c\!K_1 = -14.0 + 0.03 + 0.03 = -13.94$$

and

$$\log {}^c\!K_2 = \log K_2 + \log \gamma_{\mathrm{OH^-}} - \log \gamma_{\mathrm{HCO_3^-}} (+ \log \gamma_{\mathrm{CO_2}} = 0)$$
$$\log {}^c\!K_2 = 7.7 - 0.03 + 0.03 = 7.7$$

The system of equations to be solved is Equations 32–36 in which the equilibrium constants for the mass laws Equations 35 and 36 now have the values $10^{-13.94}$ and $10^{-6.24}$ respectively ($-6.24 = -13.94 + 7.7$). The reasoning given in Section 2.2 to solve the problem is unaffected by our change of constants and the concentrations of the dominant species are unchanged:

$$[\mathrm{Ca}^{2+}] = 10^{-3}$$
$$[\mathrm{HCO_3^-}] = 2 \times 10^{-3}$$
$$[\mathrm{CO_2}] = 10^{-4}$$

Only the concentration of the hydrogen ion is affected by our change of constants:

$$[\mathrm{H^+}] = 10^{-7.54}$$

For the hydroxide ions the net effect is null:

$$[\mathrm{OH^-}] = 10^{-6.40}$$

Note that the pH $(= -\log\{\mathrm{H^+}\})$ of this solution is affected by nonideal effects:

$$\mathrm{pH} = -\log\{\mathrm{H^+}\} = 7.54 + 0.03 = 7.57$$

instead of 7.6.

If the ionic strength of this system is increased to $0.5\,M$ by adding a background electrolyte, the pH decreases to 7.35 but the system otherwise retains the same composition. Clearly, ionic strength effects are relatively picayune in a system where the major species' concentrations are fixed by the recipe through mole balances.

A very different result is obtained in systems where the concentrations of the major solutes are controlled by the solubility of an ionic solid. Consider for example the solubility product of calcite:

$$CaCO_3(s) = Ca^{2+} + CO_3^{2-}; K_{CaCO_3}$$

$$at\ I = 0\ M, pK_{CaCO_3} = 8.35$$

$$at\ I = 0.001\ M, pK_{CaCO_3} = 8.23$$

$$at\ I = 0.1\ M, pK_{CaCO_3} = 7.47$$

$$at\ I = 0.5\ M, pK_{CaCO_3} = 7.15$$

Thus, if the concentration of Ca^{2+} is in large excess and somehow fixed, the concentration of CO_3^{2-} in a system at equilibrium with calcite can vary by more than one order of magnitude just because of nonideal effects. Clearly activity coefficient corrections must be made to obtain an accurate description of the composition of the system when the ionic strength exceeds $0.01\ M$.

5.5   Effects of Pressure on Equilibrium

To discuss the effects that changes in pressure or temperature may have on equilibrium conditions in chemical systems, we must now belatedly introduce fundamental thermodynamic quantities, other than molar free energy, such as *molar volume, entropy,* and *enthalpy.* We do so by relying on intuitive and molecular interpretations, a convenient if very "unthermodynamic" palliative.

Consider that a chemical system of volume V is subjected to an increase in pressure dP (fixed temperature; fixed composition). The increase in the free energy of the system is written

$$dG = VdP \tag{78}$$

thus defining the volume as the partial differential of the free energy with respect to the pressure. (See Section 5.7.2 for an intuitive justification of this result. Note that an increase in pressure at constant volume necessitates no mechanical energy and corresponds to no change in the *internal energy* of a system; it corresponds, however, to a *free energy change.*) To account for this change in the total free energy of the system as a sum of changes in the molar free energies of individual molecules, dG must also be given by

$$dG = \sum_i n_i \frac{\partial \mu_i}{\partial P} dP \tag{79}$$

To evaluate the pressure differential of the molar free energies, it is then necessary to break down the total volume of the system into molar volumes:

$$V = \sum_i n_i \bar{V}_i \tag{80}$$

Each (partial) molar volume $\bar{V}_i$ is defined as the increase in the total volume of the system upon the addition of an infinitesimal number of moles of the species S_i:

$$\bar{V}_i = \frac{\partial V}{\partial n_i} \tag{81}$$

For solids, $\bar{V}_i$ is simply the ratio of the molecular weight M_i of S_i to the density ρ:

$$\bar{V}_i = \frac{M_i}{\rho}$$

For gases, $\bar{V}_i$ is the ratio of the total gas volume V_g to the total number of gas molecules n_g, and thus it is the same for all the gaseous species in the same phase:

$$\bar{V}_i = \frac{V_g}{n_g} = \frac{RT}{P}$$

For solutes, $\bar{V}_i$ does not correspond to such a simple notion. In fact the partial molar volumes of aqueous solutes are often negative since addition of a salt usually results in a decrease in the total volume of the solution. Thus $\bar{V}_i$ for a solute is also dependent, in principle, upon the composition of the solution.

Substituting Equation 80 in Equation 78, comparing with Equation 79, and recognizing that the identity must be verified for any composition of the system, we find

$$\frac{\partial \mu_i}{\partial P} = \bar{V}_i = \frac{\partial V}{\partial n_i} \tag{82}$$

With the pressure dependence of the molar free energy being entirely accounted for in the standard molar free energy,*

$$\frac{\partial \mu_i}{\partial P} = \frac{\partial}{\partial P}(\mu_i^o + RT \ln X_i) = \frac{\partial \mu_i^o}{\partial P}$$

---

*As is the case for ionic strength, the effects of pressure on molar free energy can also be accounted for with an appropriate activity coefficient.

the final expression is

$$\frac{\partial \mu_i^\circ}{\partial P} = \bar{V}_i \tag{83}$$

On the basis of Equation 83 we can calculate the effect of pressure on equilibrium constants by differentiating Equation 26:

$$\frac{\partial \ln K}{\partial P} = -\frac{1}{RT}\frac{\partial}{\partial P}\left[\sum_i v_i \mu_i^\circ\right]$$

therefore,

$$\frac{\partial \ln K}{\partial P} = -\frac{\Delta V^\circ}{RT} \tag{84}$$

where

$$\Delta V^\circ = \sum_i v_i \bar{V}_i \tag{85}$$

The term ΔV° represents the change in volume of the system due to the advancement of the reaction. Equation 84 has a qualitatively pleasing aspect: a reaction that results in an expansion of the system is favored (pushed to the right) by a decrease in pressure, and vice versa, a classic example of Le Chatelier's principle. If ΔV° is approximately constant over the range of pressures considered, often not a bad approximation, the value of the equilibrium constant at a pressure P can be calculated from that at a pressure P_0:

$$\ln \frac{K_P}{K_{P_0}} = -\frac{\Delta V^\circ}{RT}(P - P_0) \tag{86}$$

For example, let us consider again the reaction for calcite dissolution:

$$CaCO_3(s) = Ca^{2+} + CO_3^{2-}$$

with the data

$$\bar{V}_{CaCO_3} = +36.9 \, cm^3 \, mol^{-1}$$
$$\bar{V}_{Ca} = -17.7 \, cm^3 \, mol^{-1}$$
$$\bar{V}_{CO_3} = -3.7 \, cm^3 \, mol^{-1}$$

This reaction is chosen as an example because it has a rather large negative volume change:

$$\Delta V^\circ = -3.7 - 17.7 - 36.9 = -58.3 \, cm^3 \, mol^{-1}$$

$$\frac{\Delta V^\circ}{RT} = \frac{58.3 \, cm^3 \, mol^{-1}}{8.31 \times 298 \, J \, mol^{-1}} = 2.35 \times 10^{-2} \, cm^3 \, J^{-1} \simeq 2.38 \times 10^{-3} \, atm^{-1}$$

Then for every 1 atm increase in pressure, the log of the solubility product is increased by

$$\log \frac{K_P}{K_{P_0}} = (2.38 \times 10^{-3}/2.3) \simeq 0.001 \tag{87}$$

corresponding to only 0.2% increase in K_{CaCO_3}. Thus despite the relatively large negative volume change of this reaction, the effect of pressure on equilibrium is rather small. In aquatic chemistry such pressure effects can generally be neglected, except in the deep ocean. At a depth of 10,000 m (1,000 atm) the calculated increase in log K_{CaCO_3} becomes approximately 1 (i.e., a factor of 10 in K_{CaCO_3}). In that case, however, the partial molar volumes cannot be considered constant as in Equation 86. They are markedly affected by pressure and ionic strength. The increase in calcite solubility in deep oceanic waters is in fact less than predicted by Equation 87.

5.6 Effect of Temperature

To study the effect of temperature on equilibrium we need to introduce the notions of enthalpy, H, and entropy, S, as distinct constituents of the free energy

$$G = H - TS \tag{88}$$

and molar enthalpies and entropies as constituents of standard molar free energies

$$\mu_i^\circ = \bar{H}_i^\circ - T\bar{S}_i^\circ \tag{89}$$

Imprecise as the molecular images may be, we conceive of the enthalpy as the free energy corresponding to intra- and intermolecular forces. Thus the energy involved in breaking (or making) chemical bonds is part of the enthalpy, as is the energy involved in overcoming the forces holding solutes to water molecules. To a first approximation, the enthalpy of a chemical species is independent of temperature. The entropy term ($-TS$), on the other hand, represents the energy involved in confining the molecules (restricting their degrees of freedom) or, as we know, in concentrating them. Entropy itself represents the degree of "mixed-up-ness" (Gibbs's term) of the system. The corresponding free energy is proportional to the absolute temperature whose scale is defined by zero entropic free energy. We note in passing that the solvation of ions in water, which results in an ordering of the solvent, involves a relatively large entropic contribution.

The expression of the molar free energy thus becomes

$$\mu_i = \bar{H}_i^o - T\bar{S}_i^o + RT \ln X_i \tag{90}$$

For the purpose of this simplified derivation, we take both $\bar{H}_i^o$ and $\bar{S}_i^o$ to be constant over the range of temperatures considered.

Dividing Equation 90 by T and differentiating with respect to T,

$$\frac{\partial(\mu_i/T)}{\partial T} = \frac{\partial(\mu_i^o/T)}{\partial T} = \frac{\partial(\bar{H}_i^o/T)}{\partial T} = -\frac{\bar{H}_i^o}{T^2} \tag{91}$$

The effect of temperature on equilibrium constants is then obtained by differentiating Equation 26:

$$\frac{\partial \ln K}{\partial T} = \frac{\partial}{\partial T}\left[-\frac{1}{RT}\sum v_i \mu_i^o\right]$$

which results in

$$\frac{\partial \ln K}{\partial T} = +\frac{\Delta H^o}{RT^2} \tag{92}$$

where

$$\Delta H^o = \sum_i v_i \bar{H}_i^o \tag{93}$$

If the enthalpy change of the reaction, ΔH^o, remains constant, the relationship between equilibrium constants at two temperatures T_0 and T is obtained by integrating Equation 92:

$$\ln\left[\frac{K_T}{K_{T_0}}\right] = -\frac{\Delta H^o}{R}\left[\frac{1}{T} - \frac{1}{T_0}\right] \tag{94}$$

This formula, which is known as the van't Hoff equation, holds rather well over the range of temperatures considered in aquatic chemistry. It is particularly useful because ΔH^o is in fact a quantity measurable a priori: it is the heat absorbed by the chemical reaction as it proceeds to the right at constant pressure and is called the *change in enthalpy* of the reaction. The change in enthalpy of the reaction corresponds chiefly to the energy involved in breaking and making bonds, unless there is also a large change in intermolecular forces, such as may be the case upon dissolution of an ionic solid. Equation 92 is also intuitively pleasing if we have Le Chatelier's intuition: an exothermic reaction ($\Delta H° < 0$) is favored (K increases) by a decrease in temperature, and vice versa.

Figure 2.5 illustrates the magnitude of the temperature effect on various equilibrium constants. The effect is rather large, with some equilibrium constants changing by a factor of 10 over the 35°C span typically encountered in natural waters. For example, a cooling to 4°C increases the solubility product of $CaCO_3$ by almost 0.1 log units and warming to 40°C decreases it by more than 0.1 log units compared to its standard value at 25°C. Unlike pressure corrections, temperature corrections of equilibrium constants cannot in general be neglected in aquatic chemistry. Although in this text we assume a constant temperature of 25°C, it is important to make appropriate temperature corrections when studying a particular water body at another temperature. Note that some of the lines on Figure 2.5 exhibit noticeable curvature and thus do not obey Equation 94. This can be accounted for in more complete formulae by considering the variations in the enthalpy (i.e., in the heat capacity of the system) due to the advancement of the reaction.

5.7   Concentration Gradients in Equilibrium Systems

We have seen in Section 1.5 that one of the consequences of thermodynamic equilibrium in a solution phase is the uniform concentration of species throughout the phase. This result is not strictly correct, but the very small or very large distances over which equilibrium concentration gradients exist in aquatic chemistry usually obscure or preclude their observation.

Temperature gradients are not possible in equilibrium systems, but pressure or electrical potential gradients are quite common. Atmospheric or hydrostatic pressure gradients do not require a continuous input of energy and do not result in perpetual motion. The same is true of the variations in electrical potential in the neighborhood of a charged object. The result of uniform concentrations in Section 1.5 was obtained by ignoring such gradients, an omission that we wish to repair here.

5.7.1   Extended Expression of Molar Free Energy   To study the effects of pressure and electrical potential gradients on chemical equilibrium, we extend the expression of molar free energies (Equation 10) to include explicitly the contribution of each species to the pressure–volume term and to the electrostatic term of the free energy of the system. The first is $\bar{V}_i(P - P_0)$, as seen in Section 5.5, P_0 being the pressure of the reference state. The second is $Z_i F(\Psi - \Psi_0)$, where Z_i is the charge number of the species, F is the Faraday constant (thus $Z_i F$ is the electrical charge of one mole of i), and Ψ_0 is the electrical potential of the reference state (usually taken at zero). Other types of energies could be included if we wished to consider other processes such as surface tension. In the study of tall water columns and their pressure gradients we must include gravitational

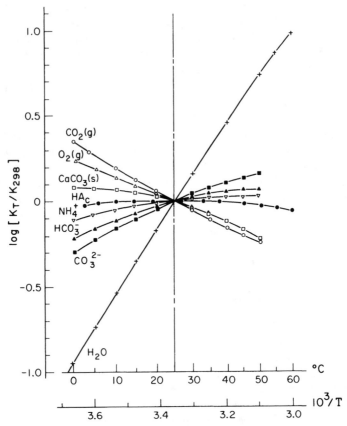

Figure 2.5   Variations of a few equilibrium constants with temperature. The reactions, the equilibrium constants at 25°C, and the corresponding data sources are as follows:

$CO_2(g):CO_2(g) = CO_2(aq)$; $\log K_{298} = -1.464$ (Harned and Davis, 1943[12]).

$O_2(g):O_2(g) = O_2(aq)$; $\log K_{298} = -3.066$ (Carpenter, 1966[13]).

$CaCO_3(s):CaCO_3(s) = Ca^{2+} + CO_3^{2-}$; $\log K_{298} = -8.475$ (Jacobson and Langmuir, 1974[14]).

$HAc:HAc = H^+ + Ac^-$; $\log K_{298} = -4.756$ (Harned and Ehlers, 1933[15]).

$NH_4^+:NH_4^+ = H^+ + NH_3(aq)$; $\log K_{298} = -9.249$ (*CRC Handbook of Chemistry and Physics*, 62nd ed. 1981[16]).

$HCO_3^-:CO_2 + H_2O = H^+ + HCO_3^-$; $\log K_{298} = -6.351$ (Harned and Davis, 1943[12]).

$CO_3^{2-}:HCO_3^- = H^+ + CO_3^{2-}$; $\log K_{298} = -10.330$ (Harned and Scholes, 1941[17]).

$H_2O:H_2O = H^+ + OH^-$; $\log K_{298} = -13.9965$ (*CRC Handbook of Chemistry and Physics*, 62nd ed., 1981[16]).

energy. Since the molar contribution of a species to the total mass m of a system is simply its molecular weight $M_i (m = \sum n_i M_i)$, the molar gravitational energy is given by $M_i g(z - z_0)$ where g is the gravity of the earth (taken here as constant for justifiable simplicity) and $(z - z_0)$ is some elevation above the reference elevation z_0.

The extended expression of molar free energy is thus written

$$\mu_i = \mu_i^o + RT \ln X_i + \bar{V}_i(P - P_0) + Z_i F(\Psi - \Psi_0) + M_i g(z - z_0) \tag{95}$$

Like the simpler expression, Equation 10, this equation satisfies the Gibbs–Duhem relation. The condition of uniformity of the molar free energy in an equilibrium system is thus applicable:

$$\mu_i = \text{constant throughout the system} \tag{96}$$

as is the equation

$$\sum_i v_i \mu_i = 0 \tag{97}$$

if the species are involved in a chemical reaction with stoichiometric coefficients v_i. Equation 95 provides a particularly convenient way to study the effects of pressure or electrical potential gradients on species concentrations—or vice versa—as in the case of osmotic pressure. As an illustration of the methodology, we now calculate the equilibrium distribution of chemical species in a tall water column, omitting the electrical term. For convenience we replace the mole fraction by the activity of the species:

$$\mu_i = \mu_i^o + RT \ln \{S_i\} + \bar{V}_i(P - P_0) + M_i g(z - z_0) \tag{98}$$

where

$\{S_i\} = 1$ for solids

$\quad\;\; = [S_i]$ for ideal solutes

$\quad\;\; = P_i$ for ideal gases

5.7.2 Distribution of Species in a Tall Water Column

Water Consider first the distribution of the solvent (the water is taken as incompressible):

$$\{\text{water}\} = 1; \qquad M_w = \rho_w \bar{V}_w$$

therefore

$$\mu_w = \mu_w^o + \bar{V}_w(P - P_0) + \rho_w \bar{V}_w g(z - z_0) \tag{99}$$

Taking the reference pressure to be that of the atmosphere ($P_0 = P_{atm}$) at the surface of the water, which is itself taken as the origin of elevations, $z_0 = 0$, we derive the hydrostatic pressure formula from the condition of uniformity of molar free energy:

$$\mu_w = \text{constant} = \mu_w^o$$

therefore

$$P = P_{atm} - \rho_w gz \tag{100}$$

(Remember that z is negative at depth.)

Conversely, on the basis of the hydrostatic pressure formula, which is merely an Archimedean statement of mechanical equilibrium, the form of the pressure term in the free energy expression can be justified. By describing the equilibrium between two adjacent layers of an incompressible fluid as both a balance of mechanical forces and an equality of free energies, one can derive Equation 78.

Solids An equally familiar result is obtained when the equation is applied to a solid species:

$$\{S_i\} = 1; \quad M_i = \rho_i \bar{V}_i \quad [P_0 = P_{atm}; \ z_0 = 0]$$

therefore

$$\mu_i = \mu_i^o + \bar{V}_i(P - P_{atm}) + \rho_i \bar{V}_i gz$$

Introducing Equation 100,

$$\mu_i = \mu_i^o + \bar{V}_i(\rho_i - \rho_w)gz \tag{101}$$

The molar free energy of the solid cannot in general be uniform in the water column:

1. $\rho_i > \rho_w$, the solid is denser than water and sinks (at equilibrium, μ_i must be minimum and thus z has the maximum negative value possible).
2. $\rho_i < \rho_w$, the solid is lighter than water and floats (at equilibrium, μ_i must be minimum and thus z has its maximum value $z = 0$).

Solutes Let us finally apply Equation 98 to a solute S_i and obtain a less commonly known result:

$$\mu_i = \mu_i^o + RT \ln[S_i] + \bar{V}_i(P - P_{atm}) + M_i gz \tag{102}$$

Introducing Equation 100,

$$\mu_i = \mu_i^o + RT \ln[S_i] + (M_i - \rho_w \bar{V}_i)gz \tag{103}$$

The condition of uniformity of the molar free energy then yields

$$\mu_i = \text{constant} = \mu_i^o + RT \ln [S_i]_0 \tag{104}$$

where $[S_i]_0$ is the concentration of the solute at the surface; therefore

$$RT \ln \frac{[S_i]}{[S_i]_0} = - [M_i - \rho_w \bar{V}_i] gz \tag{105}$$

or

$$[S_i] = [S_i]_0 \exp - \left[\frac{M_i - \rho_w \bar{V}_i}{RT} gz \right] \tag{106}$$

The distribution of solutes at equilibrium in a tall water column depends on the size and magnitude of the quantity $(M_i - \rho_w \bar{V}_i)$. The concentrations of the many species that have negative partial molar volumes in water (e.g., Ca^{2+} and CO_3^{2-}, see Section 5.5) increase exponentially with depth ($z < 0$).

5.7.3 Chemical Equilibrium in a Tall Water Column   The fundamental condition of equilibrium for a chemical reaction

$$\sum v_i \mu_i = 0 \tag{107}$$

applied to Equation 98 yields

$$\{S_1\}^{v_1} \{S_2\}^{v_2} \{S_3\}^{v_3} \ldots = K_z \tag{108}$$

where

$$- RT \ln (K_z) = \sum v_i \mu_i^o + \sum v_i \bar{V}_i (P - P_{atm}) + \sum v_i M_i gz \tag{109}$$

Introducing

$$- RT \ln K_0 = \sum v_i \mu_i^o \tag{110}$$

$$\sum v_i \bar{V}_i = \Delta V^o \tag{111}$$

$$P = P_{atm} - \rho_w gz \tag{112}$$

$$\sum v_i M_i = 0 \text{ (conservation of mass)} \tag{113}$$

we find

$$K_z = K_0 \exp \left[\frac{\rho_w gz}{RT} \Delta \bar{V}^o \right] \tag{114}$$

In reactions involving only solutes, the concentration variations due to depth in fact compensate for the effect of pressure on the equilibrium constant, and the reaction is at equilibrium throughout the water column.

For a reaction involving a solid, the situation is different. At equilibrium the solid must either float at the top or rest at the bottom. Consider, for example, a tall water column containing a saturating concentration of $CaCO_3$. In this case the solid sinks. To calculate its equilibrium, we take the origin of elevation and pressures to be at the bottom of the water column, that is, at $z = 0$, $P = P_{max}$:

$$[Ca^{2+}]_z = [Ca^{2+}]_0 \exp\left[\frac{-M_{Ca} + \rho_w \bar{V}_{Ca}}{RT} gz\right] \tag{115}$$

with $\rho_w = 1$ and z in kilometers,

$$[Ca^{2+}]_z = [Ca^{2+}]_0 \exp(-0.236z) \tag{116}$$

$$[CO_3^{2-}]_z = [CO_3^{2-}]_0 \exp\left[\frac{-M_{CO_3} + \rho_w \bar{V}_{CO_3}}{RT} gz\right] \tag{117}$$

$$[CO_3^{2-}]_z = [CO_3^{2-}]_0 \exp(-0.260z) \tag{118}$$

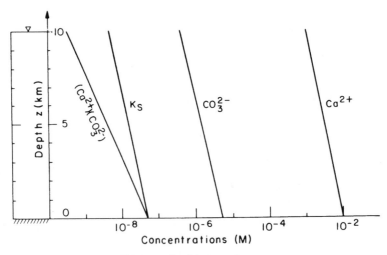

Figure 2.6 Theoretical equilibrium distribution of calcium and carbonate ions in an unmixed tall water column saturated with $CaCO_3(s)$. In the absence of mixing, the solid calcium carbonate rests on the bottom where the ion product $[Ca^{2+}][CO_3^{2-}]$ is equal to the solubility product K_s. At shallower depths the decrease in concentrations of the individual ions due to the effects of pressure on free energies more than makes up for the decrease in the solubility product K_s. The solution is thus undersaturated with respect to $CaCO_3(s)$ at all depths except the bottom.

$$K_z = K_0 \exp\left[\frac{\rho_w gz}{RT}(\bar{V}_{Ca} + \bar{V}_{CO_3} - \bar{V}_{CaCO_3})\right] \quad (119)$$

$$K_z = K_0 \exp(0.0238z) \quad (120)$$

Introducing $M_{Ca} + M_{CO_3} = M_{CaCO_3} = \bar{V}_{CaCO_3}\rho_{CaCO_3}$, into (117), (118), and (119), we get:

$$\frac{[Ca^{2+}]_z[CO_3^{2-}]_z}{K_z} = \exp\frac{gz}{RT}\bar{V}_{CaCO_3}(\rho_w - \rho_{CaCO_3}) \quad (121)$$

Since $\rho_w < \rho_{CaCO_3}$, this ratio is less than one at all depths except at the bottom (see Figure 2.6). The water is thus at equilibrium with $CaCO_3(s)$ at the bottom and undersaturated throughout the water column despite the decrease in solubility of the solid at lower pressures.

To obtain a cycle of precipitation at the surface and of dissolution at depth as is observed in the oceans, a mixing process must be invoked in addition to the effect of pressure on the solubility product of $CaCO_3$.

REFERENCES

1. G. N. Lewis and M. Randall, *Thermodynamics*, 2nd ed., revised by K. S. Pitzer and L. Brewer, McGraw-Hill, New York, 1961.

2. H. S. Harned and B. B. Owen, *The Physical Chemistry of Electrolytic Solutions*, 3rd ed., Van Nostrand-Reinhold, New York, 1958.

3. R. A. Robinson and R. H. Stokes, *Electrolyte Solutions*, 2nd ed., Butterworths, London, 1959.

3a. MINEQL, J. C. Westall, J. L. Zachary and F. M. M. Morel (1976), Technical Note #18, R. M. Parsons Laboratory, M. I. T. Cambridge MA; GEOCHEM, G. Sposito and S. Mattigod (1980), Technical Note, Department of Soil Sciences, U. California Riverside CA; MINTEQ, A. R. Felmy, D. C. Girvin and E. A. Jenne (1984), EPA-600/3-84-032 US EPA, Athens GA; MINEQL+, available from W. D. Schecher, Environmental Research Software, 16 Middle St., Hallowell ME. 04347.

4. J. N. Brønsted, *J. Am. Chem. Soc.*, **44**, 877 (1922).

5. E. A. Guggenheim, *Philos. Mag.*, **19**, 588 (1935).

6. G. Scatchard, *Chem. Rev.*, **19**, 309 (1936).

7. J. E. Mayer, *J. Chem. Phys.*, **18**, 1426 (1950).

8. G. Scatchard, *J. Am. Chem. Soc.*, **90**, 3124 (1968).

9. K. S. Pitzer, *J. Phys. Chem.*, **77**, 268 (1973).

10. L. G. Sillen and A. E. Martell, *Stability Constants of Metal Ion Complexes*, special Publication 17, Chemical Society, London, 1964; Supplement 1, Special Publication 25, Chemical Society, London, 1971.

11. A. E. Martell and R. M. Smith, *Critical Stability Constants*, Vol. 1, *Amino Acids*, Plenum, New York, 1974; R. M. Smith and A. E. Martell, Vol. 2, *Amines*, Plenum,

New York, 1975; A. E. Martell and R. M. Smith, Vol. 3, *Other Organic Ligands*, Plenum, New York, 1977; R. M. Smith and A. E. Martell, Vol. 4, *Inorganic Ligands*, Plenum, New York, 1976.

12. H. S. Harned and R. Davis, *J. Am. Chem. Soc.*, **65**, 2030 (1943).

13. J. H. Carpenter, *Limnol. Oceanogr.*, **11**, 264 (1966).

14. R. L. Jacobson and D. Langmuir, *Geochim. Cosmochim. Acta*, **38**, 301 (1974).

15. H. S. Harned and R. W. Ehlers, *J. Am. Chem. Soc.*, **55**, 652 (1933).

16. *CRC Handbook of Chemistry and Physics*, 62nd ed., CRC Press, New York, 1981.

17. H. S. Harned and S. R. Choles, *J. Am. Chem. Soc.*, **63**, 1706 (1941).

PROBLEMS

2.1 Discuss from a thermodynamic point of view the question of coexistence of two solid phases of the same composition, such as $CaCO_3$ (calcite) and $CaCO_3$ (aragonite).

2.2 Consider a compound X (acid, base, or salt) that dissociates according to the reaction

$$X = Y + Z$$

What should the standard energy change of the reaction be at the minimum for the dissociation to be considered complete ($> 99.9\%$) for all possible compositions in dilute solutions (concentrations $\leqslant 0.5\ M$)?

2.3 Consider a $10^{-3}\ M$ solution of acetic acid, the dissociation reaction $HAc = H^+ + Ac^-$, and the standard free energies on the molar scale $\mu_{H^+}^\circ = 0$, $\mu_{HAc}^\circ = -396.5\ kJ\ mol^{-1}$, $\mu_{Ac^-}^\circ = -369.3\ kJ\ mol^{-1}$, $\mu_{H_2O}^\circ = -237.1\ kJ\ mol^{-1}$.

a. What is the equilibrium constant for the reaction on the molar scale? On the mole fraction scale?

b. Neglecting all other reactions, make a plot of the free energy of the system (G) and of the free energy change of the reaction (ΔG), as the dissociation proceeds from $[Ac^-] = 0$ to $[Ac^-] = 10^{-3}\ M$.

c. Calculate the equilibrium composition of the solution.

2.4 Consider the reactions

$$CO_2(g) + H_2O = H_2CO_3^*, \qquad pK_H = 1.5$$

$$H_2CO_3^* = H^+ + HCO_3^-; \qquad pK_{a1} = 6.3$$

$$HCO_3^- = H^+ + CO_3^{2-}; \qquad pK_{a2} = 10.3$$

$$CaCO_3(s) = Ca^{2+} + CO_3^{2-}; \qquad pK_s = 8.3$$

Choose the principal components, and write the right-hand side (numerical value) of the corresponding mole balances for each of the following systems:

a. Recipe $[NaHCO_3]_T = 10^{-3} M, [HCl]_T = [NaOH]_T = 10^{-2} M$, no gas or solid phase

b. Recipe $[NaHCO_3]_T = [HCl]_T = 10^{-3} M, [NaOH]_T = 10^{-2} M$, no gas or solid phase

c. Recipe $[NaHCO_3]_T = [NaOH]_T = 10^{-3} M, [HCl]_T = 10^{-2} M$, no gas or solid phase

d. Recipe $[Na_2CO_3]_T = 10^{-3} M, [HCl]_T = [NaOH]_T = 10^{-2} M$, no gas or solid phase

e. Recipe $[NaHCO_3]_T = 10^{-3} M, P_{CO_2} = 10^{-3.5}$ atm, no solid phase

f. Recipe $[NaHCO_3]_T = 10^{-3} M, [Ca(OH)_2]_T = 10^{-2} M, CaCO_3(s)$ precipitates, no gas phase

g. Recipe $[NaHCO_3]_T = 10^{-3} M, [Ca(OH)_2]_T = 10^{-2} M, P_{CO_2} = 10^{-3.5}$ atm, $CaCO_3(s)$ precipitates

2.5 Sketch $\log C$–pH diagrams, and find approximate graphical solutions for the following equilibrium systems:

a. Recipe $[Na_2HPO_4]_T = 10^{-3} M, [HCl]_T = 10^{-3} M$

   Species $H_2O, H^+, OH^-, Na^+, Cl^-, H_3PO_4, H_2PO_4^-, HPO_4^{2-}, PO_4^{3-}$

   Reactions $H_3PO_4 = H_2PO_4^- + H^+, pK_{a1} = 2.1$

   $H_2PO_4^- = HPO_4^{2-} + H^+, pK_{a2} = 7.2$

   $HPO_4^{2-} = PO_4^{3-} + H^+, pK_{a3} = 12.3$

b. Recipe $[NH_4Cl]_T = 10^{-3} M, [NaOH]_T = 10^{-3} M$

   Species $H_2O, H^+, OH^-, Na^+, Cl^-, NH_4^+, NH_3(aq)$

   Reactions $NH_4^+ = NH_3(aq) + H^+; pK_a = 9.2$

c. Recipe   Same as system b but add $P_{NH_3} = 10^{-6}$ atm

   Species   Same as system b but add $NH_3(g)$

   Reactions Same as system b but add $NH_3(g) = NH_3(aq)$; $pK_H = -1.8$

2.6 Using a $\log C$–pH diagram, find the pH of vinegar. You may model vinegar as a 5% volume/volume solution of acetic acid.

2.7 Solve the following equilibrium problems:

a. Recipe $[Na_2CO_3]_T = 10^{-3} M, [HCl]_T = 2 \times 10^{-3} M,$
   $[NaCl]_T = 10^{-2} M, [NH_4Cl]_T = 10^{-4} M$

Species and reaction are the same as in Problems 2.4 and 2.5, but there is no gas or solid phase.

b. Recipe $P_{CO_2} = 10^{-3.5}$ atm, $P_{NH_3} = 10^{-6}$ atm, $[NaOH]_T = 10^{-3} M$,
$[Na_2HPO_4]_T = 10^{-5} M$, $[NaCl]_T = 10^{-2} M$

Species and reactions are the same as in Problems 2.4 and 2.5, but without a solid phase.

2.8 Given the following thermodynamic data, calculate the equilibrium composition for the two recipes that follow:

	μ^o $(kJ\, mol^{-1})$
H^+	0
OH^-	-157.2
H_2O	-237.1
Na^+	-261.9
Cl^-	-131.3
$NaCl(aq)$	> -370
$HCl(aq)$	$\simeq -115$
$NaOH(aq)$	$\simeq -410$
CN^-	$+172.4$
$HCN(aq)$	$+119.7$

	Recipe 1	Recipe 2
$[HCN]_T$	$10^{-3} M$	$10^{-3} M$
$[NaCl]_T$	$10^{-1} M$	$10^{-1} M$
$[NaOH]_T$	0	$10^{-3} M$

a. Calculate equilibrium constants and decide what species should be considered in the system.

b. Make appropriate ionic strength corrections.

c. Choose appropriate components, write tableaux, and solve.

2.9 Using the Davies equations, calculate the ionic strength effect

a. on the solubility of $BaSO_4$ ($pK_s = 9.7$), as NaCl is added to pure water up to 0.5 M [only species $Na^+, Cl^-, Ba^{2+}, SO_4^{2-}, BaSO_4(s)$].

b. on the pH ($= -\log\{H^+\}$), as NaOH is added to pure water up to 1 M (only species H^+, OH^-, Na^+).

c. on the pH, as HCl is added to pure water up to 1 M (only species H^+, OH^-, Cl^-).

2.10 A $10^{-3} M$ solution of acetic acid (see Problem 2.3) is transported from the warm surface of a river ($T = 30°C$, $P = 1$ atm, $I = 10^{-2} M$) to the bottom of the ocean ($T = 4°C$, $P = 400$ atm, $I = 0.7 M$). Given $\bar{H}^o_{H^+} = 0$, $\bar{H}^o_{HAc} = 486.6 \,kJ\, mol^{-1}$, $\bar{H}^o_{Ac^-} = 488.2 \,kJ\, mol^{-1}$, $\bar{V}_{H^+} = 0$, $\bar{V}_{HAc} = 50.7 \,cm^3$

PROBLEMS   **97**

mol^{-1}, $\overline{V}_{Ac^-} = 41.5\,cm^3\,mol^{-1}$, calculate the total change in the acidity constant.

2.11 Consider an intracellular fluid at a potential Ψ with respect to the extracellular medium ($\Psi_{out} = 0$).

    a. Determine the ratio of concentrations inside to outside for a freely diffusing ion of charge number Z.

    b. Define the equation for equilibrium (outside, inside, and across the membrane) for a weak acid $HA = A^- + H^+$; K_a.

CHAPTER 3

KINETICS

Thermodynamic descriptions of natural waters provide insight into the behavior of chemical species by quantifying their tendencies to undergo transformations or to partition among various phases. In many instances, a reasonable approximation of the composition of natural waters can be obtained by considering many reactions, such as acid dissociation and neutralization, complex formation, and sorption reactions, to be at chemical equilibrium. This approach fails, however, for some chemical species, particularly those that undergo oxidation or reduction reactions. The chemical disequilibrium of the biosphere is illustrated by the comparison of the atmospheric compositions of the Earth, Mars, and Venus and the calculated atmospheric composition of the Earth without life (Table 3.1). Most noticeable in this comparison is the occurrence of thermodynamically unstable combinations of constituents, such as methane in the presence of oxygen, in the Earth's atmosphere.[1]

As suggested by Table 3.1, the nonequilibrium conditions of the Earth's atmosphere are maintained by the biota, particularly photosynthetic organisms. In photosynthesis, solar energy is captured by plants and photosynthetic bacteria, carbon from CO_2 is *fixed* into the reduced organic compounds that constitute biomass, and a reservoir of oxygen is generated. This activity depends on the constant input of sunlight to the Earth's surface.

The reaction of reduced organic compounds with oxygen (combustion) results in the release of energy. The coexistence of such chemical species, whose reaction would be energetically favorable, and thus the persistence of nonequilibrium conditions depend on the *rate* at which these species react with each other. This highlights the major limitation of the equilibrium approach—thermo-

TABLE 3.1 **Composition of Planetary Atmospheres**

	Venus	Mars	Earth Without Life	Earth Actual
Carbon dioxide	96.5%	95%	98%	0.03%
Nitrogen	3.5%	2.7%	1.9%	79%
Oxygen	Trace	0.13%	0%	21%
Argon	70 ppm	1.6%	0.1%	1%
Methane	0%	0%	0%	1.7 ppm

Source: From Lovelock (1988).[1]

dynamics can tell us whether or not a reaction is favorable, but not how fast a favorable reaction will occur. Questions on the rates of chemical reactions fall into the domain of kinetics rather than thermodynamics.

We may characterize a given system or mixture of chemical species as thermodynamically *stable* if it is at equilibrium or *unstable* if it is not. The rate of a thermodynamically favored reaction, however, depends on the kinetic *inertness* or *lability* of the chemical species involved in the reaction. Unfortunately, these kinetic terms cannot be as easily defined as their thermodynamic counterparts. Kinetic inertness or lability can only be defined relative to a particular time scale. But what is the time scale of interest against which we should compare the rates of chemical reactions? For a geologist, it could be as long as the age of the Earth (over 4 billion years). An oceanographer might wish to compare reaction rates to the residence time of water in the ocean (a few thousand years). And a prominent chemist has suggested that completion of a reaction within 1 minute at 25°C be the basis for the distinction between labile and inert.[2] For our purposes, it is convenient to define a time scale based on some biological, physical, or geological process of interest and to use that time scale to characterize chemical reactions as fast or slow (Figure 3.1). Chemical reactions that are fast relative to the process defining our time scale may then be considered to be at equilibrium and reactions that are relatively slow may be ignored. Merging thermodynamic and kinetic terminology, we may then describe as *metastable* a combination of chemical species that, for practical purposes, may be regarded as being at equilibrium because of its nonreactivity even though it is *not* truly at equilibrium.

Although not mentioned explicitly above (except by our chemist), another important aspect of the kinetic barrier to a thermodynamically favorable reaction is the amount of energy available to the system. By this we usually mean thermal energy. In some cases, however, a photochemical, or light-dependent, pathway for a reaction may be kinetically favorable where a thermal pathway is not. A combination of chemical species may then be metastable at ambient temperature in the dark yet labile in the presence of light.

An aspect of carbonate chemistry that illustrates the importance of the relative

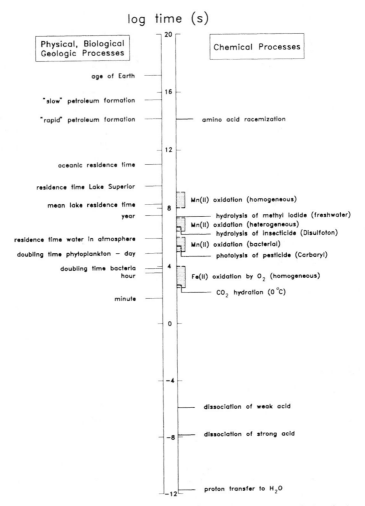

Figure 3.1 Time scale illustrating the comparison between rates of chemical processes and physical, biological, and geologic processes.

rates of physical and chemical processes is the preservation of calcite, $CaCO_3$, in ocean sediments.[3] Figure 3.2a shows the distribution of calcite-rich sediments in the world oceans. Since the carbonate sediments are composed mainly of plankton microfossils, the heterogeneous distribution of carbonate sediments is due in part to variations in biological productivity of the overlying waters. Deep ocean waters, however, are undersaturated with respect to calcium carbonate and calcite dissolves rather than being buried in the sediments. Thus the calcite distribution in Figure 3.2a is a composite record of the input of calcite, the rate of calcite dissolution (which depends on the degree of calcite saturation at the depth of the ocean floor), and the rate of sediment accumulation.

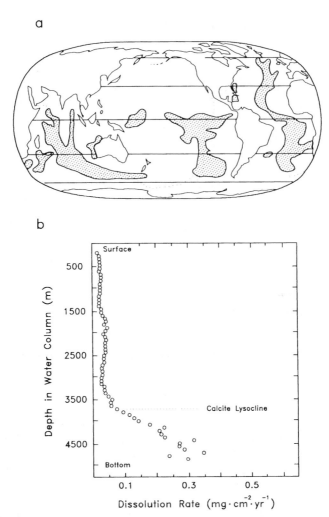

Figure 3.2 (*a*) Map showing distribution of calcite-rich sediments. Shaded areas indicate sediments containing more than 75% $CaCO_3$. From Broeker and Peng (1982).[3] (*b*) Rate of calcite dissolution as a function of depth in a 5000 m water column in the central Pacific. Data shown are the average dissolution rates determined for calcite spheres at a given depth. Water is undersaturated with respect to calcite at all depths below a few hundred meters. A sharp increase in dissolution rate is observed at 3700 m, corresponding to the depth of the calcite lysocline. Adapted from Peterson (1984).[4]

Rates of calcite dissolution have been measured in situ; increase in the dissolution rate corresponds to the depth of the calcite *lysocline*, the depth at which calcite preservation in the sediments begins to be affected by dissolution[4] (Figure 3.2*b*).

Unfortunately, it is often not practical to measure reaction rates directly in the environment. Rather, the rates of chemical reactions must be estimated for a given set of environmental conditions. To accomplish this, we must know how the reaction rates depend on the concentrations of reactants, of products in some cases, and of *catalysts* (substances that increase the rate of reaction but do not influence the thermodynamics of the reaction) and on variables such as temperature, pressure, and ionic strength. The dependence of the reaction rate on these parameters is determined by the reaction pathway or mechanism and need not correspond in any simple way to the overall stoichiometry of the reaction. Although reaction mechanisms are usually derived from empirical observations of reaction kinetics, theories of chemical kinetics provide some basis for estimating reaction rates and some insight into the effects of temperature, pressure, and ionic strength. Semiempirical correlations of kinetic rate constants and equilibrium constants, which are based on transition state concepts, may also be used to estimate reaction rates.[5]

In this chapter we examine the empirical and theoretical bases for estimating rates of chemical reactions under environmental conditions and the incorporation of kinetics into our description of the chemical composition of natural waters. Some aspects of kinetics relevant to specific reactions, such as precipitation–dissolution, complexation, and redox reactions, are covered in the chapters on those topics.

1. REACTION RATES, RATE CONSTANTS, AND MECHANISMS: ANALYSIS OF KINETIC DATA

The data from kinetic experiments commonly consist of observations of reactant and product concentrations over time during the course of the reaction with varying initial reaction conditions. The observation or trapping of reactive intermediates can be an extremely powerful tool in kinetic analysis. In the study of chemical kinetics in natural waters, this approach has been applied predominantly to photochemical reactions (and is discussed in Chapter 7). The goal of kinetic experiments is to derive a reaction mechanism and rate constants from the data. We must remember, however, that it is difficult to interpret data or even to design experiments without reference to a mechanism and also that interpretations of kinetic experiments may be rather poorly constrained by data. Thus interpretation of kinetic data often centers on demonstrating consistency of the data with a proposed mechanism.

1.1   A Few Definitions

First, we must distinguish between the overall stoichiometry of a reaction, such as

$$A + 2B \rightarrow P$$

and the reaction mechanism. The reaction mechanism consists of a set of *elementary steps,* consistent with the stoichiometry of the reaction, such as

$$A + B \rightleftarrows I$$
$$I + B \rightarrow P$$

The reaction mechanism corresponds to the actual (or supposed) reaction of molecules to give the observed products and not merely to the ratio of reactants and products described by the stoichiometric relation. The reaction mechanism may also involve *intermediates,* such as *I,* that do not appear in the overall stoichiometry of the reaction. Commonly, elementary steps are either *unimolecular,* involving only a single reacting species, or *bimolecular,* involving two reacting species. The reaction mechanism above shows the bimolecular reactions of *A* and *B* to form *I* and of *I* and *B* to form *P* and, in the back reaction of the first step, the unimolecular decomposition of *I* to yield *A* and *B*.

The *rate of reaction* is defined as the rate of formation of products or disappearance of reactants. For the overall reaction shown above:

$$\text{rate} = \frac{d[P]}{dt} = -\frac{d[A]}{dt} = -\frac{1}{2}\frac{d[B]}{dt} \tag{1}$$

The *rate law* is an expression of the empirical dependence of the rate on the concentrations of the reactants. For the rate law,

$$\text{rate} = \frac{d[P]}{dt} = k[A]^a[B]^b \tag{2}$$

k is the *rate constant.* The *order of the reaction in a reactant* corresponds to the exponent for that reactant in the rate law and the *overall reaction order* to the sum of the exponents, that is, $(a + b)$ in Equation 2. As we shall see, the observed (or overall) rate constant, the term *k* in Equation 2, derives from the intrinsic rate constants of the elementary steps.

1.2   Extraction of Rate Constants

As has already been mentioned, the interpretation of kinetic data usually presumes some reaction mechanism which is then tested by examining the agreement between the observed rate law and that derived from the proposed mechanism. In the next two sections the rate laws for irreversible and reversible reactions are described. Although these rate laws are derived for elementary reactions, the rate laws describing complex reactions (i.e., a series of elementary steps) are often of the same form.

1.2.1   *Irreversible Reactions.*   *First Order:*   A first-order elementary reaction

$$A \rightarrow P$$

has the rate law

$$\text{rate} = \frac{d[P]}{dt} = \frac{-d[A]}{dt} = k[A] \tag{3}$$

The integrated form of the rate law, in this case,

$$\ln \frac{[A]_t}{[A]_0} = \ln \left[\frac{[A]_0 - x}{[A]_0} \right] = -kt \tag{4}$$

(where $[A]_0$ is the initial concentration of A, $[A]_t$ is the concentration of A at time t, and x is the amount of A reacted at some time t) is usually used to extract the rate constant k.

With the exception of radioactive decay, not many reactions of environmental interest are truly first order. Under appropriate reaction conditions, however, many reactions appear to follow first-order kinetics. Such *pseudo* first-order kinetics may be observed if one (or more than one) reactant in a higher-order reaction is in sufficiently large excess that its concentration remains unchanged during the reaction.

Carbonate chemistry provides us with an example of pseudo first-order kinetics. In the hydration of CO_2, the reactant H_2O is in large excess; $[H_2O] \simeq 55.4\,M \gg [CO_2(aq)]$. The CO_2 hydration reaction occurs through two parallel pathways

$$CO_2(aq) + H_2O \rightarrow H_2CO_3(aq)$$

$$CO_2(aq) + H_2O \rightarrow HCO_3^- + H^+$$

These pathways are kinetically indistinguishable because of the extremely rapid acid–base equilibrium between carbonic acid and bicarbonate. Thus both pathways are included in the observed rate law,[6] which is first order in $[CO_2(aq)]$,

$$\text{rate} = \frac{-d[CO_2(aq)]}{dt} = k_{obs}[CO_2(aq)] \tag{5}$$

where

$$k_{obs} = k[H_2O] \tag{6}$$

Figure 3.3 shows a plot of $\ln([CO_2]_t/[CO_2]_0)$ vs. time; the slope of this graph corresponds to a rate constant, k_{obs}, of $2 \times 10^{-3}\,s^{-1}$ at $0°C$. Some additional rate constants for CO_2 hydration and carbonic acid dehydration[7–9] are given in Sidebar 3.1.

Another useful quantity related to the reaction rate constant is the *half-life*,

SIDEBAR 3.1

CO_2 Hydration Kinetics

The hydration of CO_2 and dehydration of H_2CO_3 are physiologically important reactions. The kinetics of these reactions have been studied intensively. Recent studies have employed relaxation techniques, in which a system at equilibrium is perturbed and the return to equilibrium, or relaxation, is followed.[7] As mentioned in the text, the hydration–dehydration reactions occur through the parallel mechanisms

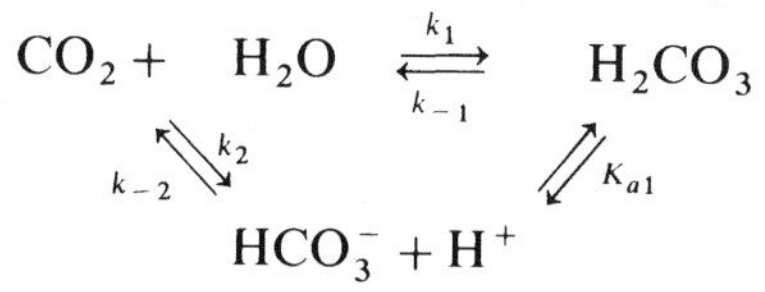

$$CO_2 + H_2O \underset{k_{-1}}{\overset{k_1}{\rightleftharpoons}} H_2CO_3$$

$$\underset{k_{-2}}{\overset{k_2}{\nwarrow}} \qquad \qquad \overset{K_{a1}}{\nearrow}$$

$$HCO_3^- + H^+$$

where the acid–base reaction is rapid compared with other reactions. Including both forward (hydration) and reverse (dehydration) in the rate expression (for detailed discussion see Section 1.2.2), we write

$$\text{rate} = \frac{-d[CO_2]}{dt} = (k_1 + k_2)[CO]_2 - k_{-1}[H_2CO_3]$$

$$- k_{-2}[HCO_3^-][H^+] \qquad \qquad \text{(i)}$$

Substituting from the mass law equation

$$[HCO_3^-] = \frac{K_{a1}[H_2CO_3]}{[H^+]} \qquad \qquad \text{(ii)}$$

we obtain

$$\frac{-d[CO_2]}{dt} = (k_1 + k_2)[CO_2] - (k_{-1} + k_{-2}K_{a1})[H_2CO_3] \qquad \text{(iii)}$$

or, equivalently,

$$\frac{-d[CO_2]}{dt} = (k_1 + k_2)[CO_2] - \frac{(k_{-1} + k_{-2}K_{a1})}{K_{a1}}[HCO_3^-][H^+] \qquad \text{(iv)}$$

(*continued*)

The individual rate constants cannot be experimentally resolved and are reported simply as k_{CO_2} and $k_{H_2CO_3}$ where

$$k_{CO_2} = (k_1 + k_2) \tag{v}$$

and

$$k_{H_2CO_3} = (k_{-1} + k_{-2}K_{a1}) \tag{vi}$$

(see below). At higher pH (pH $\geqslant$ 9), an alternative reaction pathway

$$CO_2 + OH^- \underset{k_{-4}}{\overset{k_4}{\rightleftharpoons}} HCO_3^-$$

becomes dominant. The values for the rate constants reported below show that the CO_2 hydration and H_2CO_3 dehydration reactions are strongly temperature dependent. (Temperature dependence of reaction rate constants is discussed in Sections 2.1.2 and 2.2.2.)

T (°C)	k_{CO_2} (s^{-1})	$k_{H_2CO_3}$ (s^{-1})	k_4 (M^{-1}s^{-1})	k_{-4} (s^{-1})	Ref.
0	0.0021	2.0			8
15	0.014	11.0			8
25	0.032	26.6			8
	0.040	28			7
			8500	0.0002	9
38	0.12	89.0			8

$t_{1/2}$, of a reacting species; this is the time required for the concentration of a reactant to be reduced to half of its original value. The specific relationship between the half-life and the rate constant depends on the order of the reaction. For a first-order reaction,

$$t_{1/2} = \frac{\ln 2}{k} = \frac{0.69}{k} \tag{7}$$

A very similar parameter is the *relaxation time* (sometimes called the *life-time* or *characteristic time*) for the reaction, τ. At time τ, where

$$\tau = \frac{1}{k} \tag{8}$$

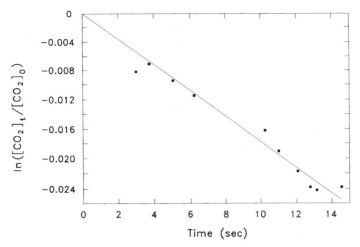

Figure 3.3 CO_2 hydration kinetics at $0°C$. Plot of $\ln([CO_2]_t/[CO_2]_0)$ vs. time (s) for the reaction of CO_2 with H_2O. The slope of the graph corresponds to the pseudo first-order rate constant k_{obs} (s^{-1}) for the rate law: rate $= -d[CO_2]/dt = k_{obs}[CO_2]$. The value of k_{obs} at $0°C$ is $2 \times 10^3\,s^{-1}$. Adapted from Jones et al. (1984).[6]

the concentration of the reacting species reaches $1/e$ (numerically, 0.368) of its initial value.

Second Order: Reactions that are second order overall may be either second order in one reactant

$$2A \rightarrow P$$

or first order in each of two reacting species

$$A + B \rightarrow P$$

For a reaction that is second order in a single reactant, the rate law is

$$\text{rate} = \frac{-d[A]}{dt} = \frac{2d[P]}{dt} = k[A]^2 \tag{9}$$

with the corresponding integrated form of the rate law:

$$\frac{1}{[A]_t} - \frac{1}{[A]_0} = \frac{1}{[A]_0 - x} - \frac{1}{[A]_0} = kt \tag{10}$$

The half-life of this reaction (unlike that for a first-order reaction, cf. Equation 7) depends on the initial concentration of the reacting species, that is,

$$t_{1/2} = \frac{1}{k[A]_0} \tag{11}$$

For a reaction that is first order in each of two reactants, the rate law is

$$\text{rate} = \frac{-d[A]}{dt} = \frac{-d[B]}{dt} = \frac{d[P]}{dt} = k[A][B] \tag{12}$$

with the corresponding integrated form:

$$\frac{1}{[B]_0 - [A]_0} \ln \frac{[A]_0[B]_t}{[B]_0[A]_t} = \frac{1}{[B]_0 - [A]_0} \ln \frac{[A]_0([B]_0 - x)}{[B]_0([A]_0 - x)} = kt \tag{13}$$

These integrated forms of the rate law may be used to extract second-order rate constants as shown below.

One example of second-order kinetics previews the subject of Chapter 7, oxidation–reduction (redox) reactions. In natural waters, iron occurs in two oxidation states, $+ \text{II}$ and $+ \text{III}$. The cycling of iron between these two oxidation states is due partly to the photochemical reduction of Fe(III) to Fe(II). Reoxidation of Fe(II) to Fe(III) can occur by reaction with a variety of oxidants including oxygen. Some natural waters, particularly fog water in polluted areas and acid-mine drainage waters, are quite acidic. At low pH, reaction with oxygen is fairly slow and reactions with alternative oxidants, such as hydrogen peroxide, may significantly affect iron redox cycling.

The reaction of Fe(II) with hydrogen peroxide has been studied in strongly acidic solution.[10] The overall reaction stoichiometry is:

$$H_2O_2 + 2Fe^{2+} + 2H^+ \rightarrow 2H_2O + 2Fe^{3+}$$

At pH 1, the concentration of H^+ changes only slightly during the course of the reaction. Thus $[H^+]$ need not be included explicitly in the rate law. The reaction is observed to be first order in $[Fe^{2+}]$ and to follow pseudo second-order kinetics with the empirical rate law

$$\text{rate} = \frac{-d[H_2O_2]}{dt} = -\frac{1}{2}\frac{d[Fe^{2+}]}{dt} = \frac{1}{2}\frac{d[Fe^{3+}]}{dt} = k[H_2O_2][Fe^{2+}] \tag{14}$$

The rate constant ($k = 3.5 \times 10^3 \ M^{-1}\text{min}^{-1}$) can be obtained from the slope of the graph shown in Figure 3.4. Here the concentration expression plotted against time is slightly different from that given in Equation 13 because the stoichiometric coefficients (for Fe^{2+} and Fe^{3+}) are not equal to one.[10]

The linearity observed in the graphs shown in Figures 3.3 and 3.4 confirms

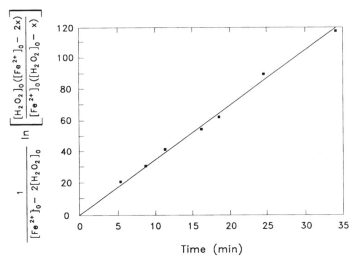

Time (min)

Figure 3.4 Kinetics of Fe^{2+} oxidation by H_2O_2 at pH 1. For the empirical rate law, $-d[H_2O_2]/dt = -1/2(d[Fe^{2+}]/dt) = 1/2(d[Fe^{3+}]/dt) = k[H_2O_2][Fe^{2+}]$, the rate constant is obtained from the slope of the graph; the value of k is $3.5 \times 10^3 \, M^{-1} \, min^{-1}$. Note: $2x = [Fe^{3+}]$. Data from Tinoco et al. (1985).[10]

the choice of first- and second-order kinetics, respectively, in modeling the experimental data. This is not, however, a sensitive test of reaction order and it is therefore important that kinetic data be obtained over a range of *initial* reactant concentrations. For example, an observation that the reaction half-life is independent of the initial concentration of the reactant (cf. Equation 7) is much stronger evidence for first-order kinetics than the linearity of the $\ln[A]$ vs. time plot for a single initial concentration of A.

1.2.2 Reversible Reactions For reactions approaching equilibrium, both the forward and back reactions must be considered (as we have already seen in Sidebar 3.1). A geochemically important example of a reversible, first-order reaction, is the *racemization* of amino acids. This reaction results in the interconversion of the two forms of an amino acid which are mirror images of each other (as shown in Figure 3.5). Of the two forms, organisms produce only one, the (L) form. After the death of the organism, the (L)-amino acids in fossil shells or bone interconvert, through abiotic chemical processes, with the (D)-form. The ratio of (D)/(L) amino acids has been used to estimate the age of sediments and fossils for samples up to one million years old[11,12] (see Figure 3.6).

As an example, consider the reversible first-order reaction for the interconversion of L-isoleucine (L) and D-alloisoleucine (D)

$$L \underset{k_b}{\overset{k_f}{\rightleftharpoons}} D$$

With initial conditions $[L]_0$ and $[D]_0$, the rate law

$$\text{rate} = \frac{d[L]}{dt} = -k_f[L] + k_b[D] \tag{15}$$

is obtained. Given that, at any time,

$$[L]_t + [D]_t = [L]_0 + [D]_0 = T \tag{16}$$

The rate law may be written as

$$\frac{d[L]}{dt} = -(k_f + k_b)[L] + k_b T \tag{17}$$

The integrated form of the rate law is then

$$[L]_t = [L]_\infty + ([L]_0 - [L]_\infty)\exp(-\lambda t) \tag{18}$$

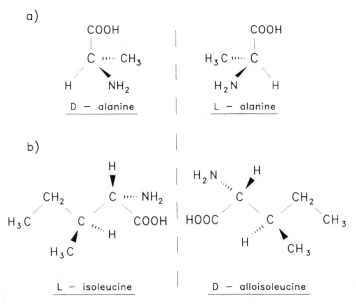

Figure 3.5 Structures of amino acids showing chirality (or handedness). Enantiomeric amino acids such as alanine (*a*) have one asymmetric carbon atom. Diastereomeric amino acids such as isoleucine (*b*) have two asymmetric carbon atoms. The wedge-shaped and dotted bonds indicate bonds projecting above and behind the plane of the paper, respectively.

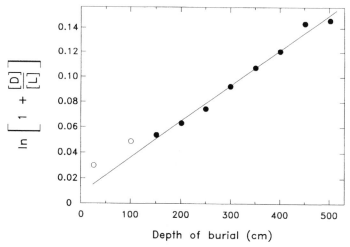

Figure 3.6 Racemization of L-isoleucine (L) to D-alloisoleucine (D) in marine sediments. The linear relationship between $\ln(1 + [D]/[L])$ and depth of burial of the sediments will fail as the reaction approaches equilibrium. The sedimentation rate (4.2 mm/1000 y) was estimated by assuming an irreversible reaction. Open circles were omitted from the calculation. From Bada et al. (1970).[11]

where

$$\lambda = k_f + k_b \tag{19}$$

and the equilibrium concentration of L, $[L]_\infty$ is

$$[L]_\infty = \frac{k_b}{k_f + k_b} T \tag{20}$$

A symmetrical solution is obtained for D:

$$[D]_t = [D]_\infty + ([D]_0 - [D]_\infty)\exp(-\lambda t) \tag{21}$$

where

$$[D]_\infty = \frac{k_f}{k_f + k_b} T \tag{22}$$

This solution to the kinetic problem introduces a *characteristic time* $1/\lambda$ or $1/(k_f + k_b)$. As illustrated in Figure 3.7, for $t \ll 1/\lambda$ the reaction has made no tangible progress, and initial conditions prevail in the system. For $t \gg 1/\lambda$ the reaction approaches completion, and the system is close to or at equilibrium.

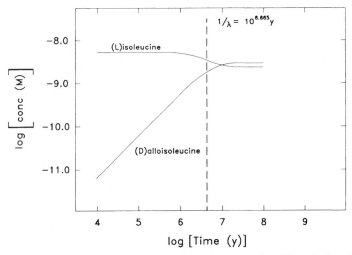

Figure 3.7 Interconversion of L-isoleucine and D-alloisoleucine. The calculated concentrations of the L- and D-forms (on a logarithmic scale) are plotted as a function of time. (also on a logarithmic scale). Values used in calculations: $[L]_0 = 5\,nM$, $[D]_0 = 0$; $k_f = 1.2 \times 10^{-7}\,y^{-1}$, $k_f/k_b = 1.25$, $k_b = 9.6 \times 10^{-8}\,y^{-1}$. Rate constants from Bada et al. (1970)[11] and Wehmiller and Hare (1971).[12]

In this example the characteristic time, $1/\lambda$, is approximately 4.6 million years. It is an important point that even if a system is far from equilibrium, the concentration of some of the reacting species may be close to their equilibrium values. In this example, the D-form of the amino acid is only slightly more stable than the L-form. Thus while the D-to-L ratio changes dramatically over geologic time, the concentration of the biogenic L-form decreases only by about a factor of two.

1.2.3 Consecutive Reactions Most overall reactions (described by a stoichiometric equation) consist of a series of elementary steps. The rate of the overall reaction (proceeding through a *single* reaction pathway or mechanism) is determined by the rate of the slowest or *rate-limiting* step.

The simplest case is that of two consecutive irreversible steps:

$$A \xrightarrow{k_1} B \xrightarrow{k_2} P$$

The rate of loss of the reactant A follows simple first-order kinetics:

$$\frac{d[A]}{dt} = -k_1[A] \qquad (23)$$

and the rate of formation of the product P exhibits a first-order dependence on the concentration of the intermediate species B:

$$\frac{d[P]}{dt} = k_2[B] \tag{24}$$

but the rate of formation of B is rather more complicated:

$$\frac{d[B]}{dt} = k_1[A] - k_2[B] \tag{25}$$

For a system where only the species A is present initially with a concentration $[A]_0$, we know, from mass conservation, that

$$[A]_t + [B]_t + [P]_t = [A]_0 \tag{26}$$

since $[B]_0 = [P]_0 = 0$.

And, for the first-order decay of A,

$$[A]_t = [A]_0 \exp(-k_1 t) \tag{27}$$

An analytical solution for the concentrations of the intermediate B and the product P over time can be obtained (as shown in Sidebar 3.2) with the result that

$$[B]_t = [A]_0 \left[\frac{k_1}{k_2 - k_1} \right] [\exp(-k_1 t) - \exp(-k_2 t)] \tag{28}$$

SIDEBAR 3.2

Kinetics of Consecutive Reactions

The analytical solution for reactant concentrations as a function of time for

$$A \xrightarrow{k_1} B \xrightarrow{k_2} P$$

May be obtained as follows for the initial conditions $[B]_0 = [P]_0 = 0$ and thus

$$[A]_t + [B]_t + [P]_t = [A]_0 \tag{i}$$

(continued)

Substituting the integrated expression for the first-order decay of A (Equation 27) into the rate expression for B (Equation 25), we obtain

$$\frac{d[B]}{dt} + k_2[B] = k_1[A]_0 \exp(-k_1 t) \tag{ii}$$

which is a first-order linear differential equation. Here, for the convenience of the less mathematically inclined readers, we solve this equation by reference to tabulated solutions, such as those in the *CRC Handbook of Chemistry and Physics*. Equation ii corresponds to the differential equation

$$(D - a)y = R(x) \tag{iii}$$

with the following substitutions to the left-hand side of the equation:

$$D = \frac{d}{dt} \tag{iv}$$

$$a = -k_2 \tag{v}$$

$$y = [B] \tag{vi}$$

The right-hand side of Equation iii corresponds to the tabulated function for $R(x)$:

$$R(x) = P(x)\exp(rx) \tag{vii}$$

where

$$P(x) = k_1[A]_0 \tag{viii}$$

$$r = -k_1 \tag{ix}$$

$$x = t \tag{x}$$

The general solution to the differential equation iii is of the form

$$y = y_p + y_c \tag{xi}$$

where y_p is obtained from the tabulated solutions and is, for our case,

$$y_p = \frac{1}{a-r}\left(P(x) + \frac{dP/dx}{a} + \frac{d^2P/dx^2}{a^2} + \cdots + \frac{d^nP/dx^n}{a^n}\right)\exp(rx) \tag{xii}$$

or, with the substitutions above,

$$y_p = \frac{1}{k_2 - k_1} k_1 [A]_0 \exp(-k_1 t) \qquad \text{(xiii)}$$

The value for y_c is obtained by setting the right-hand side of Equation iii to zero. Then for

$$(D - a)y = 0 \qquad \text{(xiv)}$$

the solution is

$$y_c = (\text{const}) \exp(ax) \qquad \text{(xv)}$$

or, in our case,

$$y_c = (\text{const}) \exp(-k_2 t) \qquad \text{(xvi)}$$

The complete solution of the differential equation is then

$$y_p + y_c = \frac{1}{k_2 - k_1} k_1 [A]_0 \exp(-k_1 t) + (\text{const}) \exp(-k_2 t) \qquad \text{(xvii)}$$

and the value of the constant can be obtained by imposing the boundary condition $[B]_0 = 0$. Then

$$\text{const} = \frac{-k_1 [A]_0}{k_2 - k_1} \qquad \text{(xviii)}$$

and

$$[B]_t = \frac{k_1 [A]_0}{k_2 - k_1} (\exp(-k_1 t) - \exp(-k_2 t)) \qquad \text{(xix)}$$

The solution for $[P]$ as a function of time may then be obtained by substitution of expressions for $[A]_t$ and $[B]_t$ in the mass conservation Equation i.

As discussed in the text, a steady-state approximation can be applied in the case where $k_2 \gg k_1$. For the initial reaction, when the concentration of A is approximately constant and equal to $[A]_0$, Equation xix then simplifies to

$$[B]_t = \frac{k_1 [A]_0}{k_2} (1 - \exp(-k_2 t)) \qquad \text{(xx)}$$

(*continued*)

> The steady-state concentration will be reached once the exponential term in Equation xx becomes negligible or approximately at $t > 1/k_2$.

and

$$[P]_t = [A]_0 \left\{ 1 - \frac{1}{k_2 - k_1} [k_2 \exp(-k_1 t) - k_1 \exp(-k_2 t)] \right\} \qquad (29)$$

Figure 3.8 shows the analytical solutions for $[A]$, $[B]$, and $[P]$ as a function of time for the conditions $[A]_0 = 1$, $[B]_0 = [P]_0 = 0$; $k_1 = 1$ (time)$^{-1}$ and $k_2 = 10$ (time)$^{-1}$ with both concentration and time in arbitrary units. The extent to

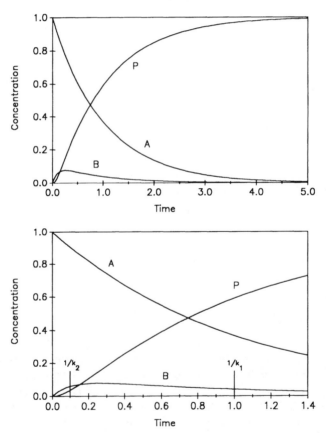

Figure 3.8 Concentrations of A, B, and P over time as the reaction $A \xrightarrow{k_1} B \xrightarrow{k_2} P$ progresses. Initial conditions: $[A]_0 = 1$, $[B]_0 = [P]_0 = 0$. Rate constants: $k_1 = 1$ (time)$^{-1}$, $k_2 = 10$ (time)$^{-1}$. Both concentration and time are expressed in arbitrary units. The initial part of the reaction is shown in detail in the lower figure, which indicates the time range over which the steady-state approximation for $[B]$ is valid, that is, $1/k_2 < t < 1/k_1$.

which the intermediate B accumulates during the reaction is determined by the relative values of the constants k_1 and k_2.

The expression for $[P]_t$ can be greatly simplified if the values of the rate constants for k_1 and k_2 are sufficiently different. For a slow first step followed by a rapid second step $(k_1 \ll k_2)$,

$$[P]_t \simeq [A]_0 [1 - \exp(-k_1 t)] \tag{30}$$

and, conversely, for a slow second step preceded by a rapid first step $(k_2 \ll k_1)$,

$$[P]_t \simeq [A]_0 [1 - \exp(-k_2 t)] \tag{31}$$

Thus we see that the rate of reaction is governed by the slower of the two steps, the rate-determining step.

If the first step is rate-determining, a small concentration of the intermediate B will be formed initially. Substantial concentrations of B cannot accumulate, however, because of the rapid decay of B to the product P. Then, except for the very beginning and very end of the reaction, the concentration of B will be approximately at *steady state* where

$$\frac{d[B]}{dt} \simeq 0 \tag{32}$$

and thus

$$k_1[A] - k_2[B] \simeq 0 \tag{33}$$

The steady-state concentrations of B, $[B]_{SS}$, is then

$$[B]_{SS} = \frac{k_1}{k_2}[A] \tag{34}$$

As mentioned above, the steady-state approximation is not valid at the very beginning of the reaction when $d[B]/dt > 0$ and $[B]$ approaches $[B]_{SS}$. This initial time period during which B accumulates is defined by the rate constant for the faster, second step (k_2). The time required to reach steady state is approximately $1/k_2$.

The steady-state approximation must also fail at the very end of the reaction where $d[B]/dt < 0$ and $[B]$ approaches 0. The approach to completion of the reaction is defined by the rate constant for the rate-limiting step, in this case k_1. The reaction approaches completion (and the steady-state approximation becomes invalid) for times greater than approximately $1/k_1$.

The steady-state approximation is very commonly applied in the analysis of enzyme kinetics. In enzymatic reactions, reversible formation of a complex between the enzyme, E, and substrate, S, is followed by irreversible formation of the product P. This is accompanied by regeneration of the enzyme, thus

$$E + S \underset{k_{-1}}{\overset{k_1}{\rightleftharpoons}} ES$$

$$ES \xrightarrow{k_2} P + E$$

The rate of formation of P (or reaction velocity, V) is

$$\frac{d[P]}{dt} = V = k_2[ES] \tag{35}$$

where

$$\frac{d[ES]}{dt} = k_1[E][S] - (k_{-1} + k_2)[ES] \tag{36}$$

Initially, the concentration of the substrate S is most often in large excess of the enzyme concentration; thus $[S]$ will be approximately constant for some time.

At the very beginning of the reaction, the intermediate ES accumulates to its steady-state concentration (see Sidebar 3.3) after which time the steady-state approximation

$$\frac{d[ES]}{dt} \simeq 0 \tag{37}$$

may be applied. Since the enzyme is regenerated, the substitution

$$[E] = E_T - [ES] \tag{38}$$

(where E_T is the total enzyme concentration) may be made, then

$$[ES]_{SS} = \frac{k_1}{k_{-1} + k_2 + k_1[S]} E_T[S] = \left[\frac{1}{\dfrac{k_{-1} + k_2}{k_1} + [S]} \right] E_T[S] \tag{39}$$

and

$$\frac{d[P]}{dt} = \left[\frac{k_2}{\dfrac{k_{-1} + k_2}{k_1} + [S]} \right] E_T[S] \tag{40}$$

SIDEBAR 3.3

Enzyme Kinetics

For the enzymatic transformation of substrate to product by the mechanism shown in the text, we may calculate the time needed to reach the steady-state concentrations of ES. With the assumption that the substrate concentration is relatively constant and using the mass conservation equation for the enzyme (Equation 38), we may write the differential equation for ES:

$$\frac{d[ES]}{dt} - (k_{-1} + k_2 + k_1[S])[ES] = k_1[S]E_T \tag{i}$$

As in Sidebar 3.2, we obtain a linear first-order differential equation which may be solved as described previously. The integrated form of this expression for $[ES] = 0$ at $t = 0$ is

$$[ES]_t = \frac{k_1 E_T[S]}{k_{-1} + k_2 + k_1[S]}\{1 - \exp[-(k_{-1} + k_2 + k_1[S])t]\} \tag{ii}$$

The rate of formation of the product P can then be written

$$\frac{d[P]}{dt} = k_2[ES] = \frac{k_2 E_T[S]}{\dfrac{k_{-1} + k_2}{k_1} + [S]}\{1 - \exp[-(k_{-1} + k_2 + k_1[S])t]\} \tag{iii}$$

By comparison of Equations ii and iii above with Equations 39 and 40 in the text, we see that the steady-state approximation is valid once the exponential terms in Equations ii and iii become negligible or approximately when $t > 1/(k_{-1} + k_2 + k_1[S])$.

This rate is often referred to as the "initial" velocity because it corresponds to the early phase of the reaction during which the substrate is not substantially depleted. We must also remember that it applies only after steady-state conditions have been established. Equation 40 corresponds to the common kinetic expression for enzymatic reactions, the Michaelis–Menten expression,

$$\frac{d[P]}{dt} = \frac{V_{\max}[S]}{K_M + [S]} \tag{41}$$

At high substrate concentrations, the system becomes saturated and the reaction proceeds at a maximum rate, V_{max}, where

$$V_{max} = k_2 E_T \tag{42}$$

The substrate concentration at which the reaction proceeds at half of its maximum rate corresponds to the Michaelis–Menten (or half-saturation) constant, K_M, where

$$K_M = \frac{k_{-1} + k_2}{k_1} \tag{43}$$

If k_2 is small compared with k_{-1}, K_M is approximately equal to k_{-1}/k_1 and thus (as discussed in Section 2.3) to the equilibrium constant for dissociation of the enzyme–substrate complex, K, where

$$K = \frac{[E][S]}{[ES]} = \frac{k_{-1}}{k_1} \tag{44}$$

In this case, the enzyme and substrate are in *pseudoequilibrium* with the complex *ES*.

The hydration of CO_2 discussed in Section 1.2.1 is an important physiological reaction which is catalyzed by carbonic anhydrase enzymes.[13,14] The *turnover number* for carbonic anhydrases ranges from 10^3 to 10^5 s^{-1}; the turnover number of an enzyme is defined as the maximum rate at which the enzymatic reaction proceeds divided by the concentration of enzyme active sites and thus corresponds to the constant k_2. (Note that some enzymes have more than one active site.) Thus the enzymatic hydration of CO_2 proceeds many orders of magnitude faster than the uncatalyzed reaction.

1.2.4 Catalysis Enzymatic reactions are examples of catalyzed reactions. Common catalysts in aqueous systems include acids, bases, and transition metals. Acid or base catalysis can be either specific or general. In specific acid catalysis, the reaction rate depends on $[H^+]$ and, in general acid catalysis, on the total acid concentration.

The effects of specific and general acid catalysis can be seen in the disproportionation of monochloramine,[15] a reaction that occurs during water chlorination. The formation of chloramines is important in the stabilization of "reactive chlorine" in chlorination systems; "breakpoint chlorination" is due to the decomposition of dichloramine.[16] Monochloramine disproportionates to give dichloramine and ammonia by the reaction

$$2NH_2Cl \xrightarrow{k_{obs}} NHCl_2 + NH_3$$

The proposed mechanism for general acid catalysis is

$$NH_2Cl + HA \rightleftharpoons NH_3Cl^+ + A^-$$

$$NH_3Cl^+ + NH_2Cl \longrightarrow NHCl_2 + NH_3 + H^+$$

The rate of monochloramine disappearance is

$$\frac{d[NH_2Cl]}{dt} = -2k_{obs}[NH_2Cl]^2 \tag{45}$$

where k_{obs} is defined with respect to dichloramine formation (thus accounting for the factor of 2). Figure 3.9a shows the observed rate constants for monochloramine disappearance at pH 3 in dilute phosphoric acid solutions where

$$k_{obs} = k_H[H^+] + k_{H_3PO_4}[H_3PO_4] + k_{H_2PO_4}[H_2PO_4^-] \tag{46}$$

(the subscripts on the rate constants indicate the catalytic species). The nonzero intercept in Figure 3.9a corresponds to specific acid catalysis (by H^+] and the slope to general acid catalysis (by H_3PO_4 and $H_2PO_4^-$). The combined effects of specific and general acid catalysis are shown in Figure 3.9b. At a given total phosphate concentration, the increase in k_{obs} with increasing proton activity (or concentration) arises both from specific acid catalysis and from the pH dependence of the ratio of the concentrations of H_3PO_4 and $H_2PO_4^-$.

For some reactions, a shift in the predominant reaction mechanism may be observed if the reaction is studied over a wide pH range. Figure 3.10 illustrates the pH dependence for reactions subject to both specific acid and specific base catalysis.[17] The predominant reaction mechanism shifts from an acid-catalyzed pathway at low pH, to an uncatalyzed pathway at neutral pH, and finally to a base-catalyzed pathway at high pH. The observed rate constant reflects the contribution of the various pathways:

$$k_{obs} = k_H[H^+] + k_N + k_{OH}[OH^-] \tag{47}$$

where k_H, k_N, and k_{OH} are the rate constants for the acid-catalyzed, uncatalyzed (neutral), and base-catalyzed pathways. This variation in the pH dependence of the reaction rate over a wide range in pH is commonly observed in the hydrolysis of organic compounds such as pesticides.

The catalytic effects of chemical species, such as acids, bases, or transition metals, must be considered if reaction rates in natural waters are to be estimated based on studies of these reactions in the laboratory. For the extrapolation to be valid the rate law must explicitly include the concentrations of catalytic species, which may vary over many orders of magnitude between the laboratory

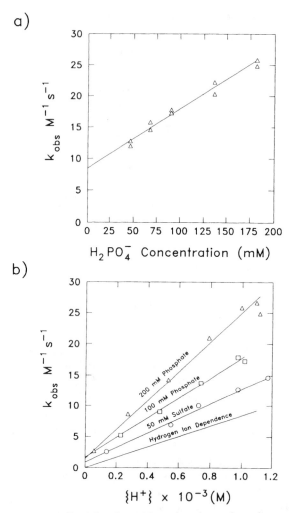

Figure 3.9 Rate constants (k_{obs}) for the acid-catalyzed reaction of monochloramine to form dichloramine as a function of (a) $H_2PO_4^-$ concentration at pH = 3 and (b) H^+ activity with varying total acid concentrations. In (a) the intercept corresponds to specific acid catalysis and the slope to general acid catalysis. In (b) the relationship between k_{obs} and $\{H^+\}$ can be seen to depend on both the type and total concentration of acid present. This behavior is a result of the combined effects of specific and general acid catalysis. From Valentine and Jafvert (1988).[15]

and the field. The difficulties involved in such extrapolations are discussed in more detail in Section 3.1.

1.2.5 Radical and Chain Reactions Many important reactions in natural waters, particularly photochemical reactions, proceed through a complex

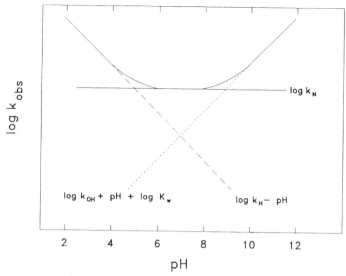

Figure 3.10 Dependence of the logarithm of the observed rate constant on pH for a reaction subject to both specific acid and specific base catalysis. At low pH, the reaction is acid catalyzed and $k_{obs} \simeq k_H[H^+]$ shown by (---). At neutral pH, the reaction is uncatalyzed and $k_{obs} \simeq k_N$ shown by (——). At high pH, the reaction is base catalyzed and $k_{obs} \simeq k_{OH}[OH^-] = k_{OH}K_w/[H^+]$ shown by ($\cdots$). The solid line corresponds to the sum of the terms $k_H[H^+] + k_N + k_{OH}[OH^-]$. Adapted from Harris (1982).[17]

array of sequential and concurrent steps. Such *chain reactions* often involve *free radicals* as intermediates. These species, which have unpaired electrons, are so reactive that they tend to combine somewhat indiscriminately with other chemical species present, resulting in a bewildering assortment of products. This leads to extremely complicated reaction mechanisms; reaction rates may exhibit fractional-order dependence on the concentration of the reacting species.

Chain reactions involve three types of steps. Reactive intermediates are generated, often photochemically, in a chain *initiation* step. Chain *propagation* occurs by the attack of radicals on other molecules to give new radicals, and chain *termination* by the recombination of radicals to give stable products.

An example of a radical chain reaction important in some water treatment processes is the decomposition of ozone.[18,19] Ozonation of drinking water may be performed for a variety of purposes, principally disinfection and removal of organic compounds.[20] In "pure water" the initiation step is the reaction of ozone with hydroxide to form a superoxide radical anion and a hydroperoxyl radical:

$$O_3 + OH^- \rightarrow O_2^- + HO_2^{\cdot}$$

In the propagation steps, the radicals formed initially react with ozone to form oxygen and hydroxyl radical and the hydroxyl radical reacts with ozone to regenerate the hydroperoxyl radical:

$$O_2^- + O_3 \xrightarrow{\ H'\ } 2O_2 + {}^{\cdot}OH$$

$${}^{\cdot}OH + O_3 \longrightarrow HO_2^{\cdot} + O_2$$

Chain termination results from the reaction of radical species with each other. These steps are outlined in Figure 3.11. In natural waters other chemical species such as bicarbonate may act as chain terminators. Organic compounds may act as chain *promoters* of ozone decomposition by increasing the efficiency of radical generation and propagation. The presence of such substances can markedly affect the apparent kinetic *chain length* (the number of chain cycles occurring per initiation step) and thus the overall efficiency of the process.

2. THEORETICAL AND SEMIEMPIRICAL ESTIMATION OF RATE CONSTANTS AND OF THEIR VARIATION WITH TEMPERATURE, PRESSURE, AND IONIC STRENGTH

Experimentally determined reaction rate constants may be used to estimate the rates of reactions in natural waters. For many reactions of environmental interest, however, rate constants have either not been determined or have been

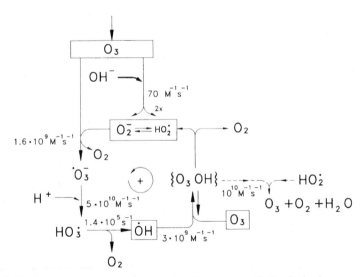

Figure 3.11   Reactions of aqueous ozone in "pure water." Catalytic decomposition of ozone occurs during the cyclic interconversion of hydroxyl radicals ($\cdot$OH) and hydroperoxyl radicals ($HO_2^{\cdot}$). From Staehelin and Hoigné (1985).[18]

determined only for conditions inappropriate to natural waters—for example, high temperature or very acidic solutions. Theories of chemical kinetics, such as *encounter theory*, *activated complex theory*, and *detailed balancing* (or *microscopic reversibility*), provide some basis for estimating rate constants and accounting for the effects of temperature, pressure, and ionic strength. *Semiempirical* estimations of rate constants are based on correlation of kinetic constants either with other kinetic constants for comparable reactions or with some thermodynamic property. This approach requires a data base of kinetic constants for similar compounds.

2.1 Encounter Theory

A fundamental constraint on the rate of a bimolecular reaction is the necessity for the reacting species to come close enough to each other for the reaction to occur as described by encounter theory for reactions in solution. A bimolecular reaction may be considered to consist of two steps: first, the reversible formation of the *encounter pair* (A,B)

$$A + B \rightleftarrows A,B$$

which occurs by diffusion of the reactant species A and B, and second, the reaction of the encounter pair to form the product P

$$A,B \rightarrow P$$

For the fastest bimolecular reactions the rate of product formation is controlled by the rate of formation of the encounter pair. These reactions are termed *diffusion-controlled* and have a rate expression of the form

$$\text{rate} = \frac{d[P]}{dt} = k[A][B] \tag{48}$$

where k is the rate constant for formation of the encounter pair, which may be derived from encounter theory. If the reaction is *activation controlled*, the rate is limited by the reaction of the encounter pair and the reaction is slower than predicted by encounter theory.

In the simplest estimation, the rate of diffusion-controlled bimolecular reaction between uncharged species of approximately equal radii depends only on the temperature and the viscosity of the solvent, water. This result is obtained by combining the Smoluchowski expression (which relates the rate constant for a bimolecular encounter to the diffusion coefficients of the reacting species and the critical distance for the reaction) with the Stokes–Einstein equation relating the diffusion coefficient of a species to its radius and the solvent viscosity.[21] The rate constant, k in cubic meters per mole per second, for a bimolecular encounter in solution, obtained from the Smoluchowski expression, is

$$k = 4\pi \mathcal{N}(D_A + D_B)r^* \tag{49}$$

where D_A and D_B are the diffusion coefficients (in square meters per second) of A and B and r^* is the critical distance (in meters) for the reaction between A and B (approximately equal to the sum of their radii). The dependence of the diffusion coefficient on the radius of the diffusing species and solvent viscosity (η) is given by the Stokes–Einstein equation:

$$D = \frac{RT}{6\pi\eta r \mathcal{N}} \tag{50}$$

For the reaction of species of approximately equal radius, $r_A = r_B = r^*/2$, we obtain, by substituting Equation 50 into Equation 49, the very simple expression

$$k \simeq \frac{8RT}{3\eta} \tag{51}$$

where the rate constant is independent of the radii and diffusion coefficients of the reacting species. In water at $25°C$ ($\eta = 8.904 \times 10^{-4} kg \cdot m^{-1} s^{-1}$), the calculated value of k is $7.4 \times 10^9 M^{-1} s^{-1}$ (with appropriate conversion of volume units from cubic meters to liters). Thus the rate of a diffusion-controlled reaction in solution is (very) approximately

$$\text{rate} = \frac{d[P]}{dt} \simeq 10^{10}[A][B] \tag{52}$$

This expression gives us a general feeling for the magnitude of the rate constants for diffusion-controlled reactions. If the diffusion coefficients and radii of the reacting species are known, a more accurate result may be obtained by using Equation 49 rather than Equation 51.

2.1.1 Diffusion-Controlled Reactions of Ions

Most chemical species in natural waters are charged rather than neutral. Electrostatic effects[22] increase the rate of diffusion-controlled reactions of oppositely charged species and decrease the rate of reaction of similarity charged species. Electrostatic effects may be incorporated into Equation 49 by including a factor P, such that

$$k \simeq P4\pi \mathcal{N}(D_A + D_B)r^* \tag{53}$$

TABLE 3.2   Electrostatic Factor (P) for Diffusion-Controlled Rate Constants

$$P = \frac{z_A z_B e^2}{4\pi\varepsilon_0\varepsilon r^* kT}\left(\frac{1}{\exp(z_A z_B e^2/4\pi\varepsilon_0\varepsilon r^* kT) - 1}\right)$$

Constants

e (charge on proton)	1.602×10^{-19} C
ε_0 (vacuum permittivity)	8.85×10^{-12} J^{-1} C^2 m^{-1}
ε (relative permittivity of medium)	78.54
k (Boltzmann constant)	1.38×10^{-23} J·deg^{-1}

Parameters with Assigned Values

r^* (critical reaction distance)	5×10^{-10} m
T (absolute temperature)	298 K
z_A or z_B (charge on ions)	See below

$z_A z_B$	P	$z_A z_B$	P
-1	1.88	$+1$	0.45
-2	3.03	$+2$	0.17
-3	4.35	$+3$	0.060
-4	5.74	$+4$	0.019

Source: From Atkins (1978).[22]

where

$$P = \left[\frac{z_A z_B e^2}{4\pi\varepsilon_0\varepsilon r^* \mathbf{k}T}\right]\left[\frac{1}{\exp(z_A z_B e^2/4\pi\varepsilon_0\varepsilon r^* \mathbf{k}T) - 1}\right] \tag{54}$$

The parameters in this equation are defined in Table 3.2. This table also gives the calculated values of P for charged species with $1 \leqslant |z_A z_B| \leqslant 4$ where z_A and z_B are the charges on the ions. From this table, we can see that the predicted acceleration in rate constants for reaction of oppositely charged species is not particularly dramatic. The predicted deceleration for like-charged species is somewhat more substantial.

Proton transfer reactions are some of the most important diffusion-controlled reactions in aquatic systems. The rate constants for these reactions[23] (given in Table 3.3) range from approximately 10^9–10^{11} M^{-1} s^{-1}. Some proton transfer reactions are even faster than predicted by encounter theory because of the participation of the solvent water molecules in the proton transfer reaction.[24]

TABLE 3.3 Rate Constants for Aqueous Proton Transfer Reactions

Reaction	$k_f(M^{-1}s^{-1})$	$k_b(s^{-1}$ or $M^{-1}s^{-1})$
$H^+ + OH^- \rightarrow H_2O$	1.4×10^{11}	2.5×10^{-5}
$H^+ + NH_3 \rightarrow NH_4^+$	4.3×10^{10}	24
$H^+ + HS^- \rightarrow H_2S$	7.5×10^{10}	8.3×10^3
$H^+ + A^- \rightarrow HA$ (A^- = acetate)	4.5×10^{10}	8.5×10^5
$H^+ + F^- \rightarrow HF$	1.0×10^{11}	7×10^7
$H^+ + SO_4^{2-} \rightarrow HSO_4^-$	1×10^{11}	1×10^9
$H^+ + H_2O \rightarrow H_3O^+$	1×10^{10}	1×10^{10}
$OH^- + NH_4^+ \rightarrow H_2O + NH_3$	3.4×10^{10}	6×10^5
$OH^- + HCO_3^- \rightarrow H_2O + CO_3^{2-}$	6×10^9	1.2×10^6
$OH^- + HPO_4^{2-} \rightarrow H_2O + PO_4^{3-}$	2×10^9	4×10^7
$OH^- + H_2O \rightarrow H_2O + OH^-$	3×10^9	3×10^9

Source: From Jolley (1984).[23]

2.1.2 Effect of Temperature on the Rate Constant

The rate constants for the proton transfer reactions between H^+ or OH^- and H_2O increase with increasing temperature[25] as shown in Figure 3.12. This exponential effect of temperature arises from the exponential temperature dependence of viscosity (η), that is,

$$\eta \propto \exp(E_a/RT) \tag{55}$$

(where E_a is an empirically derived constant, as discussed below). Thus, since the rate constant is inversely proportional to viscosity,

$$k \propto \exp(-E_a/RT) \tag{56}$$

The linear dependence of the rate constant on temperature, RT in Equation 51, can also be included, but it is much weaker than the exponential dependence from the viscosity term. The exponential temperature dependence of Equation 56 corresponds to the empirically derived *Arrhenius equation*:

$$k = A\exp(-E_a/RT) \tag{57}$$

where E_a is the *activation energy* of the reaction and A is referred to as the preexponential term. In Figure 3.12 the logarithm of the rate constant for proton transfer reactions of H^+ and OH^- with H_2O are plotted against $1/T$. The slope of this line gives values of 10 and 8.8 kJ/mole for the activation energies of these reactions. Thus the energy barrier for proton transfer reactions is comparable to the thermal energy of a solution at room temperature ($RT \approx 2.5$ kJ/mol at $T = 298$ K).

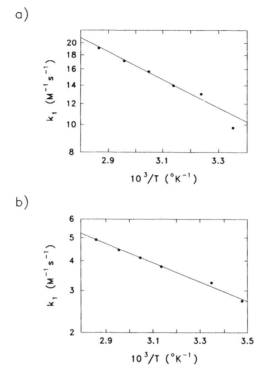

Figure 3.12   Temperature dependence of rate constants for proton transfer reactions. Rate constants (in $M^{-1}s^{-1}$ on a logarithmic scale) are plotted as a function of inverse temperature ($10^3/T$ in K^{-1}) for reactions of (a) H^+ with H_2O and (b) OH^- with H_2O. From Luz and Meiboom (1964).[25]

2.2   Activated Complex Theory

For reactions that are not diffusion controlled, encounter theory offers little insight into the reaction rates. In activated complex theory (ACT), the energy of reacting chemical species is examined along a reaction coordinate, that is, from the individual reactants A and B to the *activated complex* $(AB)^\ddagger$ to products P (as shown in Figure 3.13). Then the rate of reaction can be formulated as the product of $[(AB)^\ddagger]$, the concentration of the activated complex, and $k^\ddagger$, the rate constant for transformation of the activated complex to products, thus

$$\text{rate} = \frac{d[P]}{dt} = k^\ddagger[(AB)^\ddagger] \tag{58}$$

The activated complex may be thought of as an encounter pair organized such that a critical molecular vibration (i.e., bond distortion) corresponds to

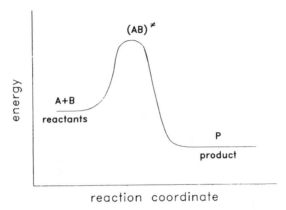

Figure 3.13 Diagram showing the energy of reacting species along the reaction coordinate for the reaction of A and B, the reactants, to $(AB)^{\ddagger}$, the activated complex, and finally to the product P.

passage through the *transition state* to products. For gas phase reactions, the distribution of molecules between reactants, A and B, and the activated complex may be expressed explicitly by the Boltzmann distribution of vibrational states. This approach, however, is not applicable to reactions in aqueous solution because of the complicated interactions between the solvent and the activated complex.

For reactions in solution, a simplification of activated complex theory is introduced in which pseudoequilibrium between reactants and the activated complex is assumed (sometimes called the "thermodynamic" approach), then

$$\frac{\{(AB)^{\ddagger}\}}{\{A\}\{B\}} = K^{\ddagger} = \left[\frac{\gamma^{\ddagger}}{\gamma_A \gamma_B}\right] \frac{[(AB)^{\ddagger}]}{[A][B]} \tag{59}$$

Neglecting the activity coefficients for the moment, we may then write

$$\frac{d[P]}{dt} = k^{\ddagger}[(AB)^{\ddagger}] = k^{\ddagger} K^{\ddagger}[A][B] \tag{60}$$

Thus we relate the rate constant of the overall reaction to the equilibrium constant for the activated complex.

Following the "thermodynamic" approach, we define a *free energy of activation* such that

$$\Delta G^{\ddagger} = -RT \ln K^{\ddagger} \tag{61}$$

Thus we may express the rate constant $k^{\ddagger} K^{\ddagger}$ in terms of $\Delta G^{\ddagger}$ or of the *entropy* and *enthalpy of activation* $\Delta G^{\ddagger} = \Delta H^{\ddagger} - T\Delta S$. Then

$$k^{\ddagger}K^{\ddagger} = k^{\ddagger}\exp\frac{(-\Delta G^{\ddagger})}{RT} = k^{\ddagger}\exp\frac{(\Delta S^{\ddagger})}{R}\exp\frac{(-\Delta H^{\ddagger})}{RT} \qquad (62)$$

These results, which relate the reaction kinetics to thermodynamic properties of the activated complex, are particularly convenient for calculating the effects of ionic strength, temperature, and pressure on reaction rate constants.

2.2.1 Effect of Ionic Strength The effect of ionic strength on reaction rates may be evaluated by including the activity coefficients for the reactants and the activated complex in Equation 60, that is,

$$\frac{d[P]}{dt} = k^{\ddagger}[(AB)^{\ddagger}] = k^{\ddagger}K^{\ddagger}\left[\frac{\gamma_A\gamma_B}{\gamma^{\ddagger}}\right][A][B] \qquad (63)$$

As discussed in Chapter 2, we can calculate activity coefficients as a function of ionic strength taking the charge of the activated complex as the the sum of the charges of the reactants. Thus there will be no ionic strength effect for

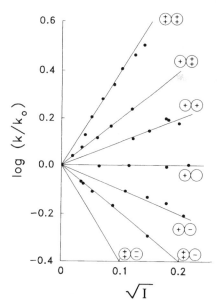

Figure 3.14 Effect of ionic strength on reaction rate constants. Rate constants at given ionic strength (normalized to the rate constant at zero ionic strength) as a function of $\sqrt{I}$ are shown for some redox and complexation reactions with species of varying charge. This behavior can be accounted for by considering the effect of ionic strength on the stability of the activated complex. From Atkins (1978).[22]

reactions with neutral molecules. For reactions between ions, increasing ionic strength decreases the rate constants for reactions of oppositely charged ions and increases the rate constant for reactions of similarly charged ions because of the decreasing electrostatic attraction or repulsion at higher ionic strength (as shown in Figure 3.14).

2.2.2 Effect of Temperature   From Equation 62, we may see that the effect of temperature on reaction rate constants is comparable to its effect on equilibrium constants (as discussed in Chapter 2). As with encounter theory, the expression relating the rate constant to temperature is of the same form as the Arrhenius equation; the preexponential term A in Equation 57 corresponds to the exponential of the entropy of activation,

$$A \simeq k^{\ddagger} \exp \frac{(\Delta S^{\ddagger})}{R} \tag{64}$$

It should be noted here that the more detailed derivation of gas-phase reactions shows that the term $k^{\ddagger}$ is also dependent on temperature—since $k^{\ddagger}$ corresponds to the transformation of the activated complex by a molecular vibration, for which more energy is available at higher temperatures—thus

$$k^{\ddagger} \simeq \frac{\mathbf{k}T}{h} \tag{65}$$

(where h is the Planck constant) but this linear temperature dependence is less important than the exponential dependence on temperature in Equation 62. By explicitly including the temperature dependence of $k^{\ddagger}$ in Equation 62, we may write

$$k^{\ddagger}K^{\ddagger} = \frac{\mathbf{k}T}{h} \exp \frac{(\Delta S^{\ddagger})}{R} \exp \frac{(-\Delta H^{\ddagger})}{RT} \tag{66}$$

or

$$\ln(k^{\ddagger}K^{\ddagger}) = \ln \frac{\mathbf{k}T}{h} + \frac{\Delta S^{\ddagger}}{R} - \frac{\Delta H^{\ddagger}}{RT} \tag{67}$$

For condensed phases $\Delta H^{\ddagger}$ and $\Delta S^{\ddagger}$ can be taken to be temperature independent. Thus

$$\frac{d\ln(k^{\ddagger}K^{\ddagger})}{dT} = \frac{\Delta H^{\ddagger} + RT}{RT^2} \tag{68}$$

By comparison of Equation 68 with the differential of the Arrhenius equation, we may see that the empirical activation energy, E_a, is related to the enthalpy of activation by the expression

$$E_a = \Delta H^{\ddagger} + RT \tag{69}$$

Thus the enthalpy of activation may be obtained from an Arrhenius plot ($\ln k$ vs. $1/T$) of experimental data.

As discussed in Section 2.1.2, reactions that are facile at room temperature (such as proton transfer reactions), have activation energies of a few kilojoules per mole. In contrast, many geochemical reactions, such as precipitation and dissolution, have significantly higher activation energies and thus a strong dependence of reaction rate on temperature. While the form of the rate laws for precipitation and dissolution will be examined in Chapter 5, here we simply use the experimental values for the rate constants for quartz precipitation as a function of temperature to obtain the activation energy for this reaction.[26] The value obtained from the slope of the Arrhenius plot in Figure 3.15 is $E_a = 53 \, kJ/mol$.

The significant decrease in precipitation rates with decreasing temperature allows the measurement of silica concentrations in thermal spring water to be used in estimating the temperature of its geothermal reservoir (*quartz geothermometer*). Application of the quartz geothermometer assumes that the silica concentration in the geothermal reservoir is controlled by quartz solubility

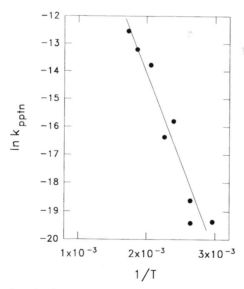

Figure 3.15 Arrhenius plot for quartz precipitation: the logarithm of the rate constant for precipitation of quartz is plotted against $1/T$. Data are from Rimstidt and Barnes (1980).[26] The activation energy E_a is obtained from the slope of the graph.

TABLE 3.4 Activation Volumes ($\Delta V^{\ddagger}$) for Aqueous Reactions

Reaction	$\Delta V^{\ddagger}$ (mL mol^{-1})
$CH_3Br + H_2O \rightarrow CH_3OH_2^+ + Br^-$	-14.5
$Co(NH_3)_5H_2^*O^{3+} + H_2O \rightarrow Co(NH_3)_5H_2O^{3+} + H_2^*O$ (isotopic exchange)	$+1.2$
$Cr(NH_3)_5H_2^*O^{3+} + H_2O \rightarrow Cr(NH_3)_5H_2O^{3+} + H_2^*O$ (isotopic exchange)	-5.8
$Co(NH_3)_5Cl^{2+} + H_2O \rightarrow Co(NH_3)_5H_2O^{3+} + Cl^-$	-10.6
$Cr(NH_3)_5Cl^{2+} + H_2O \rightarrow Cr(NH_3)_5H_2O^{3+} + Cl^-$	-10.8
Bond cleavage[a]	$+10$
Bond formation[a]	-10
Ionization[a]	-20
Charge neutralization[a]	$+20$

Source: From Moore and Pearson (1981).[27]

[a]Estimated values. Note that $\Delta V^{\ddagger}$ refers only to the volume change on the formation of the activated complex *not* the final product.

and that the concentration of silica in the thermal spring water and in the geothermal reservoir are identical. Since quartz solubility is also a function of temperature, the thermal spring water, which is cooler than the geothermal reservoir water, is supersaturated with respect to quartz. Thus the quartz geothermometer can be accurate only if slow kinetics prevent the precipitation of silica from the supersaturated thermal spring water.[26]

2.2.3 Effect of Pressure The "thermodynamic" approach also allows the calculation of the effect of pressure on rate constants. This effect is related to the change in molar volume on formation of the activated complex $\Delta V^{\ddagger}$. If $\Delta V^{\ddagger} < 0$, formation of the activated complex is favored with increasing pressure, that is, $K^{\ddagger}$ is greater. At a fixed T and ionic strength,

$$\left[\frac{d[\ln k]}{dP} \right]_{T,I} = \frac{-\Delta V^{\ddagger}}{RT} \tag{70}$$

Typical values[27] of $\Delta V^{\ddagger}$ are given in Table 3.4. For example, if $\Delta V^{\ddagger} = -10$ mL/mol and ΔP is 500 atm (corresponding roughly to the pressure at 5000 m depth in the ocean), the rate constant would increase only by $\approx 20\%$ (for $T = 298$ K and $\Delta V^{\ddagger}$ being assumed independent of pressure). With such values of $\Delta V^{\ddagger}$, pressure has only a small effect on the rate constant.

2.3 Detailed Balancing (Microscopic Reversibility)

Detailed balancing relates the forward and reverse rate constants for a reaction to the equilibrium constant for the reaction and can be used either to estimate a rate constant, from the measured values for the opposite rate constant and the equilibrium constant, or to validate measured rate constants, by comparison of their ratio with the equilibrium constant for the reaction.

At equilibrium, the net reaction rate is zero. The *principle of detailed balancing* (or *microscopic reversibility*) states that when the overall reaction is at equilibrium, each elementary step must also be at equilibrium and that, for each reversible step, the rate of the forward reaction equals the rate of the reverse reaction.

In a reaction such as proton transfer, the overall reaction corresponds to an elementary step. For the protonation of the acetate anion, A^-,

$$H^+ + A^- \underset{k_b}{\overset{k_f}{\rightleftharpoons}} HA$$

We may write the following expression for the reaction at equilibrium:

$$\frac{-d[H^+]}{dt} = \frac{-d[A^-]}{dt} = \frac{d[HA]}{dt} = 0 = k_f[H^+][A^-] - k_b[HA] \qquad (71)$$

thus

$$k_f[H^+][A^-] = k_b[HA] \qquad (72)$$

Then, at equilibrium, the ratio of the forward and reverse rate constants is equal to the equilibrium constant for the dissociation of acetic acid:

$$\frac{k_b}{k_f} = \frac{[H^+]_{eq}[A^-]_{eq}}{[HA]_{eq}} = K \qquad (73)$$

(Values for the rate constants are given in Table 3.3.) Note that while Equation 72 holds *only at equilibrium*, the relationship between the equilibrium constant and the forward and reverse rate *constants* (Equation 73) must always hold.

Although this principle strictly applies only to elementary steps, in some cases the overall equilibrium constant can be related to the constant for an individual elementary step. Thus, while the formation of complexes between alkali metals and crown ethers (Figure 3.16*a*) involves more than a single elementary step, a correlation is still observed between the rate constants for dissociation and the equilibrium stability constants of the complexes (as shown in Figure 3.16*b*). In this case k_f shows little variation with the reactants; this phenomenon is discussed in Chapter 6.

Some caution must be exercised in the application of detailed balancing over a wide range of concentrations of reacting species. For a stoichiometric reaction, detailed balancing may be used to relate the forward and reverse rate constants only if the reaction proceeds in both directions through the same mechanism. For example, experimental rate constants are often determined under conditions where the reaction is far from equilibrium. We may then attempt to estimate

a)

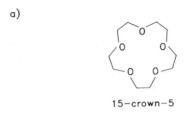

15−crown−5

b)

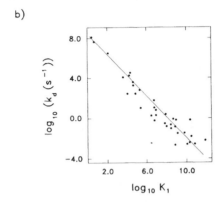

Figure 3.16   (a) Example of a crown ether. (b) Correlation of logarithms of rate constants for the dissociation of alkali metal–crown ether complexes with the logarithm of their respective equilibrium stability constants. From Burgess (1988).[28]

the opposite rate constant for this reaction by detailed balancing at significantly different concentrations of reactants and products (perhaps for conditions where the reaction is close to equilibrium). This estimation will only be valid if the reaction proceeds in both directions through the same mechanism under both concentration regimes. Variation in the predominant reaction mechanism with changes in reaction conditions is discussed in Section 3.1.

2.4   Empirical Estimations of Rate Constants

Rate constants may also be estimated empirically by correlation with other kinetic constants, with thermodynamic properties, or with other predictor variables.[29] The correlations between kinetic and equilibrium constants for a series of related reactions, termed *linear free energy relationships* (LFER), are particularly suitable for examining the effect of substituents (i.e., a small change in reactant structure) on the reaction rates. A data base of reaction rates of similar compounds is required to establish the correlation with the chosen thermodynamic parameter.

The theoretical interpretation of LFERs relates the free energy of activation

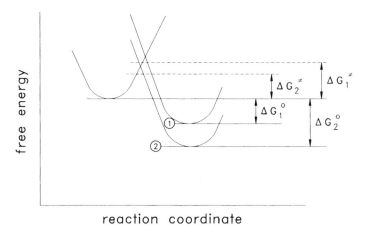

reaction coordinate

Figure 3.17 Illustration of the relationship between free energy of activation $\Delta G^{\ddagger}$ and ΔG° for a thermodynamically favorable reaction ($\Delta G^{\circ} < 0$). The activation energy $\Delta G^{\ddagger}$ is slightly below the intersection of the parabolae defining the chemical potentials of the initial and final states. After Moore and Pearson (1981).[30]

with the overall free energy of reaction.[30] This relationship is illustrated schematically in Figure 3.17. A decreasing negative free energy for the reaction corresponds to raising the right-hand parabola in Figure 3.17 (from the energy level of state 2 to that of state 1 in the figure) and is associated with an increase in $\Delta G^{\ddagger}$. In this simplified picture, the less the overall driving force for the reaction, the slower the rate.

In general, for the reaction

$$\text{reactants} \overset{k_i}{\rightleftharpoons} \text{products}$$

an LFER between the rate constant k_i and the equilibrium constant K_i would be expressed as

$$k_i = CK_i^{\alpha} \tag{74}$$

where α and C are constants. Small changes in the molar free energy of the activated state due to a substituent ($\delta\mu^{\ddagger}$) may be represented as a weighted average of the changes in the reactants and products (due to the substituent)

$$\delta\mu^{\ddagger} = \alpha\delta\mu_P^{\circ} + (1 - \alpha)\delta\mu_R^{\circ} \tag{75}$$

where the value of α is between 0 and 1 and indicates the extent to which the transition state resembles the product. Then, calculating the free energy change gives:

$$\delta\Delta G^{\ddagger} = \alpha\delta\Delta G^{\circ} \tag{76}$$

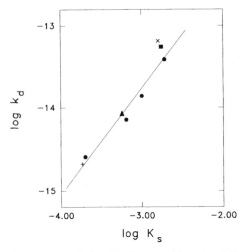

Figure 3.18 Linear free energy relationship between the rate of dissolution (k_d) and the solubility (K_s) for various silica phases. Different symbols correspond to different studies. From Wollast (1974).[31]

and thus

$$\ln k_i = \alpha \ln K_i + C \tag{77}$$

The parameters α and C may be obtained empirically for a given series of related reactions,

Linear free energy relationships have been applied to many types of reactions. The Marcus relation, an LFER for redox reactions, is discussed in Chapter 7. The LFER correlations are most often used to estimate either kinetic or thermodynamic parameters by interpolation and, in some instances, to infer mechanistic information.

The LFER shown in Figure 3.18 correlates the solubility of various solid silica phases and the rate constants for their dissolution.[31] The advantage of LFERs is that kinetic parameters, which tend to be difficult to measure, can be estimated from thermodynamic parameters, for which tabulated values are more often available. The data set on which the LFER is based must be critically examined. It is of particular importance that the LFER be applied only to an appropriate set of reactions. Reactions that involve dissimilar reactants or that occur through different reaction mechanisms should not be included in the same LFER correlation.

In practice, the application of correlation analyses has been quite broad. Reaction rate constants have been correlated with predictor variables, such as the structure or properties of reactants, even in biologically mediated reactions.[29,32] While such broad applicability is quite attractive, care must be taken not to overstep the predictive capabilities of LFERs.

3. MODELING CHEMICAL KINETICS IN NATURAL SYSTEMS

The preceding sections provide information needed to evaluate the importance of chemical kinetics in natural waters. For a specific system undergoing some perturbation, we wish to know whether the kinetics of various chemical reactions must be explicitly considered in order to describe, with some reasonable accuracy, its chemical composition. As we have already discussed, the answer to this question depends on the time scale for which we need to know the chemical composition of the system. We define this time scale by comparison with the rates of other processes, which may be physical, biological, or geological. Thus chemical kinetics will be "important" if the rate of a particular chemical reaction (under the ambient conditions of reactant concentrations, pH, temperature, etc.) is comparable to the time scale of interest. For example, if we define our time scale of interest as the doubling time of phytoplankton under optimal growth conditions (i.e., approximately one day), proton transfer reactions are fast and we may consider acid–base reactions to be at equilibrium (cf. Table 3.3). On the other hand, amino acid racemization (cf. Section 1.2.2) is clearly very slow and may be ignored. For reactions whose rates are comparable to the time scale of interest, a kinetic description becomes a necessity, and it is usually superimposed on a pseudoequilibrium model for all faster reactions. Before attempting such a model, however, it is important to examine some of the limitations on the extrapolation of kinetic data over varying reaction conditions.

3.1 Parallel and Competing Reaction Pathways

We have stated that the rate of a reaction is determined by the slowest step in the reaction mechanism. If, however, there are several *parallel* available mecha-

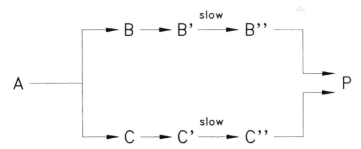

Figure 3.19 A schematic diagram showing two possible pathways for reaction of reactant A to product P either through intermediates B, B', B'' or C, C', C''. The rate-limiting step for each reaction pathway is the slowest step in that pathway ($B' \longrightarrow B''$ and $C' \longrightarrow C''$, respectively). The rate-limiting step for the overall reaction $A \longrightarrow P$ is the slowest step in the fastest pathway. Thus if the step $B' \longrightarrow B''$ is much slower than $C' \longrightarrow C''$, the overall reaction will proceed predominantly through the formation of the intermediates C, C', C'' and the transformation $C' \longrightarrow C''$ will be the rate-limiting step of the overall reaction. After Lasaga (1981).[33]

nisms (or pathways) for the reaction, then the overall reaction proceeds through the fastest available pathway.[33] Thus in Figure 3.19 the overall reaction rate is determined not by the slowest of all possible steps but by the slowest step in the fastest pathway. The availability of several parallel pathways may result in a shift in the reaction mechanism under various conditions (as was shown in Figure 3.10 for pH-dependent reactions).

This also introduces one of the great difficulties in predicting reaction rates in natural waters. Often, for analytical reasons, kinetic experiments must be conducted under conditions distinctly different from those encountered in nature, for example, at low pH or high concentrations of reactants. The extrapolation of these results to natural waters is only valid if the predominating mechanism for the reaction is the same under experimental and natural conditions. Consider the extremely simple case where transformation of some reactant A can occur either as a unimolecular reaction to product $P1$

$$A \xrightarrow{k_1} P1$$

or a bimolecular reaction to product $P2$

$$A + A \xrightarrow{k_2} P2$$

The rate expression for the loss of reactant A is

$$\frac{-d[A]}{dt} = k_1[A] + k_2[A]^2 \tag{78}$$

If the reactant concentration in the laboratory experiment is $10^{-2}\,M$ and the ambient concentration in the field is $10^{-7}\,M$, it is easy to imagine values of k_1 and k_2 for which the bimolecular pathway would overwhelmingly predominate under laboratory conditions and the unimolecular pathway under field conditions. Thus in our extrapolation from laboratory to field we can be confident only of predicting a minimum rate of reaction.

Such a shift in mechanism depending on reaction conditions has been suggested for the decomposition of the superoxide radical O_2^-, which is formed photochemically, in natural waters. This species, a key intermediate in oxygen chemistry, can decompose by the stoichiometric reaction

$$2O_2^- + 2H^+ \rightarrow H_2O_2 + O_2$$

The rate of loss of superoxide by this reaction exhibits a second-order dependence on the superoxide concentration. In oligotrophic (i.e., low productivity) seawater, this is the only observed mechanism for superoxide decomposition. In river and estuarine waters, however, an additional pathway for superoxide decomposition, with a first-order dependence on $[O_2^-]$, becomes important. The following rate law for superoxide decomposition has been observed in coastal seawater:

$$\frac{-d[O_2^-]}{dt} = k[O_2^-]^2 + k^*[O_2^-] \tag{79}$$

where k^* is the pseudo first-order rate constant for reaction of superoxide with unidentified trace constituents of the water; this pathway accounted for between 24 and 41% of the superoxide removal. This pathway is not observed in oligotrophic seawater because, at lower concentrations of the trace constituents which could scavenge superoxide, the first-order scavenging pathway cannot compete with the second-order pathway.[34,35]

3.2 The Pseudoequilibrium Approach

The appropriate mechanisms for a reaction under field conditions may be elucidated and rate constants determined in laboratory experiments or estimated from theory or by empirical methods. A kinetic description of the reaction may then be used to predict the chemical composition of a natural system as equilibrium is (re)established after some perturbation. The following discussion illustrates a simplifying approach to such a kinetic description. We model the progress of two competing chemical reactions which proceed at different rates. To simplify the calculations we assume, first, that over a short, initial time period the slower reaction can be neglected and, second, that after some time period has elapsed the faster reaction may be considered to be at equilibrium. That is, we define time domains so that a kinetic description for each reaction is included only when the characteristic time for the reaction is comparable to the time interval considered; the kinetics of the faster reaction are considered only for shorter times and the kinetics of the slower reaction only at longer times. As we shall see, the validity of this approach depends on the relative rates of the two reactions.

Metal ions in solution can interact with *ligands*, inorganic or organic species that have a free pair of electrons to share with the metal. This interaction, termed complexation, has already been mentioned briefly in Chapter 2. The thermodynamics and kinetics of complexation reactions are discussed in detail in Chapter 6. Here we anticipate that discussion by applying the pseudoequilibrium model to the competing interactions of a metal with two organic ligands. Simply by analogy with acid–base reactions we may write the following complexation reactions (omitting charges for simplicity):

$$M + L \underset{b_1}{\overset{f_1}{\rightleftharpoons}} ML; \quad K_{ML}$$

$$M + Y \underset{b_2}{\overset{f_2}{\rightleftharpoons}} MY; \quad K_{MY}$$

where K_{ML} and K_{MY} are the complexation constants. We choose conditions such that the stronger ligand, Y, is less abundant than the weaker ligand, L: $K_{MY} \gg K_{ML}$ and $[L] = 10\,\mu M$, $[Y] = 0.1\,\mu M$. Then for $[M]_T = 1\,nM$ (a reasonable value for many metals in natural waters), both L and Y are in large excess of M, and we may write the following pseudo first-order reactions:

$$M \underset{b_1}{\overset{f_1^*}{\rightleftharpoons}} ML$$

$$M \underset{b_2}{\overset{f_2^*}{\rightleftharpoons}} MY$$

where

$$f_1^* = f_1[L] \tag{80}$$

$$f_2^* = f_2[Y] \tag{81}$$

and, based on microscopic reversibility,

$$K = f/b \tag{82}$$

for both ML and MY. The governing differential equations are written:

$$\frac{d[M]}{dt} = -(f_1^* + f_2^*)[M] + b_1[ML] + b_2[MY] \tag{83}$$

$$\frac{d[ML]}{dt} = f_1^*[M] - b_1[ML] \tag{84}$$

$$\frac{d[MY]}{dt} = f_2^*[M] - b_2[MY] \tag{85}$$

The initial concentrations are $[M]_0$, $[ML]_0$, and $[MY]_0$ and the mole balance equation for M is

$$[M]_t + [ML]_t + [MY]_t = [M]_0 + [ML]_0 + [MY]_0 = T \tag{86}$$

(The analytical solution of these differential equations is given in Appendix 3.1.)

Figure 3.20 shows the concentrations of the three species with time for the total concentrations already stated, $[ML]_0 = [MY]_0 = 0$, and $f_1^* = 10^3\,s^{-1}$, $K_{ML} = 10^7$, $f_2^* = 10^{-3}\,s^{-1}$, $K_{MY} = 10^{11}$; in this case the reaction $M \rightleftharpoons ML$ is very much faster than $M \rightleftharpoons MY$. The early part of Figure 3.20 requires the exact

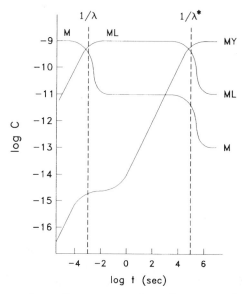

Figure 3.20 Concentrations of reactants and products over time for the reaction of metal M with ligands L and Y. Initial conditions: $[M]_0 = 1$ nM, $[L]_0 = 10\,\mu M$, $[Y]_0 = 0.1\,\mu M$. Since ligand concentrations are in large excess of $[M]_0$, the ligand concentrations are incorporated into the pseudo first-order rate constants $f_1^* = 10^3\,s^{-1}$ for formation of ML and $f_2^* = 10^{-3}\,s^{-1}$ for formation of MY. The dissociation rate constants are $b_1 = 10\,s^{-1}$ for ML and $b_2 = 10^{-7}\,s^{-1}$ for MY. The equilibrium formation constants (where $K = f/b$) are 10^7 for ML and 10^{11} for MY.

solution, but we may obtain most of the results plotted in the figure through the following approximate solution. We define two time domains: (1) an initial (short) time interval for which we neglect the slower reaction and (2) a later time for which we consider the faster reaction to be at pseudoequilibrium.

Thus for *time domain 1*, we follow the progress of the first (faster) reaction:

$$M \underset{b_1}{\overset{f_1^*}{\rightleftharpoons}} ML$$

and ignore the second (slower) one:

$$M \rightleftharpoons MY$$

Thus we hold (for this time domain) that the concentration of MY is equal to its initial value: $[MY]_t = 0$. For completeness, we may also redefine the total concentration of M as

$$T^* = T - [MY]_0 \tag{87}$$

Then (as described in Section 1.2.2), we obtain the solution

$$[M]_t = [M]_{PE} + ([M]_0 - [M]_{PE})\exp(-\lambda t) \tag{88}$$

where $[M]_{PE}$ is the pseudoequilibrium concentration of M:

$$[M]_{PE} = \frac{b_1}{f_1^* + b_1} T^* \tag{89}$$

and, as previously,

$$\lambda = f_1^* + b_1 \tag{90}$$

with a symmetrical solution for $[ML]_t$ (compare Equations 21 and 22). For species M and ML, this approximate result is indistinguishable from the exact solution for times shorter than the characteristic time for the second (slower) reaction as calculated below. We may see (from Figure 3.20) that our assumptions for time domain 1 give us a reasonable solution for $[M]_t$ and $[ML]_t$ up to at least 10^2 s. The concentration of MY can clearly not be obtained from this approximate solution (which assumes $[MY]_t = 0$) and the values shown in Figure 3.20 are those obtained from the exact solution.

We continue our approximate solution by considering the reaction over *time domain 2*, for which we take the first reaction

$$M \underset{b_1}{\overset{f_1^*}{\rightleftharpoons}} ML$$

to be at pseudoequilibrium, thus

$$[M]_t/[ML]_t = b_1/f_1^* \tag{91}$$

and follow the progress of the second (slower) reaction

$$M \underset{b_2}{\overset{f_2^*}{\rightleftharpoons}} MY$$

Substitution of the mole balance equation

$$[M]_t + [ML]_t + [MY]_t = T = \left(\frac{b_1 + f_1^*}{b_1}\right)[M]_t + [MY]_t \tag{92}$$

into the rate equation yields

$$\frac{d[MY]}{dt} = -\lambda^*[MY] + \left(\frac{b_1 f_2^*}{b_1 + f_1^*}\right)T \tag{93}$$

where

$$\lambda^* = \frac{f_2^* b_1 + b_1 b_2 + f_1^* b_2}{f_1^* + b_1} \qquad (94)$$

Then with "initial" conditions for this time domain (which correspond to the concentrations obtained from time domain 1) of $[M]_{PE}$, $[ML]_{PE}$, and $[MY]_0$, we obtain the solution for $[MY]_t$:

$$[MY]_t = [MY]_\infty + ([MY]_0 - [MY]_\infty)\exp(-\lambda^* t) \qquad (95)$$

$$[MY]_\infty = \frac{b_1 f_2^*}{b_1 f_2^* + b_1 b_2 + f_1^* b_2} T \qquad (96)$$

Note that this solution implies that we are well past the time required to establish pseudoequilibrium, which is achieved at times longer than the characteristic time of the first (fast) reaction. The concentrations of M and ML are obtained from the mass balance equation and Equation 91. Thus

$$[M]_t = \frac{b_1}{f_1^* + b_1}(T - [MY]_t) \qquad (97)$$

$$[ML]_t = \frac{f_1^*}{f_1^* + b_1}(T - [MY]_t) \qquad (98)$$

and the equilibrium concentrations are:

$$[M]_\infty = \frac{b_1 b_2}{f_1^* b_2 + b_1 b_2 + f_2^* b_1} T \qquad (99)$$

$$[ML]_\infty = \frac{f_1^* b_2}{f_1^* b_2 + b_1 b_2 + f_2^* b_1} T \qquad (100)$$

In time domain 2 the approximate solution is indistinguishable from the exact solution. For this example, the only discrepancy between the approximate and exact solutions is in the concentration of MY in the initial time period (i.e., in time domain 1) where $[MY]_t$ is held at $[MY]_0$ in the approximate solution.

In this system there are two characteristic reaction times: $1/\lambda$, which derives from the faster reaction $M \rightleftarrows ML$, and $1/\lambda^*$, which derives from the slower reaction $M \rightleftarrows MY$. The approximate solution holds in this example because these two reaction times are sufficiently different. For the conditions of Figure 3.20:

$$\frac{1}{\lambda} = \frac{1}{f_1^* + b_1} = 10^{-3}\,\text{s} \tag{101}$$

$$\frac{1}{\lambda^*} = \frac{f_1^* + b_1}{f_2^* b_1 + b_1 b_2 + f_1^* b_2} = 10^5\,\text{s} \tag{102}$$

Obviously, if the two reaction times are comparable, the approximate solution will fail to describe the system accurately. Note that the second reaction is very significantly retarded (i.e., its characteristic time lengthened) by the pseudoequilibrium of the first reaction. The second reaction, if it occurred alone, would have a characteristic time $1/\lambda_2 = 1/(f_2^* + b_2) = 10^3\,\text{s}$.

Apart from illustrating the application of the pseudoequilibrium model, this example demonstrates some of the important features of complexation kinetics in natural waters. The conditions of this example were not chosen entirely to provide a simple solution. It is often the case in natural waters that the concentrations of reacting species vary over several orders of magnitude. The initial predominance of the complex between the metal and the weaker ligand, ML, is partly due to this difference in concentration between the weaker ligand L and the stronger ligand Y, since the concentration term enters into the pseudo first-order rate constant for complex formation. The choice of the rate constants f_1 and f_2 such that the weaker ligand reacts more rapidly than the stronger ligand is explained in Chapter 6. Finally, let us note that the characteristic time for the formation of the thermodynamically favored species MY in the presence of the ligands L and Y (both in excess) in this example is about 25 hours. This equilibration time is significant on the time scale of metal uptake by microorganisms.

In natural waters, many time-dependent processes occur simultaneously. To describe the chemical composition of a natural system, we may have to consider the rates of physical and biological processes as well as the rates of chemical reactions. Even for a simple case such as a single reversible reaction in an open system, a complete description of the time-dependent chemical composition may be rather complicated (the exact solution for this case is derived in Appendix 3.2). We may, however, obtain approximate solutions by employing the partial equilibrium and pseudoequilibrium or steady-state models for chemical reactions, an approach often applied in computer algorithms for modeling transport and chemical kinetics.[36-39]

In later chapters the kinetics of specific reactions or processes, such as gas exchange, precipitation and dissolution, complexation, oxidation and reduction, and sorption and desorption, are addressed. These discussions focus on how the rates of such reactions are controlled, what mechanisms are likely to predominate, and what magnitude of reaction rates can be expected under environmental conditions. With such information we may begin to combine equilibrium and kinetic calculations to obtain a better understanding of the chemistry of natural waters.[40]

REFERENCES

1. J. Lovelock, *The Ages of Gaia*, Norton, New York, 1988, p. 9.
2. H. Taube, *Chem. Rev.*, **50**, 69 (1952).
3. W. S. Broeker and T. H. Peng, *Tracers in the Sea*, Eldigio Press, New York, 1982, Chapter 2.
4. M. N. A. Peterson, *Science*, **154**, 1542 (1966).
5. The material in this chapter is adapted in large part from the following texts: A. W. Adamson, *A Textbook of Physical Chemistry*, Academic, New York, 1972, Chapter 15; P. W. Atkins, *Physical Chemistry*, Freeman, San Francisco, CA, 1978, Chapters 26 and 27; R. S. Berry, S. A. Rice, and J. Ross, *Physical Chemistry*, Wiley, New York, 1980, Chapter 30; I. Tinoco Jr., K. Saver, and J. C. Wang, *Physical Chemistry: Principles and Applications in Biological Sciences*, 2nd ed., Prentice-Hall, New Jersey, 1985, Chapter 7.
6. P. Jones, M. L. Haggett, and J. L. Longridge, *J. Chem. Ed.*, **41**, 610 (1964).
7. W. Knoche, in *Biophysics and Physiology of Carbon Dioxide*, C. Bauer, G. Gros, and H. Bartels, Eds., Springer-Verlag, Berlin, 1980, pp. 3–11.
8. J. T. Edsall and J. Wyman, *Biophysical Chemistry*, Vol. 1, Academic, New York, 1958, p. 585.
9. W. Stumm and J. J. Morgan, *Aquatic Chemistry*, 2nd ed., Wiley-Interscience, New York, 1981, pp. 210–212.
10. I. Tinoco, Jr., K. Saver, J. C. Wang, *Physical Chemistry: Principles and Applications in Biological Sciences*, 2nd ed., Prentice-Hall, New Jersey, 1985, pp. 292–294.
11. J. L. Bada, B. P. Luyendyk, and J. B. Maynard, *Science*, **170**, 730 (1970).
12. J. Wehmiller and P. E. Hare, *Science*, **173**, 907 (1971).
13. H. F. Deutsch, *Int. J. Biochem.*, **19**, 101 (1987).
14. R. G. Khalifah and D. N. Silverman, in *The Carbonic Anhydrases*, S. J. Dodgson, R. E. Tashian, G. Gros, and N. D. Carter, Eds., Plenum, New York, 1991, Chapter 4.
15. R. L. Valentine and C. T. Jafvert, *Environ. Sci. Technol.*, **22**, 691 (1988).
16. James M. Montgomery, Consulting Engineers Inc., *Water Treatment Principles and Design*, Wiley-Interscience, New York, 1985, pp. 272–274.
17. J. C. Harris, in *Handbook of Chemical Property Estimation Methods*, W. J. Lyman, W. F. Reehl, and D. H. Rosenblatt, Eds., McGraw-Hill, New York, 1982, Chapter 7.
18. J. Staehelin and J. Hoigné, *Environ. Sci. Technol.*, **19**, 1206 (1985).
19. R. E. Buehler, J. Staehelin, and J. Hoigné, *J. Phys. Chem.*, **88**, 2560 (1984).
20. D. W. Ferguson, J. T. Gramith, and M. J. McGuire, *J. AWWA*, 32, (May 1991).
21. W. Stumm and J. J. Morgan, *Aquatic Chemistry*, 2nd ed., Wiley-Interscience, New York, 1981, pp. 96–98.
22. P. W. Atkins, *Physical Chemistry*, Freeman, San Francisco, CA, 1978, Chapter 27.
23. W. L. Jolley, *Modern Inorganic Chemistry*, McGraw-Hill, New York, 1984, p. 194.
24. A. W. Adamson, *A Textbook of Physical Chemistry*, Academic, New York, 1973, Chapters 12 and 15.
25. Z. Luz and S. Meiboom, *J. Am. Chem. Soc.*, **86**, 4768 (1964).
26. J. D. Rimstidt and H. L. Barnes, *Geochim. Cosmochim. Acta*, **44**, 1683 (1980).

27. J. W. Moore and R. G. Pearson, *Kinetics and Mechanism*, 3rd ed., Wiley-Interscience, New York, 1981, p. 277.

28. J. Burgess, *Ions in Solution*, Ellis Horwood Ltd., Chichester, 1988, p. 108.

29. P. L. Brezonik, in *Aquatic Chemical Kinetics*, W. Stumm, Ed., Wiley-Interscience, New York, 1990, Chapter 4.

30. J. W. Moore and R. G. Pearson, *Kinetics and Mechanism*, 3rd ed., Wiley-Interscience, New York, 1981, pp. 357–359.

31. R. Wollast, in *The Sea*, Vol. 5, E. D. Goldberg, Ed., Wiley-Interscience, New York, 1974, Chapter 11.

32. J. F. Kirsch, in *Advances in Linear Free Energy Relationships*, N. B. Chapman and J. Shorter, Eds., Plenum, London, 1972, Chapter 8.

33. A. C. Lasaga, in *Kinetics of Geochemical Processes, Reviews in Mineralogy*, Vol. 8, A. C. Lasaga and R. J. Kirkpatrick, Eds., Mineral Society of America, Washington, DC, 1981, Chapter 1.

34. R. G. Petasne and R. G. Zika, *Nature*, **325**, 516 (1987).

35. O. C. Zafiriou, *Mar. Chem.*, **30**, 31 (1990).

36. G. Furrer, J. C. Westall, and P. Sollins, *Geochim. Cosmochim. Acta*, **53**, 595 (1989).

37. G. Furrer, P. Sollins, and J. C. Westall, *Geochim. Cosmochim. Acta*, **54**, 2363 (1990).

38. A. A. Jennings, D. J. Kirkner, and T. L. Theis, *Water Res. Res.*, **18**, 1089 (1982).

39. G. A. Cederberg, R. L. Street, and J. O. Leckie, *Water Res. Res.*, **21**, 1095 (1985).

40. For additional discussion the reader is referred to: J. J. Morgan and A. T. Stone, in *Chemical Processes in Lakes*, W. Stumm, Ed., Wiley-Interscience, New York, 1985, Chapter 17; *Aquatic Chemical Kinetics*, W. Stumm, Ed., Wiley-Interscience, New York, 1990.

APPENDIX 3.1   TWO COMPETING FIRST-ORDER REACTIONS IN A CLOSED SYSTEM

A closed system contains a species A that reacts reversibly to form two products, B and C, according to the first-order, or pseudo first-order, reactions

$$A \underset{b_1}{\overset{f_1}{\rightleftharpoons}} B$$

$$A \underset{b_2}{\overset{f_2}{\rightleftharpoons}} C$$

The governing differential equations are written

$$\frac{d[A]}{dt} = -(f_1 + f_2)[A] + b_1[B] + b_2[C] \tag{1}$$

$$\frac{d[B]}{dt} = f_1[A] - b_1[B] \tag{2}$$

$$\frac{d[C]}{dt} = f_2[A] - b_2[C] \tag{3}$$

The initial conditions are $[A] = [A]_0$, $[B] = [B]_0$, $[C] = [C]_0$. Adding the three equations together leads to the mole balance equation

$$[A]_t + [B]_t + [C]_t = [A]_0 + [B]_0 + [C]_0 = T \tag{4}$$

This problem may be treated in a general way. Focusing first on the concentration of A, we obtain a second-order differential equation by substituting $[C]$ from Equation 4 into Equation 1, solving for $[B]$, then substituting $[B]$ and its differential into Equation 2:

$$\frac{d^2[A]}{dt^2} + \frac{\alpha d[A]}{dt} + \beta[A] = \beta[A]_\infty \tag{5}$$

where

$$\alpha = f_1 + f_2 + b_1 + b_2 \tag{6}$$

$$\beta = b_1 b_2 + b_1 f_2 + b_2 f_1 \tag{7}$$

$$[A]_\infty = \frac{b_1 b_2 T}{\beta} \tag{8}$$

The characteristics rates λ_1 and λ_2 corresponding to the solution of this differential equation are the solutions of the quadratic

$$\lambda^2 - \alpha\lambda + \beta = 0 \tag{9}$$

Since f_1, f_2, b_1, and b_2 are reaction rate constants and must be greater than zero, λ_1 and λ_2 can be shown to be both real and positive, guaranteeing a stable and nonoscillating solution of the form

$$[A]_t = K_1 \exp(-\lambda_1 t) + K_2 \exp(-\lambda_2 t) + [A]_\infty \tag{10}$$

The constants K_1 and K_2 are determined by the initial conditions $[A]_0, [B]_0$, and $[C]_0$. The corresponding solutions for $[B]_t$ and $[C]_t$ are found to be

$$[B]_t = \frac{f_1 K_1}{b_1 - \lambda_1} \exp(-\lambda_1 t) + \frac{f_1 K_2}{b_1 - \lambda_2} \exp(-\lambda_2 t) + [B]_\infty \tag{11}$$

$$[C]_t = \frac{f_2 K_1}{b_2 - \lambda_1} \exp(-\lambda_1 t) + \frac{f_2 K_2}{b_2 - \lambda_2} \exp(-\lambda_2 t) + [C]_\infty \tag{12}$$

where

$$[B]_x = \frac{f_1}{b_1}[A]_x = \frac{f_1 b_2 T}{\beta} \tag{13}$$

$$[C]_x = \frac{f_2}{b_2}[A]_x = \frac{f_2 b_1 T}{\beta} \tag{14}$$

This system is thus characterized by two reaction times: $1/\lambda_1$ and $1/\lambda_2$. For times much shorter than these reaction times, initial conditions prevail in the system; for much longer times, equilibrium is achieved. At intermediate times, it is possible that one reaction may reach pseudoequilibrium while the other reaction is still progressing. If $1/\lambda_2$ is the longer of the two characteristic times, a necessary and sufficient condition is given by

$$\lambda_2 \ll b_1 \tag{15}$$

Since $t \gg 1/\lambda_1$, the equations for A and B become, in this case,

$$[A]_t = K_2 \exp(-\lambda_2 t) + [A]_x \tag{16}$$

$$[B]_t = \frac{f_1 K_2}{b_1} \exp(-\lambda_2 t) + [B]_x = \frac{f_1}{b_1}[A]_t \tag{17}$$

(Symmetrically, C will be at pseudoequilibrium with A if $\lambda_2 \gg b_2$ and $t \gg 1/\lambda_2$.) Calculating λ_2 from Equation 9 then gives the condition for pseudoequilibrium:

$$b_1 \gg \tfrac{1}{2}[\alpha - (\alpha^2 - 4\beta)^{1/2}] \tag{18}$$

Since $b_1 < \alpha$, the two terms on the right must almost cancel each other for the inequality to be verified, and thus

$$\beta \ll \alpha \tag{19}$$

The square root can then be approximated:

$$b_1 \gg \frac{1}{2}\left[\alpha - \left(\alpha - \frac{2\beta}{\alpha}\right)\right] \tag{20}$$

therefore

$$b_1 \gg \frac{\beta}{\alpha} = \frac{b_1 b_2 + b_1 f_2 + b_2 f_1}{b_1 + b_2 + f_1 + f_2} \tag{21}$$

Necessary and sufficient conditions for this inequality to be verified (and thus for pseudoequilibrium of the first reaction to be obtainable) are

$$b_1 \gg b_2 \quad \text{and} \quad b_1 \gg f_2 \tag{22}$$

The characteristic reaction times are then given by

$$\lambda_1 = b_1 + f_1 \tag{23}$$

$$\lambda_2 = \frac{b_1 b_2 + b_1 f_2 + b_2 f_1}{b_1 + f_1} \tag{24}$$

For any time $t \gg 1/\lambda_1$, the reaction between A and B is then at pseudoequilibrium. Since the characteristic reaction times are widely different ($1/\lambda_1 \ll 1/\lambda_2$), this pseudoequilibrium can be reached much ahead of the final equilibrium.

APPENDIX 3.2   ONE REVERSIBLE FIRST-ORDER REACTION IN AN OPEN SYSTEM

Two species, A and B, react according to the reversible first-order, or pseudo first-order, reaction

$$A \underset{k_b}{\overset{k_f}{\rightleftharpoons}} B$$

in a well-mixed system with a flow rate Q through the system (volume $= V$; hydraulic rate $r = Q/V$). The influent concentrations are $[A]_i$ and $[B]_i$ and the effluent concentrations are $[A]$ and $[B]$. The governing differential equations are

$$\frac{d[A]}{dt} = r[A]_i - (k_f + r)[A] + k_b[B] \tag{1}$$

$$\frac{d[B]}{dt} = r[B]_i - (k_b + r)[B] + k_f[A] \tag{2}$$

For simplicity, consider the initial conditions: $[A]_0 = [B]_0 = 0$. Note that this choice is not indifferent; as noted in the text, initial conditions have a bearing on how fast equilibrium is reached.) Solving first for the total concentration $T = [A] + [B]$, we add Equations 1 and 2 and define $T_i = [A]_i + [B]_i$:

$$\frac{dT}{dt} = rT_i - rT \tag{3}$$

therefore

$$T = T_i[1 - \exp(-rt)] \tag{4}$$

The total concentration is unaffected by the chemical reaction, and its progress toward the stable value T_i (steady state) depends only on the hydraulic rate r.

Consider now Equation 1. Obtaining from it an expression for $[B]$ and, by differentiation, an expression for $d[B]/dt$, and introducing these into Equation 2, we obtain a second-order differential equation for $[A]$:

$$\frac{d^2[A]}{dt^2} + (\lambda + r)\frac{d[A]}{dt} + r\lambda[A] = r\lambda[A]_\infty \tag{5}$$

where

$$\lambda = k_f + k_b + r \tag{6}$$

and

$$[A]_\infty = \frac{k_b([A]_i + [B]_i) + r[A]_i}{\lambda} \tag{7}$$

The solution to this equation is

$$[A] = [A]_\infty + \frac{r[A]_i - \lambda[A]_\infty}{\lambda - r}\exp(-rt) - \frac{r([A]_i - [A]_\infty)}{\lambda - r}\exp(-\lambda t) \tag{8}$$

Symmetrically for $[B]$,

$$[B] = [B]_\infty + \frac{r[B]_i - \lambda[B]_\infty}{\lambda - r}\exp(-rt) - \frac{r([B]_i - [B]_\infty)}{\lambda - r}\exp(-\lambda t) \tag{9}$$

where

$$[B]_\infty = \frac{k_f([A]_i + [B]_i) + r[B]_i}{\lambda} \tag{10}$$

To analyze the conditions under which $[A]$ and $[B]$ are close to being at equilibrium with each other, let us first consider the final steady-state concentrations $[A]_\infty$ and $[B]_\infty$. The ratio

$$\frac{[A]_\infty}{[B]_\infty} = \frac{k_b([A]_i + [B]_i) + r[A]_i}{k_f([A]_i + [B]_i) + r[B]_i} \tag{11}$$

is close to the equilibrium ratio k_b/k_f in either of two cases:

Condition 1

$$\frac{[A]_i}{[B]_i} = \frac{k_b}{k_f} \tag{12}$$

In this case, $[A]_\infty = [A]_i$ and $[B]_\infty = [B]_i$.

Condition 2

$$r \ll \left[\frac{[A]_i + [B]_i}{[A]_i} \right] k_b \tag{13}$$

$$r \ll \left[\frac{[A]_i + [B]_i}{[B]_i} \right] k_f \tag{14}$$

In this case at least one of the reaction rate constants, k_f or k_b, is much greater than the hydraulic rate r:

$$r \ll k_b \quad \text{or} \quad r \ll k_f \tag{15}$$

and thus

$$r \ll k_f + k_b + r = \lambda \tag{16}$$

Note that the conditions in the second case are always met if both reaction rate constants are much greater than r, that is, $r \ll k_b$ and $r \ll k_f$.

We may show that in both cases the conditions which make the steady-state composition close to chemical equilibrium are also those that pertain to pseudoequilibrium much before the steady state is achieved.

Consider Equations 8 and 9. If Condition 1 is verified, the last terms are small, and the equations simplify to

$$[A] = [A]_i[1 - \exp(-rt)] \tag{17}$$

$$[B] = [B]_i[1 - \exp(-rt)] \tag{18}$$

so that

$$\frac{[A]}{[B]} = \frac{[A]_i}{[B]_i} = \frac{[A]_\infty}{[B]_\infty} \tag{19}$$

This is the trivial case where the reactants are effectively preequilibrated in the inflow and no (net) reaction occurs as the concentrations build up in the system.

To examine Condition 2, let us express $[A]_\infty$ and λ in the first two terms of Equation 8:

$$[A] = \frac{k_b([A]_i + [B]_i) + r[A]_i}{k_f + k_b + r} - \left[\frac{k_b([A]_i + [B]_i)}{k_f + k_b}\right]\exp(-rt) - [\cdots]\exp(-\lambda t)$$
(20)

and similarly for B,

$$[B] = \frac{k_f([A]_i + [B]_i) + r[A]_i}{k_f + k_b + r} - \left[\frac{k_f([A]_i + [B]_i)}{k_f + k_b}\right]\exp(-rt) - [\cdots]\exp(-\lambda t)$$
(21)

For any time much in excess of the characteristic reaction time $1/\lambda$, the last terms in $\exp(-\lambda t)$ can be neglected. Conditions 2 (including Equation 16) implicitly) are then necessary and sufficient conditions for $[A]$ and $[B]$ to take the form

$$[A] = [A]_\infty[1 - \exp(-rt)] \tag{22}$$

$$[B] = [B]_\infty[1 - \exp(-rt)] \tag{23}$$

where

$$[A]_\infty = \frac{k_b([A]_i + [B]_i)}{k_f + k_b} \tag{24}$$

$$[B]_\infty = \frac{k_f([A]_i + [B]_i)}{k_f + k_b} \tag{25}$$

The ratio $[A]/[B]$ then remains at the equilibrium value k_b/k_f for all times in excess of the characteristic reaction time $(t \gg 1/\lambda)$. Times that are sufficient for chemical equilibrium to be achieved may be reached well before the steady state:

$$\frac{1}{\lambda} \ll t \ll \frac{1}{r} \tag{26}$$

If Conditions 2 are met, the two very different characteristic times $1/r$ (hydraulic residence time) and $1/\lambda$ (reaction time) define three time scales:

$t \ll \dfrac{1}{\lambda}$ The concentrations of A and B slowly build up in the system because of the inflow. The chemical reaction does not proceed.

$\dfrac{1}{\lambda} \ll t \ll \dfrac{1}{r}$ Pseudoequilibrium is reached: $[A]/[B] = k_b/k_f$, but $[A]$ and $[B]$ change with time.

$t \gg \dfrac{1}{r}$ The final steady state is obtained:

$$[A] = [A]_\infty; \qquad [B] = [B]_\infty; \qquad \frac{[A]_\infty}{[B]_\infty} = \frac{k_b}{k_f}$$

PROBLEMS

3.1 Radioactive decay follows first-order kinetics. The half-life for radioactive decay of carbon-14 is 5730 y. (a) What is the rate constant for carbon-14 decay? (b) If an archaeological sample contains wood having only 73% of the carbon-14 found in living trees, what is the age of the sample?

3.2 Enzymatic reactions obeying Michaelis–Menten enzyme kinetics where

$$V = \frac{V_{max}[S]}{K_M + [S]}$$

are often analyzed using a linearized equation relating $1/V$ to $1/[S]$. Derive the linearized relationship. Use this relationship to calculate V_{max} and K_M for the following data.

$[S]\,(M)$	$V\,(Ms^{-1})$
1×10^{-2}	1.17×10^{-6}
2×10^{-3}	9.9×10^{-7}
1×10^{-3}	7.9×10^{-7}
5×10^{-4}	6.2×10^{-7}
3.3×10^{-4}	5.0×10^{-7}

3.3 Chlorination of organic compounds can occur during disinfection of drinking water. The chlorination of phenol (C_6H_5OH) occurs by direct reaction of hypochlorous acid (HOCl) on the phenolate anion ($C_6H_5O^-$)

$$C_6H_5O^- + HOCl \xrightarrow{k} ClC_6H_4O^- + H_2O$$

with the second-order rate constant, k, and follows the second-order rate law

$$\frac{-d[C_6H_5O^-]}{dt} = k[C_6H_5O^-][HOCl]$$

a. Write an expression for the rate of disappearance of phenol in terms of total phenol and total HOCl, that is,

$$\frac{-d[C_6H_5OH]_T}{dt} = k^*[C_6H_5OH]_T[HOCl]_T$$

where

$$[C_6H_5OH]_T = [C_6H_5OH] + [C_6H_5O^-]$$
$$[HOCl]_T = [HOCl] + [OCl^-]$$

b. How does k^* depend on $[H^+]$?

c. The pK_a of HOCl is 7.5 and the pK_a of phenol is 10. In what pH range would the fastest rate of chlorination be expected: (1) $pH < 7.5$, (2) $7.5 < pH < 10$, (3) $pH > 10$? Why?

d. Solve for the pH value at which the maximum k^* is expected.

3.4 The reaction $A + B \rightarrow P$ is studied at 100°C. At this temperature the measured rate constant is $100 \, M^{-1}s^{-1}$. If the activation energy for the reaction (E_a) is 56.7 kJ/mol, what will the rate constant for the reaction be at room temperature (25°C)?

3.5 A reaction can proceed by the mechanisms

$$A + B \underset{k_{-1}}{\overset{k_1}{\rightleftharpoons}} P \quad \text{or} \quad A + X \underset{k_{-2}}{\overset{k_2}{\rightleftharpoons}} Y \quad Y + B \underset{k_{-3}}{\overset{k_3}{\rightleftharpoons}} P + X$$

a. Write an expression relating the rate constants (for both mechanisms) to the equilibrium constant for the reaction.

b. What function does the species X serve?

c. Write rate expressions for P for each mechanism (neglect the back reaction of P) assuming steady-state concentrations of intermediates.

3.6 Two species, X and Y, are added to a system at initial concentrations $[X]_0$ and $[Y]_0$. A product Z is formed according to the bimolecular reaction

$$X + Y \underset{k_b}{\overset{k_f}{\rightleftharpoons}} Z$$

Supposing that $[X]_0 \ll [Y]_0$ and $k_b \ll k_f[Y]_0$:

a. Find the (approximate) equilibrium $(t = \infty)$ composition of the system.

b. Find the (approximate) composition of the system at all times.

CHAPTER 4

ACIDS AND BASES:
ALKALINITY AND pH IN NATURAL WATERS

The composition of natural waters is controlled by both geochemical and geobiological processes. From a chemical standpoint we can regard the exogenic cycle—the process by which mountains are slowly dissolved and transported to the bottom of the oceans—as consisting of a gigantic acid–base reaction in the flowing water, continuously retransported atop the hills by a sisyphean sun. The water, made corrosive by its acid content, mostly CO_2, dissolves the basic rock minerals it encounters, thus acquiring most of its solutes which are ultimately precipitated in the ocean sediments. The pH of the water is primarily determined by a balance between the dissolution of the weakly acidic CO_2 and that of basic rocks, primarily silicates, aluminosilicates, and carbonates. More precisely, the pH is determined by the extent of dissociation of the dissolved carbonic acid, and other weak acids such as water, whose net negative charge (HCO_3^-, CO_3^{2-}, OH^-) has to balance exactly the net positive charge from the strong mineral bases (Na^+, K^+, Ca^{2+}). Each of these two equal quantities provides a definition of *alkalinity*, one of the most central but perhaps not the best understood concept in aquatic chemistry.

By making a formal distinction between the weak and the strong acids and bases, the concept of alkalinity allows us to study in two separate steps the mechanisms that control pH in natural waters. In this chapter we examine only the weak acid–base side of the alkalinity equation, considering as given the total concentration of excess strong base—or excess strong acid in a few instances, such as acid mine drainage or acid rain. The dissolution–precipitation mechanisms controlling that concentration of excess strong base—the alkalinity—are examined in the next chapter.

157

Because in most natural waters carbon dioxide far exceeds the other weak acids, this chapter deals primarily with the carbonate system. A complication that cannot be avoided is that of CO_2 exchange between the water and the atmosphere. This we treat in three stages, considering first a closed aqueous system with no gas phase, second, an atmosphere of fixed CO_2 partial pressure at equilibrium with the water, and third, the kinetics of CO_2 exchange at the air–water interface at the end of the chapter.

Among the major aquatic weak acids, other than CO_2, are organic compounds. Because the bulk of these compounds consists of ill-defined humic acids, we use an empirical description of their acid-base properties to quantify their role in controlling the alkalinity and pH of organic-rich streams, lakes, and wetlands.

Many of the other weak acid–base components of natural waters, such as phosphate, silicate, ammonia, sulfide, or borate, contain elements essential to the formation of living matter (C, H, O, P, Si, N, S). The cycles of these elements are partially controlled by organisms, and the corresponding weak acids and bases are thus involved in numerous biological processes. In a few examples, we examine the effects of some of these biological processes on the alkalinity and pH of natural waters.

To obtain some understanding of buffering (homeostatic processes) in aquatic systems we take advantage of the mathematical simplicity of equilibrium models. Having examined the processes controlling the pH in natural waters, we ask the question: How resistant to change is that pH? The answer is merely a generalization of the familiar subject of pH buffers, and we derive an equation applicable to both simple solutions and complex model systems: the "minor species formula."

Finally, we examine the kinetics of gas exchange between the water and atmosphere, both through diffusion and surface exchange and through ebullition (bubble formation).

1. NATURAL WEAK ACIDS AND BASES

Natural waters contain a number of weak acids and bases from a variety of sources. As seen in Table 4.1, the most abundant by far is carbonate, which originates from dissolution of carbonate rock, atmospheric CO_2 transfer, and respiration of aquatic organisms. By comparison to the total carbonate concentration, which averages about $1\,mM$ in freshwater and more than twice that in seawater, all other natural weak acids and bases are relatively unimportant except in a few particular aquatic systems. Dissolution of silicate minerals contributes a total silicate concentration to freshwaters that is 5–10 times lower than the total carbonate. Both concentrations tend to be relatively higher in arid regions. Like other essential plant nutrients (phosphorus and nitrogen), silicate is very depleted in warm surface oceanic waters and increases in concentration from the deep Atlantic to the deep Pacific. This is also true, to a less dramatic degree, for carbonate. In temperate freshwaters, uptake by

TABLE 4.1   Weak Acids and Bases in Natural Waters

	Freshwater[a] Mean	Seawater[b]		
		Warm Surface	Deep Atlantic	Deep Pacific
Carbonate	0.97 mM	2.1 mM	2.3 mM	2.5 mM
Silicate	220 μM	< 3 μM	30 μM	150 μM
Ammonia	0–10 μM	< 0.5 μM	< 0.5 μM	< 0.5 μM
Phosphate	0.7 μM	< 0.2 μM	1.7 μM	2.5 μM
Borate	1 μM	0.4 mM	0.4 mM	0.4 mM
	Typical Anoxic Hypolimnion[c]	Black Sea (Deep Water)[d]		Cariaco Trench[e]
Sulfide	50–150 μM	330 μM		20 μM
Ammonia	10–40 μM	53 μM		10 μM

Source: From Refs. 1–5.
[a] Ref. 1.
[b] Ref. 2.
[c] Ref. 3.
[d] Ref. 4.
[e] Ref. 5.

plants typically keeps both ammonia and phosphate surface concentrations below detectable limits in the summer. Apart from their regeneration by the decomposition of organic matter, these two algal nutrients originate from fixation of atmospheric nitrogen and dissolution of phosphate rocks, respectively.

Hydrogen sulfide, which is virtually absent from oxygenated systems, can be found at concentrations up to 1 mM in anoxic waters such as the hypolimnions (bottoms) of stratified lakes as a result of the activity of sulfate reducing bacteria. In groundwaters contacting sulfide minerals (e.g., iron mines), the concentration of sulfide typically does not exceed 10 μM.

Borate, which is present in trace quantities in continental rock and as a result in freshwaters as well, is the second most abundant weak base in the oceans. Its much higher concentration in seawater compared to rivers is apparently the result of inefficient and poorly understood oceanic removal processes.

The log concentration versus pH diagrams for these various weak acid–weak base systems are shown in Figure 4.1. Bicarbonate (HCO_3^-) is the dominant carbonate species near neutral pH, silicate and borate are essentially undissociated, and the ammonium ion is in great excess of ammonia. Both phosphate and sulfide have acidity constants near 7. The graphs are constructed using mixed acidity constants applicable at an ionic strength of 10^{-3} M ($P = 1$ atm, $T = 25°C$). Along with rounded-off pK_a's of 6.3 and 10.3 for carbonate, these are the constants that we shall use throughout this chapter unless otherwise indicated. Other constants are given in Table 4.2, which lists the various

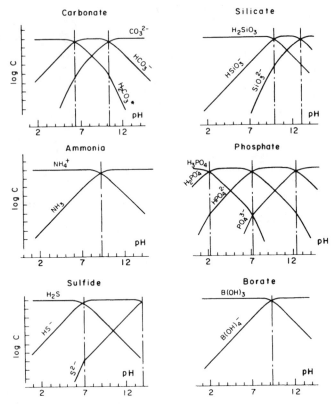

Figure 4.1 Log C–pH diagrams for weak acids–bases that are most common in natural waters.

acid–base reactions for each system. Note that for convenience the aqueous concentration of carbon dioxide and that of carbonic acid are always summed:

$$[H_2CO_3^*] = [CO_2 \cdot aq] + [H_2CO_3] \tag{1}$$

and that equilibrium constants are defined for the inclusive species $H_2CO_3^*$.

Note. The soluble hydrolysis species of some metals, particularly those of iron and aluminum [e.g., $Fe(OH)_2^+$, $Fe(OH)_4^-$, $Al(OH)_2^+$, $Al(OH)_3$] also behave as weak acids and bases and they may be important for the pH control of some aquatic systems. However, because of the predominance of insoluble hydroxide and oxide forms [$Fe(OH)_3(s)$, $Fe_2O_3(s)$, $Al(OH)_3(s)$] these particular weak acids and bases are studied in Chapter 5 which deals with solids.

Weak Acids

TABLE 4.2 Acid–Base Reactions (Mixed Acidity Constants)

	$-\log K$	
Reactions	$I = 0$	$I = 0.5\,M$
$H_2O = H^+ + OH^-$	14.00	13.89
$CO_2(g) + H_2O = H_2CO_3^*$	1.46	1.51
$H_2CO_3^* = HCO_3^- + H^+$	6.35	6.30
$HCO_3^- = CO_3^{2-} + H^+$	10.33	10.15
$H_2SiO_3 = HSiO_3^- + H^+$	9.86	9.61
$HSiO_3^- = SiO_3^{2-} + H^+$	13.1	12.71
$H_3PO_4 = H_2PO_4^- + H^+$	2.15	1.87
$H_2PO_4^- = HPO_4^{2-} + H^+$	7.20	6.72
$HPO_4^{2-} = PO_4^{3-} + H^+$	12.35	11.89
$NH_3(g) = NH_3(aq)$	-1.76	-1.64
$NH_4^+ = NH_3(aq) + H^+$	9.24	9.47
$H_2S(g) = H_2S(aq)$	0.99	0.99
$H_2S(aq) = HS^- + H^+$	7.02	6.98
$HS^- = S^{2-} + H^+$	13.9	13.45
$B(OH)_3 + H_2O = B(OH)_4^- + H^+$	9.24	8.97
$SO_2(g) = SO_2(aq)$	-0.1	—
$SO_2(aq) + H_2O = HSO_3^- + H^+$	1.9	—
$HSO_3^- = SO_3^{2-} + H^+$	7.2	—

Source: From Ref. 6.

2. ALKALINITY AND RELATED CONCEPTS

2.1 Pure Solutions of CO_2: The Equivalence Point

Before considering natural waters in all their complexity, let us examine the simplest possible case of a carbonate bearing water, that of a pure CO_2 solution with a background electrolyte.

Example 1 Consider a $10^{-3}\,M$ NaCl solution into which we introduce $10^{-4}\,M$ CO_2 by bubbling CO_2 gas through the water. The system is then closed to the atmosphere (no gas exchange) and we are interested in its equilibrium composition.

Recipe $[NaCl]_T = 10^{-3}\,M$

$[CO_2]_T = 10^{-4}\,M$

Species (H_2O), H^+, OH^-, Na^+, Cl^-, $H_2CO_3^*$, HCO_3^-, CO_3^{2-}

Reactions $H_2O = H^+ + OH^-$; $pK_w = 14.0$ $\hspace{2cm}$ (2)

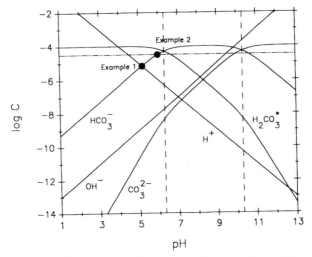

Figure 4.2 Log C–pH diagram for carbonate; graphical solutions of Example 1 (equivalence point) and Example 2.

$$H_2CO_3^* = H^+ + HCO_3^-; \quad pK_1 = 6.3 \tag{3}$$

$$HCO_3^- = H^+ + CO_3^{2-}; \quad pK_2 = 10.3 \tag{4}$$

The corresponding log C–pH diagram is given in Figure 4.2. Anticipating that $H_2CO_3^*$ is the dominant carbonate species, we choose it as one of the components. The rest of the principal components are obvious: H^+, Na^+, Cl^-. The pH of the system is obtained by solving the mole balance equation for H^+ (see Tableau 4.1):

$$TOTH = [H^+] - [OH^-] - [HCO_3^-] - 2[CO_3^{2-}] = 0 \tag{5}$$

The solution to this equation is given by

$$[H^+] = [HCO_3^-] \gg [OH^-], [CO_3^{2-}] \tag{6}$$

The pH can be read on the graph (Figure 4.2); it can also be obtained algebraically: in this pH region, the mole balance for carbonate simplifies to

$$TOTH_2CO_3 = [H_2CO_3^*] + [HCO_3^-] = [CO_2]_T = 10^{-4} \tag{7}$$

Introduction of Equation 6 and the mass law for reaction 3 into Equation 7 yields

$$10^{6.3}[H^+]^2 + [H^+] = [CO_2]_T = 10^{-4}$$

TABLEAU 4.1

	H^+	Na^+	Cl^-	$H_2CO_3^*$	$\log K$
H^+	1				
OH^-	-1				-14.0
Na^+		1			
Cl^-			1		
$H_2CO_3^*$				1	
HCO_3^-	-1			1	-6.3
CO_3^{2-}	-2			1	-16.6
Example 1					
NaCl		1	1		$10^{-3} M$
CO_2				1	$10^{-4} M$
Example 2					
NaOH	-1	1			$10^{-4.5} M$

This quadratic can now be solved for $[H^+]$ and by substitution into the mass laws all the concentrations can be calculated.

$$[H^+] = 10^{-5.17} = 6.8 \times 10^{-6}$$

$$[HCO_3^-] = 10^{-5.17} = 6.8 \times 10^{-6}$$

$$[CO_3^{2-}] = 10^{-10.3} = 5.0 \times 10^{-11}$$

$$[OH^-] = 10^{-8.84} = 1.5 \times 10^{-9}$$

$$[H_2CO_3^*] = 10^{-4.04} = 9.12 \times 10^{-5}$$

In this case, because of the round-off errors in the exponents, a more precise value of $[H_2CO_3^*]$ is obtained by difference from the mole balance equation 7:

$$[H_2CO_3^*] = 9.32 \times 10^{-5}$$

Also

$$[Na^+] = [Cl^-] = 10^{-3}$$

The pH of such a solution containing no strong acid or strong base is called the CO_2 equivalence point. It is defined as the pH of a pure solution of CO_2 in water and is given mathematically by the solution of Equation 5, which usually simplifies to $[H^+] = [HCO_3^-]$. The equivalence point depends on the total carbonate concentration and, to a lesser degree, on the ionic strength of the solution. Figure 4.3 gives the CO_2 equivalence point for two different ionic strengths, as a function of the total concentration of carbonate in solution. The symbol C_T is used to designate this concentration. It differs from $TOTH_2CO_3$

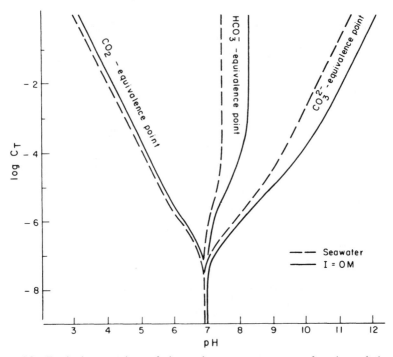

Figure 4.3 Equivalence points of the carbonate system as a function of the total carbonate in solution. The seawater lines are calculated using apparent acidity constants that include the effects of ion pairing; see Chapter 6.

by including neither the concentration of the gas phase $[CO_2(g)]$, nor the concentrations of the various carbonate solids when those are considered.

2.2 Alkalinity: Preliminary Notion

As mentioned earlier, natural waters gain strong base from the dissolution of rocks. *The resulting net concentration of strong base in excess of strong acid is, by definition, the alkalinity of the water.* For didactic purposes, let us simply add NaOH to our closed carbonate system of Example 1.

Example 2

$$\text{Recipe} \quad [NaCl]_T = 10^{-3} M$$

$$[CO_2]_T = 10^{-4} M$$

$$[NaOH]_T = 10^{-4.5} M$$

The reactions and species are identical to those in the previous system. The addition of NaOH at the bottom of Tableau 4.1 yields only two changes in the mole balances:

$$TOTH = [H^+] - [OH^-] - [HCO_3^-] - 2[CO_3^{2-}]$$

$$= -[NaOH]_T = -10^{-4.5} \tag{8}$$

$$TOTNa = [Na^+] = [NaCl]_T + [NaOH]_T = 10^{-2.99} \tag{9}$$

According to the definition, the alkalinity of this system, $[NaOH]_T$, is then equal to the negative value of $TOTH$ as expressed in Equation 8:

$$Alk = -[H^+] + [OH^-] + [HCO_3^-] + 2[CO_3^{2-}] = [NaOH]_T \tag{10}$$

It is also equal to the difference between the sodium and the chloride mole balances:

$$TOTNa - TOTCl = Alk = [Na^+] - [Cl^-] = [NaOH]_T \tag{11}$$

The presence of the strong base NaOH provides an excess of positive charge from the strong electrolytes ($[Na^+] > [Cl^-]$), which is balanced by the net negative charge from the dissociation of carbon dioxide and water. As discussed in the next section, this interpretation of alkalinity as a balance of charges is not strictly correct; yet it provides a convenient preliminary notion:

$$Alk = [\text{excess negative charge from weak acids}]$$

$$= [\text{excess positive charge from strong bases}] \tag{12}$$

Both the left- and the right-hand sides of this formula (top and bottom) correspond to useful conceptual and mathematical definitions of alkalinity. Because in most natural waters carbonate is much more abundant than other weak acids and bases, and Na^+, K^+, Ca^{2+}, Mg^{2+}, Cl^-, and SO_4^{2-} are the other major ions, the alkalinity formula can usually be expressed explicitly as

$$Alk = [OH^-] - [H^+] + [HCO_3^-] + 2[CO_3^{2-}]$$

$$= [Na^+] + [K^+] + 2[Ca^{2+}] + 2[Mg^{2+}] - [Cl^-] - 2[SO_4^{2-}] \tag{13}$$

The separation of the weak and strong acids and bases on each side of the equal sign permits a description of the pH buffering of the aqueous phase in natural waters: as more and more strong base is added to the system, the weak acids dissociate increasingly ($H_2CO_3^* \rightarrow HCO_3^- \rightarrow CO_3^{2-}$) to balance the excess positive charge. The degree of dissociation of these weak acids is what determines the pH of the water (as given by the mass laws for HCO_3^- and CO_3^{2-}).

In our example, the addition of NaOH raises the pH and $[HCO_3^-]$ becomes larger than $[H^+]$. The approximate solution to Equation 10 is given by

$$[HCO_3^-] = [NaOH]_T = 10^{-4.5} \qquad \text{(see Figure 4.2)} \qquad (14)$$

Introducing Equation 14 and the mass law for HCO_3^- into the carbonate mole balance yields

$$TOTH_2CO_3 = 10^{6.3}[H^+]10^{-4.5} + 10^{-4.5} + \frac{10^{-10.3} \times 10^{-4.5}}{[H^+]} = 10^{-4.0}$$

therefore pH = 5.97 and

$$[H_2CO_3^*] = 6.85 \times 10^{-5}; \qquad [HCO_3^-] = 3.16 \times 10^{-5}$$

2.3 Alkalinity: Mathematical Definition

Although convenient, the conceptualization of alkalinity as a charge balance is not strictly correct; the proper concept is that of an acid–base balance. The actual definition of alkalinity is thus founded on a conservation equation for hydrogen ions ($TOTH$) that may differ from the electroneutrality equation, as seen in the following mathematical definition:

The alkalinity of a solution is the negative of the TOTH expression when the components are the principal components of the solution at the CO_2 equivalence point (H^+, $H_2CO_3^*$ (or, equivalently, CO_2), Na^+, Cl^-, etc. since the equivalence point is typically in the range of pH 4 to 5).

Weak acids and bases whose principal components are charged at the CO_2 equivalence point (e.g., NH_4^+ or $H_2PO_4^-$) make the difference between this precise definition of alkalinity and the previous notion of balance of charges. The more exact concept is that of a proton deficiency (with respect to a zero level defined by the principal components at the CO_2 equivalence point) for weak acids and bases which is balanced (or caused) by an excess of strong base over strong acid. In fact, from the point of view of defining alkalinity, strong acids and strong bases can now be characterized precisely as H^+ and OH^- containing compounds that are totally dissociated at the CO_2 equivalence point. For example, as can be seen in Figure 4.1, H_3PO_4 ($pK_{a1} = 2.2$, $pK_{a2} = 7.2$, $pK_{a3} = 12.4$) is a strong monoprotic acid and a weak diprotic acid; NH_3 ($= NH_4OH$; $pK_a = 9.2$) is a strong base.

Note: The definition of alkalinity is normally restricted to the aqueous phase. However, it can be easily extended to heterogeneous systems including gases or solids by considering the proper principal components (in any of the phases of the system) at the CO_2 equivalence point.

Let us now write explicit expressions of alkalinity for aquatic systems containing other weak acids and bases than CO_2 and water. Focusing on the left-hand side of the $TOTH$ equation, we find for a simple carbonate system:

$$Alk = -[H^+] + [OH^-] + [HCO_3^-] + 2[CO_3^{2-}] \qquad (15)$$

and for a system that contains other natural weak acids and bases:

$$Alk = \underbrace{-[H^+] + [OH^-] + [HCO_3^-] + 2[CO_3^{2-}]}_{\text{C-Alk}} + \underbrace{[NH_3]}_{\text{N-Alk}} + \underbrace{[HS^-] + 2[S^{2-}]}_{\text{S-Alk}}$$

$$\underbrace{+[HSiO_3^-] + 2[SiO_3^{2-}]}_{\text{Si-Alk}} + \underbrace{[B(OH)_4^-]}_{\text{B-Alk}} \underbrace{-[H_3PO_4] + [HPO_4^{2-}] + 2[PO_4^{3-}]}_{\text{P-Alk}} \qquad (16)$$

The species, H^+, $H_2CO_3^*$, NH_4^+, H_2S, H_2SiO_3, $B(OH)_3$, and $H_2PO_4^-$ are the principal components at the CO_2 equivalence point; but for H^+, they are the major species dissolved in the pH range 4 to 6 (see Figure 4.1 and Tableau 4.2). For convenience we use the expressions C-Alk, S-Alk, and P-Alk to designate the contributions of the various weak acids to the overall alkalinity. (The effect of organic acids on the alkalinity and acid–base behavior of natural waters is discussed in Section 6.)

The contribution of complexes such as $NaCO_3^-$ or $CaCO_3$ to the alkalinity expression is easily obtained from the knowledge of the principal components at the CO_2 equivalence point (H^+, $H_2CO_3^*$, Na^+, $H_2PO_4^-$; see Tableau 4.2):

$$Alk = -[H^+] + [OH^-] + [CaOH^+] + [HCO_3^-] + 2[CO_3^{2-}] + 2[NaCO_3^-]$$
$$+ 2[CaCO_3] - [H_3PO_4] + [HPO_4^{2-}] + 2[PO_4^{3-}] + [CaHPO_4]$$
$$+ \text{N-Alk} + \text{S-Alk} + \text{Si-Alk} + \text{B-Alk} \qquad (17)$$

A completely systematic method for writing out the alkalinity expression is thus provided by its mathematical definition, even for very complex chemical systems. If the definition of alkalinity were purely mathematical, however, it would be of limited usefulness; an essential property of alkalinity is that it is a measurable quantity.

2.4 Alkalinity: Experimental Definition

The alkalinity of a solution is its acid neutralizing capacity when the end point of the titration is the CO_2 equivalence point.

This definition focuses on the right-hand side of the $TOTH$ equation and is most easily understood by referring to our previous example: if HCl is added to the system at a concentration equal to that of NaOH, the alkalinity goes to zero, which is the condition for the CO_2 equivalence point. Vice versa, if HCl

TABLEAU 4.2

	H^+	$H_2CO_3^*$	NH_4^+	H_2S	H_2SiO_3	$B(OH)_3$	$H_2PO_4^-$	Na^+	Ca^{2+}
H^+	1								
OH^-	−1								
$CaOH^+$	−1								1
$H_2CO_3^*$		1							
HCO_3^-	−1	1							
CO_3^{2-}	−2	1							
$NaCO_3^-$	−2	1						1	
$CaCO_3$	−2	1							1
NH_4^+			1						
NH_3	−1		1						
H_2S				1					
HS^-	−1			1					
S^{2-}	−2			1					
H_2SiO_3					1				
$HSiO_3^-$	−1				1				
SiO_3^{2-}	−2				1				
$B(OH)_3$						1			
$B(OH)_4^-$	−1					1			
H_3PO_4	1						1		
$H_2PO_4^-$							1		
HPO_4^{2-}	−1						1		
PO_4^{3-}	−2						1		
$CaHPO_4$	−1						1		1

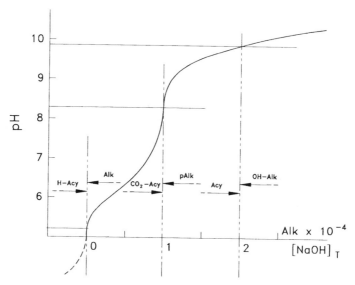

Figure 4.4 Variation of pH with alkalinity (alkalinity titration) for a $10^{-4}\,M$ carbonate solution. The arrows indicate the endpoints of the alkalimetric and acidimetric titrations corresponding to the various base- and acid-neutralizing capacities of the solution: H-Acy = mineral acidity; CO_2-Acy = CO_2 acidity; Acy = acidity; OH-Alk = caustic alkalinity; p-Alk = p-alkalinity; Alk = alkalinity.

is added until the pH reaches the CO_2 equivalence point, the original alkalinity of the system is equal to the concentration of HCl added. This provides an experimental method for determining the alkalinity of a solution by acidimetric titration. The end point is usually recognizable as a sharp inflection point in the titration curve (see Figure 4.4) and small errors in the end point affect the alkalinity determination negligibly. The major source of inaccuracy is commonly due to the loss of CO_2 to the atmosphere when the titration is carried to a fixed pH as the end point. (Remember that the CO_2 equivalence point is a function of C_T and that alkalinity is defined for the aqueous phase exclusively!). This can be minimized in various ways and several techniques have been developed to measure alkalinity with great accuracy (see Sidebar 4.1 on Gran Titrations). Note that the presence of a weak acid with a pK_a near the CO_2 equivalence point will create discrepancies between the mathematical and experimental definitions of alkalinity since the acid will be incompletely titrated. It will also complicate the experimental determination of alkalinity by buffering the pH near the end point.

2.5 Other Related Definitions and Quantities

In the same manner that we have defined the CO_2 equivalence point as the pH of a pure CO_2 solution, we can define the bicarbonate and carbonate equivalence

SIDEBAR 4.1

Gran Titrations

In low alkalinity, relatively low pH waters, the alkalinity cannot be obtained accurately by acidimetric titration to a predetermined end-point. The error introduced in such an approach can easily be as large as the intended measurement. Detection of the appropriate inflection points in the titration curve is obviously also very inaccurate. The method of choice to measure alkalinity in such waters is the Gran titration,[7] which consists of incremental acid additions past the CO_2 equivalence point where $[H^+]$ becomes the dominant term in the $TOTH$ equation.

$$TOTH = [\text{acid added}] - \text{Alk} = [H^+] - [OH^-] - [HCO_3^-] + \cdots$$

$$[\text{Acid added}] \simeq [H^+] + \text{Alk}$$

Thus at low pH (usually in the range 4–3) the plot of $[H^+]$ ($= 10^{-pH}$) vs. concentration of acid added is linear with a slope of unity and an intercept equal to the alkalinity of the sample. A calculated example of such a titration for a sample with $\text{Alk} = 0.3 \, \text{meq L}^{-1}$ is shown in Figure 4.5. Above a hydrogen ion concentration of $2 \times 10^{-4} \, M$, the maximum deviation from the unit slope asymptote due to HCO_3^- is less than $2 \, \mu M$ (for $C_T \leqslant 10^{-2} \, M$). Larger errors are introduced by the presence of other weak acids (see Section 6). The Gran titration method is in no way restricted to low alkalinity, low pH water and is in fact standard for most water samples.

points as the pH's of pure bicarbonate (e.g., $NaHCO_3$) and carbonate (e.g., Na_2CO_3) salt solutions, respectively. Mathematically, these equivalence points are most conveniently obtained by equating to zero the proton balance equations corresponding to the respective principal components.

Bicarbonate Equivalence Point (Pure NaHCO₃ Solution) The components H^+, HCO_3^-, and Na^+ yield Tableau 4.3, from which we obtain the $TOTH$ equation:

$$TOTH = [H^+] - [OH^-] + [H_2CO_3^*] - [CO_3^{2-}] = 0 \qquad (18)$$

For example, with $[NaHCO_3]_T = 10^{-3} \, M$, the $TOTH$ equation simplifies approximately to

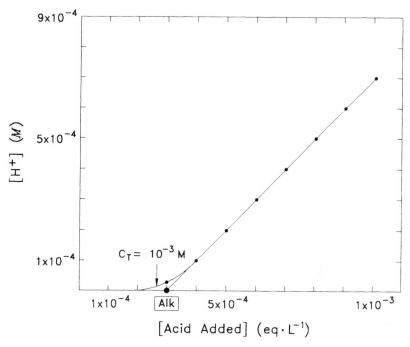

Figure 4.5 Calculated Gran titrations of water samples with an alkalinity of $3 \times 10^{-4} \, \text{eq L}^{-1}$. The total carbonate concentration was varied from 0 to 10 mM.

TABLEAU 4.3

	H^+	HCO_3^-	Na^+
H^+	1		
OH^-	-1		
$H_2CO_3^*$	1	1	
HCO_3^-		1	
CO_3^{2-}	-1	1	
Na^+			1
$NaHCO_3$		1	1

$$[H_2CO_3^*] = [CO_3^{2-}] \tag{19}$$

therefore pH = 8.3 (see Figures 4.3 and 4.6).

Carbonate Equivalence Point (Pure Na₂CO₃ Solution) The components H^+, CO_3^{2-}, and Na^+ yield Tableau 4.4.

$$TOTH = [H^+] - [OH^-] + 2[H_2CO_3^*] + [HCO_3^-] = 0 \tag{20}$$

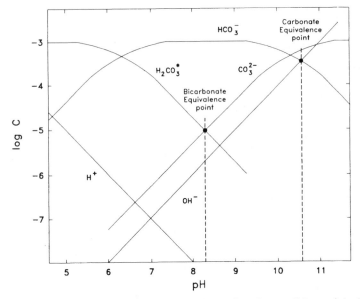

Figure 4.6 Log C–pH diagram for carbonate illustrating the conditions of the bicarbonate and carbonate equivalence points.

TABLEAU 4.4

	H^+	CO_3^{2-}	Na^+
H^+	1		
OH^-	-1		
$H_2CO_3^*$	2	1	
HCO_3^-	1	1	
CO_3^{2-}		1	
Na^+			1
Na_2CO_3		1	2

For example, with $[Na_2CO_3]_T = 10^{-3} M$, the $TOTH$ equation simplifies to

$$[OH^-] = [HCO_3^-] \tag{21}$$

Therefore pH = 10.55 (see Figures 4.3 and 4.6).

Acid and Base Neutralizing Capacity We have defined the alkalinity of a solution as its acid neutralizing capacity when the end point of the titration is the CO_2 equivalence point. Similar quantities can be defined as the acid or base neutralizing capacity of a solution for HCO_3^- and CO_3^{2-} equivalence points as titration end points. The mathematical expressions of these quantities are the $TOTH$ equations corresponding to the respective principal components.

For CO_2 equivalence point as end point,

$$TOTH_1 = [H^+] - [OH^-] - [HCO_3^-] - 2[CO_3^{2-}] \qquad (22)$$

For HCO_3^- equivalence point as end point,

$$TOTH_2 = [H^+] - [OH^-] + [H_2CO_3^*] - [CO_3^{2-}] \qquad (23)$$

For CO_3^{2-} equivalence point as end point,

$$TOTH_3 = [H^+] - [OH^-] + 2[H_2CO_3^*] + [HCO_3^-] \qquad (24)$$

Each of these quantities is given a different name depending on whether it is positive or negative; that is, whether the corresponding end point is obtained by acid or base addition.

For acid neutralizing quantities $(TOTH < 0)$:

$$\text{Alkalinity} = -TOTH_1$$

$$p\text{-alkalinity*} = -TOTH_2$$

$$\text{Caustic alkalinity} = -TOTH_3$$

For base neutralizing quantities $(TOTH > 0)$:

$$\text{Mineral acidity} = TOTH_1$$

$$CO_2 \text{ acidity} = TOTH_2$$

$$\text{Acidity} = TOTH_3$$

Each of these quantities is illustrated in the calculated titration curve of Figure 4.4.*

Note 1. In simple carbonate systems the various acid and base neutralizing quantities are related to each other by the equation

$$TOTH_3 = TOTH_2 + C_T = TOTH_1 + 2C_T \qquad (25)$$

where C_T is again the total dissolved carbonate. This equation is useful to obtain values of $TOTH$ when switching among carbonate components.

---

*In the older literature (and in the previous version of this text) the term "carbonate alkalinity" is sometimes used to designate-$TOTH_2$.

Note 2. The definitions of all the acid and base neutralizing quantities can be generalized to complex systems by considering the $TOTH$ equations corresponding to the principal components of all acids and bases at the various equivalence points. For example, in the presence of small concentrations of ammonia, sulfide, silicate, borate, and phosphate, the p-alkalinity is obtained by choosing the principal components H^+, HCO_3^-, NH_4^+, HS^-, H_2SiO_3, $B(OH)_3$, HPO_4^{2-}:

$$p\text{-Alk} = -TOTH_2 = -[H^+] + [OH^-] - [H_2CO_3^*] + [CO_3^{2-}]$$
$$+ [NH_3] - [H_2S] + [S^{2-}] + [HSiO_3^-] + 2[SiO_3^{2-}] + [B(OH)_4^-]$$
$$- 2[H_3PO_4] - [H_2PO_4^-] + [PO_4^{3-}] \tag{26}$$

and the acidity is obtained by choosing the principal components H^+, CO_3^{2-}, NH_3, HS^-, $HSiO_3^-$, $B(OH)_4^-$, HPO_4^{2-}:

$$\text{Acy} = TOTH_3 = [H^+] - [OH^-] + 2[H_2CO_3^*] + [HCO_3^-]$$
$$+ [NH_4^+] + [H_2S] - [S^{2-}] + [H_2SiO_3] - [SiO_3^{2-}] + [B(OH)_3]$$
$$+ 2[H_3PO_4] + [H_2PO_4^-] - [PO_4^{3-}] \tag{27}$$

3. ACID–BASE CALCULATIONS FOR NATURAL WATERS

Consider the usual situation of a natural water whose acid–base chemistry is dominated by the carbonate system (Alk $\simeq$ C-Alk). The canonical equilibrium problem described in Chapter 2 consists of calculating the composition of the system, including its pH, given the recipe of the system. This recipe may often be replaced by the equivalent conservative parameters C_T and Alk. Alternatively, the calculation may be constrained, not by a recipe but by analytical information, most often the pH and Alk of the solution. Any two of the three parameters C_T, Alk, and pH completely define the acid–base chemistry of the system and the third can be calculated from the other two. This is illustrated in Figure 4.7 which shows the dependence of Alk on C_T for various pH's.

3.1   pH of a Carbonate System Given C_T and Alk

A minor complication arises in making such acid–base calculations. This concerns the choice of appropriate $TOTH$ equations for solutions with a pH above 6.3. We use the following simple example as an illustration.

Example 3   Consider a natural water defined by

$$C_T = 1.4 \times 10^{-3} M; \qquad \text{Alk} = 1.6 \times 10^{-3} \, \text{eq L}^{-1}$$

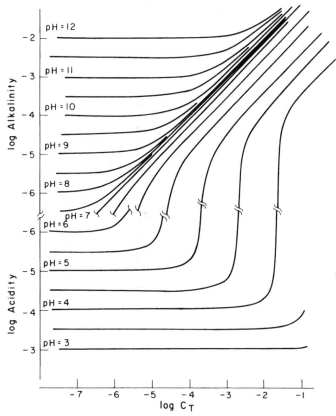

Figure 4.7   General relationship between total carbonate (C_T), alkalinity $(=\text{-acidity})$, and pH in natural waters when the acid–base chemistry is dominated by the carbonate system. (Many graphs similar in concept to this one and others in this chapter and the next were pioneered by various authors including notably Sillén, Butler, Deffeyes, and Stumm and Morgan.)

Calculate the composition of the water assuming carbonate to be the only weak acid.

The governing mole balance equations are:

$$C_T = [H_2CO_3^*] + [HCO_3^-] + [CO_3^{2-}] = 1.4 \times 10^{-3} \tag{28}$$

$$\text{Alk} = -[H^+] + [OH^-] + [HCO_3^-] + 2[CO_3^{2-}] = 1.6 \times 10^{-3} \tag{29}$$

To be systematic, let us recast this problem in its canonical form, defining it by a proper recipe. For this purpose we may consider that the alkalinity has been introduced in the form of a strong base (e.g., NaOH) and the carbon as CO_2 so that the alkalinity is not affected by it.

Recipe $[CO_2]_T = C_T = 1.4 \times 10^{-3} M$

$[NaOH]_T = Alk = 1.6 \times 10^{-3} M$

Species (H_2O), H^+, OH^-, $H_2CO_3^*$, HCO_3^-, CO_3^{2-} (and Na^+)

Reactions $H_2O = H^+ + OH^-$; $pK_w = 14.0$ (30)

$H_2CO_3^* = H^+ + HCO_3^-$; $pK_1 = 6.3$ (31)

$HCO_3^- = H^+ + CO_3^{2-}$; $pK_2 = 10.3$ (32)

Since in this system the acid (CO_2) and the base (NaOH) are roughly in balance, we expect the pH to be circumneutral, and we choose HCO_3^- as the principal carbonate component. As can be seen by inspection of Equations 28 and 29, this is true of any system where C_T and Alk are close to each other.

The corresponding mole balance equations (see Tableau 4.5, where Na^+ is omitted as irrelevant) are then given by the difference between the C_T and Alk expressions and by the C_T expression itself:

$$TOTH = [H^+] - [OH^-] + [H_2CO_3^*] - [CO_3^{2-}] = C_T - Alk = -2 \times 10^{-4}$$
(33)

$$TOTHCO_3^- = [H_2CO_3^*] + [HCO_3^-] + [CO_3^{2-}] = C_T = 1.4 \times 10^{-3}$$
(34)

The solution of the problem in this form is straightforward.

The dominant term in Equation 33, which has a negative right-hand side, must be $-[CO_3^{2-}]$.

$$\therefore [CO_3^{2-}] \simeq 2 \times 10^{-4} = 10^{-3.70}$$

Introducing this result in Equation 34 and neglecting $[H_2CO_3^*]$ gives

$$[HCO_3^-] = 1.4 \times 10^{-3} - 2 \times 10^{-4} = 1.2 \times 10^{-3} = 10^{-2.92}$$

TABLEAU 4.5

	H^+	HCO_3^-
H^+	1	
OH^-	-1	
$H_2CO_3^*$	1	1
HCO_3^-		1
CO_3^{2-}	-1	1
$C_T = [CO_2]_T$	1	1
$Alk = [NaOH]_T$	-1	0

The pH is then calculated from the mass law (Reaction 32):

$$\frac{[H^+][CO_3^{2-}]}{[HCO_3^-]} = 10^{-10.3}$$

$$\therefore [H^+] = \frac{10^{-10.3} \times 10^{-2.92}}{10^{-3.70}} = 10^{-9.52}$$

$$pH = 9.52$$

By substituting into the mass law of Reaction 31, we can verify that $[H_2CO_3^*]$ is indeed neglible in Equations 33 and 34:

$$[H_2CO_3^*] = 10^{+6.3}[H^+][HCO_3^-] = 10^{-6.14}$$

3.2 pH of a Carbonate System Containing Phosphate

A more complete description of natural waters includes other weak acids and bases, whose presence affects both alkalinity and pH. As an illustration, we calculate the composition of the system in Example 3, considering that the measured alkalinity is determined by the concentration of phosphate, as well as carbonate, species.

Example 4 Calculate the composition of the same system $(C_T = 1.4\,mM;$ Alk $= 1.6\,meq\,L^{-1})$ assuming that it now contains $0.2\,mM\,[PO_4]_T$.
The three governing mole balance equations are:

$$C_T = [H_2CO_3^*] + [HCO_3^-] + [CO_3^{2-}] = 1.4 \times 10^{-3} = 10^{-2.85} \tag{35}$$

$$P_T = [H_3PO_4] + [H_2PO_4^-] + [HPO_4^{2-}] + [PO_4^{3-}] = 2 \times 10^{-4} = 10^{-3.70} \tag{36}$$

$$\text{Alk} = -[H^+] + [OH^-] + [HCO_3^-] + 2[CO_3^{2-}] - [H_3PO_4] + [HPO_4^{2-}]$$
$$+ 2[PO_4^{3-}] = 1.6 \times 10^{-3} = 10^{-2.80} \tag{37}$$

If we assume that the pH of this system is not too different from that calculated in Example 3, then the principal components are H^+, HCO_3^-, and HPO_4^{2-}. Thus, rather than the alkalinity equation, the appropriate $TOTH$ equation is again obtained by differences among Equation 35, 36, and 37, as can be seen in Tableau 4.6.

$$TOTH = [H^+] - [OH^-] + [H_2CO_3^*] - [CO_3^{2-}] + 2[H_3PO_4]$$
$$+ [H_2PO_4^-] - [PO_4^{3-}] = C_T + P_T - \text{Alk} = 0 \tag{38}$$

Again this result can be justified by considering that the recipe of the system

TABLEAU 4.6

	H^+	HCO_3^-	HPO_4^{2-}
H^+	1		
OH^-	-1		
$H_2CO_3^*$	1	1	
HCO_3^-		1	
CO_3^{2-}	-1	1	
H_3PO_4	2		1
$H_2PO_4^-$	1		1
HPO_4^{2-}			1
PO_4^{3-}	-1		1
$C_T = [CO_2]_T$	1	1	
$P_T = [NaH_2PO_4]_T$	1		1
$Alk = [NaOH]_T$	-1		

consists of NaOH $(= Alk)$, CO_2 $(= C_T)$ and NaH_2PO_4 $(= P_T)$ since neither CO_2 nor NaH_2PO_4 contribute to the alkalinity.

At a pH around 9, the dominant terms in Equation 38 are clearly $[H_2CO_3^*]$, $[CO_3^{2-}]$, and $[H_2PO_4^-]$. Expressing these concentrations as a function of the principal components by substitution of the appropriate mass law equations, we obtain

$$10^{6.3}[H^+][HCO_3^-] - 10^{-10.3}\frac{[HCO_3^-]}{[H^+]} + 10^{7.16}[H^+][HPO_4^{2-}] = 0$$

The problem is solved by introducing the approximations $[HCO_3^-] = C_T$ and $[HPO_4^{2-}] = P_T$, from the mole balances (Equations 35 and 36):

$$\therefore 10^{6.3}10^{-2.85}[H^+] - \frac{10^{-10.3}\,10^{-2.85}}{[H^+]} + 10^{7.16}10^{-3.70}[H^+] = 0$$

$$\therefore [H^+]^2 = \frac{10^{-13.15}}{10^{3.45} + 10^{3.46}} = 10^{-16.91}$$

$$\therefore pH = 8.45$$

The minor species can now be calculated from the appropriate mass laws and the approximation, verified.

3.3 Use of Ionization Fraction Parameters

Considering again a pure carbonate system, let us now generalize the problem and see how we can calculate the pH of a solution for any alkalinity. That is, given a total inorganic carbon concentration $[CO_2]_T = C_T$, we wish to calculate the pH of the solution for any strong base addition, $[NaOH]_T = B_T$. In other words, we wish to calculate the acid–base titration curve for any given concentration, C_T, of carbonic acid.

By its very nature this problem spans a wide pH range and each of the three carbonate species is a principal component in its own pH domain of dominance. For the sake of generality, we must then forego the simplification afforded by the choice of the proper principal components. The governing equations are again:

Mole Balances

$$\text{Alk} = -[H^+] + [OH^-] + [HCO_3^-] + 2[CO_3^{2-}] = B_T \qquad (39)$$

$$C_T = [H_2CO_3^*] + [HCO_3^-] + [CO_3^{2-}] \qquad (40)$$

Mass Laws

$$[OH^-] = 10^{-14}[H^+]^{-1} \qquad (41)$$

$$[HCO_3^-] = 10^{-6.3}[H^+]^{-1}[H_2CO_3^*] \qquad (42)$$

$$[CO_3^{2-}] = 10^{-16.6}[H^+]^{-2}[H_2CO_3^*] \qquad (43)$$

Since we wish to derive a general formula relating pH to Alk, we cannot make any a priori approximations. The key to this problem is to exploit the universal functionality that relates the concentrations of individual weak acid species to their pK_a's, their total concentrations, and the pH, as exhibited in the universal shape of log C–pH diagrams.

By substitution of Equations 42 and 43 into Equation 40, an explicit formula for $[H_2CO_3^*]$ is obtained:

$$[H_2CO_3^*] = \alpha_0 C_T \qquad (44)$$

with

$$\alpha_0 = (1 + 10^{-6.3}[H^+]^{-1} + 10^{-16.6}[H^+]^{-2})^{-1} \qquad (45)$$

Substituting Equation 44 back into Equations 42 and 43 gives us similar formulae for $[HCO_3^-]$ and $[CO_3^{2-}]$:

$$[HCO_3^-] = \alpha_1 C_T \qquad (46)$$

$$[CO_3^{2-}] = \alpha_2 C_T \qquad (47)$$

with

$$\alpha_1 = (10^{6.3}[H^+] + 1 + 10^{-10.3}[H^+]^{-1})^{-1} \qquad (48)$$

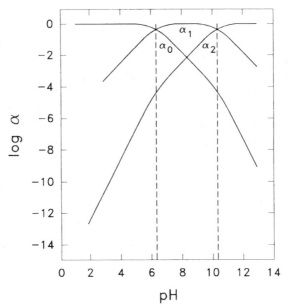

Figure 4.8 Ionization fractions of the carbonate species as a function of pH: $\alpha_0 = [H_2CO_3^*]/C_T$; $\alpha_1 = [HCO_3^-]/C_T$; $\alpha_2 = [CO_3^{2-}]/C_T$.

and

$$\alpha_2 = (10^{16.6}[H^+]^2 + 10^{10.3}[H^+] + 1)^{-1} \tag{49}$$

The *ionization fractions* α_0, α_1, and α_2 represent the fractions of the total carbonate present as each of the particular carbonate species at any pH. By necessity the sum of the three ionization fractions must always equal unity, as can be verified by adding Equations 45, 48 and 49:

$$\alpha_0 + \alpha_1 + \alpha_2 = 1 \tag{50}$$

A plot of the log of the ionization fractions as a function of pH, according to Equations 45, 48 and 49, exhibits the precise characteristics of a log C–pH graph for weak acids with a normalization to $0[= \log(1)]$ on the vertical axis (Figure 4.8). The ionization fraction formulation is thus general for all weak acids; the α parameters for diprotic acids are obtained by substituting the appropriate constants in Equations 45, 48, and 49, or by using the corresponding expressions for monoprotic or triprotic acids.

The solution to our problem is now derived by introducing Equations 41, 46, and 47 into the Alk Equation 39:

$$-[H^+] + 10^{-14}[H^+]^{-1} + \alpha_1 C_T + 2\alpha_2 C_T = \text{Alk} \tag{51}$$

Equation 51 is an implicit function of $[H^+]$ which can theoretically be solved for any given pair of values for C_T and Alk. While we have obtained a compact formulation for the sought for alkalimetric titration curve, the problem of calculating the pH has really not been resolved. Equation 51 is a third-degree polynomial in $[H^+]$, and the easy solutions that come from making the appropriate simplifications in each pH range have been all but obscured. The easy way to solve Equation 51 is to consider the inverse problem and calculate Alk for a given pH and C_T. This is indeed the most straightforward way to obtain the graph of a titration curve: for a series of hydrogen ion concentrations, $[H^+]$, both the ionization fractions and the right-hand side of Equation 51 are readily calculated. The results of such calculations are shown in Figures 4.4 and 4.7.

4. EQUILIBRIUM WITH THE GAS PHASE

4.1 Carbon Dioxide

So far we have considered the aqueous phase to be closed to the atmosphere, assuming no gas exchange at all. Many aquatic solutes are volatile, however, and the atmosphere contains a host of interesting gases, including CO_2 (Table 4.3). As discussed later, gas exchange kinetics at the air–water interface result in equilibration times on the order of days. Aqueous systems of fixed total carbonate concentration at acid–base equilibrium are thus useful models of natural waters for "fast" processes taking place on a time scale of hours or less.

TABLE 4.3   Typical Partial Pressures of Gases in the Atmosphere and Henry's Law Constants

Gas	P (atm)	$\log K_H$ ($M \, atm^{-1}$)
N_2	0.78	-3.18
O_2	0.21	-2.90
CO_2	3.5×10^{-4}	-1.46
CO	$1 - 5 \times 10^{-7}$	-3.02
CH_4	1.7×10^{-6}	-2.89
O_3	$1 - 9 \times 10^{-8}$	-2.03
SO_2	$0.1 - 2 \times 10^{-9}$	$+0.09$
H_2S	$\leqslant 2 \times 10^{-10}$	-0.99
NO_2	$1 - 5 \times 10^{-9}$	-2.00
NO	$1 - 5 \times 10^{-10}$	-2.72
N_2O	3×10^{-7}	-1.59
NH_3	$0.1 - 5 \times 10^{-9}$	$+1.76$
H_2	6×10^{-7}	-3.10

Source: From Refs. 8–12.

At the other extreme, if we are interested in long time scales (i.e., average annual composition of surface waters), it is then a good approximation to consider CO_2 to be at equilibrium between the water and the atmosphere ($P_{CO_2} \simeq 0.03\% = 10^{-3.5}$ atm)*. In the first case, the one phase model system is defined by two mole balance equations, Alk and C_T, which determine the pH of carbonate dominated waters as illustrated in Figure 4.7. Any two of the three parameters Alk, C_T, and pH provide the third one. In the second case, the acid–base composition of the aqueous phase (its pH) is determined by the alkalinity and the equilibrium partial pressure of CO_2 in the atmosphere. The three interdependent parameters are then Alk, P_{CO_2}, and pH. The mole balance equation for carbonate is no longer useful and is replaced by the mass law solubility equation (Henry's law) for CO_2 dissolution. Let us consider a quantitative example.

Example 5 Calculate the composition of a natural water with an alkalinity of 1 meq L^{-1} at equilibrium with the atmosphere. To be systematic let us recast this problem in its canonical form:

Recipe $\quad [NaOH]_T = Alk = 10^{-3} M$

Species $\quad (H_2O)$, H^+, OH^-, Na^+, $CO_2(g)$, $H_2CO_3^*$, HCO_3^-, CO_3^{2-}

Reactions $\quad H_2O = H^+ + OH^-$; $\qquad pK_w = 14.0$ $\qquad$ (52)

$\qquad\qquad\quad CO_2(g) + H_2O = H_2CO_3^*$; $\quad pK_H = 1.5$ $\qquad$ (53)

$\qquad\qquad\quad H_2CO_3^* = H^+ + HCO_3^-$; $\qquad pK_{a1} = 6.3$ $\qquad$ (54)

$\qquad\qquad\quad HCO_3^- = H^+ + CO_3^{2-}$; $\qquad pK_{a2} = 10.3$ $\qquad$ (55)

Although we expect again (as we usually do for natural waters) the pH to be circumneutral, HCO_3^- is not the principal carbonate component in this system. The activity of the species CO_2 (g) is fixed at $P_{CO_2} = 10^{-3.5}$ atm, and its total concentration (in moles per liter of solution) is arbitrary since it depends on the volume of the gas phase which is not specified. Thus, we choose the species CO_2 as a component in order to "isolate" its concentration in one equation only and benefit from its fixed activity. [Note that if the arbitrary volume of the gas phase is taken sufficiently large compared to that of the aqueous phase, $CO_2(g)$ is in fact the dominant inorganic carbon species.] The resulting principal component set is then H^+, CO_2 (and Na^+ which is irrelevant), as shown in Tableau 4.7.

---

*As noted in Chapter 1, the concentration of CO_2 in the atmosphere is increasing, largely as a result of fossil fuel combustion. A CO_2 concentration of approximately 280 ppmV ($10^{-3.55}$ atm) has been estimated for the modern, pre-industrial atmosphere. Recently measured values are significantly higher—from 315 ppmV in 1958 to 355 ppmV ($10^{-3.45}$ atm) in 1990. The current rate of increase in atmosphere CO_2 is approximately 1.5 ppmV (or 0.4%) per year.

TABLEAU 4.7

	H^+	CO_2	$\log K$
H^+	1		
OH^-	-1		-14
$CO_2(g)$		1	
$H_2CO_3^*$		1	-1.5
HCO_3^-	-1	1	-7.8
CO_3^{2-}	-2	1	-18.1
$[CO_2]_T = ?$	0		
$Alk = [NaOH]_T$	-1	0	

Mole Balances

$$TOTH = [H^+] - [OH^-] - [HCO_3^-] - 2[CO_3^{2-}] = -[NaOH]_T$$
$$= -Alk = -10^{-3} \tag{56}$$
$$TOTCO_2 = [CO_2 \cdot g] + [H_2CO_3^*] + [HCO_3^-] + [CO_3^{2-}] = [CO_2]_T = ? \tag{57}$$

Mass Laws

$$[OH^-] = 10^{-14}[H^+]^{-1} \tag{58}$$
$$[H_2CO_3^*] = 10^{-1.5} P_{CO_2} = 10^{-5.0} \tag{59}$$

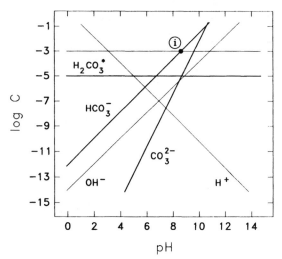

Figure 4.9 Log C–pH diagram for carbonate at equilibrium with $CO_2(g)$; graphical solution of Example 5 (point i).

$$[HCO_3^-] = 10^{-7.8}[H^+]^{-1} P_{CO_2} = 10^{-11.3}[H^+]^{-1} \qquad (60)$$

$$[CO_3^{2-}] = 10^{-18.1}[H^+]^{-2} P_{CO_2} = 10^{-21.6}[H^+]^{-2} \qquad (61)$$

The mass law equations are explicit functions of H^+ and are easily plotted on a log C–pH diagram (Figure 4.9). We expect $[HCO_3^-]$ to be the dominant term in the $TOTH$ equation 56:

$$[HCO_3^-] = 10^{-3.0}$$

Thus, by substitution into the mass law (Equation 60):

$$10^{-11.3}[H^+]^{-1} = 10^{-3.0}$$

$$\therefore [H^+] = 10^{-8.3}$$

$$pH = 8.3$$

All other species are then obtained:

$$[OH^-] = 10^{-5.7}$$

$$[CO_3^{2-}] = 10^{-5.0}$$

$$* \qquad * \qquad *$$

As is apparent from its definition, the alkalinity of a solution is independent of any gain or loss of CO_2 by the solution. (CO_2 is a component and thus has a coefficient of zero in the H^+ column.) Conceptually, this may be seen most easily by noting that in the $TOTH$ expression the excess of strong bases is unaffected by CO_2 exchange. Nonetheless, the concentrations of each of the species on the left-hand side of the $TOTH$ expression (H^+, OH^-, HCO_3^-, etc.) do vary; it is only their algebraic combination, the alkalinity, which is invariant upon CO_2 exchange. The total dissolved carbonate (C_T) is, of course, directly dependent on CO_2 gain or loss by the solution.

A solution to the general problem of finding the pH, knowing P_{CO_2} and the alkalinity of a system where carbonate is the only weak acid, is given directly by the alkalinity equation. The various species concentrations are replaced by their mass law expressions as a function of H^+ and P_{CO_2}:

$$Alk = -[H^+]$$

$$+ \underbrace{K_w[H^+]^{-1}}_{[OH^-]} + \underbrace{K_H K_{a1} P_{CO_2}[H^+]^{-1}}_{[HCO_3^-]} + \underbrace{2K_H K_{a1} K_{a2} P_{CO_2}[H^+]^{-2}}_{[CO_3^{2-}]} \qquad (62)$$

K_H, K_{a1}, K_{a2} are the Henry's law and the first and second acidity constants of the carbonate system.

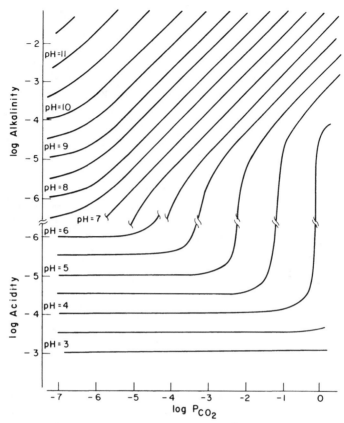

Figure 4.10 General relationship between partial pressure of $CO_2(P_{CO_2})$, alkalinity ($= -$ acidity), and pH in natural waters at equilibrium with a gas phase when the acid–base chemistry is dominated by the carbonate system.

This is an explicit third-degree polynomial of $[H^+]$ and it can be solved for any combination of P_{CO_2} and Alk, including negative alkalinities. To obtain a general graph of the relation between the three parameters P_{CO_2}, Alk, and pH, as illustrated in Figure 4.10 (compare with Figure 4.7), it is of course more convenient to calculate the alkalinity given $[H^+]$ and P_{CO_2}.

4.2 Other Volatile Species

Volatile species other than CO_2 are also present in natural waters. Some have principally an atmospheric source (N_2), some are biogenic (CO_2, O_2, H_2S, CH_4, NH_3, N_2O, NO_2), and some are pollutants with direct or indirect anthropogenic sources (SO_2, NO_x, HCl). As seen in Table 4.2, some of these volatile species, namely H_2S, NH_3 (NH_4^+), and SO_2 (H_2SO_3), are weak acids whose properties should be taken into account when studying the acid–base behavior of natural waters.

Hydrogen sulfide is a particularly important species formed by reduction of sulfate in anoxic waters (see Chapter 7). The effect of gain and loss of H_2S on the alkalinity and pH of a system is much like that of carbon dioxide since the principal component of the sulfide system at the CO_2 equivalence point is H_2S. Loss of H_2S to the atmosphere (which has a very low partial pressure of the gas; ca. 10^{-10} atm) will not affect the alkalinity of the water. The total soluble concentration of sulfide $[H_2S]_T$ will decrease and the pH will increase accordingly.

Example 6   Consider a water whose carbonate system is initially at equilibrium with the atmosphere ($P_{CO_2} = 10^{-3.5}$ atm). This water contains sulfide: $[H_2S]_T = 2 \times 10^{-4} M$ and its initial pH is 8.0. What will the final pH of this water be when hydrogen sulfide is lost to the atmosphere? We assume that atmospheric $CO_2(g)$ remains at equilibrium with the solution and that $P_{H_2S} = 0$ atm.

1. Initially

$$P_{CO_2} = 10^{-3.5} \text{ atm}$$

$$S_T = 10^{-3.7} M$$

$$pH = 8.0$$

The concentrations of the various species in the system are easily calculated:

$$[H_2CO_3^*] = 10^{-1.5} \, 10^{-3.5} = 10^{-5}$$

$$[HCO_3^-] = \frac{10^{-6.3} \times 10^{-5}}{10^{-8}} = 10^{-3.3} = 5 \times 10^{-4}$$

$$[CO_3^{2-}] = \frac{10^{-10.3} \times 10^{-3.3}}{10^{-8}} = 10^{-5.6} = 2.5 \times 10^{-6}$$

Since the pK_a for the H_2S–HS^- reaction is 7.0, at pH 8 most of the sulfide will be HS^- and about one tenth will be H_2S (see Figure 4.1).

$$S_T = [H_2S] + [HS^-] + [S^{2-}] = 10^{-3.7}$$

$$= 10^7 \, 10^{-8}[HS^-] + [HS^-] + 10^{-13.9}\frac{[HS^-]}{10^{-8}} = 10^{-3.7}$$

therefore

$$[HS^-] = 10^{-3.74} = 1.82 \times 10^{-4}$$

$$[H_2S] = 10^7 \times 10^{-8} \times 10^{-3.78} = 10^{-4.74} = 1.82 \times 10^{-5}$$

$$[S^{2-}] = \frac{10^{-13.9} \times 10^{-3.74}}{10^{-8}} = 10^{-9.64} = 2.3 \times 10^{-10}$$

We can obtain the value of the alkalinity for this system:

$$\text{Alk} = -[H^+] + [OH^-] + [HCO_3^-] + 2[CO_3^{2-}] + [HS^-] + 2[S^{2-}] \qquad (63)$$

$$\cdots \simeq [HCO_3^-] + [HS^-] = 10^{-3.16} = 6.9 \times 10^{-4}\,\text{eq L}^{-1}$$

2. Eventually

$$P_{CO_2} = 10^{-3.5}\,\text{atm (unchanged)}$$

$$S_T = 0\,M$$

(All sulfide must escape from the system since we assume $P_{H_2S} = 0$) Alk = $10^{-3.16}$ eq L^{-1} (remains unchanged as CO_2 and H_2S exchange with the atmosphere)

Developing the alkalinity expression in the absence of sulfide:

$$\text{Alk} = -[H^+] + [OH^-] + [HCO_3^-] + 2[CO_3^{2-}] = 10^{-3.16} \qquad (64)$$

Assuming all terms to be negligible except $[HCO_3^-]$:

$$[HCO_3^-] = 10^{-3.16} = \frac{10^{-6.3}\,10^{-1.5}\,10^{-3.5}}{[H^+]}$$

$$[H^+] = 10^{-8.14}$$

After verifying that other terms in the alkalinity expression are indeed negligible, we have a final pH = 8.14. The volatilization of H_2S has resulted in only a small pH increase as CO_2 has replaced the lost H_2S.

$$*\qquad*\qquad*$$

The situation for ammonia is different from that of CO_2 or H_2S since NH_4^+, not NH_3, is the principal component at the CO_2 equivalence point. Ammonia is a strong base, and loss of NH_3 to the atmosphere ($P_{NH_3} \simeq 10^{-9}$ atm) thus results in a decrease of the alkalinity of the system. Concomitantly, the pH of the solution decreases. Ammonia is however much more soluble than carbon dioxide or hydrogen sulfide and ammonia losses are accordingly much slower.

While NH_3 originates chiefly from human and animal waste, sulfur dioxide (SO_2) and nitrogen oxides ($NO_x = NO$ and NO_2) emanate from fossil fuel combustion. Because they are released initially to the atmosphere, these species have a strong impact on the chemistry of a particular type of aquatic system: atmospheric waters, that is, rain and fog.

Ground fogs are formed by condensation of water vapor on aerosols—fine particles suspended in the air. Both gaseous and particulate pollutants can be efficiently scavenged by fog which thus contributes to their deposition on leaves of trees or crops. The chemical composition of fog water is influenced not only by the composition of the condensation nuclei which include aerosols such as $(NH_4)_2SO_4$ or NH_4NO_3 or trace-metal-rich dust or soil, but also by the gases absorbed by the fog droplets.[13,14]

Example 7. Effect of Volatile Species on the Acid–Base Chemistry of Fog Droplets To provide a quantitative example, consider a fog formed in a Swiss cow pasture by condensation of water vapor on inert particles. We wish to calculate the composition and pH of the fog water, assuming equilibrium between the fog water and the atmosphere containing a relatively high concentration of $SO_2(g)$:

$$P_{CO_2} = 10^{-3.5}\,\text{atm}; \qquad P_{SO_2} = 10^{-7.7}\,\text{atm}; \qquad P_{NH_3} = 10^{-9.2}\,\text{atm}$$

Since the fog droplets are considered at equilibrium with atmospheric gases, we choose CO_2, SO_2, and NH_3 as components, and obtain from Tableau 4.8 the following $TOTH$ equation:

$$TOTH = [H^+] - [OH^-] - [HCO_3^-] - 2[CO_3^{2-}] - [HSO_3^-] - 2[SO_3^{2-}]$$
$$+ [NH_4^+] = 0 \qquad (65)$$

TABLEAU 4.8

	H^+	CO_2	SO_2	NH_3	$\log K$
H^+	1				
OH^-	-1				-14
$H_2CO_3^*$		1			-1.5
HCO_3^-	-1	1			-7.8
CO_3^{2-}	-2	1			-18.1
$SO_2(aq)$			1		$+0.1$
HSO_3^-	-1		1		-1.8
SO_3^{2-}	-2		1		-9.0
$NH_3(aq)$				1	1.8
NH_4^+	1			1	11.0
$[CO_2]_T$		1			
$[SO_2]_T$			1		
$[NH_3]_T$				1	

In this system the volatile strong base (NH_3) is neutralized by the volatile acids SO_2 and CO_2. Since SO_2 is both more soluble and more acidic than CO_2, we hypothesize that it is dominant. We thus make the approximation that all species other than HSO_3^- and NH_4^+ can be neglected in Equation 65. Then

$$[NH_4^+] - [HSO_3^-] = 0$$

Expressing these concentrations in terms of $[H^+]$ and the partial pressures of the gases, we obtain

$$10^{11.0} P_{NH_3}[H^+] = \frac{10^{-1.8} P_{SO_2}}{[H^+]} \qquad (66)$$

or

$$[H^+]^2 = \frac{10^{-1.8} P_{SO_2}}{10^{11.0} P_{NH_3}}$$

For $P_{SO_2} = 10^{-7.7}$ atm and $P_{NH_3} = 10^{-9.2}$ atm, we obtain:

$$[H^+]^2 = 10^{-11.3}; \qquad pH = 5.65$$

The species concentrations are obtained by substitution into the mass law equations:

$$[H_2CO_3^*] = 10^{-5.0} \qquad [SO_2 \cdot aq] = 10^{-7.6} \qquad [NH_3 \cdot aq] = 10^{-7.4}$$
$$[HCO_3^-] = 10^{-5.65} \qquad [HSO_3^-] = 10^{-3.85} \qquad [NH_4^+] = 10^{-3.85}$$
$$[CO_3^{2-}] = 10^{-10.3} \qquad [SO_3^{2-}] = 10^{-5.4}$$

By substitution into the $TOTH$ equation 65, we can verify that our approximations are tolerable.

The alkalinity of the fog is given by the expression

$$Alk = -[H^+] + [OH^-] + [HCO_3^-] + 2[CO_3^{2-}] - [SO_2 \cdot aq]$$
$$+ [SO_3^{2-}] + [NH_3 \cdot aq] \qquad (67)$$

Thus

$$Alk \simeq 10^{-5.1} \text{ eq L}^{-1}$$

We may then ask how the chemistry of the fog would change if the partial pressure of NH_3 were increased locally 100-fold as a result of bovine biological activity. With the same assumption as before ($[NH_4^+] = [HSO_3^-]$), we obtain:

$$[H^+] = 10^{-6.65}; \quad pH = 6.65$$

This assumption, however, gives a pH for which the concentration of SO_3^{2-} is not negligible in Equation 65. (Note that $[HSO_3^-] = [SO_3^{2-}]$ at pH 7.2). Thus neglecting the SO_3^{2-} term results in an overestimation of the pH. Solving the equation

$$[NH_4^+] = [HSO_3^-] + 2[SO_3^{2-}]$$

by trial and error, we obtain the more accurate result

$$[H^+] = 10^{-6.57}; \quad pH = 6.57$$

As a consequence of acid neutralization by additional SO_2 dissolution and dissociation, the 100-fold increase in P_{NH_3} results in only a very small pH increase. Concentrations of both ammonia and sulfur species in the fog are

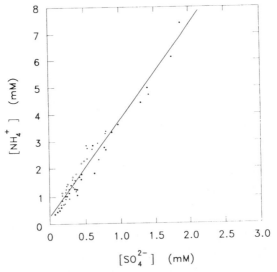

Figure 4.11 Concentrations of ammonium and sulfate in fog waters collected in 1985 and 1986 in an urban area (near Zurich, Switzerland). The strong observed correlation between $[NH_4^+]$ and $[SO_4^{2-}]$ may be due to several factors. With higher NH_4^+ concentrations, both absorption of $SO_2(g)$ into the fog droplets and its oxidation to SO_4^{2-} are favored. Formation of fog droplets by condensation of water on ammonium sulfate aerosols could also contribute to the observed correlation. Adapted from Sigg et al. (1987).[14]

higher:

$$[H_2CO_3^*] = 10^{-5.0} \quad [SO_2 \cdot aq] = 10^{-7.6} \quad [NH_3 \cdot aq] = 10^{-5.4}$$

$$[HCO_3^-] = 10^{-4.73} \quad [HSO_3^-] = 10^{-2.93} \quad [NH_4^+] = 10^{-2.77}$$

$$[CO_3^{2-}] = 10^{-8.46} \quad [SO_3^{2-}] = 10^{-3.56}$$

and Alk $\simeq [HCO_3^-] + [SO_3^{2-}] + [NH_3 \cdot aq] = 10^{-3.5}$ eq L^{-1}.

* * *

Further SO_2 oxidation reactions (see Chapter 7) increase the acidity of fog water by generating sulfuric acid, one of the two major causes of acid rain. The NH_3-partial pressure and resulting fog water $[NH_4^+]$ can have several effects on the extent of SO_2 oxidation in fog water. First, at higher $[NH_4^+]$ the equilibrium $[SO_2]_T$ is higher, thus more sulfur is available for oxidation. Second, higher $[NH_4^+]$ and higher Alk neutralize the acidity produced in the SO_2 oxidation reaction and thus favor progress of the reaction. Constant ratios of NH_4^+/SO_4^{2-} have been observed in fog waters (see Figure 4.11). This correlation may be due in part to the effects mentioned above, though fog formation on $(NH_4)_2SO_4$ aerosols may also be important.

5. MIXING OF TWO WATERS

In estuaries, river confluences, and wastewater disposal sites, waters are mixed. Given sufficient knowledge of the chemical characteristics of the original waters and a description of the mixing process, one should be able to predict the chemical characteristics of the mixture. The basic approach to solving such a problem is to consider the mixing of "conservative" quantities; that is, the total concentrations of the components. If two solutions of 1 L each are mixed, the recipe for the mixture is obviously the sum of the two original recipes divided by two; the pH of the mixture, however, is not the average of the two original pH's. To perform such a "mixing" operation, the numerical values (right-hand side) of the mole balances describing the original waters have to be obtained based on the same set of components. Depending on the original and final constraints (e.g., P_{CO_2} vs. C_T) that are imposed on the system, we may have to consider, serially, different sets of components.

Figure 4.12 illustrates such a solution scheme as applied to the pH of mixtures and to the carbonate system that usually controls it. The conservative quantities, Alk and C_T, are obtained from the given chemical definition of the original waters. As there is no instantaneous CO_2 exchange with the atmosphere, these quantities are "mixed" conservatively and yield the initial composition of the mixture. Over long times of equilibration with the atmosphere, CO_2 equilibrium can be assumed and the mixture is then defined by a fixed partial pressure of

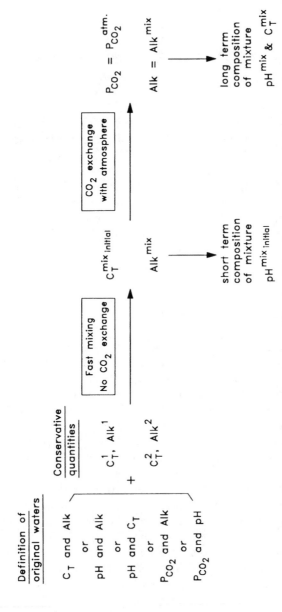

Figure 4.12 General factors controlling the pH in mixtures of carbonate-bearing waters, and schematic of the general method of calculation.

CO_2 and the "mixed" alkalinity. These two equilibrium models idealize gas exchange processes as complete or infinitely slow and mixing processes as instantaneous in given proportions. A more detailed description of the hydrodynamics and of the gas exchange would permit a chemical definition of the system that varies in time and space:

$$\text{Alk } (x, y, z, t); \ C_T(x, y, z, t) \ \rightarrow \ \text{pH}(x, y, z, t)$$

Example 8

River A:	pH = 8.2	Alk = 10^{-3} eq L^{-1}
River B:	pH = 5.7	Alk = 10^{-5} eq L^{-1}
Mixture:	2 volumes of A for 3 volumes of B	

To obtain the total carbonate content of the water after mixing, we need to calculate C_T for each of the rivers.

River A At a pH of 8.2, HCO_3^- is roughly 100 times more abundant than the other carbonate species:

$$\text{Alk}^A = -[H^+] + [OH^-] + [HCO_3^-] + 2[CO_3^{2-}] = 10^{-3.0} \qquad (68)$$

therefore

$$[HCO_3^-] + 2 \times 10^{-10.3} \ 10^{+8.2}[HCO_3^-] = 10^{-3.0}$$

From the mass laws we then obtain

$$[HCO_3^-] = 10^{-3.01}$$
$$[H_2CO_3^*] = 10^{+6.3} \ 10^{-8.2} \ 10^{-3.01} = 10^{-4.91}$$
$$[CO_3^{2-}] = 10^{-10.3} \ 10^{+8.2} \ 10^{-3.01} = 10^{-5.11}$$
$$C_T^A = [HCO_3^-] + [H_2CO_3^*] + [CO_3^{2-}] = 10^{-3.00} \qquad (69)$$

River B At a pH of 5.7, $H_2CO_3^*$ is the principal carbonate component:

$$\text{Alk}^B = -[H^+] + [OH^-] + [HCO_3^-] + 2[CO_3^{2-}] = 10^{-5.0} \qquad (70)$$

Since $[OH^-]$ and $[CO_3^{2-}]$ are negligible:

$$[HCO_3^-] = 10^{-5} + 10^{-5.7} = 10^{-4.92}$$
$$[H_2CO_3^*] = 10^{+6.3} \ 10^{-5.7} \ 10^{-4.92} = 10^{-4.32}$$
$$C_T^B = [H_2CO_3^*] + [HCO_3^-] = 10^{-4.22} = 6.0 \times 10^{-5} \qquad (71)$$

Initial Mix   Before CO_2 exchange:

$$Alk^{mix} = \tfrac{2}{5}Alk^A + \tfrac{3}{5}Alk^B = 10^{-3.39} = 4.06 \times 10^{-4} \, eq \, L^{-1} \qquad (72)$$

$$C_T^{mix} = \tfrac{2}{5}C_T^A + \tfrac{3}{5}C_T^B = 10^{-3.36} = 4.36 \times 10^{-4} \, M \qquad (73)$$

The pH of the mixture is expected to be approximately neutral and HCO_3^- is thus expected to be the principal component for the carbonate system. Choosing HCO_3^- as a component:

$$TOTH = [H^+] - [OH^-] + [H_2CO_3^*] - [CO_3^{2-}] = C_T - Alk = 3 \times 10^{-5}$$
$$= 10^{-4.52} \qquad (74)$$

$$TOTHCO_3 = [H_2CO_3^*] + [HCO_3^-] + [CO_3^{2-}] = C_T = 4.36 \times 10^{-4} \qquad (75)$$

In the neutral pH range, $[H_2CO_3^*]$ is the major positive term in the $TOTH$ equation. Neglecting the other terms as a first approximation,

$$[H_2CO_3^*] = 10^{-4.52}$$

and

$$[HCO_3^-] = C_T - [H_2CO_3^*] = 4.06 \times 10^{-4} = 10^{-3.39}$$

therefore

$$[H^+] = \frac{[H_2CO_3^*]}{[HCO_3^-]} 10^{-6.3} = 10^{-7.43}$$

$$pH_{mix} = 7.43, \text{ initially}$$

The approximations can be verified to be valid at this point.

Mix Equilibrated with Atmosphere   After CO_2 equilibration:

$$Alk^{mix} = 10^{-3.39} \, eq \, L^{-1}$$

$$P_{CO_2} = 10^{-3.5} \, atm$$

With the components H^+ and CO_2, the $TOTH$ equation is

$$TOTH = [H^+] - [OH^-] - [HCO_3^-] - 2[CO_3^{2-}] = -Alk = -10^{-3.39}$$

Trial and error, or graphical examination, shows that $[HCO_3^-]$ is the major term of the $TOTH$ equation:

$$[HCO_3^-] = 10^{-3.39} = \frac{10^{-1.5}\,10^{-6.3}P_{CO_2}}{[H^+]}$$

therefore

$$[H^+] = \frac{10^{-1.5}\,10^{-6.3}\,10^{-3.5}}{10^{-3.39}} = 10^{-7.91}$$

$$pH_{mix} = 7.91, \text{ eventually}$$

The solution of a mixing problem such as this one is made particularly convenient by repeated use of generalized graphs relating C_T, Alk, and pH (Figure 4.7) and P_{CO_2}, Alk, and pH (Figure 4.10).

Note The major point to be remembered from this example is that only conservative quantities can be added to each other. To find the composition of a mixture, concentrations of particular species cannot be averaged out, much less their logarithms. (For example, one has to be careful when defining an "average pH" for rain water.) However, any set of components can be chosen to perform the "mixing operation" as long as it is the same for each part of the system. In the preceding example, we could have calculated $TOTH$ for each of the rivers using HCO_3^- as a component and directly obtained a convenient $TOTH$ equation for the mix, without ever explicitly calculating its alkalinity. However, it is mandatory to use CO_2 as a component and alkalinity as the proton balance equation whenever the system is controlled by a fixed partial pressure of CO_2 rather than a total carbonate concentration.

6. THE ROLE OF ORGANIC ACIDS IN THE ACID–BASE CHEMISTRY OF NATURAL WATERS

The relationships between Alk, C_T or P_{CO_2}, and pH exhibited in Figures 4.7 and 4.10 provide a convenient means of quantifying the acid–base chemistry of a natural water in which inorganic carbon is the dominant weak acid. As illustrated in the previous examples we can easily calculate the effects of CO_2 exchange or the result of mixing two water types in given proportions. The presence of other weak acids, particularly phosphate, sulfide in anoxic waters, borate in seawater, and sulfite and ammonia in atmospheric water make such problems more complicated. Nevertheless the specification of the relevant total concentrations (S_T, P_T, B_T, etc.) allows us to completely define the system and perform the necessary calculations.

A more complicated problem is posed by the presence of organic acids which have a marked influence on the acid–base chemistry of many low alkalinity waters—those of greatest interest in the context of acid rain. We can always obtain the total dissolved organic carbon concentration, but we usually cannot quantify all the individual organic acids in a body of water. (The organic carbon concentration is normally expressed in grams of carbon per liter: Org_T in $g\,C\,L^{-1}$; carbon typically represents one half of the total weight of dissolved organic compounds.) The difficulty then is to describe the acid–base properties of these compounds.

6.1   Acid–Base Chemistry of Humic Acids

Natural organic acids are dominated by so-called humic acids, the yellow-brown dissolved material that gives some rivers and swamps their characteristic color. These complex compounds exhibit pK_a's that span the range on both sides of the CO_2-equivalence point, thus confusing both the definition and the experimental determination of a water's alkalinity. Further, these pK_a's are poorly defined, because of electrostatic interactions among neighboring functional groups on the same macromolecule. Thus a proper quantitative modeling of the acid–base reactions of these organic acids is relatively complicated and we defer it to Chapter 6.

For now we take an approximate empirical approach and assume that the average degree of deprotonation of the mixture of organic acids found in natural waters is given, as a function of pH, by the data of Figure 4.13 and Table 4.4. The total organic acid concentration HA_T is obtained by multiplying the organic carbon concentration by the average number of ionizable acid groups per unit weight of organic carbon, approximately 10^{-2} eq g^{-1}:

$$HA_T(\text{eq L}^{-1}) \simeq 10^{-2} \times Org_T(g\,C\,L^{-1}) \tag{76}$$

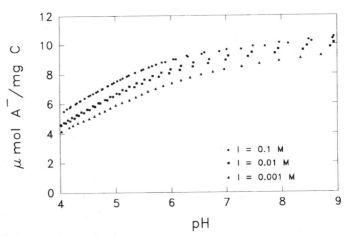

Figure 4.13   Alkalimetric titrations of a fulvic acid sample.[16] The low ionic strength results are in accord with the data given in Table 4.4. The effects of ionic strength are discussed in Chapter 6.

TABLE 4.4 Ionization of Humic Acids

pH	$\alpha = A^-/A_T$	pH	$\alpha = A^-/A_T$	pH	$\alpha = A^-/A_T$
3.0	0.32	5.0	0.73	7.0	0.970
3.1	0.33	5.1	0.75		
3.2	0.35	5.2	0.77	7.2	0.977
3.3	0.37	5.3	0.79		
3.4	0.38	5.4	0.81	7.4	0.983
3.5	0.40	5.5	0.83		
3.6	0.42	5.6	0.84	7.6	0.987
3.7	0.44	5.7	0.86		
3.8	0.47	5.8	0.87	7.8	0.990
3.9	0.49	5.9	0.88		
4.0	0.51	6.0	0.90	8.0	0.993
4.1	0.53	6.1	0.91		
4.2	0.55	6.2	0.92		
4.3	0.58	6.3	0.926		
4.4	0.60	6.4	0.935		
4.5	0.62	6.5	0.942	8.5	0.997
4.6	0.65	6.6	0.949		
4.7	0.67	6.7	0.955		
4.8	0.69	6.8	0.960		
4.9	0.71	6.9	0.965		
				9.0	0.999

Source: From Ref. 16.

Although there are of course variations in the natural organic acid chemistry of various water bodies, these empirical descriptions have been found to have wide applicability. With few exceptions (such as bogs) the role of organic matter in the acid–base chemistry of natural waters is well described quantitatively by this "average" humic acid.

6.2 Defining Alkalinity in the Presence of Organic Acids

To avoid the complications presented by the variable degree of protonation of organic acids near the CO_2 equivalence point, it is convenient to choose the completely protonated organic acids as a component, in addition to the usual principal components at the CO_2-equivalence point. We can then define the "charge-balance alkalinity" (CB-Alk)[17] as

$$- \text{CB-Alk} = TOTH = [H^+] - [OH^-] - [HCO_3^-] - 2[CO_3^{2-}] \cdots - [A^-]$$

$$(77)$$

CB-Alk is the value obtained by difference between the analytically measured

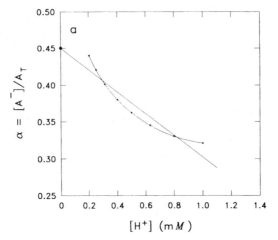

Figure 4.14a The fraction of deprotonated humic acid sites as a function of $[H^+]$. In the range of interest to Gran titrations, the data can be approximated by a straight line with a y-intercept at 0.45.

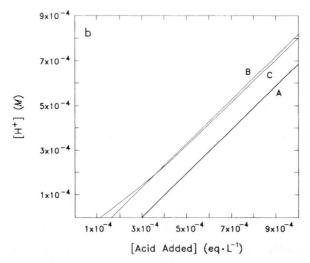

Figure 4.14b Calculated Gran titrations of three water samples:
(A) $C_T = 3 \times 10^{-4} M$; C-Alk $= 3 \times 10^{-4} M$; $pH_{init} = 8.3$. (B) $C_T = 3 \times 10^{-4} M$; C-Alk $= 1.65 \times 10^{-4} M$; $pH_{init} = 6.39$. (C) $C_T = 3 \times 10^{-4} M$; CB-Alk $= 3 \times 10^{-4} M$; $HA_T = 3 \times 10^{-4} M$. $Org_T = 30\,mg\,CL^{-1}$; $pH_{init} = 5.5$.

strong base cations and strong acid anions:

$$CB\text{-Alk} = [Na^+] + [K^+] + 2[Ca^{2+}] + 2[Mg^{2+}] - [Cl^-] - 2[SO_4^-] \cdots \quad (78)$$

Thus CB-Alk is an experimentally determinable quantity. In practice, as discussed in Sidebar 4.2, 45% of the organic acids are measured as strong acids in a Gran titration, thus:

$$CB\text{-Alk} = Gran\text{-Alk} + 4.5 \times 10^{-3} \, Org_T(g \, C \, L^{-1}) \quad (79)$$

This approximate formula allows one to calculate one of the quantities, CB-Alk, Gran-Alk, or Org_T, from the other two.

Note 1. The expression "charge balance alkalinity" is not strictly correct since Equation 77 should also include the total phosphate concentration (a "strong" monoprotic acid) as discussed before, as well as any other strong acid.

Note 2. The component set chosen for CB-Alk results in different implicit definitions of weak and strong acids for inorganic and organic species.

Note 3. The approach taken here can be generalized to other species such as $Al(OH)_3$, which also becomes protonated at low pH.

6.3 Acid–Base Calculations Involving Organic Acids

Like all other properly defined values of $TOTH$ equations, CB-Alk is a conservative quantity which can be appropriately "diluted," or "mixed". The chemistry of a system, including its pH, is determined completely from a knowledge of its CB-Alk and of the relevant analytical concentrations, including C_T, and HA_T (or Org_T). When the organic acid concentration is large, the calculation is somewhat complicated because the proportion of organic acid that is deprotonated must be obtained from an empirical (tabular) expression, rather than from a mass law equation. Graphical methods have been proposed;[17] they involve drawing a family of graphs of CB-Alk vs. Org_T for a series of pH's at a given value of P_{CO_2} or C_T. These graphs can be used to study conveniently the effects of dilution, acid or base addition, organic carbon elimination, or water mixing on the acid–base chemistry or organic bearing waters.

To illustrate how the presence of organic acids may influence not only our acid–base calculations but also our data interpretation, we reexamine Example 8 with a new set of assumptions.

Example 9 According to the calculations of Example 8, both River A and River B are initially supersaturated with $CO_2(g)$. Having measured the given pH's and alkalinities (through Gran titrations) we might take a different interpretation of the data and consider that both rivers are in fact at equilibrium with the atmosphere and that organic acids (and no other weak acids) are the

SIDEBAR 4.2

Gran Titration of Humic Acids

Gran titrations are routinely used to determine the alkalinity of organic-rich water samples, for which, as explained earlier, it is difficult to define the acidimetric titration end point. As seen in Figure 4.13, a sizable fraction of the acid groups on humic acids is titrated in the pH range 4–3 where Gran titrations are normally performed.

$$[\text{acid added}] = \text{CB-Alk} + [H^+] - [A^-]$$

$$[\text{acid added}] = \text{CB-Alk} + [H^+] - \alpha A_T$$

An arithmetic plot of the proportion of deprotonated organic acids, α (see Table 4.4), as a function of $[H^+]$ shows that the linear fit in the range $[H^+] = 200{-}1000\,\mu M$ (which is typically used in Gran titrations) has an intercept at 0.45 and a slope of $-0.15\,\text{m}M^{-1}$ (Figure 4.14a). The linear approximation of the acidimetric titration is then

$$[\text{acid added}] = \text{CB-Alk} - 0.45\,A_T + [H^+](1 + 0.15 \times 10^3\,A_T)$$

Thus the alkalinity determined from a Gran titration includes about 55% of the organic acid content of the sample, as if 45% of the organic acids were "strong acids" (Figure 4.14b).

$$\text{Gran-Alk} = \text{CB-Alk} - 0.45\,A_T$$

$$\text{Gran-Alk} \simeq \text{CB-Alk} - 4.5 \times 10^{-3}\,\text{Org}_T \qquad (\text{Org}_T \text{ in g C L}^{-1})$$

For most natural concentrations of organic acid, the slope of the linear fit of the $[H^+]$ vs. [acid added] graph,

$$[1 + 0.15 \times 10^3\,A_T]^{-1} \simeq [1 + 1.5\,\text{Org}_T]^{-1} \qquad (\text{Org}_T \text{ in g C L}^{-1})$$

is negligibly different from unity. Only for high organic concentrations $(\text{Org}_T = 0.1\,\text{g L}^{-1})$ does the slope of the Gran titration decrease measurably (0.87), but this effect may then be hidden by the curvature of the titration graph.

reasons why the pH's of these waters are lower than expected from their alkalinities and atmospheric P_{CO_2}.

Let us recalculate the compositions of A, B, and their mix under these new assumptions.

River A   Equilibrium with atmosphere:

$$P_{CO_2} = 10^{-3.5} \, \text{atm}; \quad \text{pH} = 8.2$$

$$\therefore [H_2CO_3^*] = 10^{-5.0}$$

$$[HCO_3^-] = \frac{10^{-6.3} \cdot 10^{-5.0}}{10^{-8.2}} = 10^{-3.1}$$

$$[CO_3^{2-}] = \frac{10^{-10.3} \cdot 10^{-3.1}}{10^{-8.2}} = 10^{-5.2}$$

$$\text{C-Alk}^A \simeq [HCO_3^-] + 2[CO_3^{2-}] = 8.06 \times 10^{-4} \, \text{eq L}^{-1} \tag{80}$$

$$C_T^A = [H_2CO_3^*] + [HCO_3^-] + [CO_3^{2-}] = 8.10 \times 10^{-4} \, M^{-1} \tag{81}$$

Concentration of organic acid:

Since the organic acids are almost 100% deprotonated at pH = 8.2 (see Table 4.4), the measured alkalinity is given by the equation

$$\text{Gran-Alk} = \text{C-Alk} + (1.00 - 0.45) \, HA_T \tag{82}$$

$$\therefore HA_T = \frac{\text{Gran-Alk} - \text{C-Alk}}{0.55} = 3.53 \times 10^{-4} \, M \tag{83}$$

Thus, River A contains approximately 35.3 mg C L^{-1} of organic acids and

$$\text{CB-Alk}^A = 8.06 \times 10^{-4} + 3.53 \times 10^{-4} = 11.59 \times 10^{-4} \, \text{eq L}^{-1} \tag{84}$$

River B   Equilibrium with atmosphere:

$$P_{CO_2} = 10^{-3.5} \, \text{atm}; \quad \text{pH} = 5.7$$

$$\therefore [H_2CO_3^*] = 10^{-5.0}$$

$$[HCO_3^-] = \frac{10^{-6.3} \cdot 10^{-5.0}}{10^{-5.7}} = 10^{-5.6}$$

$$[CO_3^{2-}] = \frac{10^{-10.3} \cdot 10^{-5.6}}{10^{-5.7}} = 10^{-10.2}$$

$$\text{C-Alk}^B \simeq [HCO_3^-] - [H^+] = 5.2 \times 10^{-7} \, \text{eq L}^{-1} \, (\simeq 0!) \tag{85}$$

$$C_T^B = [H_2CO_3^*] + [HCO_3^-] + [CO_3^{2-}] = 1.25 \times 10^{-5} \tag{86}$$

Concentration of organic acid:
At a pH = 5.7, the organic acids are 86% dissociated, thus

$$HA_T = \frac{\text{Gran-Alk} - \text{C-Alk}}{0.86 - 0.45} = 2.31 \times 10^{-5} M \tag{87}$$

River B thus contains approximately 2.31 mg C L^{-1} of organic acids and

$$CB\text{-}Alk^B = 5.2 \times 10^{-7} + 2.31 \times 10^{-5} = 2.36 \times 10^{-5} \text{ eq L}^{-1}$$

Note: The previous calculations, which are based on the differences between measured alkalinities and calculated C-alk, are quite sensitive to the values chosen for the equilibrium constants of the carbonate system.

Initial Mix

$$CB\text{-}Alk^{\text{mix}} = \tfrac{2}{5}CB\text{-}Alk^A + \tfrac{3}{5}CB\text{-}Alk^B = 4.78 \times 10^{-4} \text{ eq L}^{-1}$$

$$C_T^{\text{mix}} = \tfrac{2}{5}C_T^A + \tfrac{3}{5}C_T^B = 3.31 \times 10^{-4} M$$

$$HA_T^{\text{mix}} = \tfrac{2}{5}HA_T^A + \tfrac{3}{5}HA_T^B = 1.55 \times 10^{-4} M$$

Consider as an initial guess that the pH is near the value calculated initially in Example 8: pH ≃ 7.40. At such pH, 98% of the organic acids are deprotonated ([A$^-$] ≃ HA$_T$) and [HCO$_3^-$] dominates the other terms in the C-Alk expression:

$$CB\text{-}Alk^{\text{mix}} = -[H^+] + [OH^-] + [HCO_3^-] + 2[CO_3^{2-}] + [A^-] \tag{88}$$

$$\simeq [HCO_3^-] + HA_T$$

$$\therefore C\text{-}Alk^{\text{mix}} \simeq CB\text{-}Alk - HA_T = 3.23 \times 10^{-4} = [HCO_3^-]$$

The pH is then obtained by solving the carbonate mole balance equation:

$$C_T^{\text{mix}} \simeq [H_2CO_3^*] + [HCO_3^-] = 3.31 \times 10^{-4} \tag{89}$$

$$\therefore [H_2CO_3^*] = 8 \times 10^{-6}$$

$$[H^+] = 10^{-6.3}\frac{[H_2CO_3^*]}{[HCO_3^-]} = 10^{-7.91}$$

$$pH_{\text{mix}} = 7.91, \text{ initially}$$

We can now verify that our assumptions ([A$^-$] ≃ HA$_T$ and [CO$_3^{2-}$] negligible in Equation 88) are indeed appropriate. Note that this calculation is made particularly easy by the assumption that all the organic acids are deprotonated. At lower pH's where this assumption does not hold, the solution would require a trial and error iterative method.

Equilibrium with Atmosphere   By sheer coincidence this initial mix is nearly at equilibrium with the atmosphere:

$$P_{CO_2} = 10^{-3.5} \, atm; \qquad [H_2CO_3^*] = 10^{-5}$$

$$C\text{-}Alk^{mix} = 3.23 \times 10^{-4} \simeq [HCO_3^-]$$

$$\therefore [H^+] = \frac{10^{-6.3} \cdot 10^{-5}}{10^{-3.49}} = 10^{-7.81}$$

$$pH_{mix} = 7.81, \text{ eventually}$$

The organic acids remain fully deprotonated in this pH range and do not buffer the pH decrease resulting from the invasion of CO_2.

Oxidation of Organic Acids   Let us assume now that we UV-oxidize all the organic acids (to H_2O and CO_2) in a sample of the mixed river water, while keeping it at equilibrium with the atmosphere. What is the new pH of the sample? In the absence of organic acids the only alkalinity is provided by the carbonate system:

$$CB\text{-}Alk = C\text{-}Alk = -[H^+] + [OH^-] + [HCO_3^-] + 2[CO_3^{2-}]$$

$$= 4.78 \times 10^{-4} \qquad\qquad (90)$$

$$\therefore [HCO_3^-] \simeq 4.78 \times 10^{-4} = 10^{-3.32}$$

$$[H_2CO_3^*] = 10^{-1.5} P_{CO_2} = 10^{-5}$$

$$\therefore [H^+] = \frac{10^{-6.3} \cdot 10^{-5}}{10^{-3.32}} = 10^{-7.98}$$

$$pH \simeq 7.89$$

Eliminating the organic acid from the water has resulted in a slight pH increase.

7. EFFECTS OF BIOLOGICAL PROCESSES ON pH AND ALKALINITY

7.1 Photosynthesis and Respiration

So far we have considered only the atmosphere to be a source or a sink for CO_2 in aquatic systems. In fact, photosynthesis and respiration within the water column and in the benthos often dominate the utilization and production of carbon dioxide in natural waters. A very simplified chemical representation of these processes is given by the reaction

$$CO_2 + H_2O \underset{\text{respiration}}{\overset{\text{photosynthesis}}{\rightleftharpoons}} \text{``CH}_2O\text{''} + O_2 \qquad\qquad (91)$$

where "CH_2O" is a symbol for organic matter. Despite its crudity, this chemical representation gives a reasonable approximation of the stoichiometry of total photosynthetic and respiratory activities as they affect water chemistry.

According to the reaction shown above, photosynthesis and respiration do not affect the alkalinity of the water. In its simplicity, Reaction 91 does not include the relatively minor effects of nitrogen uptake and release on alkalinity (which are discussed below). Alkalinity is also slightly affected if the organic compound respired is an organic acid. The magnitude of this effect is indicated in the preceding example. From the point of view of the carbonate system, however, it makes no difference whether the CO_2 is gained or lost in autochthonous processes or by exchange with the atmosphere. In both cases, the alkalinity of the system remains unchanged (consider the right-hand side of the $TOTH$ equation, with $H_2CO_3^*$ as the carbonate component) and the pH decreases with increasing total carbonate. To obtain quantitatively the effect of photosynthesis and respiration on pH, we must measure the net loss or gain of CO_2 and calculate the new pH given a constant alkalinity and the new total carbonate concentration

$$C_T^{new} = C_T^{old} + \text{gain of } CO_2 \tag{92}$$

A semiquantitative estimate of the effect of CO_2 gain or loss on pH is obtained by considering the reactions written with the principal components of the system:

$$pH < 6.3 \qquad CO_2 + H_2O = \text{"}CH_2O\text{"} + O_2 \tag{93}$$

$$6.3 < pH < 10.3 \quad HCO_3^- + H^+ = \text{"}CH_2O\text{"} + O_2 \tag{94}$$

$$pH > 10.3 \qquad CO_3^{2-} + 2H^+ = \text{"}CH_2O\text{"} + O_2 \tag{95}$$

At low pH, the effect of photosynthesis and respiration on pH is minimal (below pH = 6.3 the reaction shows no proton to be consumed); at high pH the effect is large (above pH = 10.3 two protons are consumed per carbon fixed). In the carbonate and the bicarbonate dominance regions, CO_2 acts effectively as a strong diprotic and monoprotic acid, respectively. This intuitive result can be rationalized mathematically by considering the $TOTH$ equation corresponding to the principal components of the system: what is basic or acidic at any pH can be defined as a negative or positive contribution to the $TOTH$ equation. In a first approximation, the effect of a change in composition on pH is given by the sign and the magnitude of the change in the numerical value of the $TOTH$ equation (right-hand-side) corresponding to the principal components of the system (see Section 8).

Plant growth requires more than just carbon, water, and light. Other nutrients that are necessary in relatively large quantities are phosphorus and nitrogen, chiefly provided in aquatic systems by phosphate and nitrate or ammonium.

The average proportions of these major elements in algal biomass is described by the Redfield formula[18]:

$$\text{Protoplasm} = C_{106}H_{263}O_{110}N_{16}P_1 \qquad (96)$$

A more complete stoichiometric description of photosynthetic and respiratory processes in natural waters is thus provided by the following reactions:

$$106CO_2 + 16NO_3^- + H_2PO_4^- + 122H_2O + 17H^+$$

$$\underset{R}{\overset{P}{\rightleftharpoons}} \text{Protoplasm} + 138O_2 \qquad (97)$$

$$106CO_2 + 16NH_4^+ + H_2PO_4^- + 106H_2O$$

$$\underset{R}{\overset{P}{\rightleftharpoons}} \text{Protoplasm} + 106O_2 + 15H^+ \qquad (98)$$

The overall reactions permit an accurate study of the role of photosynthesis and respiration in modifying the pH and the alkalinity of natural waters despite their inapplicability to particular organisms under particular conditions. As written, they provide a direct measure of alkalinity changes since the reactants are the principal components at the CO_2 equivalence point:

1. When nitrate is the nitrogen source, the alkalinity increases by 0.16 equivalent ($= 17/106$) per mole of carbon fixed.
2. When ammonium is the nitrogen source, the alkalinity decreases by 0.14 equivalent ($= 15/106$) per mole of carbon fixed.

The inclusion of nitrogen and phosphorus in the reactions for photosynthesis and respiration does not markedly modify our previous results with respect to the net pH effects of these processes. Consider, for example, the reactions written with the principal components in the pH range 7–9:

$$106HCO_3^- + 16NO_3^- + HPO_4^{2-} + 16H_2O + 124H^+$$

$$\underset{R}{\overset{P}{\rightleftharpoons}} \text{Protoplasm} + 138O_2 \qquad (99)$$

$$106CO_2 + 16NH_4^+ + HPO_4^{2-} + 92H^+$$

$$\underset{R}{\overset{P}{\rightleftharpoons}} \text{Protoplasm} + 106O_2 \qquad (100)$$

Clearly, in all situations the uptake of CO_2 dominates the acid–base effect of photosynthesis. For example, if bicarbonate is the major inorganic carbon species, uptake of CO_2 results in an increase in pH. This result is independent of whether $CO_2(aq)$, H_2CO_3, or HCO_3^- is the actual species taken up by the algae as long as the stoichiometry of the reaction is correct (i.e., H^+, is the counter ion for HCO_3^-). However, the magnitude of the pH effect is influenced by the nature of the nitrogen source. Photosynthesis drives the pH higher when nitrate rather than ammonium is the nitrogen source for aquatic plants. At low pH where $H_2CO_3^*$ is the major carbonate species, the nitrogen uptake can dominate the acid–base chemistry of photosynthesis. For example, uptake of ammonia in bog water (pH $\simeq 4$) can result in a net pH decrease, as seen in Reaction 98.

7.2   Other Microbial Processes

Microbial processes other than photosynthesis and respiration are also involved in the cycles of elements. Sulfate reduction, nitrogen fixation, nitrification, denitrification, and methanogenesis, are all locally and globally important in governing the cycles of sulfur, nitrogen, and carbon.

Till we examine the energetics of these oxidation–reduction processes in Chapter 7, we can use the stoichiometric approach illustrated above to estimate their effects on the pH and alkalinity of aquatic systems. As before, the methodology consists of writing overall reactions describing the stoichiometry of the various processes as a function of the principal components for the alkalinity expression (i.e., at the CO_2 equivalence point; pH $\simeq 5$). The production or consumption of protons in such reactions then yields the alkalinity changes. The effects on pH are obtained by considering the gains or losses of protons when reactants and products are written as the principal components under ambient conditions.

Table 4.5 presents a list of such reactions and the corresponding qualitative effects on alkalinity and pH. A quantitative description of the pH effect can be obtained when chemical conditions are specified by considering the effects of the various reactions on the mole balance expressions. Note that in all the reactions of Table 4.5, the changes in alkalinity are precisely the changes in the electrical charge of the species of specific interest (S, N, C). For example, from an acid–base point of view, denitrification and sulfate reduction are strictly equivalent to removal of nitric and sulfuric acid, respectively.

Example 10   Consider the effect on pH and alkalinity of reducing all the sulfate to sulfide (without any loss of gas) in a lake characterized initially by

$$Alk^0 = 10^{-3}\,eq\,L^{-1}$$

$$[SO_4^{2-}]_T^0 = 10^{-4}\,M$$

$$pH^0 = 7.5$$

Choice of Components   The principal components at a pH of 7.5 are H^+, HCO_3^-, SO_4^{2-}, and HS^- (see Figure 4.1).

TABLE 4.5 **Effects of Aquatic Microbial Processes on Alkalinity and pH in Closed Systems**

Sulfate Reduction[a]

$$SO_4^{2-} + 2\text{“}CH_2O\text{”} + 2H^+ \rightarrow H_2S + 2H_2O + 2CO_2 \tag{1}$$

Alkalinity increases by 2 eq. per mole of sulfate reduced
pH $< 6.2 \rightarrow$ pH increases
$7.0 >$ pH $> 6.2 \rightarrow$ pH $\sim$ constant
pH $> 7.0 \rightarrow$ pH decreases

Nitrogen Fixation (with concomitant photosynthesis)

$$106CO_2 + 8N_2 + H_2PO_4^- + 130H_2O + H^+ \rightarrow \text{Protoplasm} + 118O_2 \tag{2}$$

Alkalinity increases by 0.13 eq. per mole of nitrogen fixed but this effect is due to P uptake and is in fact negligible
pH increases throughout pH range

Nitrification

$$NH_4^+ + \tfrac{3}{2}O_2 \rightarrow NO_2^- + 2H^+ + H_2O \tag{3}$$

Alkalinity decreases by 2 eq. per mole of ammonium oxidized
pH decreases throughout pH range

$$NO_2^- + \tfrac{1}{2}O_2 \rightarrow NO_3^- \tag{4}$$

No effect on alkalinity or pH

Denitrification

$$4NO_3^- + 5\text{“}CH_2O\text{”} + 4H^+ \rightarrow 2N_2 + 5CO_2 + 7H_2O \tag{5}$$

Alkalinity increases by 1 eq. per mole of nitrate reduced
pH $< 6.3 \rightarrow$ pH increases
pH $> 6.3 \rightarrow$ pH decreases slightly

Methane Fermentation

$$\text{“}CH_2O\text{”} + \text{“}CH_2O\text{”} \rightarrow CH_4(g) + CO_2 \tag{6}$$

Alkalinity remains constant
pH $< 6.3 \rightarrow$ pH $\sim$ constant
pH $> 6.3 \rightarrow$ pH decreases

[a]“CH_2O” represents organic matter.

TABLEAU 4.9

	H^+	HCO_3^-	SO_4^{2-}	HS^-
H^+	1			
OH^-	-1			
$H_2CO_3^*$	1	1		
HCO_3^-		1		
CO_3^{2-}	-1	1		
SO_4^{2-}			1	
H_2S	1			1
HS^-				1
S^{2-}	-1			1
Initial composition	$10^{-4.21}$	$10^{-2.97}$	$10^{-4.0}$	0
Change in composition	$+10^{-4.0}$	$+10^{-3.7}$	$-10^{-4.0}$	$+10^{-4.0}$
Final composition	$10^{-3.79}$	$10^{-2.90}$	0	$10^{-4.0}$

Initial Composition   We need to obtain the right-hand side of the mole balance equations corresponding to our principal components. The last two columns of Tableau 4.9 give

$$TOTSO_4 = 10^{-4}$$

$$TOTHS = 0$$

To obtain the values of $TOTH$ and $TOTHCO_3$ we have to calculate the concentrations of all the carbonate species given the alkalinity (10^{-3}) and the pH (7.5):

$$Alk = -[H^+] + [OH^-] + HCO_3^-] + 2[CO_3^{2-}] = 10^{-3}$$

$$= -10^{-7.5} + 10^{-6.5} + [HCO_3^-] + 2\frac{10^{-10.3}}{10^{-7.5}}[HCO_3^-] = 10^{-3} \quad (101)$$

therefore

$$[HCO_3^-] = 10^{-3}$$

$$[CO_3^{2-}] = \frac{10^{-10.3}}{10^{-7.5}}10^{-3} = 10^{-5.8}$$

$$[H_2CO_3^*] = 10^{6.3}\,10^{-7.5}\,10^{-3} = 10^{-4.2}$$

From this calculation we obtain (see Tableau 4.9)

$$TOTH = 10^{-7.5} - 10^{-6.5} + 10^{-4.2} - 10^{-5.8} = 10^{-4.21}$$

$$TOTHCO_3 = 10^{-4.2} + 10^{-3.0} + 10^{-5.8} = 10^{-2.97}$$

Change in Composition   Let us write the reaction for sulfate reduction with the principal components of the system (see Table 4.5):

$$SO_4^{2-} + 2\text{"}CH_2O\text{"} \rightarrow HS^- + 2HCO_3^- + H^+ \tag{102}$$

Given $\Delta TOTSO_4^{2-} = -10^{-4}$ (sulfate entirely reduced), we obtain, according to the stoichiometry of Reaction 102,

$$\Delta TOTH = +10^{-4}$$

$$\Delta TOTHCO_3 = +2 \times 10^{-4}$$

$$\Delta TOTHS = +10^{-4}$$

therefore

$$\text{new } TOTH = 10^{-3.79}$$

$$\text{new } TOTHCO_3 = 10^{-2.90}$$

$$\text{new } TOTHS = 10^{-4}$$

Final Composition   Following the columns of Tableau 4.9, we can write the mole balance equations:

$$TOTH = [H^+] - [OH^-] + [H_2CO_3^*] - [CO_3^{2-}] + [H_2S] - [S^{2-}]$$
$$= 10^{-3.79} \tag{103}$$

$$TOTHCO_3 = [H_2CO_3^*] + [HCO_3^-] + [CO_3^{2-}] = 10^{-2.90} \tag{104}$$

$$TOTHS = [H_2S] + [HS^-] + [S^{2-}] = 10^{-4.0} \tag{105}$$

Since the right-hand side of the $TOTH$ equation has increased compared to its original value, the pH is expected to be lower than before, and $[CO_3^{2-}]$ and $[S^{2-}]$ can be neglected in all the mole balances, as can $[H^+]$ and $[OH^-]$ in Equation 103:

$$TOTH = [H_2CO_3^*] + [H_2S] = 10^{6.3}[H^+][HCO_3^-]$$
$$+ 10^{7.0}[H^+][HS^-] = 10^{-3.79}$$

$$TOTHCO_3 = [H_2CO_3^*] + [HCO_3^-] = (10^{6.3}[H^+] + 1)[HCO_3^-] = 10^{-2.90}$$

$$TOTHS = [H_2S] + [HS^-] = (10^{7.0}[H^+] + 1)[HS^-] = 10^{-4.0}$$

Substituting $[HCO_3^-]$ and $[HS^-]$ into the first equation yields an explicit equation for $[H^+]$:

$$\frac{10^{3.4}[H^+]}{1 + 10^{6.3}[H^+]} + \frac{10^{3.0}[H^+]}{1 + 10^{7.0}[H^+]} = 10^{-3.79}$$

The resulting quadratic can be solved:

$$[H^+] = 10^{-7.25}$$

Substituting back into the other mole balances

$$[HCO_3^-] = 10^{-2.95}$$
$$[HS^-] = 10^{-4.19}$$

The other species concentrations are obtained from the mass laws:

$$[H_2CO_3^*] = 10^{-3.9}; \quad [CO_3^{2-}] = 10^{-6.00}$$
$$[H_2S] = 10^{-4.44}; \quad [S^{2-}] = 10^{-10.84}$$

Note that the reduction of the sulfate has resulted in a slight increase in the alkalinity of the system:

$$\text{Alk}' = -TOTH + TOT\text{HCO}_3 + TOT\text{HS} = 10^{-2.92}$$

Despite this consumption of sulfuric acid, the pH actually decreases from 7.5 to 7.25. This is due to the production of CO_2 accompanying the sulfate reduction process.

8. BUFFER CAPACITY

From the point of view of *Homo polluens* what matters most among the characteristics of the environment is its resistance to change, its homeostasis, its buffering processes. Staying in the chemical realm, the problem is to understand how and by how much the composition of the environment will change for a given input. Attacking such a problem is quite a leap from the traditional study of buffers that are used to stabilize pH in reaction vessels. Yet, if we narrow our focus to the question of pH in natural waters—a particularly topical question in the face of a quasi-ubiquitous acid rain problem in the Northern Hemisphere— the same principles that apply in the laboratory can be used to ascertain the behavior of lakes, rivers, groundwaters, and oceans.

Before we develop a general formula for the "buffer capacity" of chemical systems, it is worth remembering the limitations of our approach. Although the systems that we consider may be quite complex indeed, they are but very simplified idealizations of real systems; the most tenuous simplification is assuredly that of chemical equilibrium. We have to recall here our earlier arguments of time scales and partial equilibrium. To the traditional buffering questions regarding how much the pH will change for a given addition of acid or base and what species or reactions control the pH, we have to add: over what time scale? The homeostatic behavior of the real system is not the same over hours as it is over geological times; it is dependent on a number of relatively slow reactions such as equilibration with gas or solid phases. We shall see that our corresponding partial equilibrium models have different buffering capacities. These buffering capacities pertain to reality only to the extent that the equilibrium models do.

8.1 Definition

The buffer capacity of a chemical system is typically defined as the ratio of an infinitesimal base addition to the resulting infinitesimal increase in pH. It is given by the slope of the titration curve—base concentration minus acid concentration vs. pH—which may be obtained theoretically or experimentally:

$$\beta_H = -\frac{\Delta TOTH}{\Delta pH} \tag{106}$$

The buffer capacity so defined has dimensions of moles per liter and provides a measure of the ability of the system to resist pH changes upon addition of strong acid or base. For example, if we consider the titration curve of Figure 4.4, we can see that the buffer capacity of the carbonate system is largest when the pH is close to the pK_a's (6.3 and 10.3).

To proceed with calculations of buffer capacities in complex systems, we need a more precise mathematical definition. Let us define an "ideal" strong acid or base as one that affects a chemical system exclusively through hydrogen or hydroxide ion additions (e.g., an acid HA that dissociates completely into H^+ and A^-, the latter being strictly unreactive). In practice, most strong acids and bases are good approximations of ideal ones. Consider now the addition of an infinitesimal quantity $(\partial TOTH)$ of an ideal strong acid or base, and the resulting change (∂pH) in pH. A strict mathematical definition of the buffer capacity β_H of the system is given by

$$\beta_H = -\left[\frac{\partial pH}{\partial TOTH}\right]^{-1} \tag{107}$$

The partial derivative indicates that all total component concentrations, except H^+, are kept constant. Note that the order of the differentiation is consistent with the causality of the real or imaginary titration process: the pH change results from an acid or base addition and not vice versa, as implied in the preceding formula.

8.2 Minor Species Formula

An approximate value of the pH buffering capacity of any chemical system is given by the minor species formula:[19]

$$\beta_H = 2.3 \sum_i \lambda_i^2 [S_i] \tag{108}$$

where λ_i is the coefficient of the species S_i in the $TOTH$ equation when the components are the principal components of the system. This can be rephrased in a more compact but perhaps more esoteric form: where λ_i is the proton level of i compared to the principal components of the system. Because $\lambda_i = 0$ for the principal components, the species S_i which contribute to the value of β_H are by definition the "minor species."

A simplified derivation of this formula can be given as a direct differentiation of the $TOTH$ equation: Consider the generalized tableau involving the principal components of the system (Tableau 4.10). From the coefficients in the H^+ column, the proton conservation equation can be written:

$$TOTH = \sum_i \lambda_i [S_i] \tag{109}$$

All the components (here the principal components) but H^+ have identically null λ_i. Consider the mass laws of the other species (the "minor species") as a function of the components:

$$[S_i] = \frac{K_i}{\gamma_i} \{H^+\}^{\lambda_i} (a_1)^{v_{i1}} (a_2)^{v_{i2}} \cdots \tag{110}$$

K_i are the equilibrium constants and γ_i the activity coefficients. If the ionic

TABLEAU 4.10 Principal Components

	Hydrogen Ion H^-	Solids		Gases		Major Dissolved Species		
Component Activities	$\{H^-\}$	Z_1 $a_1 = 1$	$Z_2 \cdots$ $a_2 = 1 \cdots$	G_1 P_1	$G_2 \cdots$ $P_2 \cdots$	C_1 $C_2 \cdots$ $\cdots$		$C_k \cdots$ $a_k = \gamma_k(C_k) \cdots$
Major Species C_1	0	0	$0 \cdots$	0	$0 \cdots$	1	$0 \cdots$	
C_2	0	0	$0 \cdots$	0	$0 \cdots$	0	$1 \quad 0 \cdots$	0
C_k	0	0	$0 \cdots$	0	$0 \cdots$	0	0	1
Minor Species $S_1 = H^+$	$\lambda_1 = 1$	$v_{11} = 0$	$v_{12} = 0 \cdots$					$v_{1k} = 0$
$S_2 = OH^-$	$\lambda_2 = -1$	$v_{21} = 0$	$v_{22} = 0$					$v_{2k} = 0$
S_3	λ_3	v_{31}	v_{32}					v_{3k}
S_i	λ_i	v_{i1}	v_{i2}	$\cdots$				v_{ik}

strength is given, the activity coefficients are fixed. The activities (a_k) of the principal components, except that of H^+, are also approximately independent of $TOTH$:

1. For solids $a_k = 1$.
2. For gases $a_k = P_k$, always taken as constant.
3. For solutes $a_k = \gamma_k[C_k]$ where the activity coefficient γ_k is constant for a given ionic strength.

Consider the mole balance equation corresponding to C_k (see Tableau 4.10):

$$TOTC_k = [C_k] + v_{3k}[S_3] + v_{4k}[S_4] + \cdots \tag{111}$$

By definition of the principal components, $[C_k]$ is the largest concentration in this equation. If all the other concentrations can be neglected

$$[C_k] \simeq TOTC_k = \text{constant}$$

From our own experience, we know that this approximation holds quite well under many circumstances. There are of course some cases in which two concentrations have approximately the same numerical value, for example, near the pK_a of a weak acid. This is the situation that makes the minor species formula only an approximation of the buffer capacity of a chemical system. We shall examine later the magnitude of the errors that result from it.

Grouping all constant terms into one (X_i), the mass law expression for the minor species S_i becomes a function of $\{H^+\}$ only:

$$[S_i] = X_i \cdot \{H^+\}^{\lambda_i} \tag{112}$$

where X_i is a constant independent of $TOTH$ (and pH). The $TOTH$ equation is then given as an explicit function of $\{H^+\}$ only and the partial differentials can be replaced by total differentials:

$$TOTH = \sum_i X_i \lambda_i \{H^+\}^{\lambda_i} \tag{113}$$

therefore

$$\beta_H = -\left[\frac{\partial pH}{\partial TOTH}\right]^{-1} = -\left[\frac{dpH}{dTOTH}\right]^{-1}$$

$$= -\frac{dTOTH}{dpH} = +2.3\{H^+\}\frac{dTOTH}{d\{H^+\}} \tag{114}$$

The factor of 2.3 is introduced by the differentiation of a base 10 logarithm.

$$\beta_H = 2.3\{H^+\}\sum_i X_i\lambda_i^2\{H^+\}^{\lambda_i - 1} = 2.3\sum_i \lambda_i^2 X_i\{H^+\}^{\lambda_i} \tag{115}$$

$$= 2.3\sum_i \lambda_i^2 [S_i] \quad \text{QED} \tag{116}$$

This intuitive derivation of the minor species formula provides also a definition of what is acidic (proton donor: $\Delta TOTH > 0$) or basic (proton acceptor: $\Delta TOTH < 0$) in a given system. An added chemical constituent behaves as an acid (or a base) in a chemical system if it has a positive (or negative) proton level compared to the principal components of the system. Thus the application of the minor species formula need not be restricted to strong acids and bases; it can be used to calculate small pH variations brought about by any change in the composition of a system:

$$\Delta pH = -\frac{\Delta TOTH}{\beta_H}$$

In this formula $\Delta TOTH$ is calculated as the change in the right-hand side of the $TOTH$ equation corresponding to the principal components of the system. In effect, the minor species formula provides a justification, a generalization, and a quantification of the qualitative reasoning that we carried out in our discussion of the effects of biological processes on pH (Section 7). This is elaborated on further in Section 8.5.

8.3   Buffer Capacity of Weak Acid–Base Solutions

Consider a solution of a weak acid:

$$HA = H^+ + A^-; \qquad K_a \tag{117}$$

For pH's below the pK_a the major species, and hence the principal component, is HA. The buffer capacity is then given according to the minor species theorem:

$$pH < pK_a: \beta_H = 2.3\{[A^-] + [H^+](+ [OH^-]; \text{ small at low pH})\} \tag{118}$$

Conversely for pH's above the pK_a, the buffer capacity is given by:

$$pH > pK_a: \beta_H = 2.3\{[HA] + [OH^-](+ [H^+]; \text{ small at high pH})\} \tag{119}$$

For pH equal to the pK_a, either HA or A^- can be taken as the major or minor species:

$$pH = pK_a: \beta_H = 2.3([HA] + [H^+] + [OH^-]) = 2.3([A^-] + [H^+] + [OH^-]) \tag{120}$$

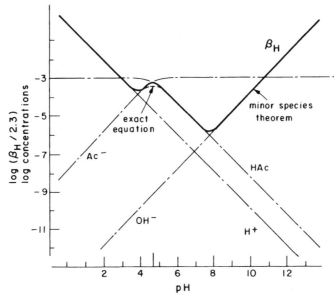

Figure 4.15 Buffer capacity of the acetate system. Except at the pK_a, where the buffer capacity is maximum, the minor species theorem gives values of β_H that are indistinguishable from those of the exact equation.

For such a simple system an exact expression of β_H can actually be derived, valid at any pH:

$$\beta_H = 2.3\left[[H^+] + [OH^-] + \frac{[HA][A^-]}{[HA] + [A^-]} \right] \tag{121}$$

As can be seen in Figure 4.15, where both this equation and the approximation given by the minor species formula are plotted on a $\log C$–pH graph, the minor species formula provides an excellent approximation of the buffer capacity over the whole pH range. The fit of the two formulae is worst when there is no dominant species, that is, when the pH is equal to the pK_a. In this case the minor species formula overestimates the buffer capacity by a factor of 2. This is true even in very complex cases. For applications in which more precision is needed, a correction factor can be applied to the value of β_H obtained from the formula. Note that the buffer capacity of a weak acid–base solution is maximum at its pK_a where it is equal to one quarter the total acid–base concentration ($\times 2.3$). One pH unit away from the pK_a, the buffer capacity is reduced to one tenth of the total acid–base concentration ($\times 2.3$). This provides a convenient rule of thumb for choosing appropriate buffers in the laboratory.

8.4   Buffer Capacity of Heterogeneous Systems

The buffer capacity of heterogeneous systems has been studied relatively little, partly because slow reactions make experimental studies difficult, partly because the mathematical complexity of the theoretical calculations appears formidable. Paradoxically, the minor species formula is especially easy to apply to such systems as the choice of principal components is made more obvious by the presence of solids and gases.

Consider, for example, a carbonate-bearing water at equilibrium with a fixed partial pressure of CO_2. With CO_2 as principal component, the buffering capacity of such a system is given directly by

$$\beta_H = 2.3([HCO_3^-] + 4[CO_3^{2-}] + [H^+] + [OH^-]) \tag{122}$$

For many natural waters this quantity is practically equal to the alkalinity ($\times 2.3$). It is much larger than the buffering capacity of the aqueous phase alone which, in the neutral pH range, is obtained by applying Equation 108 to the carbonate system with HCO_3^- as the principal component:

$$\beta_H = 2.3([H_2CO_3^*] + [CO_3^{2-}] + [H^+] + [OH^-]) \tag{123}$$

At pH = 8 (ca. seawater pH) the buffer capacity of the carbonate system at equilibrium with the gas phase is roughly a hundred times greater than that of the isolated aqueous phase. Note that the buffer capacity of the system is also increased by the presence of carbonate complexes, since the concentration of such species are also included in the expression of β_H.

Consider now a carbonate system at equilibrium not only with a fixed partial pressure of CO_2 but also with a calcium carbonate solid. Taking both CO_2 and $CaCO_3$ as principal components, we have

$$\beta_H = 2.3(4[Ca^{2+}] + [HCO_3^-] + 4[CO_3^{2-}] + [H^+] + [OH^-]) \tag{124}$$

If more heterogeneous reactions are considered, the buffering capacity of the system increases as more soluble species are included in the expression of β_H, in this case $[Ca^{2+}]$.

At the limit, if all dissolved species in an aquatic system are controlled by the solubility of solid and gas phases, an expression of the maximum possible buffering capacity is obtained. Consider a complete set of principal components including only H^+, gases, and solids. (It is shown in Chapter 5 that this set contains the maximum number of solid and gas phases that can be at equilibrium with the solution.) Then the proton level λ_i of each species is exactly equal to its electrical charge, and β_H is given by

$$\beta_H^{max} = 4.6I \tag{125}$$

where $I = \frac{1}{2}\sum \lambda_i^2 [S_i]$ is the ionic strength of the system. Of course this theoretical maximum value of the buffering capacity of an aquatic system may not be attained in fact, even if we consider the geochemical processes that control the composition of the water over geological time. For example, in most waters the concentration of the chloride ion is probably not controlled by the solubility of any mineral phase. Nonetheless, the expression of β_H^{max} shows that the buffer capacity of a system cannot be larger than its ionic strength ($\times 4.6$) and this value is a rough estimate of the buffer capacity of the system in which all heterogeneous geochemical processes are considered operative.

8.5 Applications and Limitations of the Minor Species Theorem

Recall Examples 6 and 10 where we calculated changes in pH due to changes in the composition of a system. We shall apply the minor species formula to these examples to illustrate the methodology and gain some insight into the applicability of this differential approach to calculating pH changes.

In Example 6 the system is at equilibrium with the $CO_2(g)$ in the atmosphere, but not with $H_2S(g)$. The principal components at pH $= 8$, H^+, CO_2, HS^-, and the initial composition, $[HCO_3^-] = 10^{-3.3} M$, $[CO_3^{2-}] = 10^{-5.6} M$, $[H_2S] = 10^{-4.74} M$, $[S^{2-}] = 10^{-9.64} M$, yield the buffer capacity:

$$\beta_H = 2.3([HCO_3^-] + 4[CO_3^{2-}] + [H_2S] + [S^{2-}] = 1.22 \times 10^{-3} M$$

The reaction of escape of hydrogen sulfide gas written with the principal components

$$HS^- + H^+ \rightarrow H_2S(g)$$

yields

$$\Delta TOTH = -S_T = -2 \times 10^{-4} M$$

$$\therefore \Delta pH \simeq +\frac{2 \times 10^{-4}}{1.22 \times 10^{-3}} \simeq +0.16$$

By comparing with the result of Example 6 ($\Delta pH = +0.14$), we can see that the application of the minor species theorem gives a reasonably accurate answer in this case.

In Example 10 where the pH change is brought about by the reduction of sulfate to sulfide, at the initial pH of 7.5, the principal components of the system are H^+, HCO_3^-, HS^-, and SO_4^{2-}. Thus

$$\beta_H = 2.3([H_2CO_3^*] + [CO_3^{2-}] + \text{no sulfide initially}) = 1.49 \times 10^{-4} M$$

The sulfate reduction reaction with the principal components

$$SO_4^{2-} + 2\text{"}CH_2O\text{"} \rightarrow HS^- + 2HCO_3^- + H^+$$

yields

$$\Delta TOTH = -\Delta[SO_4^{2-}]_T = +10^{-4}$$

$$\Delta pH = -\frac{10^{-4}}{1.49 \times 10^{-4}} = -0.67$$

This calculated change in pH is markedly larger than the actual one (-0.25), a reflection of the fact that a sizeable fraction of the sulfide formed remains undissociated as H_2S (hence $\Delta TOTH$ is effectively smaller than $10^{-4} M$) and that the buffering capacity that is provided by the sulfide itself as it forms has not been accounted for (i.e., β_H is effectively larger than $1.49 \times 10^{-4} M$).

Overall it is apparent that the calculation of pH changes on the basis of the buffer capacity of a system provides accurate answers only when these pH changes are relatively small; say $|\Delta pH| < 0.2$. One should also be wary of changes in compositions where weak acids and bases (e.g., H_2S) are introduced into a system, thus changing its buffer capacity. Nonetheless, the direction of pH change and its rough magnitude can always be calculated conveniently by applying the methodology outlined above. In particular, buffer capacity calculations provide a quick method to estimate whether or not the pH will change appreciably.

9. EXCHANGE OF GASES AT THE AIR–WATER INTERFACE

At any given time, most natural waters are not at equilibrium with the atmosphere because the aquatic processes that consume or produce volatile compounds are often faster than gas exchange. For example, on a summer day, the rate of photosynthesis by phytoplankton is usually fast enough to result in supersaturation of oxygen and undersaturation of CO_2 in the euphotic (well lit) zones of lakes and oceans. Conversely, the rate of respiration during the night can cause O_2 depletion and CO_2 supersaturation. As a result of these and other processes (microbial activity, temperature changes, etc.), there is normally a diurnal and seasonal cycle of gas exchange in natural water bodies.

The general expression for the exchange of a nonreactive volatile species between a gaseous and liquid phase is of the form

$$J_g = k_g(C_g^s - C_g) \tag{126}$$

where J_g is the rate of transfer of the gas per unit area $(\text{mol cm}^{-2} \text{s}^{-1})$, k_g is the rate transfer coefficient (cm s^{-1}), C_g^s is the dissolved concentration of the gas (mol L^{-1}) at equilibrium with the partial pressure P_g in the gas phase, and C_g is the dissolved gas concentration in the liquid phase (mol L^{-1}).

This empirical rate law can be rationalized by physical models of the exchange process. Various models provide different interpretations for the transfer coefficient k_g as a function of the properties of the volatile species (primarily its molecular diffusion coefficient D_g) and the characteristics of the fluid motion, usually restricted to the liquid phase unless the species is extremely volatile. These models are of two general types:

1. *The surface film model* in which gas exchange is viewed as limited by molecular diffusion through a quiescent surface film. The transfer coefficient is then interpreted as the ratio of the molecular diffusion coefficient and the thickness Z of the boundary layer:

$$k_g = \frac{D_g}{Z} \tag{127}$$

2. *The surface renewal model(s)* in which turbulent eddies periodically mix the surface layer with the bulk solution. The transfer coefficient is then interpreted as the square root of the ratio of the molecular diffusion coefficient D_g and a characteristic time θ, representing the frequency of renewal of the surface film by bulk solution due to turbulence:

$$k_g = \left(\frac{D_g}{\theta} \right)^{1/2} \tag{128}$$

Various surface renewal models differ primarily in the underlying assumptions for the distribution of surface renewal times, leading to different formulae for the effective parameter θ.

Conclusive experimental evidence proving the correctness or incorrectness of either model is still lacking. Although available data may be construed to favor slightly the surface renewal model, the great conceptual simplicity of the surface film model makes it a better heuristic tool, and it is the one that we are presenting succinctly in the following sections.

9.1 Governing Equation

By assuming that the transfer of a gas between water and atmosphere is limited by molecular diffusion through a water interfacial laminar layer, we obtain a mathematical description of gas transfer kinetics. For steady state conditions under a number of simplifying assumptions, the rate of gas transfer may be described by an equation of the form

$$J_g = 10^3 \frac{D}{Z} (C_g^s - C_g) \tag{129}$$

where J_g is the rate of transfer (flux) of the gas per unit area ($\mathrm{mol\,cm^{-2}\,s^{-1}}$), D is the molecular diffusion coefficient of the gas ($\mathrm{cm^2\,s^{-1}}$), C_g^s is the soluble concentration ($\mathrm{mol\,L^{-1}}$) of the gas in equilibrium with the given partial pressure

P_g, C_g is the concentration of the gas in the bulk aqueous phase (mol L^{-1}), and Z is the average depth of the water laminar boundary layer through which the diffusion is taking place (cm). The factor of 10^3 is for consistency among the various units.

With the exception of NH_3 and SO_2, which are extremely soluble, the gases of interest in aquatic systems ($O_2, N_2, CO_2, H_2S, CH_4, NO_x$) are sufficiently volatile that no boundary layer in the gas phase need be considered. For example, in the case of transfer from air to water, there is no microzone of gas depletion on the air side of the interface, as the rate of mixing in the air is much faster than the rate of dissolution into the water.

Many of the assumptions leading to Equation 129 are rarely, if ever, satisfied in natural waters. Still, the form of the equation is worth keeping as it conveniently separates parameters describing the hydrodynamic regime of the system (Z) from those pertaining to the particular gas of interest. Although the parameter Z may not strictly correspond to the thickness of an actual laminar boundary layer, it describes the influence of the mixing regime—the turbulence of the water—on the gas exchange kinetics as a single parameter: the thickness of an equivalent boundary layer. In principle, Z should be the same for all gases in a body of water at a particular point in time. For a given gas only D/Z is measurable directly as the ratio of the gas flux to the concentration gradient, as described by Equation 129.

Because the molecular diffusion coefficients of typical aquatic solutes span a relatively narrow range of values (2 to 5×10^{-5} cm^{2}s^{-1}), the transfer of gases between the atmosphere and surface waters is dominated by the hydrodynamic characteristics of the water and is almost independent of the nature of the gas. Chemical oceanographers who have a propensity for simple physical images like to speak of the transfer coefficient k_g as a "piston velocity," with values in the range of 10^{-4}–10^{-2} cm s^{-1}. For surface water depths on the order of 1–10 m, this represents a characteristic exchange time on the order of 10^4–10^7 s, that is, 2 h–100 days.

9.2 Chemical Enhancement

Equation 129 provides a convenient method of describing gas exchange kinetics for unreactive gases such as O_2, N_2, or CH_4. However, for gases that dissociate into weak acids and bases (CO_2, H_2S, NH_3), it has sometimes been observed experimentally that the gas transfer is more rapid than expected on the basis of the equation, using an independent estimate of the boundary layer thickness. This is interpreted as the effect of chemical reactions within the boundary layer. To take such an effect into account, one can define a chemical enhancement factor (E_g)[20-22] such that the actual gas flux is given by

$$J_g = 10^3 \, E_g \frac{D}{Z}(C_g^s - C_g) \tag{130}$$

Theoretical values of E_g can be obtained by considering the kinetics of the various chemical reactions and the diffusion of all chemical species in infinitesimal sublayers within the laminar boundary layer. The problem is particularly complex as several competing reactions have to be considered simultaneously and such effects as the electrical interactions among ions of different mobilities have to be taken into account. Such sophistication does not seem warranted in face of the intrinsic crudeness of the laminar boundary layer model itself and the imprecision of the experimental measurements. In addition, the parameter E_g is dependent upon the particular chemistry of the system; it varies as the water equilibrates with the atmosphere.

We shall examine here limiting cases to this problem, focusing on the question of CO_2 transfer across the air–water interface. Consider the usual boundary layer model (see Figure 4.16), consisting of an unmixed water layer of thickness Z between an atmosphere of a given fixed P_{CO_2} and a well-mixed bulk solution of known chemistry. As a molecule of CO_2 is dissolved into the upper boundary of the layer, it can either diffuse downward or react with water according to

$$CO_2 + H_2O \underset{}{\overset{k_1}{\rightleftharpoons}} HCO_3^- + H^+ \tag{131}$$

or
$$CO_2 + OH^- \underset{}{\overset{k_2}{\rightleftharpoons}} HCO_3^- \tag{132}$$

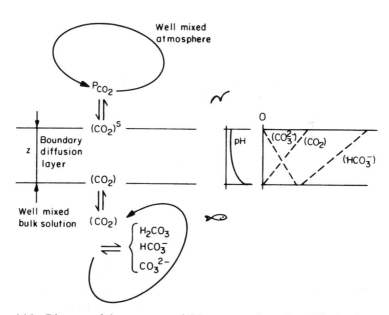

Figure 4.16 Diagram of the transport of CO_2 across a boundary diffusion layer at the air–water interface. The carbonate species concentration profiles are drawn assuming constant alkalinity throughout the diffusion layer. In this illustration the water is undersaturated with respect to the atmosphere, and the net carbon flux is toward the water. Note that the $[CO_3^{2-}]$ gradient is opposite to those $[CO_2]$ and $[HCO_3^-]$, corresponding to the increase in pH with depth.

The forward kinetics of these two reactions, which are known to be the slowest of all those involved in the carbonate system, are described by the equation[23]

$$\frac{d[CO_2]}{dt} = -k_1[CO_2] - k_2[CO_2][OH^-] + \cdots \tag{133}$$

where $k_1 = 3 \times 10^{-2} s^{-1}$ and $k_2 = 8.5 \times 10^3 M^{-1} s^{-1}$ at 25°C.

If the resulting reaction rate is much slower than the diffusion process across the layer, we can effectively ignore the reactivity of CO_2 and treat the gas exchange problem according to Equation (129). If, at the other extreme, the chemical reactions are much faster than diffusion ($D_{CO_2} = 2 \times 10^{-5} cm^2 s^{-1}$), then we can treat the problem by considering equilibrium at every point in the layer (equilibrium enhancement). If the chemical and transport rates are comparable, we then have no choice but to consider both simultaneously in a rather complex model. The critical parameter determining which of these cases is applicable is the depth Z of the boundary layer:

1. The characteristic time for diffusion through the boundary layer is given by the ratio of the total number of moles of CO_2 in a given volume of the boundary layer to the flux through that volume:

$$t_{diffusion} \simeq \frac{[CO_2]Z(\text{unit area})}{J_{CO_2}(\text{unit area})} \simeq \frac{Z^2}{D} \tag{134}$$

2. The characteristic time for the chemical reaction of CO_2 is given directly by the forward kinetic coefficients:

$$t_{reaction} \simeq (k_1 + k_2[OH^-])^{-1} \tag{135}$$

3. Equating the two time scales ($t_{diffusion} = t_{reaction}$) yields the critical depth of the diffusion boundary layer:

$$Z_{crit} \simeq \left[\frac{D}{k_1 + k_2[OH^-]}\right]^{1/2} = \left[\frac{1}{1 + 10^{-8.55}/[H^+]}\right]^{1/2} \times 250\,\mu m \tag{136}$$

A logarithmic plot of this equation is shown in Figure 4.17. Only in very quiescent lakes with deep surface boundary layers or in alkaline waters are the chemical reactions faster than the diffusion through the surface layer. In most natural waters, the enhancement of the exchange kinetics of CO_2 due to chemical reactions can be neglected.

9.3   Kinetics of CO_2 Equilibrium in a Water Column

Let us consider the time course of CO_2 equilibration in a well-mixed water column with the typical conditions of negligible chemical enhancement. Carbon

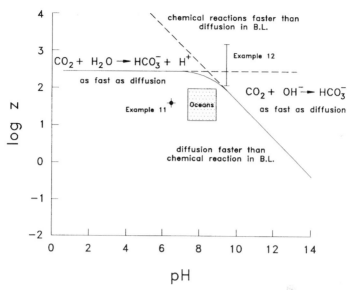

Figure 4.17   Chemical enhancement of CO_2 transport across the air–water interface. In most bodies of water the effective thickness of the diffusion boundary layer (given here as $\log Z$, Z in μm) is sufficiently small that CO_2 diffusion is rapid compared to the kinetics of HCO_3^- formation and chemical enhancement is negligible.

dioxide is then the only species transported across the boundary layer; chemical reactions in the bulk solution are much faster than the CO_2 transport and equilibrium can be assumed among the carbonate species.

Example 11   Given a water body of average depth, $h = 10\,m$, with sufficient mixing to ensure uniform concentrations throughout, and the following initial conditions, how will the carbonate concentration in the water change as a function of time?

$$P_{CO_2} = 10^{-3.5}\,\text{atm}$$

$$\text{Alk} = 10^{-3.5}\,\text{eq}\,L^{-1}$$

$$pH^0 = 6.5$$

Boundary layer thickness: $Z = 40\,\mu m$ (typical range: 20–1000 μm)

$$D_{CO_2} = 2 \times 10^{-5}\,\text{cm}^2\,\text{s}^{-1}$$

Using our previous notation to study the chemistry, let us make the approximation

$$[CO_2]_{aq} \simeq [H_2CO_3^*]$$

In this system for pH < 9, the following approximations are easily justified:

$$Alk = [HCO_3^-]$$

$$C_T = [H_2CO_3^*] + [HCO_3^-]$$

$$C_T^0 = Alk(10^{6.3} \times 10^{-6.5} + 1) = 10^{-3.3}$$

$$[CO_2] = [H_2CO_3^*] = C_T - Alk$$

Using dm ($= 10^{-1}$ m) throughout as unit of length since $1\,dm^3 = 1\,L$:

$$\frac{dC_T}{dt} = J_{CO_2} \frac{Area}{Volume} = \frac{J_{CO_2}}{h} = \frac{J_{CO_2}}{100}$$

therefore

$$\frac{dC_T}{dt} = 10^{-2} \frac{D_{CO_2}}{Z} ([CO_2]^s - [CO_2])$$

where

$$10^{-2} \frac{D_{CO_2}}{Z} = 10^{-5.3}\,s^{-1} = 10^{-1.75}\,h^{-1}$$

$$[CO_2]^s = 10^{-1.5} P_{CO_2} = 10^{-5}\,M$$

$$[CO_2] = C_T - Alk = C_T - 10^{-3.5}$$

$$\frac{dC_T}{dt} = 10^{-1.75}[10^{-3.5} - C_T]$$

The solution of this differential equation for the initial condition $C_T^0 = 10^{-3.3}\,M$ is given by

$$C_T = 10^{-3.5} + 10^{-3.7} e^{-0.018t} \qquad (t\text{ in hours}) \qquad (137)$$

This solution is plotted in Figure 4.18 where it can be seen that about five days are necessary to approach equilibrium with the atmosphere. At each point in time, the concentration of all solution species (including H^+) can be computed from C_T and Alk by assuming chemical equilibrium in the bulk solution phase.

9.4   Equilibrium Enhancement of CO_2 Exchange

As seen in Figure 4.17, it is possible to find conditions where the chemical reactions of the carbonate system in the diffusion layer are in fact faster than the diffusion itself. In such a case a very simplified treatment consists of assuming chemical equilibrium at each boundary of the layer and adding together the

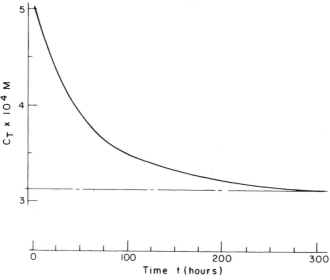

Figure 4.18 Kinetics of CO_2 equilibration (with no chemical enhancement) in a well-mixed 10-m water column with an effective diffusion boundary layer of $40\,\mu m$ (Example 11).

diffusion fluxes of each of the carbonate species:

$$J_i = \frac{D_i}{Z}([S_i]^s - [S_i])$$ (138)

where i refers to the various carbonate species.

If we consider the constraint of electroneutrality and neglect the problem of electrostatic interactions among diffusing ions, it is reasonable to assume that the alkalinity is constant throughout the diffusion layer and equal to the alkalinity of the bulk solution. This assumption is particularly good if the ionic strength is markedly higher than the carbonate species concentrations.

All the diffusion coefficients being approximately equal ($\simeq 2 \times 10^{-5}\,cm^2\,s^{-1}$), the enhancement factor is simply the ratio of the C_T gradient to the CO_2 gradient:

$$J_T = \sum_i J_i = \frac{D}{Z}[C_T^s - C_T]$$ (139)

$$J_{CO_2} = \frac{D}{Z}([CO_2]^s - [CO_2])$$ (140)

$$E_g = \frac{J_T}{J_{CO_2}} = \frac{C_T^s - C_T}{[CO_2]^s - [CO_2]}$$ (141)

Example 12 Consider the conditions in a quiescent lake:

$$P_{CO_2} = 10^{-3.5} \text{ atm}$$
$$\text{Alk} = 10^{-4} \text{ eq L}^{-1}$$
$$\text{pH} = 9.5$$

The boundary conditions at the air interface are defined by

$$P_{CO_2} = 10^{-3.5} \text{ atm}$$
$$\text{Alk} = 10^{-4} \text{ eq L}^{-1}$$

therefore

$$\text{pH}^s = 7.3$$
$$[H_2CO_3^*]^s = 10^{-5} M$$
$$[HCO_3^-]^s = 10^{-4} M$$
$$[CO_3^{2-}]^s = 10^{-7} M$$

On the bulk water side of the interface, the boundary conditions are given by

$$[H^+] = 10^{-9.5}$$
$$\text{Alk} = 10^{-4} \text{ eq L}^{-1}$$

therefore

$$[H_2CO_3^*] = 10^{-7.49} M$$
$$[HCO_3^-] = 10^{-4.29} M$$
$$[CO_3^{2-}] = 10^{-5.09} M$$

Note that the CO_3^{2-} gradient is opposite to the $H_2CO_3^*$ and HCO_3^- gradients. The total carbonate and carbon dioxide gradients are obtained by difference

$$C_T^s - C_T = 5 \times 10^{-5}$$
$$[CO_2]^s - [CO_2] = [H_2CO_3^*]^s - [H_2CO_3^*] = 10^{-5}$$

yielding a maximum enhancement factor $E_g = 5$. Actual experimental laboratory measurements of such a system[21] have yielded $E_g \simeq 3.7 - 6.5$.

* * *

Although this discussion is limited to the coupling of chemical reactions with the surface film model, it has been demonstrated that the influence of chemical reactions on the rate of gas transfer are practically the same in all physical models of gas transfer.[24] In the case of equilibrium enhancement where the chemical reactions are considered infinitely fast, the very same result is applicable regardless of the model: the volatile species gradient can simply be replaced by the total concentration gradient in the transfer rate equation. This provides a convenient way to evaluate an upper limit on the possible role of chemical reactions in the transport of reactive solutes.

9.5 Ebullition

Formation of bubbles by air entrainment in breaking waves may greatly enhance the kinetics of gas exchange at the air–water interface. Volatile species may also be released to the atmosphere through the formation of bubbles which rise through the water column and escape at the surface. Such bubbles are formed spontaneously from supersaturated gases in the water. This process, known as ebullition, provides an efficient means of gas release from the water to the air, particularly for volatile species such as methane that result from biotic fermentation in anoxic sediment layers.

A necessary condition for a bubble to form in water is that the sum of the partial pressures of the volatile species be in excess of the ambient hydrostatic pressure. For example, in a system where the principal volatile species besides H_2O itself are N_2, CO_2, and CH_4, this condition can be written

$$P_{N_2} + P_{CO_2} + P_{CH_4} + P_{H_2O} > P_z \tag{142}$$

where $P_z = (P_{atm} + \rho g z)$ is the hydrostatic pressure at the given depth z. This inequality may become verified either because the hydrostatic pressure on the right-hand side is decreased mechanically, or because the partial pressures of gases are increased by the formation in situ of the volatile species. The former case is commonly observed when bringing up sediment cores from anoxic sediments. It also occurs in littoral zones subjected to tidal regimes.[25] At ebb tide, a relatively rapid decrease in ambient pressure creates a large supersaturation of the volatile species that were previously at or below saturation in shallow sediments. This supersaturation overcomes the energetic barrier created by the large surface energies of small bubbles (see the discussion of nucleation energy in Chapter 5) and leads to rapid bubble formation.

In contrast, in methanogenic sediments (see Chapter 7) the gradual buildup in methane partial pressure leads to the formation of bubbles at the surface of solids through heterogeneous rather than homogeneous nucleation. In such systems the sum of the partial pressures of the volatile species is thus effectively maintained at or near the ambient hydrostatic pressure corresponding to the depth z:

$$P_{N_2} + P_{CO_2} + P_{CH_4} + P_{H_2O} = P_z = P_{atm} + \rho g z \tag{143}$$

The composition of the bubbles is given by the relative magnitude of each of the terms in Equation 143 and thus reflects the volatility of each species as well as the composition of the water. Since methane fermentation produces CH_4 and CO_2 roughly in equal proportions (see Chapter 8), and methane is about 20 times more volatile than carbon dioxide, the resulting bubbles contain a large excess of methane. Thus ebullition is an effective means for CH_4 release. In contrast, surface exchange normally dominates the release of CO_2, which is much more soluble than CH_4 and thus more effectively transported by diffusion to the surface as a dissolved species.

Example 13. Ebullition in a Methanogenic Marsh   To illustrate quantitatively the effect of ebullition on gas transfer and water composition, consider a 50-cm deep anoxic acidic marsh. Methane fermentation in the sediments produces CH_4 and CO_2 in equimolar proportions at a rate $m = 5 \times 10^{-11} \, mol \, cm^{-2} \, s^{-1}$. At the sediment surface, which can be considered a point source of these two gases, heterogeneous bubble formation maintains the sum of the partial pressures at the ambient hydrostatic pressure (1.05 atm) according to Equation 143. In the water column the bubbles rise so fast to the surface that no dissolution occurs. Transport of solutes in the water column is described by a dispersion coefficient $D = 10^{-3} \, cm^2 \, s^{-1}$, valid for all dissolved species. Chemical reactions such as bicarbonate formation from CO_2 or methane oxidation are ignored. The atmospheric partial pressures of CO_2, CH_4, and N_2 are $10^{-3.5}, 10^{-5.7}$, and $10^{-0.1}$ atm, respectively.

For each gas, the rate of ebullition ε_i ($mol \, cm^{-2} \, s^{-1}$) is proportional to its partial pressure at the sediment surface

$$\varepsilon_i = \frac{P_i}{P_z} \varepsilon$$

where ε is the total rate of ebullition for all the gases together. A steady state mass balance equation for each species at the sediment surface can be written by expressing that the sum of its transport by diffusion and by ebullition must be equal to its rate of formation from methanogenesis:

$$\frac{D}{z}([CH_4]_b - [CH_4]_s) + \varepsilon K_{CH_4}^{-1}[CH_4]_b P_z^{-1} = m \qquad (144)$$

$$\frac{D}{z}([CO_2]_b - [CO_2]_s) + \varepsilon K_{CO_2}^{-1}[CO_2]_b P_z^{-1} = m \qquad (145)$$

$$\frac{D}{z}([N_2]_b - [N_2]_s) + \varepsilon K_{N_2}^{-1}[N_2]_b P_z^{-1} = 0 \qquad (146)$$

where z is the depth of the marsh, the subscripts b and s designate the bottom and surface concentrations, and K represents the Henry's law constants (mol L^{-1} atm^{-1}) for the species. The concentrations at the water surface are calculated to be at equilibrium with the atmosphere (see Tables 4.2 and 4.3 for constants):

$$[CH_4]_s \simeq P_{CH_4}^{atm} K_{CH_4} = 2.9 \times 10^{-9} M$$

$$[CO_2]_s \simeq P_{CO_2}^{atm} K_{CO_2} \simeq 10^{-5} M$$

$$[N_2]_s = P_{N_2}^{atm} K_{N_2} \simeq 5.2 \times 10^{-4} M$$

Given the rate of methanogenesis, m, we wish to calculate the concentrations of the species at the bottom of the marsh and their rates of diffusion and ebullition. The unknowns are thus $[CH_4]_b, [CO_2]_b, [N_2]_b$ and ε. A fourth equation is given by the pressure condition at the sediment surface:

$$P_{CH_4} + P_{CO_2} + P_{N_2} + P_{H_2O} = 1.05 \, atm \tag{147}$$

Replacing P_{H_2O} by the saturating water pressure (ca. 0.05 atm) and the other partial pressures by the corresponding Henry's law expressions:

$$K_{CH_4}^{-1} [CH_4]_b + K_{CO_2}^{-1} [CO_2]_b + K_{N_2}^{-1} [N_2]_b \simeq 1 \, atm \tag{148}$$

With the approximations $[CH_4]_s \ll [CH_4]_b$ and $[CO_2]_s \ll [CO_2]_b$, we can substitute the bottom concentrations from Equations 144–146 into Equation 148 and obtain

$$\frac{m}{K_{CH_4} D/z + \varepsilon/P_z} + \frac{m}{K_{CO_2} D/z + \varepsilon/P_z} + \frac{(D/z)[N_2]_s}{K_{N_2} D/z + \varepsilon/P_z} = 1 \, atm \tag{149}$$

This equation can be solved for ε and by substitution back into Equations 144–146 we obtain the concentrations of the species at the bottom of the marsh. An approximate solution for ε is obtained by neglecting the middle term in Equation 149 and replacing K_{CH_4} and K_{N_2} by the average $K \simeq 10^{-6} \, mol \, cm^{-3}$ atm^{-1} (instead of $10^{-2.84}$ and $10^{-3.18} \, M \, atm^{-1}$ respectively).[26]

$$\varepsilon = [m + (D/z)[N_2]_s - KD/z] P_z = 4.2 \times 10^{-11} \, mol \, cm^{-2} \, s^{-1}$$

By substituting this approximate solution in the original equation, we can verify that this solution is actually quite good.
Then

$$[CH_4]_b \simeq \frac{m}{D/z + \varepsilon K_{CH_4}^{-1} P_z^{-1}} \simeq 1.05 \times 10^{-6} \, mol \, cm^{-3} = 1.05 \, mM$$

$$[CO_2]_b \simeq \frac{m}{D/z + \varepsilon K_{CO_2}^{-1} P_z^{-1}} \simeq 2.35 \times 10^{-6} \, mol \, cm^{-3} = 2.35 \, mM$$

$$[N_2]_b \simeq \frac{(D/z)[N_2]_s}{D/z + \varepsilon K_{N_2}^{-1} P_z^{-1}} \simeq 1.29 \times 10^{-7} \, mol \, cm^{-3} = 0.13 \, mM$$

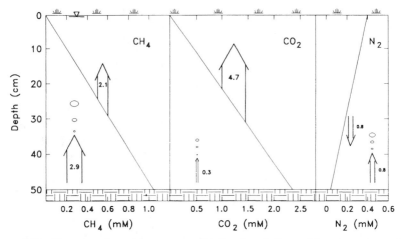

Figure 4.19 Concentrations (shown as solid lines) and fluxes (shown as arrows) of volatile species in anoxic, acidic marsh (Example 13). Methane and carbon dioxide are produced in the sediments; their concentrations are highest at the bottom (depth = 50 cm). Magnitude of fluxes through ebullition (upward for all species) and through diffusion and surface exchange (upward for methane and carbon dioxide, downward for nitrogen) are indicated by the width and direction of the arrows; values for fluxes in units of 10^{-11} mol·cm^{-2}s^{-1}. [Note that the compressed concentration scale for carbon dioxide is not reflected in the width of the flux arrows.]

Thus the diffusive fluxes of the dissolved species to the surface are

$$D/z\,([CH_4]_b - [CH_4]_s) = 2.10 \times 10^{-11} \text{ mole } CH_4\,cm^{-2}\,s^{-1}$$

$$D/z\,([CO_2]_b - [CO_2]_s) = 4.70 \times 10^{-11} \text{ mole } CO_2\,cm^{-2}\,s^{-1}$$

$$D/z\,([N_2]_b - [N_2]_s) \simeq -0.78 \times 10^{-11} \text{ mole } N_2\,cm^{-2}\,s^{-1}$$

and the ebullition fluxes are

$$\varepsilon K_{CH_4}^{-1}[CH_4]_b\,P_z^{-1} = 2.9 \times 10^{-11} \text{ mol } CH_4\,cm^{-2}\,s^{-1}$$

$$\varepsilon K_{CO_2}^{-1}[CO_2]_b\,P_z^{-1} = 0.30 \times 10^{-11} \text{ mol } CO_2\,cm^{-2}\,s^{-1}$$

$$\varepsilon K_{N_2}^{-1}[N_2]_b\,P_z^{-1} = 0.78 \times 10^{-11} \text{ mol } N_2\,cm^{-2}\,s^{-1}$$

$$\varepsilon P_{H_2O}\,P_z^{-1} = 0.20 \times 10^{-11} \text{ mol } H_2O\,cm^{-2}\,s^{-1}$$

Thus CO_2 is vented chiefly through diffusion and surface exchange with the atmosphere but more than half the CH_4 is vented by ebullition. The bubbles are made up of 69% CH_4, 19% N_2, 5% H_2O, and only 7% CO_2. In this example, the major gases in the bubbles emanating from the methane-fermenting sediments are CH_4 and, paradoxically, N_2 which originates from the atmosphere and diffuses downward as a dissolved species (Figure 4.19). In deeper waters the diffusive fluxes of dissolved gases would be slower, thus decreasing the contribution of N_2 to ebullition and increasing that of CO_2.

Other trace gases that may be formed in such reduced systems include H_2, CO, H_2S, N_2O, and NH_3. Hydrogen sulfide has not been found as a dominant gaseous species in bubbles in natural waters. As discussed in Chapter 7, the high pH conditions that result from H_2S formation are also conducive to H_2S dissociation (to HS^-) rather than volatilization.

REFERENCES

1. H. D. Holland, *The Chemistry of the Atmosphere and Oceans*, Wiley, New York, 1978.
2. J. P. Riley and G. Skirrow, Eds., *Chemical Oceanography*, Academic, New York, 1965.
3. G. E. Hutchinson, *A Treatise on Limnology*, Wiley, New York, 1974.
4. J. P. Riley and C. Skirrow, *Chemical Oceanography*, 2nd ed., Academic, London, 1975.
5. F. A. Richards and B. B. Benson, *Deep Sea Res.*, **7**, 254 (1961).
6. R. M. Smith and A. E. Martell, *Critical Stability Constants*, Vol. 4, *Inorganic Ligands*, Plenum, New York, 1976.
7. G. Gran, *Analyst*, **77**, 661 (1952).
8. R. E. Dickinson and R. J. Cicerone, *Nature*, **319**, 109 (1986).
9. J. H. Seinfeld, *Atmospheric Chemistry and Physics of Air Pollution*, Wiley, New York, 1986, pp. 8, 99.
10. R. F. Weiss, *Deep Sea Res.*, **17**, 721 (1970).
11. D. A. Wiesenburg and N. L. Guinasso, Jr., *J. Chem. Eng. Data*, **24**, 356 (1979).
12. W. Stumm and J. J. Morgan, *Aquatic Chemistry*, 2nd ed., Wiley, New York, 1981, p. 109
13. W. Stumm, L. Sigg, and J. L. Schnoor, *Environ. Sci. Technol.*, **21**, (1987).
14. L. Sigg, W. Stumm, J. Zobrist, and F. Zurcher, *Chimia*, **41**, 159 (1987).
15. B. G. Oliver, E. M. Thurman, and R. L. Malcolm, *Geochim. Cosmochim. Acta*, **47**, 2031 (1983).
16. B. M. Bartschat, S. E. Cabaniss, and F. M. M. Morel, *Environ. Sci. Technol.*, **26**, 284 (1992).
17. H. F. Hemond, *Environ. Sci. Technol.*, **24**, 1486 (1990).
18. A. C. Redfield, *James Johnston Memorial Volume*, Liverpool University Press, Liverpool, UK, 1934.
19. F. M. M. Morel, R. E. McDuff, and J. J. Morgan, *J. Mar. Chem.*, **4**, 1 (1976)
20. J. A. Quinn and N. C. Otto, *J. Geophys. Res.*, **76**, 1539 (1971).
21. S. Emerson, *Limnol. Oceanogr.*, **20**, 743 (1975).
22. S. Emerson, *Limnol. Oceanogr.*, **20**, 754 (1975).
23. W. Stumm and J. J. Morgan, *Aquatic Chemistry*, 2nd ed., Wiley, New York, 1981, p. 211.
24. P. V. Danckwerts, *Gas–Liquid Reactions*, McGraw-Hill, New York, 1970.
25. C. S. Martens, G. W. Kipphut, and J. Val Klump, *Science*, **208**, 285 (1980).
26. H. F. Hemond, T. P. Army, W. K. Nuttle, and D. G. Chen, "Element Cycling in

Wetlands: Interaction with Physical Mass Transport," in *Sources and Fates of Aquatic Pollutants*, H. Hites and S. Eisenreich, Eds., *Advances in Chemistry Series*, Vol. 216, American Chemical Society, Washington, DC, 1987.

PROBLEMS

4.1   What is the alkalinity in each of the following systems?

    a. $[NaOH]_T = 10^{-3} M$; $[NaHCO_3]_T = 10^{-2} M$; $P_{CO_2} = 10^{-2.5}$ atm

    b. $[Na_2CO_3]_T = 10^{-4} M$; $[NH_3]_T = 10^{-2} M$; $P_{CO_2} = 10^{-2.5}$ atm

    c. Carbonate system: $P_{CO_2} = 10^{-3.5}$ atm; pH = 7.3

    d. $[H_2SO_4]_T = [NaHCO_3]_T = 10^{-3} M$

4.2   Calculate the pH given:

    a. $Alk = C_T = 10^{-3} M$

    b. $[NaOH]_T = [CO_2]_T = 10^{-3} M$

    c. $P_{CO_2} = 10^{-3.5}$ atm; $Alk = 10^{-3} M$

    d. $Alk = -10^{-2} M$ (mineral acidity)

4.3   Consider the approximate photosynthetic reaction: $CO_2 + H_2O \rightarrow$ "CH_2O" $+ O_2$. What are the qualitative changes in pH in the following systems?

    a. $C_T = 10^{-3} M$; pH = 7.3

    b. $P_{CO_2} = 10^{-3.5}$ atm; pH = 7.3

    c. $CaCO_3(s)$ at equilibrium; $P_{CO_2} = 10^{-3.5}$ atm

    d. $CaCO_3(s)$ at equilibrium; pH = 11.0; $[Ca^{2+}] = 10^{-3} M$

4.4   What are the alkalinity changes introduced by each of the following reactions, each proceeding to the right by $10^{-3} M$ of the first reactant (solids precipitate and gases escape the system)?

    a. $HS^- + H^+ \rightarrow H_2S(g)$

    b. $SO_4^{2-} + 2$ "CH_2O" $\rightarrow HS^- + 2HCO_3^- + H^+$

    c. $NH_4^+ \rightarrow NH_3(g) + H^+$

    d. $Ca^{2+} + 2HCO_3^- \rightarrow CaCO_3(s) + CO_2(g) + H_2O$

4.5   **a.** Using the appropriate data from Table 4.5, provide the alkalinity expression (left-hand side) for a carbonate system containing dissolved aluminum and ferric hydroxide species.

    b. What are the effects of $Al(OH)_3(s)$ and $Fe(OH)_3(s)$ precipitation on alkalinity? What are the qualitative effects on pH, for pH $\simeq 7.0$? For pH $\simeq 9.0$?

4.6   Planet Thethys has a nitrogen and hydrogen sulfide atmosphere ($P \simeq 1$ atm), liquid water, and photosynthetically supported life. Thethysian aquatic chemists have thus defined the alkalinity expression with respect to the H_2S equivalence point.

a. Write this alkalinity expression (left-hand side) including necessary carbonate and phosphate species.

b. What is the alkalinity of the system

$$[Na_2S]_T = 10^{-3}\,M; \qquad [CaCO_3]_T = 10^{-4}\,M$$

c. What is the effect on alkalinity of $CO_2(g)$ exchange with the atmosphere?

d. What is the effect on alkalinity of the photosynthetic reaction

$$CO_2(g) + \tfrac{2}{3}H_2S(g) + \tfrac{1}{3}H_2O \rightarrow \text{``}CH_2O\text{''} + \tfrac{2}{3}SO_2(g)$$

e. For alkalinity $\cong 10^{-3}\,M$ and $P_{H_2S} = 10^{-3}$ atm, what is the average pH in Thethysian lakes? What are the qualitative effects of $CO_2(g)$ exchange and photosynthesis on this pH?

4.7 An acid mine drainage ($[H_2SO_4]_T = 10^{-4}\,M$, $P_{CO_2} = 10^{-1.5}$ atm) mixes 1:2 with a stream characterized by $P_{CO_2} = 10^{-1.5}$ atm, pH = 6.0. What are the alkalinity and the pH of (a) the acid mine drainage, (b) the stream, (c) the mix at the point of confluence, (d) the mix after cascading down the hill and equilibrating with the atmosphere, and (e) the mix after standing in a small lake and supporting the growth of 3 mg L^{-1} of algal biomass (NO_3^- as N source)?

4.8 Yesterday you prepared a $10^{-2}\,M$ NaOH solution in distilled water.

a. What pH did you measure right away?

b. The pH measured today is only 10; why?

c. What will be the ultimate pH?

4.9 Using apparent constants appropriate for seawater (see Chapter 6), derive a plot of pH versus C_T, given Alk = $2.5 \times 10^{-3}\,M$ (C_T from 0 to $3 \times 10^{-3}\,M$). How good a measure of photosynthesis and respiration would a surface seawater pH be? Discuss.

4.10 Calculate the small change in pH due to the addition of $10^{-5}\,M$ $FeCl_3$ in a system at equilibrium with $Fe(OH)_3(s)$ at pH = 7.3 and $C_T = 10^{-3}\,M$.

4.11 Your backyard pond (a nice little water body with a stable groundwater source) had a pH of 8.2 in April. The pH measured now (November) is 7.5, and when you bubble a sample with air, the pH goes to 7.7.

a. Why do these pH differences occur?

b. What would have been the measured pH if you had bubbled a sample in April?

c. Calculate the alkalinity and the total carbonate content of the pond in April and November.

4.12   An industry discharges 700 kg of ammonium chloride (NH_4Cl) per hour into a stream. For the purpose of this problem, the stream is divided into three regions (flow rate = $2m^3 s^{-1}$):

1. Upstream from the discharge where the water is at equilibrium with the atmosphere (0.032% CO_2; 0.0000% NH_3).
2. Some distance downstream from the discharge where the waste has been mixed and equilibrated with the water, but air–water equilibrium has not had time to be reestablished.
3. Farther from the discharge where air–water equilibrium has been reestablished. Questions:

   a. The pH measured in region 1 is 9.4. Supposing that sodium is the main cation in the stream, what is the concentration of sodium?
   b. Which way will the pH vary as we measure farther downstream: region 1? 2? 3?
   c. List species, write appropriate mole balance equations, and compute the pH in region 2.
   d. Compute the pH in region 3.

4.13   The epilimnion of your backyard pond has a pH $\simeq 7.0$ and alkalinity $\simeq 10^{-3} M$.

   a. Assuming that the carbonate system dominates the chemistry of your pond, calculate the concentrations of all species. Is your pond at equilibrium with the atmosphere? Explain.
   b. You just washed the bricks of your house walls with HCl. How much acid can you add to your pond before changing the pH by more than 0.2 units (get an approximate answer by considering the buffering capacity).
   c. The hypolimnion of your pond is anoxic. Its composition is derived from that of the epilimnion by reducing $3 \times 10^{-4} M$ of SO_4^{2-} to H_2S. What are the alkalinity, the pH, and the concentrations of all species in the hypolimnion?
   d. In the fall the epilimnion (1 m depth) and the hypolimnion (2 m depth) mix with minimal gas exchange. What is the composition (Alk, pH, concentrations) of the resulting water?
   e. What would the composition of the mixed water be if you forced mixing by bubbling air at the bottom of the pond (assume gas exchange, not oxidation).

4.14   Consider a carbonate solution: pH = 8.0; $C_T = 5.0 \times 10^{-4} M$.

   a. Calculate the alkalinity and the detailed composition of the system.
   b. Calculate and draw a precise acidimetric titration curve (pH versus $[HCl]_T$), considering C_T = constant.
   c. Calculate and draw a precise acidimetric titration curve for $P_{CO_2} = 10^{-3.51}$ atm = constant.

d. Suppose that the solution is contained in a 1-L stoppered bottle and that the pressure of the enclosed gas (approximately 10 mL) is monitored. Draw an acidimetric titration curve (P_{CO_2} vs. $[HCl]_T$).

e. Discuss the relative merits of the three corresponding methods for alkalinity determination.

4.15 The following data were obtained from analysis of a rainwater sample collected in Pittsburgh, Pennsylvania, in September 1979:

Ion	mg L^{-1}	mol L^{-1} ($\times 10^5$)
		Concentration
H$^+$	(pH = 4.19)	6.46
NH$_4^+$	1.75	12.5
Na$^+$	0.337	1.46
K$^+$	0.741	1.90
Ca^{2+}	3.00	7.50
Mg^{2+}	0.855	3.56
SO$_4^{2-}$	13.0	13.5
NO$_3^-$	2.15	15.4
Cl$^-$	0.44	1.23

a. Compute the $\sum$ cation equivalent/$\sum$ anion equivalent; what does this reveal?

b. Compute the mineral acidity of this sample.

c. Sketch an *approximate* titration curve assuming:

That the titration is carried out with 0.01 N NaOH on a 100-mL sample in a closed jacketed beaker.

That high-purity nitrogen gas is bubbled through the sample so that the titration is performed under a nitrogen atmosphere.

That the given pH remains essentially unchanged after the onset of N$_2$(g) bubbling and prior to the addition of NaOH.

Indicate the points of inflection and explain their significance.

d. Explain the purpose of performing the titration in a closed jacketed beaker under a nitrogen atmosphere.

4.16 For three diffusion boundary layer thicknesses (30–60–120 μm) calculate the rate of CO$_2$ exchange across the surface of the ocean ($T = 20°C$) as a function of pH in the range of 7.5–9 (use "apparent" carbonate acidity constants: $K_{a1} = 10^{-6.0}$; $K_{a2} = 10^{-8.9}$).

CHAPTER 5

SOLID DISSOLUTION AND PRECIPITATION: ACQUISITION AND CONTROL OF ALKALINITY

With an average water runoff of $30 \, \text{cm} \, \text{y}^{-1}$, the hydrologic cycle erodes the continents at an overall rate of about $60 \, \mu\text{m} \, \text{y}^{-1}$. Approximately $0.5 \, \text{g} \, \text{L}^{-1}$ of weathered continental rock is thus transported by rivers to the oceans, roughly 80% as suspended solids (physical erosion) and only about 20% as dissolved species (chemical erosion).[1] The suspended material of rivers is not completely inert chemically. For example, detrital clays have ion exchange properties and serve as an important sink for sodium in seawater.[2] The organic fraction of the suspended material in rivers contains the essential elements of biomass (chiefly carbon, nitrogen, phosphorus) which become remineralized upon decomposition in estuaries. Especially for phosphorus, this organic particulate transport contributes a sizable part of the overall continental input to the oceans.

In this chapter, however, we largely ignore the existence of the suspended material and focus on the smaller part of the exogenic material flux from the continents to the ocean, that which undergoes phase changes from the continental rock to the aqueous solution and finally to the sediments. Dissolution and precipitation reactions control these phase changes; they are the chemical processes that we wish to study here as an idealization of weathering and sedimentation phenomena.

There are wide differences in the conditions under which dissolution of continental rocks and precipitation of oceanic sediments take place. In the first case very low ionic strength rainwater infiltrates through the soil where the effective partial pressure of CO_2 is high (10^{-3} to 10^{-1} atm; typically $10^{-2.5}$ atm), because of to intense microbial activity. Some of the solids are highly unstable in this corrosive and continually renewed groundwater environment; the chemical driving forces are rather large and reasonably well understood. At the other extreme is seawater, which contains about half the chloride of the earth's surface and is a medium of high ionic strength (about $0.67\,M$). The residence time of the water is long (ca. 3,000 years) and the surface ocean is practically at equilibrium with the atmosphere ($P_{CO_2} = 10^{-3.5}$ atm). Even for seawater constituents which are removed by precipitation, the resulting solids are for the most part barely, and perhaps only locally, oversaturated. In general, the precipitation of ocean sediments is either biogenic, as for calcium carbonate and silica, or occurs only in restricted environments, as in evaporite formation. The oceanic distribution of many of the minor elements, such as the trace metals, is more strongly influenced by biological removal and adsorption phenomena than by precipitation of corresponding solid phases.

In Chapter 4 we examined the concept of alkalinity in part as a convenient way to study the chemistry of dissolved weak acids—particularly CO_2—without worrying about the source of the balancing strong bases. The question of how natural waters acquire their alkalinity by dissolution of continental rock is one that we wish to address here. Owing to the complexity and multiplicity of rock and soil types and to the sometimes sluggish and poorly characterized dissolution kinetics, we may not be able to develop precise quantitative models of alkalinity in freshwaters. Yet we can use limiting cases to understand chemically how the nature of the rocks controls the alkalinity of the water.

Our understanding of alkalinity control in the oceans is both stronger and more tenuous. On relatively short time scales, the alkalinity of the oceans is rather constant in time and space, and the small variations that exists are well studied and easily accounted for. Over geological times, the difficult question is paradoxically not why seawater alkalinity is so high (about $2.3\,\text{meq}\,L^{-1}$ compared with $0.1\,\text{meq}\,L^{-1}$ for the average river), but why it is so low. If seawater were simply river water concentrated by evaporation, its alkalinity would be almost a thousand times what it is (by normalization to chloride concentrations). To understand the control of alkalinity in seawater, we thus need to understand quantitatively the removal processes for the major cations Ca^{2+}, Na^+, K^+, and Mg^{2+}. These do not involve simply equilibration with solid phases in the sediments; ion exchange in detrital clays, high and low temperature reactions with volcanic material at oceanic ridges, and biological activity all seem to play important roles as well.

The questions addressed in this chapter are somewhat narrower and simpler than those posed by the study of the exogenic cycle. After a brief examination of the chemical nature of rocks, we develop some simple methodologies—diagrams and calculations—to study equilibrium between solids and the aqueous phase. This allows us to predict whether or not a particular constituent is saturated with respect to a particular phase and what concentrations of dissolved constituents a certain solid may contribute to a contacting water. Generalizing our approach to include several chemical compounds, we then examine the question of the coexistence of solid phases, and determine the conditions under which solids involving common constituents can coexist at equilibrium and which solid is thermodynamically most stable when they cannot coexist. These questions bring us to discuss Gibbs' phase rule and its significance in natural waters in the context of the mathematical and conceptual framework we have developed. Going back to a more direct study of the exogenic cycle, we then examine a few simple examples of weathering reactions to show how freshwaters acquire their alkalinity, how the composition of the water is controlled by the types of rocks it encounters. We also look at the mechanisms that control the alkalinity of seawater. Since equilibrium concepts are clearly insufficient in these studies of the acquisition and control of alkalinity, we close the chapter with a brief discussion of precipitation–dissolution kinetics in aquatic systems.

A number of caveats are in order before embarking on the study of solubility relationships in aquatic systems:

1. Many precipitation–dissolution reactions are sluggish. For example, a large supersaturation of $CaCO_3$ is often observed in natural waters before precipitation of the solid actually occurs. High temperature, the presence of nucleating surfaces, and biological activity can all dramatically enhance precipitation–dissolution kinetics. The proper application of equilibrium relationships depends on one's knowledge of such processes.

2. Even when the kinetics of precipitation are fast, the solid formed is often not the most stable solid thermodynamically. For example, opal—a cryptocrystalline form of silica—is typically precipitated by organisms where quartz (crystalline silica) is thermodynamically the stable form of SiO_2. The evolution of the solid to its more stable form (e.g., through dehydration or recrystallization) is usually very slow and often requires high temperatures.

3. The metastable solids that are initially precipitated are often "nonstoichiometric." For example, a common form of manganese oxide, "γ-MnO_2," has the approximate stoichiometry of $MnO_{1.3}$. Many natural solids also typically contain impurities, foreign ions incorporated in the matrix. The equilibrium constants we use for such solids have operational value, but probably little true thermodynamic significance.

4. The formation of pure solid phases is not the only, or perhaps even the dominant, process by which many solutes are removed from solution. In Chapter 8 we see how surface adsorption and solid solution formation can remove a solute from solution much below saturation conditions for pure solids. Such processes can be viewed as a way to describe thermodynamically the formation of solid species with activities different from unity.

1. THE CHEMICAL NATURE OF ROCKS

In order to understand the composition of natural waters, it is indispensable to understand something of the composition of the rocks that they contact, transform, and partially dissolve. These necessary excursions into the field of geology are often particularly frustrating to the aquatic chemist. It sometimes seems as if xenophobic mineralogists had arranged their multidimensional taxonomy with the intention of forming a maze impenetrable to outsiders.

At the risk of tediousness and oversimplification, let us recall here a few fundamentals of geological terminology. Rocks (which are assemblages of minerals) are broadly classified as igneous, sedimentary, or metamorphic. Igneous rocks are solidified magma, the molten material from below the earth's crust. Their weathering produces sedimentary rocks, which cover some 80% of the surface of the continents. Metamorphic rocks result from pressure and heat transformations of the other rocks and are often classified, as they are here, with their parent material by geochemists who are more interested in composition than appearances. Igneous rocks come in two basic shades (though many colors), black and gray. The darker shade is characteristic of lavas and described by an assortment of adjectives: dark, basaltic, extrusive, volcanic, and mafic (rich in magnesium and iron). The light-colored material is typically formed by intrusions of cooled magma into the crust and variously described as light, granitic, intrusive, plutonic, or sialic (rich in silica and aluminum). Sedimentary rocks comprise (1) sandstones, the familiar stuff on beaches; (2) shales, which represent more than 50% of all exposed rocks and include the clays; (3) evaporites, which are salt deposits formed by evaporation of seawater; and (4) limestones such as chalk and marbles.

Only four elements account for about 89% of the igneous rock mass: oxygen, silicon, aluminum, and iron (Table 5.1). The four elements that provide the

TABLE 5.1 Major Elements in Continental Rock[a]

Element	Igneous Rock	Sedimentary Rock
O	46.8	49.0
Si	29.7	27.4
Al	8.4	7.6
Fe	4.6	4.4
Ca	3.5	3.3
Mg	1.8	1.6
Na	2.5	0.7
K	2.7	2.6
C		1.3
H		0.4
Cl		1.7

Source: Adapted from Garrels and MacKenzie (1971).[3]

[a]The figures given are in weight percent; rocks dried to 110°C.

major cation content of natural waters—calcium, sodium, potassium, and magnesium—roughly account for the remaining 11%. The elemental composition of average sedimentary rock is of course very similar to that of the parent igneous rock, the major difference being the addition of some CO_2 (0.05 g/g), H_2O (0.04 g/g), and HCl (0.03 g/g). Although their elemental composition is similar, the mineralogy of igneous and sedimentary rocks is quite different. Tables 5.2 and 5.3 list minerals—taken here as defined chemical solid phases— that predominate in the various types of rocks. Silicates and silicate-rich alumino-silicates (chiefly feldspars) are characteristic of igneous rocks, while carbonates and silicate-poor alumino-silicates (mostly clays) are abundant in sedimentary material. Note that abraded but unweathered igneous rocks also constitute a variable fraction of the sedimentary rock mass and that quartz is ubiquitous. In these tables, the formulae of complex alumino-silicates are given as the "oxide formulae" to emphasize the basic nature of the minerals and to simplify their stoichiometric decomposition into the components SiO_2 and Al_2O_3 (or any combination thereof, such as kaolinite: $Al_2O_3 \cdot 2SiO_2 \cdot 2H_2O = Al_2Si_2O_5(OH)_4$).

By itself, the stoichiometric information provides a rough understanding of the chemistry of weathering. The major constituents of rocks are rather insoluble: aluminum and iron oxides contribute no significant solute concentrations to natural waters and dissolved silica concentrations average only 0.15 mM. In addition to the relative enrichment of aluminum over silicon in individual rocks (not in the average rock composition), weathering is thus largely the process of progressively stripping the four major cations, Ca^{2+}, Na^+, K^+, and Mg^{2+}, from the alumino-silicates. Note the relative sodium impoverishment in sedimentary rock. In the process, the water acquires part of its dissolved load, most of which actually originates from the dissolution of carbonates, sulfates, sulfides, and chlorides in sedimentary material (plus inputs from gases and aerosols).

To go further in our understanding of weathering processes and to make it more quantitative, thermodynamic or kinetic information is necessary. Table 5.4 provides solubility constants for most of the major minerals listed in Tables 5.2 and 5.3 and for some other minerals such as sulfides and hydroxides that are not particularly abundant in continental rocks but play an important role in aquatic chemistry. Also listed in Table 5.4 are some copper and lead solid phases to exemplify the chemical control of trace elements in natural waters. It should be noted that thermodynamic data for solids that react very slowly at ordinary pressures and temperatures are difficult to obtain (they are often extrapolated from high pressure and temperature conditions), and may thus be unreliable. Also, minerals that exist in continuously variable compositions (e.g., montmorillonites) are in principle difficult to define thermodynamically; the corresponding constants listed in Table 5.4 are provided for illustrative purposes, not for exact thermodynamic calculations. Finally, the presence of foreign ions may affect markedly the stability of natural minerals compared to their idealized pure chemical counterparts.

TABLE 5.2 Some Important Minerals in Continental Igneous Rocks

	Silicates		Alumino-Silicates: Micas	Alumino-Silicates: Feldspars	
Basaltic ↑	**Olivines**[a]			**Plagioclases**[c]	
	Forsterite	$2MgO \cdot SiO_2$		Anorthite	$CaO \cdot Al_2O_3 \cdot 2SiO_2$
	Fayalite	$2FeO \cdot SiO_2$			
	Pyroxenes				
	Enstatite	$MgO \cdot SiO_2$			
	Wollastonite	$CaO \cdot SiO_2$			
	Hedenbergite	$CaO \cdot FeO \cdot 2SiO_2$			
	Diopside	$CaO \cdot MgO \cdot 2SiO_2$			
	Amphiboles[b]		Biotite or Phlogopite	Albite	$Na_2O \cdot Al_2O_3 \cdot 6SiO_2$
	Tremolite	$2CaO \cdot 5MgO \cdot 8SiO_2 \cdot H_2O$	$K_2O \cdot 6MgO \cdot Al_2O_3 \cdot 6SiO_2 \cdot 2H_2O$		
	Iron tremolite	$2CaO \cdot 5FeO \cdot 8SiO_2 \cdot H_2O$			
	Quartz		Muscovite	**Orthoclases**	
↓	α-quartz	SiO_2	$K_2O \cdot 3Al_2O_3 \cdot 6SiO_2 \cdot 2H_2O$	K. feldspar	$K_2O \cdot Al_2O_3 \cdot 6SiO_2$
Granitic	Cristobalite	SiO_2			

[a] Continuously variable composition from Mg to Fe.
[b] Variable composition; Na often part of composition; in the hornblende variety, Al substitutes for part of the Si.
[c] Continuously variable composition from Ca to Na.

TABLE 5.3 Some Important Minerals in Continental Sedimentary Rocks

Silicates (Predominant in Sandstones)		Alumino-Silicates (Predominant in Shales: Clays)	Carbonates (Predominant in Limestone)		Chlorides (Predominant in Evaporites)		Sulfates (Predominant in Evaporites)	
Quartz		**Montmorillonites**[a]	Calcite	$CaCO_3$	Halite	NaCl	Gypsum	$CaSO_4 \cdot 2H_2O$
α-quartz	SiO_2	$Na_2O \cdot 7Al_2O_3 \cdot 22SiO_2 \cdot nH_2O$	Aragonite	$CaCO_3$			Anhydrite	$CaSO_4$
Cristobalite	SiO_2	$CaO \cdot 7Al_2O_3 \cdot 22SiO_2 \cdot nH_2O$	Dolomite	$CaMg(CO_3)_2$				
Amorphous Silica	$SiO_2 \cdot nH_2O$	**Illite** (Hydrated mica)	Magnesite	$MgCO_3$				
		$3K_2O \cdot 2MgO \cdot 9Al_2O_3 \cdot 28SiO_2 \cdot 8H_2O$						
		Chlorites						
		$5MgO \cdot Al_2O_3 \cdot 3SiO_2 \cdot 4H_2O$						
		Kaolinite						
		$Al_2O_3 \cdot 2SiO_2 \cdot 2H_2O$						

[a] Extremely variable composition; often contains K and Mg.

TABLE 5.4 Solubility Products of Various Minerals

	Log K^a	Reference
Chlorides		
Halite	$NaCl(s) = Na^+ + Cl^-$	1.54	3
Sylvite	$KCl(s) = K^+ + Cl^-$	0.98	3
Chlorargyrite	$AgCl(s) = Ag^+ + Cl^-$	-9.74	4
Sulfates		
Gypsum	$CaSO_4 \cdot 2H_2O(s) = Ca^{2+} + SO_4^{2-} + 2H_2O$	-4.62	4
Celestite	$SrSO_4(s) = Sr^{2+} + SO_4^{2-}$	-6.50	4
Barite	$BaSO_4(s) = Ba^{2+} + SO_4^{2-}$	-9.96	4
Oxides and Hydroxides[b]		
Calcium hydroxide	$Ca(OH)_2(s) = Ca^{2+} + 2OH^-$	-5.19	4, 5
	$CaOH^+ = Ca^{2+} + OH^-$	-1.15	5
Brucite	$Mg(OH)_2(s) = Mg^{2+} + 2OH^-$	-11.1	4, 5
	$MgOH^+ = Mg^{2+} + OH^-$	-2.6	5
	$Mg_4(OH)_4^{4+} = 4Mg^{2+} + 4OH^-$	-16.3	4, 5
Gibbsite	$Al(OH)_3(s) = Al^{3+} + 3OH^-$	-33.5	4, 5
	$AlOH^{2+} = Al^{3+} + OH^-$	-9.0	5
	$Al(OH)_2^+ = Al^{3+} + 2OH^-$	-18.7	5
	$Al(OH)_3 = Al^{3+} + 3OH^-$	-27.0	5
	$Al(OH)_4^- = Al^{3+} + 4OH^-$	-33.0	5
	$Al_2(OH)_2^{4+} = 2Al^{3+} + 2OH^-$	-20.3	5
	$Al_3(OH)_4^{5+} = 3Al^{3+} + 4OH^-$	-42.1	5
Manganous hydroxide	$Mn(OH)_2(s) = Mn^{2+} + 2OH^-$	-12.8	5
	$MnOH^+ = Mn^{2+} + OH^-$	-3.4	5
Ferrous hydroxide	$Fe(OH)_2(s) = Fe^{2+} + 2OH^-$	-15.1	5
	$FeOH^+ = Fe^{2+} + OH^-$	-4.5	5
	$Fe(OH)_2 = Fe^{2+} + 2OH^-$	-7.4	5
	$Fe(OH)_3^- = Fe^{2+} + 3OH^-$	-11.0	5
	$Fe(OH)_4^{2-} = Fe^{2+} + 4OH^-$	-10.0	5

243

TABLE 5.4 (*Continued*)

		Log K^a	Reference
Goethite	$\alpha \cdot FeOOH(s) + H_2O = Fe^{3+} + 3OH^-$	-41.5	4, 5
Ferric hydroxide	$am \cdot Fe(OH)_3(s) = Fe^{3+} + 3OH^-$	-38.8	4
Hematite	$\frac{1}{2}\alpha \cdot Fe_2O_3(s) + \frac{3}{2}H_2O = Fe^{3+} + 3OH^-$	-42.7	4
	$FeOH^{2+} = Fe^{3+} + OH^-$	-11.8	5
	$Fe(OH)_2^+ = Fe^{3+} + 2OH^-$	-22.3	5
	$Fe(OH)_4^- = Fe^{3+} + 4OH^-$	-34.4	5
	$Fe_2(OH)_2^{4+} = 2Fe^{3+} + 2OH^-$	-25.1	5
	$Fe_3(OH)_4^{5+} = 3Fe^{3+} + 4OH^-$	-49.7	5
Tenorite	$CuO(s) + H_2O = Cu^{2+} + 2OH^-$	-20.4	5
Cupric hydroxide	$Cu(OH)_2(s) = Cu^{2+} + 2OH^-$	-19.4	5
	$Cu(OH)_4^{2-} = Cu^{2+} + 4OH^-$	-16.4	5
	$Cu_2(OH)_2^{2+} = 2Cu^{2+} + 2OH^-$	-17.6	5
Litharge (red)	$PbO(s) + H_2O = Pb^{2+} + 2OH^-$	-15.3	5
Massicot (yellow)	$PbO(s) + H_2O = Pb^{2+} + 2OH^-$	-15.1	5
	$PbOH^+ = Pb^{2+} + OH^-$	-6.3	5
	$Pb(OH)_2 = Pb^{2+} + 2OH^-$	-10.9	5
	$Pb(OH)_3^- = Pb^{2+} + 3OH^-$	-13.9	5
	$Pb_2(OH)^{3+} = 2Pb^{2+} + OH^-$	-7.6	5
	$Pb_3(OH)_4^{2+} = 3Pb^{2+} + 4OH^-$	-32.1	5
	$Pb_4(OH)_4^{4+} = 4Pb^{2+} + 4OH^-$	-35.1	5
	$Pb_6(OH)_8^{4+} = 6Pb^{2+} + 8OH^-$	-68.4	5
Carbonates			
Aragonite	$CaCO_3(s) = Ca^{2+} + CO_3^{2-}$	-8.22	4
Calcite	$CaCO_3(s) = Ca^{2+} + CO_3^{2-}$	-8.35	4
Magnesite	$MgCO_3(s) = Mg^{2+} + CO_3^{2-}$	-7.46	4
Nesquehonite	$MgCO_3 \cdot 3H_2O(s) = Mg^{2+} + CO_3^{2-} + 3H_2O$	-4.67	4
Dolomite	$CaMg(CO_3)_2(s) = \text{calcite} + \text{magnesite}$	-1.70	3

Mineral	Reaction	log K	
Disordered dolomite	$CaMg(CO_3)_2(s) = $ calcite $+$ magnesite	-0.08	3
Strontianite	$SrCO_3(s) = Sr^{2+} + CO_3^{2-}$	-9.0	4
Rhodochrosite	$MnCO_3(s) = Mn^{2+} + CO_3^{2-}$	-9.3	4
Siderite	$FeCO_3(s) = Fe^{2+} + CO_3^{2-}$	-10.7	4
Malachite	$Cu_2CO_3(OH)_2(s) = 2Cu^{2+} + CO_3^{2-} + 2OH^-$	-33.8	4
Azurite	$Cu_3(CO_3)_2(OH)_2(s) = 3Cu^{2+} + 2CO_3^{2-} + 2OH^-$	-46.0	4
Cerussite	$PbCO_3(s) = Pb^{2+} + CO_3^{2-}$	-13.1	4
Phosphates			
Brushite	$CaHPO_4 \cdot 2H_2O(s) = Ca^{2+} + HPO_4^{2-} + 2H_2O$	-6.6	4
Hydroxylapatite	$Ca_5(PO_4)_3OH(s) = 5Ca^{2+} + 3PO_4^{3-} + OH^-$	-55.6	8
Newberyite	$MgHPO_4 \cdot 3H_2O(s) = Mg^{2+} + HPO_4^{2-} + 3H_2O$	-5.8	4
Bobierrite	$Mg_3(PO_4)_2 \cdot 8H_2O(s) = 3Mg^{2+} + 2PO_4^{3-} + 8H_2O$	-25.2	4
Vivianite	$Fe_3(PO_4)_2 \cdot 8H_2O(s) = 3Fe^{2+} + 2PO_4^{3-} + 8H_2O$	-36.0	4
Strengite	$FePO_4 \cdot 2H_2O(s) = Fe^{3+} + PO_4^{3-} + 2H_2O$	-26.4	4
Berlinite	$AlPO_4(s) = Al^{3+} + PO_4^{3-}$	-20.6	6
Sulfides			
Pyrrhotite	$FeS(s) = Fe^{2+} + S^{2-}$	-18.1	4
Pyrite	$FeS_2(s) = $ pyrrhotite $+ S^0$	-10.4	3
Alabandite	$MnS(s) = Mn^{2+} + S^{2-}$	-13.5	4
Covellite	$CuS(s) = Cu^{2+} + S^{2-}$	-36.1	4
Galena	$PbS(s) = Pb^{2+} + S^{2-}$	-27.5	4
Chalcopyrite	$CuFeS_2(s) = $ covellite $+$ pyrrhotite	-6.0	3
Silicates			
Quartz ($\alpha + \beta$)	$SiO_2(s) + H_2O = H_2SiO_3$	-4.00	3
Cristobalite ($\alpha + \beta$)	$SiO_2(s) + H_2O = H_2SiO_3$	-3.45	3
Amorphous silica	$SiO_2(s) + H_2O = H_2SiO_3$	-2.71	3
Forsterite	$\frac{1}{4}(2MgO \cdot SiO_2)(s) + H^+ = \frac{1}{2}Mg^{2+} + \frac{1}{4}H_2SiO_3 + \frac{1}{4}H_2O$	7.11	3
Fayalite	$\frac{1}{4}(2FeO \cdot SiO_2)(s) + H^+ = \frac{1}{2}Fe^{2+} + \frac{1}{4}H_2SiO_3 + \frac{1}{4}H_2O$	4.21	3
Enstatite	$\frac{1}{2}(MgO \cdot SiO_2)(s) + H^+ = \frac{1}{2}Mg^{2+} + \frac{1}{2}H_2SiO_3$	5.82	3
Wollastonite	$\frac{1}{2}(CaO \cdot SiO_2)(s) + H^+ = \frac{1}{2}Ca^{2+} + \frac{1}{2}H_2SiO_3$	6.82	3

245

TABLE 5.4 (*Continued*)

		Log K^a	Reference
Hedenbergite	$\frac{1}{4}(CaO \cdot FeO \cdot 2SiO_2)(s) + H^+$ $= \frac{1}{4}Ca^{2+} + \frac{1}{4}Fe^{2+} + \frac{1}{2}H_2SiO_3$	4.60	3
Diopside	$\frac{1}{4}(CaO \cdot MgO \cdot 2SiO_2)(s) + H^+$ $= \frac{1}{4}Ca^{2+} + \frac{1}{4}Mg^{2+} + \frac{1}{2}H_2SiO_3$	5.30	3
Tremolite	$\frac{1}{14}(2CaO \cdot 5MgO \cdot 8SiO_2 \cdot H_2O)(s) + H^+$ $= \frac{1}{7}Ca^{2+} + \frac{5}{14}Mg^{2+} + \frac{4}{7}H_2SiO_3$	4.46	3
Alumino-silicates			
Gibbsite	$Al(OH)_3(s) + H_2SiO_3 = \frac{1}{2}\text{kaolinite} + \frac{3}{2}H_2O$	4.25	3
Phlogopite	$\frac{1}{14}(K_2O \cdot 6MgO \cdot Al_2O_3 \cdot 6SiO_2 \cdot 2H_2O)(s) + H^+$ $= \frac{1}{14}\text{kaolinite} + \frac{1}{7}K^+ + \frac{3}{7}Mg^{2+} + \frac{2}{7}H_2SiO_3 + \frac{3}{14}H_2O$	5.01	3
Muscovite	$\frac{1}{2}(K_2O \cdot 3Al_2O_3 \cdot 6SiO_2 \cdot 2H_2O)(s) + H^+ + \frac{3}{2}H_2O$ $= \frac{3}{2}\text{kaolinite} + K^+$	3.51	3
Anorthite	$\frac{1}{2}(CaO \cdot Al_2O_3 \cdot 2SiO_2)(s) + H^+ + \frac{1}{2}H_2O$ $= \frac{1}{2}\text{kaolinite} + \frac{1}{2}Ca^{2+}$	9.83	3
Albite	$\frac{1}{2}(Na_2O \cdot Al_2O_3 \cdot 6SiO_2)(s) + H^+ + \frac{5}{2}H_2O$ $= \frac{1}{2}\text{kaolinite} + 2H_2SiO_3 + Na^+$	-0.68	3
K-feldspar	$\frac{1}{2}(K_2O \cdot Al_2O_3 \cdot 6SiO_2)(s) + H^+ + \frac{5}{2}H_2O$ $= \frac{1}{2}\text{kaolinite} + 2H_2SiO_3 + K^+$	-3.54	3
Na-montmorillonite[c]	$\frac{1}{2}(Na_2O \cdot 7Al_2O_3 \cdot 22SiO_2 \cdot 6H_2O)(s) + H^+$ $+ \frac{15}{2}H_2O = \frac{2}{7}\text{kaolinite} + 4H_2SiO_3 + Na^+$	-9.1	8
Ca-montmorillonite[c]	$\frac{1}{2}(CaO \cdot 7Al_2O_3 \cdot 22SiO_2 \cdot 6H_2O)(s) + H^+$ $+ \frac{15}{2}H_2O = \frac{2}{7}\text{kaolinite} + 4H_2SiO_3 + \frac{1}{2}Ca^{2+}$	-7.7	8

[a] Except where noted, all constants are valid for 25°C and ionic strength $I = 0$ M.

[b] See also oxides formed in redox reactions, Table 7.1.

[c] As noted by Helgeson et al.,[4] thermodynamic properties of minerals with continuously variable stoichiometric composition (e.g., montmorillonites and illites) are not simply defined. The values reported here allow for representative calculations of water composition not rigorous thermodynamic geochemical analysis. Note also that the solubilities of anorthite and albite reported by Helgeson et al.,[4] are markedly higher than those reported by Stumm and Morgan[8] and that they may thus be inconsistent with the tabulated values for Na- and Ca-montmorillonites.

2. SOLUBILITY OF SOLIDS: EFFECTS OF CONCENTRATION

Let us now examine the problem of determining whether a specific solid is or is not present at equilibrium given a particular recipe for the system. We may solve this problem numerically by initially assuming the solid to be present, carrying out the calculation of equilibrium composition, and finally checking that all components are present in sufficient concentrations to insure the presence of the solid; that is, the solid must have a positive concentration in the system. Alternatively, the calculation of equilibrium composition can be carried out by assuming that the solid does not form in the system and ultimately verifying whether or not the solubility product is exceeded. We may then generalize the problem and determine the conditions (e.g., concentration and pH) under which a given solid may form or the composition of a solution at equilibrium with a given solid phase.

2.1 Solubility of Hydroxides (and Oxides)

The formation and dissolution of hydroxide and oxide solids (oxides can be considered simply as dehydrated hydroxides) are important in the aquatic chemistry of many metal ions, particularly iron, aluminum, and manganese. Since the hydroxide ion concentration is given directly as a function of pH (the precipitation reactions can in fact be written with H^+ production rather than OH^- consumption), the mathematical problem is straightforward, the only complication being the existence of hydrolysis species, such as $Fe(OH)_2^+$ or $MnOH^+$. These hydrolysis species often play a dominant role in controlling the solubility of hydroxide-forming metals; thus their equilibrium constants are listed along with those of the solids in Table 5.4.

Example 1. Solubility of Fe(OH)$_3$(s)   First, let us determine the distribution of dissolved iron species obtained by mixing solutions of $FeCl_3$ and NaOH. We choose as the solubility product that of amorphous $Fe(OH)_3(s)$, which is most applicable to a system containing a fresh iron precipitate, and ignore, for simplicity, ionic strength effects and possible iron chloride complexes (see Chapter 6).

Recipe $10^{-4}\,M\ FeCl_3$
              $10^{-4}\,M\ NaOH$

Species $(H_2O), H^+, OH^-, Na^+, Cl^-, Fe^{3+}, FeOH^{2+}, Fe(OH)_2^+,$
            $Fe(OH)_4^-, Fe_2(OH)_2^{4+}, Fe_3(OH)_4^{5+}, Fe(OH)_3(s)$ (one solid phase)

Reactions

$$H_2O = H^+ + OH^- \qquad pK_w = 14.0$$
$$FeOH^{2+} = Fe^{3+} + OH^- \qquad pK = 11.8$$
$$Fe(OH)_2^+ = Fe^{3+} + 2OH^- \qquad pK = 22.3$$
$$Fe(OH)_4^- = Fe^{3+} + 4OH^- \qquad pK = 34.4$$
$$Fe_2(OH)_2^{4+} = 2Fe^{3+} + 2OH^- \qquad pK = 25.1$$
$$Fe_3(OH)_4^{5+} = 3Fe^{3+} + 4OH^- \qquad pK = 49.7$$
$$Fe(OH)_3(s) = Fe^{3+} + 3OH^- \qquad pK_s = 38.8$$

As we have seen in previous examples, the correct choice of components makes solving equilibrium problems much simpler. If the solution is saturated with respect to $Fe(OH)_3(s)$ (a hypothesis that we shall verify later), then addition of more $Fe(OH)_3(s)$ to the system will leave the aqueous solution unchanged. Because we are interested primarily in the composition of the aqueous phase, the amount of solid is arbitrary and the concentration $[Fe(OH)_3 \cdot s]$ (in moles per liter of solution) is not a necessary variable to describe the system. Note also that the activity of the solid is fixed at unity and is unrelated to the "concentration" of the solid in the aqueous phase. For these reasons we eliminate the solid "concentration" from all mole balance equations but one. This is achieved directly in the manner used previously to eliminate the water concentration: $Fe(OH)_3(s)$ is chosen as a component. The choice of principal components is then straightforward: $(H_2O), H^+, Na^+, Cl^-$, and $Fe(OH)_3(s)$, as listed in Tableau 5.1.

The corresponding mole balance equations are

$$TOTH = [H^+] - [OH^-] + 3[Fe^{3+}] + 2[FeOH^{2+}] + [Fe(OH)_2^+]$$
$$- [Fe(OH)_4^-] + 4[Fe_2(OH)_2^{4+}] + 5[Fe_3(OH)_4^{5+}]$$
$$= 3[FeCl_3]_T - [NaOH]_T = 2 \times 10^{-4} \tag{1}$$

$$TOTNa = [Na^+] = [NaOH]_T = 10^{-4} \tag{2}$$

$$TOTCl = [Cl^-] = 3[FeCl_3]_T = 3 \times 10^{-4} \tag{3}$$

$$TOTFe(OH)_3 = [Fe^{3+}] + [FeOH^{2+}] + [Fe(OH)_2^+] + [Fe(OH)_4^-]$$
$$+ 2[Fe_2(OH)_2^{4+}] + 3[Fe_3(OH)_4^{5+}] + [Fe(OH)_3 \cdot s]$$
$$= [FeCl_3]_T = 10^{-4} \tag{4}$$

TABLEAU 5.1

	H^+	Na^+	Cl^-	$Fe(OH)_3(s)$	$\log K$
H^+	1				
OH^-	-1				-14.0
Na^+		1			
Cl^-			1		
Fe^{3+}	3			1	3.2
$FeOH^{2+}$	2			1	1.0
$Fe(OH)_2^+$	1			1	-2.5
$Fe(OH)_4^-$	-1			1	-18.4
$Fe_2(OH)_2^{4+}$	4			2	3.5
$Fe_3(OH)_4^{5+}$	5			3	3.3
$Fe(OH)_3(s)$				1	
$[FeCl_3]_T$	3		3	1	$10^{-4} M$
$[NaOH]_T$	-1	1			$10^{-4} M$

The equilibrium constants given in the last column of the tableau are obtained, with some algebraic manipulation, by rewriting all the reactions in terms of the components used in the tableau. Thus, the unit activity of the solid phase is conveniently included in the various mass law equations. The last mole balance given by Equation 4 is not necessary to obtain the equilibrium composition of the aqueous phase and can be solved after the others to obtain the solid concentration $[Fe(OH)_3 \cdot s]$.

Solution Equations 2 and 3 are trivial:

$$[Na^+] = 10^{-4}$$

$$[Cl^-] = 3 \times 10^{-4}$$

For a hydroxide solid such as $Fe(OH)_3$, the most convenient way to display the solubility as a function of concentration and pH is in the form of a log C–pH diagram where the mass law expressions for the iron hydrolysis species are plotted as straight lines of various slopes. For example, since the mass law expression for $[Fe^{3+}]$ at equilibrium with the hydroxide solid is

$$[Fe^{3+}] = 10^{3.2}[H^+]^3 \tag{5}$$

we plot the line

$$\log[Fe^{3+}] = 3.2 - 3\,pH \tag{6}$$

for this species on the log C–pH diagram (see Figure 5.1). Other species are plotted in a similar manner.

We may now compare the $TOTH$ equation with the log C–pH diagram. According to the hypothesis of large inequalities among concentrations, the solution to the $TOTH$ equation is most probably given by one of the following possibilities:

(i) $\quad\quad [H^+] = 2 \times 10^{-4} \gg$ all other species

(ii) $\quad\quad [Fe^{3+}] = 0.67 \times 10^{-4} \gg$ all other species

(iii) $\quad [FeOH^{2+}] = 10^{-4} \gg$ all other species

(iv) $\quad [Fe(OH)_2^+] = 2 \times 10^{-4} \gg$ all other species

(v) $\quad [Fe_2(OH)_2^{4+}] = 5 \times 10^{-5} \gg$ all other species

(vi) $\quad [Fe_3(OH)_4^{5+}] = 4 \times 10^{-5} \gg$ all other species

Other possibilities could be considered in which two terms of opposite signs in the $TOTH$ equation would be equal and $\gg 2 \times 10^{-4}$. However, the intersections of the $[OH^-]$ and $[Fe(OH)_4^-]$ lines with the others on Figure 5.1 all yield concentrations $\ll 2 \times 10^{-4}$.

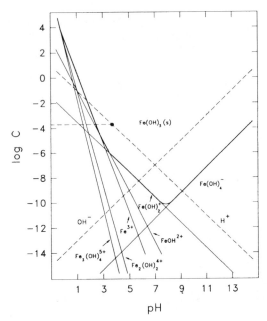

Figure 5.1 Log C–pH diagram for Fe(III) at equilibrium with amorphous Fe(OH)$_3$(s), (Example 1). Any system whose total Fe(III) concentration (log C) and pH yield a point within the hatched area of the graph is supersaturated with respect to am.Fe(OH)$_3$; at equilibrium the soluble Fe(III) and the pH of such a system must yield a point on the saturation line of the diagram. Note the importance of the polymeric hydroxo species at low pH and the increase in solubility at high pH due to Fe(OH)$_4^-$ formation. If a more stable form of ferric hydroxide were considered (see Table 5.4), the whole diagram would be translated to lower log C values.

It is readily seen on Figure 5.1 that (i) is the correct choice:

$$[H^+] = 2 \times 10^{-4} = 10^{-3.7}$$

therefore

$$pH = 3.7$$

Thus, according to the mass laws:

$$[OH^-] = 10^{-14}[H^+]^{-1} = 10^{-10.3}$$

$$[Fe^{3+}] = 10^{3.2} [H^+]^3 = 10^{-7.9}$$

$$[FeOH^{2+}] = 10^{1.0} [H^+]^2 = 10^{-6.4}$$

$$[Fe(OH)_2^+] = 10^{-2.5} [H^+] = 10^{-6.2}$$

$$[Fe(OH)_4^-] = 10^{-18.4} [H^+]^{-1} = 10^{-14.7}$$

$$[Fe_2(OH)_2^{4+}] = 10^{3.5} [H^+]^4 = 10^{-11.3}$$

$$[Fe_3(OH)_4^{5+}] = 10^{3.3} [H^+]^5 = 10^{-15.2}$$

From Equation 4 we can calculate the "concentration" of the solid phase, that is, the number of moles of solid per liter of solution:

$$[Fe(OH)_3 \cdot s] = 10^{-4} - 10^{-7.9} - 10^{-6.4} - 10^{-6.2} - 10^{-14.7}$$

$$- 2 \times 10^{-11.3} - 3 \times 10^{-15.2} = 9.89 \times 10^{-5} M$$

The solid is thus only 1% dissolved.

To obtain a more general solution, we may note that the total Fe(III) in solution in the presence of the solid $Fe(OH)_3(s)$ is also the maximum concentration of dissolved iron at any pH:

$$[Fe(III)]_{dissolved} = [Fe^{3+}] + [Fe(OH)^{2+}] + [Fe(OH)_2^+] + [Fe(OH)_4^-]$$

$$+ 2[Fe_2(OH)_2^{4+}] + 3[Fe_3(OH)_4^{5+}] \tag{7}$$

For any given pH and analytical concentration of Fe(III), the graph of Figure 5.1 shows whether the solid $Fe(OH)_3$ is saturated or not, and gives the concentrations of all Fe(III) species in solution. Below the line for the maximum dissolved Fe(III) concentration at a given pH, solutions are undersaturated, that is, the *ion activity product* (IAP) is less than the solubility product:

$$IAP = [Fe^{3+}][OH^-]^3 < K_s \tag{8}$$

and the *saturation quotient* Ω is less than 1, where

$$\Omega = \frac{IAP}{K_s} \tag{9}$$

When the saturation quotient is greater than 1, the solution is supersaturated and the solid must precipitate for the system to reach equilibrium.

As can be seen in Figure 5.1, the hydroxide complexes of Fe(III), not the free ion Fe^{3+}, account for most of the dissolved iron concentration throughout the pH range of natural waters. However, as discussed in Chapter 6, other dissolved complexes may dominate the hydroxide species. A graph such as Figure 5.1 is thus valid only when there are no important dissolved Fe(III) species other than the hydroxide complexes; in particular, there must be no strong organic Fe(III) complexing agent in the system.

$$* \quad * \quad * \quad *$$

It is not always possible to guess correctly whether a given solid phase should or should not be present at equilibrium. When there are few such solids, a trial and error procedure yields the correct answer. As we have seen in this example, an initial guess must be made as to whether or not the solid, in this case $Fe(OH)_3(s)$, precipitates. At the end of the calculation, the assumption can be verified by calculating the "concentration" of the solid. If the assumption is not valid, the problem must be recalculated with another choice for the Fe component. In complex situations where there are many possible solids involving common components, this trial and error procedure can become quite difficult, and short of trying all possibilities (which can number in the thousands), there is in fact no easy way to obtain the correct set of solids.

An additional complication arises from the effect of the formation of hydrolysis species and the precipitation or dissolution of $Fe(OH)_3(s)$ on the pH of the solution; ferric salts themselves are strong acids. While pH can be varied independently of $[Fe(III)]_T$ by addition of strong base or strong acid, an experimental change in $[Fe(III)]_T$ typically leads to a pH variation.

Example 2. Effects of Fe(III) Hydrolysis on pH   With the recipe

$$[FeCl_3]_T = 2 \times 10^{-3} M$$

we may again guess that ferric hydroxide precipitates and thus choose the principal components H^+, Cl^-, and $Fe(OH)_3(s)$. Thus

$$TOTH = [H^+] - [OH^-] + 3[Fe^{3+}] + 2[FeOH^{2+}] + [Fe(OH)_2^+]$$
$$- [Fe(OH)_4^-] + 4[Fe_2(OH)_2^{4+}] + 5[Fe_3(OH)_4^{5+}] = 3[FeCl_3]_T$$
$$= 6 \times 10^{-3} M \tag{10}$$

All the terms in this equation may be expressed in terms of $[H^+]$:

$$[H^+] - 10^{-14}[H^+]^{-1} + (3)10^{3.2}[H^+]^3 + (2)10^{1.0}[H^+]^2 + 10^{-2.5}[H^+]$$
$$- 10^{-18.4}[H^+]^{-1} + (4)10^{3.5}[H^+]^4 + (5)10^{3.3}[H^+]^5 = 6 \times 10^{-3} \tag{11}$$

Solving this polynomial by trial and error leads to pH = 2.31.

A pure ferric chloride solution is thus strongly acidic and at such a low pH, the iron is quite soluble (see Figure 5.1):

$$[Fe^{3+}] = 1.86 \times 10^{-4} M$$
$$[FeOH^{2+}] = 2.40 \times 10^{-4} M$$
$$[Fe(OH)_2^+] = 1.55 \times 10^{-5} M$$

therefore

$$[Fe(III)]_{dissolved} = 4.42 \times 10^{-4} M.$$

At higher concentrations of ferric salts the pH is in fact so low that the solid actually—and paradoxically—dissolves. The pH below which the solid dissolves can be calculated by taking the difference between the $TOTFe(OH)_3$ and $TOTH$ equations (neglecting OH^-, $Fe(OH)_4^-$, and $Fe_3(OH)_4^{5+}$, which are small at the pH of interest):

$$TOTH = [H^+] + 3[Fe^{3+}] + 2[FeOH^{2+}] + [Fe(OH)_2^+]$$
$$+ 4[Fe_2(OH)_2^{4+}] = 3[FeCl_3]_T \tag{12}$$

$$TOTFe(OH)_3 = [Fe(OH)_3 \cdot s] + [Fe^{3+}] + [FeOH^{2+}] + [Fe(OH)_2^+]$$
$$+ 2[Fe_2(OH)_2^{4+}] = [FeCl_3]_T \tag{13}$$

therefore

$$3TOTFe(OH)_3 - TOTH = 3[Fe(OH)_3 \cdot s] + [FeOH^{2+}] + 2[Fe(OH)_2^+]$$
$$+ 2[Fe_2(OH)_2^{4+}] - [H^+] = 0 \tag{14}$$

and

$$3[Fe(OH)_3 \cdot s] = [H^+] - [FeOH^{2+}] - 2[Fe(OH)_2^+] - 2[Fe_2(OH)_2^{4+}] \tag{15}$$

The solid dissolves when the solid concentration becomes negative:

$$[FeOH^{2+}] + 2[Fe(OH)_2^+] + 2[Fe_2(OH)_2^{4+}] \geqslant [H^+] \tag{16}$$

that is, when

$$10^{1.0}[H^+]^2 + (2)10^{-2.5}[H^+] + (2)10^{3.5}[H^+]^4 \geqslant [H^+] \tag{17}$$

or

$$[H^+] > 10^{-1.35} M.$$

Introducing this value of $[H^+]$ into the $TOTFe$ equation with substituted mass law expressions gives the minimum $FeCl_3$ concentration above which the $Fe(OH)_3$ dissolves:

$$[FeCl_3]_T \geqslant (10^{3.2})(10^{-4.05}) + (10^{1.0})(10^{-2.7}) + (10^{-2.5})(10^{-1.35})$$
$$+ (2)(10^{3.5})(10^{-5.4})$$

that is

$$[FeCl_3]_T \geqslant 0.18 \, M.$$

While such Fe(III) concentrations are extremely unlikely in natural waters, hydrolysis of iron and aluminum may significantly affect pH and alkalinity, as we shall see in Examples 11 and 12.

2.2 Solubility of Carbonates, Sulfides, and Phosphates

Studying the precipitation of hydroxide solids as a function of pH is a relatively straightforward exercise since the hydroxide concentration is known immediately at any given pH. This is not the case with other ligands such as carbonate, sulfide, or phosphate, whose free concentrations depend not only on pH but also on the total ligand concentration, and perhaps on the concentration of the precipitating metal (if it is present in excess of the ligand). In the next few examples we examine the different possible situations for precipitation of a metal with a weak acid ligand. For consistency among the examples, calcite ($CaCO_3$) is used throughout as the precipitating solid, but the concepts and the methodology are applicable to all other carbonates, sulfides, and phosphates. (FeS, a dominant solid in anoxic, i.e., oxygen free, systems, is used extensively as an example in the next section.)

It is easiest to study carbonate precipitation when the system is at equilibrium with a fixed partial pressure of CO_2. In this case, the free ligand concentration is given explicitly as a function of pH, and simple numerical or graphical solutions are obtained.

Example 3. Solubility of CaCO₃ at Fixed P_{CO₂} *Consider the typical partial pressure of carbon dioxide in the atmosphere:* $P_{CO_2} = 10^{-3.5}$ atm. From the mass laws of the carbonate system, the free carbonate concentration can be expressed as a function of pH:

$$[CO_3^{2-}] = \frac{10^{-6.3}}{[H^+]} \times \frac{10^{-10.3}}{[H^+]} \times 10^{-1.5} P_{CO_2} = 10^{-21.6}[H^+]^{-2} \tag{18}$$

When calcite precipitates, the free calcium ion also becomes an explicit function of pH:

$$[Ca^{2+}][CO_3^{2-}] = K_s = 10^{-8.3} \tag{19}$$

therefore

$$[Ca^{2+}]_{max} = 10^{+13.3} [H^+]^2 \qquad (20)$$

For any pH this result, illustrated in Figure 5.2 as a $\log C$–pH diagram, gives the maximum concentration of calcium that can be added to a solution before saturation of calcite under the given partial pressure of CO_2. Note that no dissolved calcium species other than Ca^{2+} is assumed to be present here. For example, consider the recipe

$$P_{CO_2} = 10^{-3.5} \text{ atm}$$

$$[CaCO_3]_T = 10^{-2.0} M$$

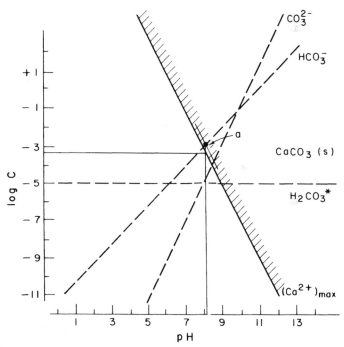

Figure 5.2 Log C–pH diagram for Ca^{2+} and carbonate at equilibrium with $P_{CO_2} = 10^{-3.5}$ atm and $CaCO_3(s)$ (Example 3). A system whose calcium concentration and pH yield a point in the hatched area of the graph is supersaturated with respect to $CaCO_3$(calcite); at equilibrium the free calcium concentration and the pH must yield a point on the saturation line of the diagram. The solution of Example 3 is given by **point a**, which satisfies the condition $[HCO_3^-] = 2[Ca^{2+}]$. The composition of the system is then obtained from the intersection of the corresponding vertical line with the graphs of the various species.

With the assumption that $CaCO_3$ precipitates, the principal components H^+, CO_2, and $CaCO_3$ lead to the equation

$$TOTH = [H^+] - [OH^-] - [HCO_3^-] - 2[CO_3^{2-}] + 2[Ca^{2+}] = 0 \quad (21)$$

Graphically (see **point a** on Figure 5.2) the solution of this equation is given by

$$2[Ca^{2+}] = [HCO_3^-] \quad (22)$$

so that

$$2 \times 10^{+13.3}[H^+]^2 = \frac{10^{-6.3}}{[H^+]} 10^{-1.5} \times 10^{-3.5} \quad (23)$$

$$pH = 8.3$$

$$[Ca^{2+}] = 10^{-3.3} = 5 \times 10^{-4}$$

Since

$$TOTCa = [Ca^{2+}] + [CaCO_3 \cdot s] = 10^{-2.0} \quad (24)$$

then

$$[CaCO_3 \cdot s] = 10^{-2.0} - 10^{-3.3} = 9.5 \times 10^{-3}$$

For concentrations of $[CaCO_3]_T$ less than $5 \times 10^{-4}\ M$, the solid would dissolve.

$$* \quad * \quad * \quad *$$

Calcium carbonate systems not at equilibrium with a fixed partial pressure of CO_2 are slightly more difficult to study since the free carbonate concentration is then not a function of pH only; $[CO_3^{2-}]$ also depends on the total dissolved carbonate which itself depends on the extent of precipitation. In the next two examples we consider systems with given total calcium and carbonate concentrations and with a fixed pH that may be varied arbitrarily by addition of strong acid or strong base. The problem is to find out if there is enough calcium and carbonate for calcite precipitation at any pH, and to determine the resulting composition of the system.

Example 4. Solubility of CaCO₃ (s) Given [CO₃]ₜ > [Ca]ₜ If the total carbonate in the system is in large excess of the total calcium, for example,

$$Ca_T = 10^{-4}\ M; \qquad [CO_3]_T = 10^{-2}\ M$$

then the speciation of carbonate can be affected only minimally by the presence

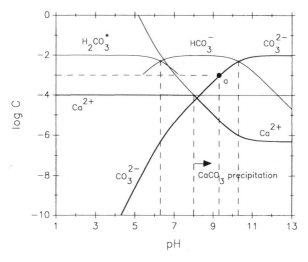

Figure 5.3 $\log C$–pH diagram for Ca^{2+} and carbonate with $[CO_3]_T = 10^{-2} M$ and $Ca_T = 10^{-4} M$ (Example 4). Two graphs for calcium are obtained, one by considering no solid species and the other by assuming precipitation of $CaCO_3(s)$. The critical pH for $CaCO_3(s)$ precipitation is obtained by the intersection of these two calcium graphs, and the equilibrium free calcium ion concentration as a function of pH is given by the lower of the graphs (heavy line).

of calcium, whether calcite precipitates or not. The various carbonate species are thus plotted in the usual $\log C$–pH diagram in Figure 5.3.

For calcium, on the other hand, different dissolved concentrations are calculated depending on whether the solid precipitates or not. If the solid precipitates, the dissolved calcium concentration is given by the solubility expression

$$[Ca^{2+}] = \frac{K_s}{[CO_3^{2-}]} = \frac{10^{-8.3}}{[CO_3^{2-}]} \tag{25}$$

If the solid does not precipitate, all the calcium remains in solution:

$$[Ca^{2+}] = [Ca]_T = 10^{-4} M \tag{26}$$

Both of the $[Ca^{2+}]$ versus pH graphs corresponding to Equations 25 and 26 are plotted in Figure 5.3. At any pH, the lower of these two graphs (smaller $[Ca^{2+}]$) gives the correct calcium concentration at equilibrium. This is because the solubility product expression (Equation 25) gives both the maximum concentration of calcium in solution and the minimum concentration of calcium necessary to obtain a precipitate at a given free carbonate concentration. Thus at low $[CO_3^{2-}]$ there is insufficient calcium to meet the solubility product, precipitation does not occur, and $[Ca^{2+}]$ is controlled by the total calcium concentration. At higher $[CO_3^{2-}]$, calcium is abundant enough to allow precipitation, and $[Ca^{2+}]$ is controlled by the solubility product expression.

The critical pH above which the solid precipitates is obtained by equating the two expressions for $[Ca^{2+}]$, Equations 25 and 26:

$$\frac{10^{-8.3}}{[CO_3^{2-}]} = [Ca^{2+}] = [Ca_T] = 10^{-4} \tag{27}$$

According to Figure 5.3, the two graphs intersect in the region where HCO_3^- is the major carbonate species, thus providing an expression for $[CO_3^{2-}]$:

$$[CO_3^{2-}] = 10^{-10.3}\frac{[HCO_3^-]}{[H^+]} = 10^{-10.3}\frac{[CO_3]_T}{[H^+]} = \frac{10^{-12.3}}{[H^+]} \tag{28}$$

and

$$\frac{10^{-8.3}}{10^{-12.3}}[H^+] = 10^{-4} \tag{29}$$

$$pH_{crit} = 8.0$$

Given a complete recipe, we can obtain the pH of the solution on the graph by examining the appropriate $TOTH$ equation, that corresponding to the principal components of the system. Consider, for instance, the following recipe for Example 4:

$$[CaCO_3]_T = 10^{-4}\,M$$

$$[Na_2CO_3]_T = 10^{-3}\,M$$

$$[NaHCO_3]_T = 8.9 \times 10^{-3}\,M \text{ (to give } [CO_3]_T = 10^{-2}\,M)$$

Guessing that the solid does precipitate and thus choosing the principal components, $H^+, HCO_3^-, CaCO_3$, and Na^+, we obtain the appropriate $TOTH$ equation:

$$TOTH = [H^+] - [OH^-] + [H_2CO_3^*] - [CO_3^{2-}] + [Ca^{2+}]$$
$$= -[Na_2CO_3]_T = -10^{-3}\,M \tag{30}$$

This equation is readily simplified to $[CO_3^{2-}] = 10^{-3}\,M$, which leads to pH = 9.3 graphically; see **point a** on Figure 5.3.

*   *   *   *

The methodology used in this example is typical of that used for all such problems:

1. Graph species for the component in excess (metal or ligand).
2. Graph species for the less abundant components under both hypotheses of the presence and absence of solid.
3. Take the "lower lines" of the less abundant component graph as giving the correct species concentration.
4. Obtain the critical pH (the threshold of precipitation) by equating the expressions for the less abundant component under conditions of precipitation and no precipitation (in the diagram the pH is given by the intersection of the two corresponding graphs).

Note that this methodology breaks down when both components involved in the precipitation of a solid are present at approximately the same concentration. In such a case, one may use the methodology to obtain the critical precipitation pH. Below that pH, in the absence of precipitation, the mole balance equations ($TOTCa$ and $TOTCO_3$ in our example) are independent of each other and provide the correct solution regardless of their relative numerical values. However, to obtain the solution composition in the precipitation region, where the solid itself must be chosen as a component, it becomes necessary to consider simultaneously several species and their corresponding mass laws, as illustrated in the next example.

Example 5. Solubility of $CaCO_3(s)$ Given $[Ca]_T \simeq [CO_3]_T$ Consider the recipe

$$[Ca]_T = 10^{-3.2}\, M$$

$$[CO_3]_T = 10^{-3.3}\, M$$

Following the methodology presented earlier, we first plot the free calcium concentration on the graph as if no solid precipitated (Figure 5.4):

$$TOTCa = [Ca^{2+}] = [Ca]_T = 10^{-3.2}\, M \tag{31}$$

For $[CO_3^{2-}]$, if $CaCO_3$ precipitates,

$$[CO_3^{2-}] = \frac{10^{-8.3}}{[Ca^{2+}]} = 10^{-5.1}\, M \tag{32}$$

If $CaCO_3(s)$ does not precipitate, $[CO_3^{2-}]$ is given by the usual log C–pH diagram. In particular, in the region where HCO_3^- is the major carbonate species,

$$[CO_3^{2-}] = 10^{-10.3}\frac{[HCO_3^-]}{[H^+]} = 10^{-10.3}\frac{[CO_3]_T}{[H^+]} = \frac{10^{-13.6}}{[H^+]} \tag{33}$$

The lower of the two $[CO_3^{2-}]$ graphs in Figure 5.4 is the pertinent one. The two lines intersect in the HCO_3^- region; the critical pH is then obtained by equating

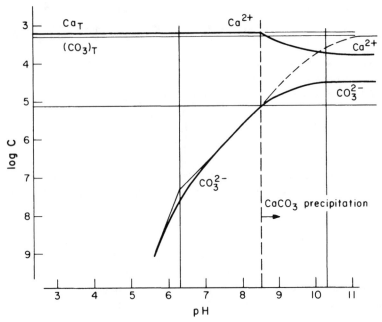

Figure 5.4   Log C–pH diagram for Ca^{2+} and carbonate with $[CO_3]_T = 10^{-3.3}\,M$ and $Ca_T = 10^{-3.2}\,M$ (Example 5). The critical pH for $CaCO_3(s)$ precipitation is obtained by the intersection of the two $[CO_3^{2-}]$ graphs according to the general methodology. Above this critical pH a significant fraction of both calcium and carbonate are precipitated as $CaCO_3$; the free calcium and carbonate concentrations are then obtained as the solution of the quadratic equation given in Example 5.

Equations 32 and 33:

$$10^{-5.1} = \frac{10^{-13.6}}{[H^+]} \tag{34}$$

therefore

$$pH_{crit} = 8.5$$

Above this pH, the precipitation of $CaCO_3$ will markedly decrease the concentrations of all species in solution. Consider the $TOTCa$ equation corresponding to the principal components H^+, Ca^{2+}, and $CaCO_3$:

$$TOTCa = [Ca^{2+}] - [CO_3^{2-}] - [HCO_3^-] - [H_2CO_3^*] = [Ca]_T - [CO_3]_T$$
$$= 10^{-3.89} \tag{35}$$

All concentrations can be expressed as a function of $[CO_3^{2-}]$ and $[H^+]$:

$$\frac{10^{-8.3}}{[CO_3^{2-}]} - [CO_3^{2-}] - 10^{10.3}[H^+][CO_3^{2-}] = 10^{-3.89} \tag{36}$$

The graph of this implicit function of $[H^+]$ is shown on Figure 5.4. The equivalent expression for $[Ca^{2+}]$ is also plotted on the figure:

$$[Ca^{2+}] - \frac{10^{-8.3}}{[Ca^{2+}]} - 10^{2.0}\frac{[H^+]}{[Ca^{2+}]} = 10^{-3.89} \tag{37}$$

At very high pH, when $[HCO_3^-]$ is negligible, $[CO_3^{2-}]$ and $[Ca^{2+}]$ are obtained by solving a simple quadratic:

$$\frac{10^{-8.3}}{[CO_3^{2-}]} - [CO_3^{2-}] = 10^{-3.89} \tag{38}$$

therefore

$$[CO_3^{2-}] = 10^{-4.50}$$
$$[Ca^{2+}] = 10^{-3.80}$$

Although the approximate equality of the total metal and total ligand concentrations makes this example more complicated than the previous ones, the appropriate choice of principal components does lead to a reasonably simple solution.

*   *   *

Analytical information for natural waters is usually limited to the total concentrations in the aqueous phase, for example, Ca_T, C_T, and Alk. If we want to know whether or not $CaCO_3$ is saturated in such systems, the problem to be solved is slightly different from, and in fact easier than, those we have considered so far. The maximum (i.e., saturating) concentrations of calcium can be calculated given C_T and pH or alkalinity and pH; the simplest approach is to use the ionization fraction expression for the free carbonate (see Section 3.3 of Chapter 4). The solution of these problems is left as an exercise (Problem 5.4).

3. COMPETITION AND COEXISTENCE AMONG SEVERAL SOLID PHASES

At high pH in a system containing calcium and carbonate the precipitation of calcium hydroxide can, in principle, become thermodynamically favorable:

$$Ca^{2+} + 2OH^- = Ca(OH)_2(s) \qquad pK_s = 5.2$$

Which of the two possible solid phases considered [$CaCO_3$ and $Ca(OH)_2$] then controls the dissolved calcium concentration as pH is increased? Can both solids coexist at equilibrium? Although these are the types of questions that we wish to address here, calcium provides a poor example because the precipitation of $Ca(OH)_2$ takes place at pH values outside of the range of interest for natural waters (pH > 12). As a more realistic simple example, let us consider the iron(II)–sulfide–carbonate system.

Example 6. Competition between Sulfide and Carbonate Fe(II) Solids   To illustrate how two ligands may compete for precipitation of the same metal, let us consider a situation in which both total ligand concentrations are in excess of the metal.

Recipe $[Fe \cdot II]_T = 10^{-6} M$

$[S \cdot -II]_T = 10^{-5} M$

$[CO_3]_T = 10^{-3} M$

Reactions $H_2O = H^+ + OH^-$ $pK_w = 14$

$H_2S = HS^- + H^+$ $pK = 7.0$

$HS^- = S^{2-} + H^+$ $pK = 13.9$

$H_2CO_3^* = HCO_3^- + H^+$ $pK = 6.3$

$HCO_3^- = CO_3^{2-} + H^+$ $pK = 10.3$

$FeS(s) = Fe^{2+} + S^{2-}$ $pK_{s1} = 18.1$

$FeCO_3(s) = Fe^{2+} + CO_3^{2-}$ $pK_{s2} = 10.7$

$Fe(OH)_2(s) + 2H^+ = Fe^{2+} + 2H_2O$ $pK_{s3} = -12.9$

$FeOH^+ + H^+ = Fe^{2+} + H_2O$ $pK = -9.5$

$Fe(OH)_2^0 + 2H^+ = Fe^{2+} + 2H_2O$ $pK = -20.6$

$Fe(OH)_3^- + 3H^+ = Fe^{2+} + 3H_2O$ $pK = -31.0$

Here we have expressed the hydrolysis reactions directly in terms of H^+ rather than OH^- since we wish to plot concentrations vs. pH.

Since the total concentrations of sulfide and carbonate are in great excess of the iron, the corresponding free ion concentrations are effectively independent of any precipitation; the graphs for $[S^{2-}]$ and $[CO_3^{2-}]$ can be calculated in the usual manner from the ionization fraction formula. Ignoring for the moment $Fe(OH)_2(s)$, we find three possible cases for the free ferrous ion concentration:

1. No solid precipitates:

$$[Fe^{2+}] = \frac{[Fe \cdot II]_T}{1 + \dfrac{10^{-9.5}}{[H^+]} + \dfrac{10^{-20.6}}{[H^+]^2} + \dfrac{10^{-31}}{[H^+]^3}} \tag{39}$$

which simplifies at low pH to $Fe^{2+} = Fe_T = 10^{-6} M$.

2. FeS(s) precipitates:

$$[Fe^{2+}] = \frac{K_{s1}}{[S^{2-}]} = \frac{10^{-18.1}}{[S^{2-}]} \tag{40}$$

3. $FeCO_3(s)$ precipitates:

$$[Fe^{2+}] = \frac{K_{s2}}{[CO_3^{2-}]} = \frac{10^{-10.7}}{[CO_3^{2-}]} \tag{41}$$

Each of the corresponding three graphs is plotted on Figure 5.5. As in the previous examples (4 and 5), the lowest Fe^{2+} concentration graph is the correct one at any pH. As seen on Figure 5.5, a simple case of competition among the two solids arises: the stable solid at a given pH is the one that most "lowers" the ferrous ion concentration. In this particular case, the ferrous carbonate solid never forms because the ferrous sulfide is too insoluble and it keeps $[Fe^{2+}]$ below saturation of $FeCO_3$ at all pH's; that is, the graph for Equation 40 is always lower than that for Equation 41.

The critical pH at which FeS precipitates is obtained in the usual way by equating Equations 39 and 40. According to Figure 5.5, this point is in the H_2S region [where Fe(II) hydrolysis is negligible]:

$$10^{-6.0} = \frac{10^{-18.1}}{[S^{2-}]} = \frac{10^{-18.1} \times [H^+]^2}{10^{-13.9} \times 10^{-7.0} \times [S \cdot -II]_T} \tag{42}$$

therefore

$$pH_{crit} = 6.9$$

This value is a bit low because it is very close to the upper limit of the H_2S region. Expressing the free sulfide by its ionization fraction,

$$[S^{2-}] = [S \cdot -II]_T (10^{20.9}[H^+]^2 + 10^{13.9}[H^+])^{-1} \tag{43}$$

yields the more exact answer,

$$pH_{crit} = 7.07$$

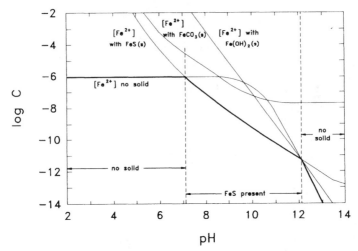

Figure 5.5   Log C–pH diagram for Fe(II) in a carbonate and sulfide bearing water with all ligands in excess: $[CO_3]_T = 10^{-3} M$, $[S \cdot -II]_T = 10^{-5} M$, $[Fe \cdot II]_T = 10^{-6} M$ (Example 6). Four lines for $[Fe^{2+}]$ are plotted, considering no precipitate, $FeCO_3(s)$, $FeS(s)$, or $Fe(OH)_2(s)$. Hydrolysis of Fe^{2+} is included, which results in decreased $[Fe^{2+}]$ at high pH in the absence of any solid precipitate. In determining $[Fe^{2+}]$ as controlled by carbonate or sulfide precipitation, $[CO_3^{2-}]$ and $[S^{2-}]$ are calculated as explicit functions of $[H^+]$ and total concentrations, using the ionization fractions $[CO_3^{2-}] = [CO_3]_T$ $(10^{16.3}[H^+]^2 + 10^{10.3}[H^+] + 1)^{-1}$ and $[S^{2-}] = [S \cdot -II]_T$ $(10^{20.9}[H^+]^2 + 10^{13.9}[H^+] + 1)^{-1}$. This approach is computationally more complicated than that described in the text and was not done by hand. The domains of stability of the possible solids are delimited by the intersection of the $[Fe^{2+}]$ lines. At each pH, the lowest $[Fe^{2+}]$ value corresponds to the controlling solid (if any) as shown by the heavy lines. In this case, $FeCO_3(s)$ and $Fe(OH)_2(s)$ are never stable while $FeS(s)$ precipitates in the mid-pH region.

For the sake of completeness, let us consider the possible precipitation of the ferrous hydroxide solid that we have neglected so far in this example. When this solid precipitates, the free ferrous ion concentration is given by

$$[Fe^{2+}] = 10^{12.9}[H^+]^2 \tag{44}$$

The corresponding graph shown on Figure 5.5 is never the lowest of the $[Fe^{2+}]$ graphs; that is, the hydroxide does not precipitate at this total Fe(II) concentration.

At very high pH, the graph for $[Fe^{2+}]$ corresponding to precipitation of FeS again crosses the graph for $[Fe^{2+}]$ in the absence of any precipitating solid. That is, hydrolysis of dissolved Fe(II) lowers $[Fe^{2+}]$ to such an extent that the solubility product is no longer exceeded and the FeS(s) redissolves.

In this pH range, if no solid precipitates, the dominant hydrolysis species is

$Fe(OH)_3^-$:

$$[Fe \cdot II]_T \simeq [Fe(OH)_3^-] = \frac{10^{-31}[Fe^{2+}]}{[H^+]^3} \qquad (45)$$

and thus

$$[Fe^{2+}] = 10^{31}[H^+]^3[Fe \cdot II]_T \qquad (46)$$

From Figure 5.5 we see that redissolution of FeS occurs in the pH domain where HS^- is the dominant sulfide species. Equation 40 may thus be written

$$[Fe^{2+}] = \frac{10^{-18.1}10^{13.9}[H^+]}{[S \cdot -II]_T} \qquad (47)$$

Equating Equations 46 and 47, we obtain

$$[H^+]^2 = \frac{10^{-31}10^{-18.1}10^{13.9}}{[S \cdot -II]_T[Fe \cdot II]_T} \qquad (48)$$

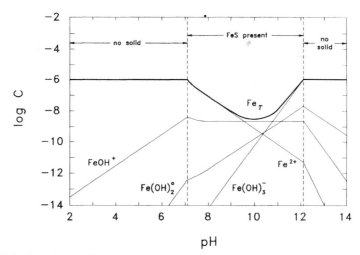

Figure 5.6 Log C–pH diagram showing the effect of hydrolysis on dissolved Fe(II) concentrations. Conditions as in Figure 5.5 (Example 6). The graph for $[Fe^{2+}]$ as a function of pH corresponds to $[Fe^{2+}]$ as determined by the controlling solid (FeS(s) in the mid–pH region) or by hydrolysis (also shown by the heavy line in Figure 5.5). The concentrations of the hydrolysis species and the sum of all dissolved Fe(II) species, Fe_T, are calculated from $[Fe^{2+}]$ and pH (Fe_T shown by heavy line).

or

$$pH_{crit} = 12.1$$

Figure 5.5 indicates which solids, if any, are present over the pH range and what the corresponding Fe^{2+} concentrations are. It also provides direct information on the total dissolved Fe (Fe_T) in the lower pH range where hydrolysis is unimportant. Figure 5.6 shows the concentrations of the various Fe(II) hydrolysis species and of the sum of these species (i.e., Fe_T). Note that $Fe(OH)_2^0$ is never the dominant species. The concentrations of the hydrolysis species may be calculated as a function of pH by using the appropriate expression for $[Fe^{2+}]$, that is, Equation 40 when FeS is present ($7.07 < pH < 12.1$) and Equation 39 for lower or higher pH values where no solid is present.

* * *

In Example 6 both solids, FeS(s) and $FeCO_3$(s), cannot coexist because, from low to high pH, the free ferrous ion $[Fe^{2+}]$ is kept below the saturation value for carbonate, first by its total concentration, then by sulfide precipitation and finally by $Fe(OH)_3^-$ formation. For the carbonate solid to form, there must be a large Fe^{2+} concentration. When FeS precipitates, the carbonate solid will form only if the total iron concentration is in excess of the total sulfide.

Example 7. Coexistence of Sulfide and Carbonate Fe(II) Solids

$$Recipe \quad [Fe \cdot II] = 10^{-4}\,M$$

$$[S \cdot -II]_T = 10^{-5}\,M$$

$$[CO_3]_T = 10^{-3}\,M$$

Reactions   Same as Example 6

Following our usual method, we plot the free ion concentration graphs in the order of the total concentrations in the system. The free carbonate ion concentration $[CO_3^{2-}]$ is not affected much by any precipitation since the total carbonate is 10 times in excess of anything else. The free ferrous ion can only be slightly affected by sulfide, and the graphs for $[Fe^{2+}]$ are obtained as usual (Figure 5.7).

1. No precipitate forms:

$$[Fe^{2+}] = \frac{[Fe \cdot II]_T}{1 + \dfrac{10^{-9.5}}{[H^+]} + \dfrac{10^{-20.6}}{[H^+]^2} + \dfrac{10^{-31}}{[H^+]^3}} \tag{49}$$

Again, at low pH, $Fe^{2+} = Fe_T = 10^{-4}\,M$.

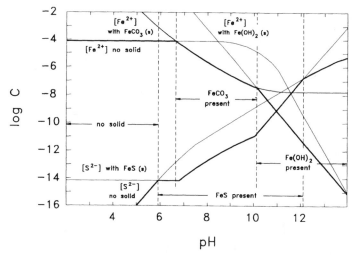

Figure 5.7 Log C–pH diagram for Fe(II) in a carbonate and sulfide bearing water with iron in excess of sulfide: $[CO_3]_T = 10^{-3} M$, $[S \cdot -II]_T = 10^{-5} M$, $[Fe \cdot II]_T = 10^{-4} M$ (Example 7). Because of the difference in total concentrations, precipitation of FeS(s) cannot have much effect on $[Fe^{2+}]$. Thus the lines for $[Fe^{2+}]$ are plotted considering no precipitate, FeCO$_3$(s), or Fe(OH)$_2$(s). The lowest values for $[Fe^{2+}]$ (heavy line) and intersections of the $[Fe^{2+}]$ lines give the domains of stability of FeCO$_3$(s) and Fe(OH)$_2$(s), which do not overlap. The domain of stability of FeS(s) is determined by plotting the lines for $[S^{2-}]$ as a function of pH. In the absence of any solid $[S^{2-}]$ is calculated from the total concentration and ionization fraction. Once FeS(s) precipitates, $[S^{2-}]$ is controlled by $[Fe^{2+}]$, which is in turn controlled by $[CO_3^{2-}]$ when FeCO$_3$(s) is present and by $[OH^-]$ when Fe(OH)$_2$(s) is present. Because the total iron is in excess of the total sulfide, the domain of stability of FeS(s) can overlap with that of the other two solids.

2. FeCO$_3$ precipitates:

$$[Fe^{2+}] = \frac{K_{s2}}{[CO_3^{2-}]} = \frac{10^{-10.7}}{[CO_3^{2-}]} \tag{50}$$

The correct $[Fe^{2+}]$ graph is the lowest one on Figure 5.7, and the critical pH for FeCO$_3$ precipitation is obtained by equating Equations 49 and 50. In the HCO$_3^-$ region

$$10^{-4} = \frac{10^{-10.7}}{[CO_3^{2-}]} = \frac{10^{-10.7}[H^+]}{10^{-10.3}[CO_3]_T} \tag{51}$$

therefore

$$pH_{crit} = 6.6$$

We can turn now to the question of the speciation of the trace sulfide that we have heretofore neglected.

If FeS does not precipitate, $[S^{2-}]$ is obtained from the usual sulfide graph (see Figure 5.7). In the H_2S region

$$[S^{2-}] = \frac{10^{-13.9} \times 10^{-7.0}[S \cdot - II]_T}{[H^+]^2} = \frac{10^{-25.9}}{[H^+]^2} \tag{52}$$

If FeS precipitates,

$$[S^{2-}] = \frac{K_{s1}}{[Fe^{2+}]} = \frac{10^{-18.1}}{[Fe^{2+}]} \tag{53}$$

In Equation 53 $[Fe^{2+}]$ is given by either Equation 49 or 50, whichever yields the lower concentration at any given pH. For pH < 6.6,

$$[S^{2-}] = \frac{10^{-18.1}}{[Fe^{2+}]} = \frac{10^{-18.1}}{10^{-4.0}} = 10^{-14.1} \tag{54}$$

The correct $[S^{2-}]$ graph is the lowest one in Figure 5.7 and we obtain the critical pH for FeS precipitation by equating Equations 52 and 54 (guessing from the graph that the critical pH is less than 6.6):

$$\frac{10^{-25.9}}{[H^+]^2} = 10^{-14.1} \tag{55}$$

therefore

$$pH'_{crit} = 5.9$$

Above this pH, FeS precipitates and controls the sulfide species concentrations. In the region where the ferrous ion concentration is controlled by precipitation of $FeCO_3$ (pH > 6.6), the free sulfide is then dependent on the free carbonate, in a sort of "second-order interaction:"

$$[S^{2-}] = \frac{10^{-18.1}}{[Fe^{2+}]} = \frac{10^{-18.1}}{10^{-10.7}}[CO_3^{2-}] = 10^{-7.4}[CO_3^{2-}] \tag{56}$$

3. Finally let us consider again the precipitation of $Fe(OH)_2(s)$:

$$[Fe^{2+}] = 10^{12.9}[H^+]^2 \tag{57}$$

Equating Equations 50 and 57, we find the critical pH above which $Fe(OH)_2(s)$

precipitates and $FeCO_3(s)$ dissolves; in the HCO_3^- region

$$10^{12.9}[H^+]^2 = \frac{10^{-10.7}}{[CO_3^{2-}]} = \frac{10^{-10.7}[H^+]}{10^{-10.3}[CO_3]_T} \tag{58}$$

therefore

$$pH''_{crit} = 10.3$$

Again, this pH is too close to the pK to be accurate; $[CO_3^{2-}]$ cannot be neglected. The more complete formula given by the ionization fraction,

$$[CO_3^{2-}] = [CO_3]_T \left(1 + 10^{10.3}[H^+] \right)^{-1} \tag{59}$$

yields

$$pH''_{crit} = 10.1$$

Also, as in Example 6, $FeS(s)$ dissolves for pH above 12.1:

$$pH'''_{crit} = 12.1$$

The graph of Figure 5.7 is thus made up of 5 regions:

pH < 5.9	Fe is totally in solution as Fe^{2+}
5.9 < pH < 6.6	Fe is mostly present as Fe^{2+} but $FeS(s)$ precipitates
6.6 < pH < 10.1	Fe is mostly present as $FeCO_3(s)$, which coexists with $FeS(s)$
10.1 < pH < 12.1	Fe is mostly present as $Fe(OH)_2(s)$, which coexists with $FeS(s)$
12.1 < pH < 14	Fe is precipitated as $Fe(OH)_2(s)$ but the dissolved species $Fe(OH)_3^-$ becomes increasingly important with increasing pH

Again, calculation of dissolved $Fe(Fe_T)$ as a function of pH must take hydrolyzed species into account, as shown in Figure 5.8. At very high pH, the hydroxide solid is almost completely redissolved.

* * *

The only difference between the last two examples is the concentration of total Fe(II) in the system. One may obtain a general description of the chemistry of Fe(II) under all combinations of Fe(II) total concentration and pH—all other conditions such as $[CO_3]_T$ and $[S \cdot - II]_T$ being fixed and given—in the form of

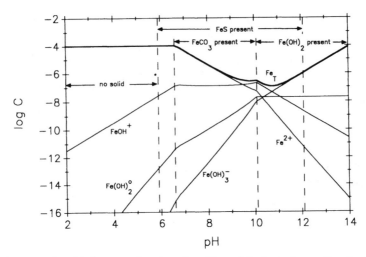

Figure 5.8 Log C–pH diagram showing the effect of hydrolysis on dissolved Fe(II) concentrations. Conditions as Figure 5.7 (Example 7). The graph for $[Fe^{2+}]$ corresponds to the heavy line in Figure 5.7. The concentrations of the hydrolysis species and the sum of all dissolved Fe(II) species, Fe_T, are calculated from $[Fe^{2+}]$ and pH (Fe_T shown by heavy line).

a $\log[Fe\cdot II]_T$ versus pH diagram, as shown in Figure 5.9. Such *stability diagrams* are obtained by a generalization of the methodology described in the preceding examples. The boundaries of the various domains are derived by equating the controlling expressions for the major species in the neighboring areas, a not very difficult but often tedious process in which the effect of hydrolysis of Fe(II) on the solubility of the various solids must also be considered (see steps 1a–c below). [Note that, for simplicity, complete equations for the ionization fractions are not used below. This introduces slight inaccuracies in Figure 5.9 near the carbonate and sulfide pK_a's. For the more complete solution refer to the calculations of Examples 6 and 7.] The various lines of Figure 5.9, where $[CO_3]_T = 10^{-3}\ M$ and $[S\cdot - II]_T = 10^{-5}\ M$, are obtained in the following eight steps:

1a. $Fe^{2+}/FeS(s)$ line (southwestern boundary of FeS stability domain). Equating $[Fe^{2+}] = [Fe\cdot II]_T$ and $[Fe^{2+}] = K_{s1}/[S^{2-}]$ and expressing $[S^{2-}]$ as a function of $[S\cdot - II]_T$ and pH yields line A in the region where hydrolysis is unimportant:

$$pH < 7; \qquad [Fe\cdot II]_T = 10^{7.8}\ [H^+]^2 \qquad (60)$$

$$pH < 9.5; \qquad [Fe\cdot II]_T = 10^{0.8}\ [H^+] \qquad (61)$$

1b. $FeOH^+/FeS(s)$ line (southern boundary of FeS stability domain). Substituting from $[FeOH^+] = 10^{-9.5}\ [Fe^{2+}][H^+]^{-1} = [Fe\cdot II]_T$ into $[Fe^{2+}] = K_{s1}/[S^{2-}]$ and expressing $[S^{2-}]$ as in step 1a above yields the continu-

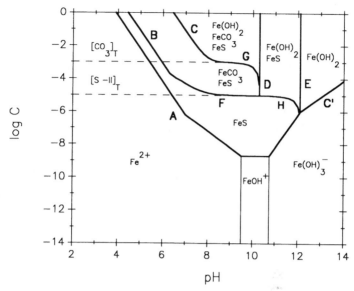

Figure 5.9   Stability diagram for Fe(II) in the presence of carbonate $(10^{-3}\ M)$ and sulfide $(10^{-5}\ M)$. The boundaries of the domains of stability of the various solids are given by the lines A and E for FeS(s); B, F, and D for $FeCO_3$(s); and C, G, D, H, and C' for $Fe(OH)_2$(s). The solids present in each domain are indicated on the diagram to illustrate the domains of solid coexistence when the iron is in excess of the precipitating ligands and the domains of competition among solids at low iron concentration.

ation of line A:

$$9.5 < pH < 10.75; \qquad [\text{Fe·II}]_T = 10^{-8.7} \tag{62}$$

1c. $Fe(OH)_3^-/FeS(s)$ line (southeastern boundary of FeS stability domain). Substituting from $[Fe(OH)_3^-] = 10^{-31}[Fe^{2+}][H^+]^{-3} = [\text{Fe·II}]_T$ into $[Fe^{2+}] = K_{s1}/[S^{2-}]$ and expressing $[S^{2-}]$ as in step 1a yields the final section of line A:

$$10.75 < pH < 12.1; \qquad [\text{Fe·II}]_T = 10^{-30.2}[H^+]^{-2} \tag{63}$$

2. $Fe^{2+}/FeCO_3$(s) line (southwestern boundary of $FeCO_3$ stability domain). Equating $[Fe^{2+}] = [\text{Fe·II}]_T$ and $[Fe^{2+}] = K_{s2}/[CO_3^{2-}]$ and expressing $[CO_3^{2-}]$ as a function of $[CO_3]_T$ and pH yields line B:

$$pH < 6.3; \qquad [\text{Fe·II}]_T = 10^{8.9}[H^+]^2 \tag{64}$$

$$pH > 6.3; \qquad [\text{Fe·II}]_T = 10^{2.6}[H^+] \tag{65}$$

3a. $Fe^{2+}/Fe(OH)_2$(s) line (southwestern boundary of $Fe(OH)_2$ stability domain). Equating $[Fe^{2+}] = [Fe \cdot II]_T$ and $[Fe^{2+}] = K_{s3}[H^+]^2$ yields line C:

$$[Fe \cdot II]_T = 10^{12.9}[H^+]^2 \tag{66}$$

3b. $Fe(OH)_3^-/Fe(OH)_2(s)$ line (southeastern boundary of $Fe(OH)_2$ stability domain). For $[Fe(OH)_3^-] = [Fe \cdot II]_T$, equating $[Fe^{2+}] = 10^{31}[Fe \cdot II]_T [H^+]^3$ and $[Fe^{2+}] = K_{s3}[H^+]^2$ yields line C':

$$[Fe \cdot II]_T = 10^{-18.1} [H^+]^{-1} \tag{67}$$

4. $FeCO_3(s)/Fe(OH)_2(s)$ line (eastern boundary of $FeCO_3$ stability domain). Equating $[Fe^{2+}] = K_{s2}/[CO_3^{2-}]$ and $[Fe^{2+}] = K_{s3}[H^+]^2$ yields line D:

$$pH = 10.3 \quad \text{(as in Example 7)} \tag{68}$$

5. $FeS(s)/Fe(OH)_2(s)$ line (eastern boundary of FeS stability domain). Equating $[Fe^{2+}] = K_{s1}/[S^{2-}]$ and $[Fe^{2+}] = K_{s3}[H^+]^2$ yields line E:

$$pH = 12.1 \quad \text{(as in Example 6)} \tag{69}$$

The remaining three lines correspond to boundaries where a ligand is exhausted by precipitation and a new iron solid forms. These are more difficult to obtain precisely than the others, but easier to determine approximately. The methodology is described in some detail for the $FeCO_3(s)$ precipitation line corresponding to exhaustion of sulfide.

6. $FeCO_3(s)/FeS(s)$ line (southern boundary of $FeCO_3$ stability domain). In the FeS stability domain with FeS and HS^- as components, the relevant mole balance is written

$$TOTHS = [HS^-] - [Fe^{2+}] = [S \cdot - II]_T - [Fe \cdot II]_T \tag{70}$$

When $FeCO_3$ precipitates, $[Fe^{2+}]$ is given by

$$[Fe^{2+}] = \frac{K_{s2}}{[CO_3^{2-}]} \tag{71}$$

and since FeS also precipitates

$$[S^{2-}] = \frac{K_{s1}}{[Fe^{2+}]} = \frac{[CO_3^{2-}]K_{s1}}{K_{s2}} \tag{72}$$

Expressing $[Fe^{2+}]$ and $[HS^-] = 10^{13.9}[H^+][S^{2-}]$ as a function of $[CO_3]_T$ and $[H^+]$ in the $TOTHS$ equation then yields line F:

$$[Fe \cdot II]_T = 10^{-5.01} + 10^{2.6}[H^+] \tag{73}$$

$[Fe \cdot II]_T = [S \cdot - II]_T = 10^{-5.0}$ is a good intuitive approximation for this boundary since iron becomes available for precipitation with carbonate when it is in excess of the sulfide.

7. $Fe(OH)_2(s)/FeCO_3(s)$ line (southern boundary of $Fe(OH)_2$ stability domain). Using $[Fe^{2+}] = K_{s3}[H^+]^2$ and $[CO_3^{2-}] = K_{s2}/[Fe^{2+}]$ in the $TOTHCO_3$ equation pertinent in the $FeCO_3$ domain $([HCO_3^-] - [Fe^{2+}] = [CO_3]_T - [Fe \cdot II]_T)$ yields line G:

$$[Fe \cdot II]_T = 10^{-3.0} - \frac{10^{-13.3}}{[H^+]} + 10^{12.9}[H^+]^2 \qquad (74)$$

$[Fe \cdot II]_T = [CO_3]_T = 10^{-3.0}$ is a good intuitive approximation.

8. $Fe(OH)_2(s)/FeS(s)$ line (southern boundary of $Fe(OH)_2$ stability domain). Using $[Fe^{2+}] = K_{s3}[H^+]^2$ and $[S^{2-}] = K_{s1}/[Fe^{2+}]$ in the $TOTHS$ equation pertinent in the FeS domain $([HS^-] - [Fe^{2+}] = [S \cdot - II]_T - [Fe \cdot II]_T)$ yields line H:

$$[Fe \cdot II]_T = 10^{-5.0} - \frac{10^{-17.1}}{[H^+]} + 10^{12.9}[H^+]^2 \qquad (75)$$

Again, $[Fe \cdot II]_T = [S \cdot - II]_T = 10^{-5}$ is a good intuitive approximation.

Note on Figure 5.9 that the stability domains of the various solids do not coincide with the domains of dominance—sections of the log $[Fe \cdot II]_T$ versus pH diagram where a solid accounts for most of the total iron concentration. For example, $[Fe^{2+}]$ is the principal iron species in parts of the domains where $FeCO_3(s)$ and $FeS(s)$ precipitate, and there are several regions where the various solids coexist. The results of Examples 6 and 7 can be checked in Figure 5.9 by following the appropriate horizontal lines on the graph.

In general, if we consider the possible precipitation of one metal with several ligands, it is rather obvious from Example 6 that solids with ligands in excess of the metal cannot coexist at equilibrium. This is true in particular for hydroxides (there is effectively a total concentration of $55.4 M$ of hydroxide in water) and usually for carbonates (with the exception of calcium and magnesium in systems such as seawater). Typically we can expect that either the hydroxide or the carbonate solid of a metal may precipitate but not both.

As illustrated in Examples 6 and 7, in anoxic waters the relative values of pH, $[S \cdot - II]_T$, $[CO_3]_T$, and $[Fe \cdot II]_T$ are of great importance in determining the sulfide chemistry and hence the chemistry of many trace metals that form insoluble sulfides. The free sulfide ion activity may be controlled by (1) $[S \cdot - II]_T$ and pH when there is no precipitate, (2) $[Fe \cdot II]_T$ when FeS precipitates but not $FeCO_3$, (3) $[CO_3]_T$ and pH when both FeS and $FeCO_3$ precipitate and $[CO_3]_T > [S \cdot - II]_T$.

4. THE PHASE RULE

Determining the possible coexistence among solid phases in aquatic systems containing a large number of components can become quite complicated. An upper limit on the number of phases that can coexist at equilibrium is given by the "Gibbs phase rule," a well-known and almost mystical theorem of chemical thermodynamics. First formulated by J. W. Gibbs in 1875,[9] the phase rule is a mere mathematical consequence of the conservation and energetic constraints in chemical systems at equilibrium. Here we state and illustrate the phase rule for the restricted conditions considered so far—dilute aqueous solutions at a given pressure and temperature. In so doing, we hope to enhance our understanding of this classic theorem and to widen the field of its applicability in aquatic systems.

Consider the first three rules we have set for choosing principal components: (1) H_2O; (2) H^+; (3) species of fixed activities such as gases at given partial pressures or solid phases. A question that readily arises is: Could it happen that there would be more species with fixed activities than there are independent components in the system? The answer to this question is our first statement of the phase rule: *The number of species with fixed activities (solids and gases at fixed partial pressure) must be smaller than the number of components in a chemical system at fixed pressure and temperature.* (Note that this statement of the rule is unaffected by whether water is or is not considered as a component; if we choose to consider water as an explicit component, we also have to consider its activity fixed at unity.)

As given here, the phase rule is reasonably intuitive. Since all species are either components or expressed as a function of components by mass laws, a system in which all components have fixed activities is strictly invariant. The number of species with fixed activities then certainly cannot exceed the number of components; we can show that it must in fact be smaller using a simple example to avoid the unnecessary complications and awkward notation of the general case. The demonstration consists of a reasoning ad absurdo in which we consider a system with as many solids and gases at fixed partial pressure as components and show that equilibrium cannot be achieved.

Example 8. Maximum Number of Phases in the Ca–CO₃ System Using our familiar calcium and carbonate system, let us ask the question: Can the solids calcium carbonate and calcium hydroxide coexist in water under a given partial pressure of carbon dioxide? According to our phase rule, this should be impossible since there would be three fixed activities ($\{CO_2\} = P_{CO_2}$, $\{CaCO_3\} = 1$, and $\{Ca(OH)_2\} = 1$) in a three-component system (e.g., H^+, Ca^{2+}, CO_3^{2-}).

Consider, for example, that the system is defined by $P_{CO_2} = 10^{-3.5}$ atm and the usual reactions:

$$CO_2(g) + H_2O = H_2CO_3^* \qquad pK = 1.5$$
$$H_2CO_3^* = 2H^+ + CO_3^{2-} \qquad pK = 16.6$$

$$CaCO_3(s) = Ca^{2+} + CO_3^{2-} \qquad pK = 8.3$$
$$Ca(OH)_2(s) + 2H^+ = Ca^{2+} + 2H_2O \qquad pK = -22.8$$

By combination of these four reactions, we can write another reaction involving exclusively solids, gases, and water:

$$CO_2(g) + Ca(OH)_2(s) = CaCO_3(s) + H_2O \qquad pK = -13.0$$

It is clear that at the given partial pressure of CO_2 this reaction cannot be at equilibrium:

$$\frac{1}{P_{CO_2}} = 10^{+3.5} \neq 10^{+13.0}$$

The free energy of this reaction, $\Delta G = -RT \ln K - RT \ln P_{CO_2} = -54.3 \, kJ$ mol^{-1}, is always negative and the reaction must proceed to the right until either $CO_2(g)$ or $Ca(OH)_2(s)$ is exhausted:

$$CO_2(g) + Ca(OH)_2(s) \rightarrow CaCO_3(s) + H_2O$$

Note that if the partial pressure of $CO_2(g)$ were allowed to drop in such a gedanken experiment, an equilibrium would be achieved at $P_{CO_2} = 10^{-13.0}$ atm (a small partial pressure!). In that situation, the reaction would have a free energy change of zero and could proceed indifferently in either direction. We can then acknowledge the notable exception that such thermodynamically degenerate situations provide to our phase rule: in addition to the number of components minus one, there *can* be any number of species with fixed activities if those can be formed from the others according to a reaction with zero free energy change. This is an unlikely occurrence in aquatic systems which have temperatures and pressures fixed independently of their chemistry. The determination of the temperature–pressure domains where such coexistence of phases is possible is of course an important aspect of high-temperature geochemistry.

The reasoning that we just applied to the Ca–CO_3 system can be applied to any other system in which we hypothesize as many species with fixed activities as we have components. By induction, we then have a demonstration of the phase rule as stated here. Note that the same reasoning can also be applied to any *subset* of solids and gases at fixed partial pressure in any given system. We thus have effectively demonstrated a much more powerful enunciation of the phase rule: *Species with fixed activities must be independent (and can always be chosen as components).*

For example, according to this rule, two solids with identical formulae such as calcite and aragonite ($CaCO_3$) cannot coexist at equilibrium no matter how many components there are, unless the reaction $CaCO_3$ (aragonite) = $CaCO_3$ (calcite) has a zero free-energy change, in which case the solids are thermodynamically indistinguishable.

Similarly, a reaction that involves exclusively solids, gases, and water (no solute) cannot be at equilibrium under arbitrary fixed partial pressures of the gases. The ability to write such a chemical reaction implies ipso facto that the solids and the gases are not independent components. The various species thus cannot coexist at equilibrium and the reaction has to proceed completely in one direction.

For example, silica, aluminum oxide (gibbsite), and kaolinite cannot coexist at equilibrium:

$$2SiO_2(s) + Al_2O_3 \cdot 3H_2O(s) \rightarrow Al_2O_3 \cdot 2SiO_2 \cdot 2H_2O + H_2O$$

(silica)   (gibbsite)   (kaolinite)

and neither can calcite, kaolinite, and anorthite under a fixed (arbitrary) pressure of CO_2:

$$CaO \cdot Al_2O_3 \cdot 2SiO_2 + CO_2(g) + 2H_2O \rightarrow Al_2O_3 \cdot 2SiO_2 \cdot 2H_2O + CaCO_3(s)$$

(anorthite)   (kaolinite)   (calcite)

Thermodynamically, these reactions have to proceed all the way in one direction—to the right under ambient P_{CO_2}. The solids may, however, coexist because of very slow kinetics. For example kaolinite results from the weathering of alumino-silicates but does not form at ordinary pressures and temperatures from silica and gibbsite or from dissolved silicic acid and aluminum.

The CO_2 of the atmosphere must be buffered by some mechanism, however. One way that P_{CO_2} may be fixed geochemically over long times (greater than 10^5 y) is in fact through an equilibrium between some calcium (or magnesium) silicate and carbonate phases similar to that written above for anorthite. The effective constant of such a reaction must then match the long-term average P_{CO_2} of the atmosphere.

5. THE ACQUISITION OF ALKALINITY IN FRESHWATER

The story of how freshwaters acquire their alkalinity is essentially the story of how freshwaters acquire their major dissolved cations, Ca^{2+}, Na^+, K^+, and Mg^{2+}, from basic rocks. Most of the important noncarbonate anions (Cl^- and SO_4^{2-}) in freshwaters originate from the dissolution of chloride and sulfate salts (e.g., $NaCl$ and $CaSO_4$), which are not acidic and contribute no net negative alkalinity. The major exception to this is the dissolution of sulfide minerals such as pyrite (FeS_2), whose global negative contribution to freshwater alkalinity is difficult to quantify but probably small.

Following Garrels and MacKenzie[3] consider, for example, the average composition of North American rivers. Table 5.5 shows how the major ions can be roughly accounted for in five steps:

TABLE 5.5 Synthesis of Average North American Freshwater from Rock Types[a]

	Average Water	From Atmosphere[b]	From Halite	From Gypsum	From Limestone	From Feldspars
HCO_3^-	1.25	0.04			1.01	0.20
SO_4^{2-}	0.42	0.03		0.39		
Cl^-	0.23	0.10	0.13			
Ca^{2+}	1.05	0.04		0.39	0.62	
Mg^{2+}	0.42	0.03			0.39	
Na^+	0.39	0.09	0.13			0.17
K^+	0.04	0.01				0.03
H_2SiO_3	0.15	0.01				0.14

Source: Adapted from Garrels and MacKenzie (1971).[3]

[a] The figures given are in milliequivalents per liter; H_2SiO_3 is given in millimolar.

[b] From rainwater analysis. Does not include the CO_2 utilized in the weathering of limestone and feldspar.

1. The atmospheric contribution is estimated on the basis of rainwater composition.

2. All remaining chloride is taken to be halite dissolution:

$$NaCl \rightarrow Na^+ + Cl^-$$

3. The sulfate is attributed to calcium sulfate (gypsum):

$$CaSO_4 \cdot 2H_2O \rightarrow Ca^{2+} + SO_4^{2-} + 2H_2O$$

4. Magnesium and the balance of Ca^{2+} are ascribed to carbonate rock:

$$CaCO_3 + CO_2(g) + H_2O \rightarrow Ca^{2+} + 2HCO_3^-$$

$$MgCO_3 + CO_2(g) + H_2O \rightarrow Mg^{2+} + 2HCO_3^-$$

5. Finally, the silicate and the monovalent cations are taken to originate from the weathering of feldspars. In order to obtain the low ratio of H_2SiO_3 to the remaining $Na^+ + K^+$ (approximately 1 compared to about 3 in the rock), we can hypothesize reprecipitation of some of the silicic acid:

$$K_2O \cdot Al_2O_3 \cdot 6SiO_2(s) + 2CO_2(g) + 2.4H_2O$$

(K-feldspar)

$$\rightarrow 2K^+ + 2HCO_3^- + Al_2O_3 \cdot 2SiO_2(s) + 2.6SiO_2(s) + 1.4H_2SiO_3(aq)$$

(kaolinite)

and

$$Na_2O \cdot Al_2O_3 \cdot 6SiO_2(s) + 2CO_2(g) + 2.4H_2O$$

(albite)

$$\rightarrow 2Na^+ + 2HCO_3^- + Al_2O_3 \cdot 2SiO_2(s) + 2.6SiO_2(s) + 1.4H_2SiO_3(aq)$$

(kaolinite)

This very simplified genealogy of average North American river water shows that some 80% of the alkalinity is attributable to carbonate rock dissolution (it makes no difference how the Mg^{2+} and Ca^{2+} are distributed between sulfate and carbonate) and 20% to the weathering of alumino-silicate rocks. More detailed calculations based on different estimations for the world average river water[1] (with a 50% higher alkalinity) show only 60% of the alkalinity to be contributed by dissolution of carbonates, with most of the remainder originating from the dissolution of calcium and magnesium silicates and aluminosilicates.

The major interest in the sort of genealogical analysis shown in Table 5.5 is not merely to correlate average water and average rock composition, but rather to understand the differences among natural waters, to relate a particular water chemistry to the regional mineralogy. The next four examples illustrate on the basis of simple equilibrium models how such important solids as carbonates, alumino-silicates, and sulfides contribute to the alkalinity of freshwaters. Although these examples are instructive and the results of the calculations are within the range of observations, they should not be taken too literally. It must be remembered that the thermodynamic properties of some of the actual mineral phases are not easily defined and that equilibrium may be a poor approximation of reality, particularly in well-drained areas where the residence time of the water in the soil is short.

Example 9. Dissolution of $CaCO_3(s)$ at a Given P_{CO_2} As the first and simplest example let us consider the dissolution of calcium carbonate under a fixed partial pressure of CO_2.
Recipe

1. $P_{CO_2} = 10^{-2.5}$ atm (note the high value typical of groundwater).
2. $CaCO_3(s)$ is present at equilibrium with the water.

The choice of principal components is straightforward, leading to Tableau 5.2:

$$TOTH = [H^+] - [OH^-] - [HCO_3^-] - 2[CO_3^{2-}] + 2[Ca^{2+}] = 0 \quad (76)$$

The major terms are $[HCO_3^-]$ and $[Ca^{2+}]$:

TABLEAU 5.2

	H^+	CO_2	$CaCO_3$	$\log K$
H^+	1			
OH^-	-1			-14.0
HCO_3^-	-1	1		-7.8
$H_2CO_3^*$		1		-1.5
CO_3^{2-}	-2	1		-18.1
Ca^{2+}	2	-1	1	9.8
CO_2		1		
$CaCO_3$			1	

$$10^{-7.8}[H^+]^{-1}P_{CO_2} = 2 \times 10^{9.8}[H^+]^2(P_{CO_2})^{-1} \tag{77}$$

$$[H^+]^3 = 10^{-17.9}(P_{CO_2})^2 = 10^{-22.9} \tag{78}$$

therefore

$$pH = 7.63$$

$$[Ca^{2+}] = (10^{9.8})(10^{-15.3})(10^{2.5}) = 10^{-2.97} = 1.07 \times 10^{-3}\,M$$

$$Alk = [HCO_3^-] = 2[Ca^{2+}] = 2.14 \times 10^{-3}$$

The alkalinity of such a system is typical of freshwaters in calcareous regions (and about twice average river water). A general relationship between alkalinity and P_{CO_2} for this system can be obtained:

$$Alk = [HCO_3^-] = 10^{-7.8}P_{CO_2}[H^+]^{-1} \tag{79}$$

Introducing $[H^+] = 10^{-5.97}(P_{CO_2})^{2/3}$ from Equation 78:

$$Alk = 10^{-1.83}(P_{CO_2})^{1/3} \tag{80}$$

*   *   *

The process by which the water acquires its alkalinity in a situation like that of Example 9 can be represented as a single reaction:

$$CO_2(g) + CaCO_3(s) + H_2O = 2HCO_3^- + Ca^{2+}$$

Half of the carbon thus dissolved into the water originates from the rock and half from the gas phase. If the effective partial pressure of CO_2 is increased, the amount of carbonate rock that is dissolved also increases. As a result, precipitation–dissolution of carbonate rock is emphatically not a homeostatic mechanism for the total carbon in the atmosphere and the hydrosphere. For example, the $CO_2(g)$ produced by fossil fuel burning cannot ultimately find its way into limestone according to the low-temperature geochemical processes of the carbonate system. In fact, the historical increase in P_{CO_2} should ultimately result in an intensified weathering of carbonate rock on the continents or in a decrease of net carbonate sedimentation on the ocean floor or both.

As illustrated in Table 5.5, sodium and potassium in freshwaters originate principally from the weathering of alumino-silicates. The reaction usually involves the formation of another solid ("incongruent dissolution") and has the general formula:

$$(Na, K)Al\cdot silicate + H^+ \rightarrow Na^+, K^+ + H_2SiO_3 + weathered\ Al\cdot silicate$$

The stoichiometry is of course variable, involving more or less water, and the hydrogen ion is usually provided by the dissolution and dissociation of CO_2. Let us consider the weathering of an igneous rock, albite (pure Na-feldspar), to the common clay Na-montmorillonite.

Example 10. Weathering of Albite to Na-Montmorillonite

Recipe

1. $P_{CO_2} = 10^{-2.5}\,atm$
2. Albite: $Na_2O\cdot Al_2O_3\cdot 6SiO_2(s)$
3. Na-montmorillonite: $Na_2O\cdot 7Al_2O_3\cdot 22SiO_2\cdot nH_2O(s)$

From Table 5.4 we can write the reaction:

$$\tfrac{7}{20}Albite + \tfrac{3}{5}H^+ + H_2O = \tfrac{1}{20}Na\text{-mtte} + H_2SiO_3 + \tfrac{3}{5}Na^+ \qquad pK = -0.43$$

Choosing H^+ and Na^+ to round out the set of components yields Tableau 5.3.

$$TOTH = [H^+] - [OH^-] - [HCO_3^-] - 2[CO_3^{2-}] + \tfrac{3}{5}[H_2SiO_3] - \tfrac{2}{5}[HSiO_3^-]$$
$$= 0 \tag{81}$$

$$TOTNa = [Na^+] - \tfrac{3}{5}[H_2SiO_3] - \tfrac{3}{5}[HSiO_3^-] = 0 \tag{82}$$

The major terms are $[Na^+]$, $[H_2SiO_3]$, and $[HCO_3^-]$:

$$\tfrac{3}{5}[H_2SiO_3] = [HCO_3^-] \tag{83}$$

$$\tfrac{3}{5}[H_2SiO_3] = [Na^+] \tag{84}$$

TABLEAU 5.3

	H^+	Albite	Na-mtte	CO_2	Na^+	$\log K$
H^+	1					
OH^-	-1					
$H_2CO_3^*$				1		-1.5
HCO_3^-	-1			1		-7.8
CO_3^{2-}	-2			1		-18.1
Na^+					1	
H_2SiO_3	$3/5$	$7/20$	$-1/20$		$-3/5$	0.43
$HSiO_3^-$	$-2/5$	$7/20$	$-1/20$		$-3/5$	-9.43
CO_2				1		
Albite		1				
Na-mtte			1			

Introducing the mass law for $[H_2SiO_3]$ in Equation 84 leads to:

$$\tfrac{3}{5} \times 10^{0.43} [H^+]^{3/5} [Na^+]^{-3/5} = [Na^+] \tag{85}$$

therefore

$$[Na^+] = (\tfrac{3}{5} \times 10^{0.43} [H^+]^{3/5})^{5/8} = 10^{0.13} [H^+]^{3/8} \tag{86}$$

$$\tfrac{3}{5}[H_2SiO_3] = 10^{0.13} [H^+]^{3/8} \tag{87}$$

Substituting Equation 87 into Equation 83 and expressing the mass law for $[HCO_3^-]$ yields

$$10^{0.13} [H^+]^{3/8} = 10^{-7.8} \times P_{CO_2}[H^+]^{-1} \tag{88}$$

therefore

$$[H^+]^{11/8} = 10^{-10.43} \tag{89}$$

$$pH = 7.59$$

$$[Na^+] = 10^{0.13}(10^{-7.59})^{3/8} = 10^{-2.71} = 1.9 \times 10^{-3} \, M$$

$$Alk = [Na^+] = [HCO_3^-] = 1.9 \times 10^{-3} \, M$$

The alkalinity of this system is slightly more than that of the average freshwater. The general relationship between alkalinity and P_{CO_2} can be obtained by combining Equations 86 and 88:

$$Alk = 10^{-2.03} \times (P_{CO_2})^{3/11} \tag{90}$$

* * *

TABLE 5.6   Mean Composition of Rain and Stream Waters at Hubbard Brook, New Hampshire

		Concentration (μM)	
	Rain	Stream (812 m) (High Elevation)	Stream (415 m) (Low Elevation)
H^+	73.9	18.4	2.1
Al(III)	—	26.3	5.6
Ca^{2+}	4.3	21.8	42.5
Na^+	5.2	16.1	37.8
Mg^{2+}	2.1	8.6	18.5
NH_4^+	12.9	2.1	0.4
SO_4^{2-}	29.9	62.5	59.3
NO_3^-	23.1	32.3	16.1
SiO_2	—	36.6	103
pH	4.13	4.74	5.68

Source: Johnson et al. (1981).[10]

The weathering of alumino-silicates usually proceeds further than the formation of montmorillonites. The cations can be completely dissolved, leading to the formation of kaolinite, the most thermodynamically stable of the clays. The alkalinity of a system in which Na-montmorillonite is weathered to kaolinite (cf. Problem 5.11) is even less sensitive to the partial pressure of CO_2 than those of the previous examples.

Many more examples of such weathering reactions could be presented. They could also be combined in the same model to obtain representative compositions of freshwater systems. Even if the absolute concentrations of ions do not correspond to an equilibrium for the weathering reactions, the ratios of major elements dissolved into the water are usually good indicators of the minerals from which they originate.

The weathering of alumino-silicates is also the long-term sink for acidity of modern "acid rain." In contrast with natural "acid rain" (where the acidity is due chiefly to CO_2), the acidity in modern "acid rain" is due to SO_4^{2-} and NO_3^- (see Table 5.6). The source of this acidity is the anthropogenic input of SO_2 and NO_x, which are oxidized in the atmosphere. Because the weathering reactions of alumino-silicates proceed rather slowly, mineral acidity of the rain is neutralized in the short term by reaction with more labile, though less abundant, materials. In "sensitive" areas (i.e., relatively poor in carbonate minerals), weathering of aluminum oxides can be significant.[10]

Example 11. "Acid Rain" and Chemical Weathering   The chemical composition of rainwater and streamwater in the Hubbard Brook Experimental Forest, New Hampshire, has been extensively monitored (see Table 5.6). The initial change in composition (in H^+ and Al) between rainwater and streamwater at the highest elevation has been ascribed to dissolution of reactive aluminum oxides in the soil zone. Later changes in streamwater composition have been attributed to chemical weathering of primary silicate minerals.

For the initial reaction of rainwater (at pH 4.13) with gibbsite, we may simply treat the rainwater as a dilute solution of a single strong acid (e.g., HNO_3), thus,

$$\text{Recipe} \quad [HNO_3]_T = 10^{-4.13} \, M$$
$$\text{Gibbsite: } Al(OH)_3(s)$$

The reactions for the dissolution of gibbsite and aluminum hydrolysis (neglecting polymeric species) are:

Reactions

$$Al(OH)_3(s) + 3H^+ = Al^{3+} + 3H_2O \qquad pK = -8.5$$

$$AlOH^{2+} + H^+ = Al^{3+} + H_2O \qquad pK = -5.0$$

$$Al(OH)_2^+ + 2H^+ = Al^{3+} + 2H_2O \qquad pK = -9.3$$

$$Al(OH)_3^0 + 3H^+ = Al^{3+} + 3H_2O \qquad pK = -15.0$$

$$Al(OH)_4^- + 4H^+ = Al^{3+} + 4H_2O \qquad pK = -23.0$$

We may calculate the equilibrium pH and total, dissolved aluminum concentration from Tableau 5.4 and the $TOTH$ equation:

$$TOTH = [H^+] - [OH^-] + 3[Al^{3+}] + 2[AlOH^{2+}] + [Al(OH)_2^+]$$
$$- [Al(OH)_4^-] = 10^{-4.13} \qquad (91)$$

Since the concentrations of the Al species are determined by gibbsite solubility, we may express all their concentrations as functions of $[H^+]$, then (neglecting

TABLEAU 5.4

	H^+	NO_3^-	Gibbsite	$\log K$
H^+	1			
OH^-	-1			-14
NO_3^-		1		
Al^{3+}	3		1	8.5
$AlOH^{2+}$	2		1	3.5
$Al(OH)_2^+$	1		1	-0.8
$Al(OH)_3^0$			1	-6.5
$Al(OH)_4^-$	-1		1	-14.5
$[HNO_3]_T$	1	1		$10^{-4.13} \, M$
$[Gibbsite]_T$			1	

$[OH^-]$ and $[Al(OH)_4^-]$),

$$TOTH = [H^+] + 3 \times 10^{8.5}[H^+]^3 + 2 \times 10^{3.5}[H^+]^2 + 10^{-0.8}[H^+] = 10^{-4.13}$$

$$(92)$$

Trial and error solution of this polynomial gives

$$pH = 4.50$$

The corresponding concentrations of all the Al species can be calculated for this pH:

$$[Al^{3+}] = 10^{-5.0} M$$

$$[AlOH^{2+}] = 10^{-5.5} M$$

$$[Al(OH)_2^+] = 10^{-5.3} M$$

$$[Al(OH)_3^0] = 10^{-6.5} M$$

$$\text{total dissolved Al} = 10^{-4.73} M = 18.5 \, \mu M$$

Comparing these calculated concentrations with the average composition of the high elevation streamwater (in Table 5.6), we see that our calculations underpredict the extent of neutralization ($pH_{pred} = 4.50$ vs. $pH_{obs} = 4.74$). This implies that other neutralization reactions are also important. This possibility is supported by the increase in $[Ca^{2+}]$, which suggests that Ca-containing minerals are also being chemically weathered. Furthermore, the observed total dissolved Al concentration exceeds the predicted value. The calculation, however, does not include all possible dissolved aluminum species; organically complexed aluminum, inorganic SO_4^{2-} and F^- complexes, and polymeric species have been neglected.

Although dissolution of gibbsite initially neutralizes some of the acidity of the rainwater, $Al(OH)_3(s)$ dissolution is not a "permanent" acidity sink because, as the pH increases, eventual reprecipitation of $Al(OH)_3(s)$ returns acidity to the water.

To account for the neutralization of the streamwater observed at lower elevation, we shall also consider weathering of alumino-silicate minerals by including anorthite in our recipe.

$$\text{Recipe} \quad [HNO_3]_r = 10^{-4.13} M,$$
Gibbsite: $Al(OH)_3(s)$
Anorthite: $CaO \cdot Al_2O_3 \cdot 2SiO_2(s)$

We may then calculate, for the pH of the lower elevation streamwater (pH 5.7), the dissolved aluminum and calcium concentrations resulting from the weathering of gibbsite and anorthite. Here, we consider the weathering of anorthite to lead directly to gibbsite and silicic acid without formation of crystalline secondary minerals such as kaolinite. The choice of this weathering reaction is consistent with the field observation that crystalline secondary minerals are not found as weathering products of granitic rocks in acidic catchments.[11] Thus we include, in addition to reactions for gibbsite dissolution and Al hydrolysis, the reactions

$$CaO \cdot Al_2O_3 \cdot 2SiO_2 + 2H^+ + 4H_2O$$

$$= 2Al(OH)_3(s) + 2H_2SiO_3 + Ca^{2+} \qquad pK = -11.16$$

$$H_2SiO_3 = HSiO_3^- + H^+ \qquad pK = 9.86$$

From Tableau 5.5, we obtain the $TOTH$ equation:

$$TOTH = [H^+] - [OH^-] + 3[Al^{3+}] + 2[AlOH^{2+}] + [Al(OH)_2^+]$$

$$- [Al(OH)_4^-] + 2[Ca^{2+}] - [HSiO_3^-] = 10^{-4.13} \qquad (93)$$

Expressing the concentrations of the Al species in terms of $[H^+]$, we can calculate $[Ca^{2+}]$ at the pH of the lower elevation streamwater, pH = 5.7. We neglect the species OH^-, $Al(OH)_4^-$, and $HSiO_3^-$, which are unimportant below neutral pH. Then

$$[Ca^{2+}] = \tfrac{1}{2}(10^{-4.13} - [H^+] - 3 \times 10^{8.5}[H^+]^3$$

$$- 2 \times 10^{3.5}[H^+]^2 - 10^{-0.8}[H^+])$$

$$[Ca^{2+}] = 10^{-4.445} M = 35.9 \,\mu M \qquad (94)$$

TABLEAU 5.5

	H^+	NO_3^-	Anorthite	H_2SiO_3	Gibbsite	$\log K$
H^+	1					
OH^-	-1					-14
NO_3^-		1				
Al^{3+}	3				1	8.5
$AlOH^{2+}$	2				1	3.5
$Al(OH)_2^+$	1				1	-0.8
$Al(OH)_3^0$					1	-6.5
$Al(OH)_4^-$	-1				1	-14.5
Ca^{2+}	2		1	-2	-2	11.16
H_2SiO_3				1		
$HSiO_3^-$	-1			1		9.86
$[HNO_3]_T$	1	1				$10^{-4.13} M$
$[Gibbsite]_T$					1	
$[Anorthite]_T$			1			

which is a reasonable approximation of the observed (mean) calcium concentration in lower elevation streamwater.

Here we have used the tableau, at least for the calcium produced by anorthite weathering, only for bookkeeping, that is, to keep track of the stoichiometry of the weathering reaction. We have calculated $[Ca^{2+}]$ for an observed streamwater pH, but these values of pH and $[Ca^{2+}]$ do not correspond to equilibrium of the water with anorthite. This bookkeeping application of the tableau introduced in Chapter 1 is useful for systems where the kinetics are sufficiently slow that equilibrium is not readily achieved (the kinetics of weathering reactions are discussed in Section 7) or for reactions where the equilibrium constant is so large that the reaction goes to completion (as illustrated in the next example).

As a drastic departure from the examples of weathering of alumino-silicate rocks, let us examine the case of an acid mine drainage: the acquisition of negative alkalinity (i.e., mineral acidity) from sulfide rocks.

Example 12. Dissolution and Oxidation of Pyrite In acid mine drainage sites, the oxidative weathering of sulfide ores and mine tailings produces acidic, metal-enriched waters. The chemical composition of acid mine drainage water for a California mining district is given in Table 5.7. We may use these data to predict the extent of pyrite dissolution and the resulting pH of the drainage water. Let us assume that the chemical composition of the drainage water results from the reaction of groundwater of zero alkalinity (unpolluted rainwater) with pyrite ($FeS_2 \cdot s$), other sulfide minerals (ZnS and CuS), gibbsite, silicates, and alumino-

TABLE 5.7 Composition of Acid Mine Drainage Waters

	Concentration (M)	Log Concentration
H^+		-2.8
total Fe	2.15×10^{-3}	-2.67
Fe(II)	2.22×10^{-4}	-3.91
Fe(III)[a]	2.03×10^{-3}	-2.69
Mn	2.37×10^{-5}	-4.63
Zn	2.66×10^{-4}	-3.58
Cu	2.58×10^{-4}	-3.59
Al	7.04×10^{-4}	-3.15
Na	1.83×10^{-4}	-3.74
K	7.42×10^{-6}	-5.13
Ca	5.24×10^{-4}	-3.28
Mg	6.58×10^{-4}	-3.18
SO_4	6.60×10^{-3}	-2.18
Cl	1.97×10^{-5}	-4.71
SiO_2	4.99×10^{-4}	-3.30

Source: Filipek et al. (1987)[12] (site 22).

[a]Fe(III) calculated from the difference between (measured) total Fe and Fe(II).

silicates. As we shall see, the dissolution of pyrite produces acidity which is, in turn, partially titrated in subsequent weathering reactions.

Recipe

FeS_2, CuS, ZnS, enstatite ($MgO \cdot SiO_2$), anorthite ($CaO \cdot Al_2O_3 \cdot 2SiO_2$), gibbsite ($Al(OH)_3$)

Reactions

$$FeOH^{2+} + H^+ = Fe^{3+} + H_2O \qquad\qquad pK = -2.2$$

$$H_2SiO_3 = HSiO_3^- + H^+ \qquad\qquad pK = 9.86$$

$$FeOH_3(s) + 3H^+ = Fe^{3+} + 3H_2O \qquad\qquad pK = -3.2$$

$$Al(OH)_3(s) + 3H^+ = Al^{3+} + 3H_2O \qquad\qquad pK = -8.5$$

$$MgO \cdot SiO_2(s) + 2H^+ = Mg^{2+} + H_2SiO_3 \qquad\qquad pK = -11.64$$

$$CaO \cdot Al_2O_3 \cdot 2SiO_2(s) + 2H^+ + 4H_2O$$
$$= Ca^{2+} + 2Al(OH)_3(s) + 2H_2SiO_3 \qquad\qquad pK = -11.61$$

$$FeS_2(s) + \tfrac{7}{2}O_2 + HH_2O \rightarrow Fe^{2+} + 2SO_4^{2-} + 2H^+$$

$$Fe^{2+} + \tfrac{1}{4}O_2 + H^+ \rightarrow Fe^{3+} + \tfrac{1}{2}H_2O$$

$$ZnS(s) + 2O_2 \rightarrow Zn^{2+} + SO_4^{2-}$$

$$CuS(s) + 2O_2 \rightarrow Cu^{2+} + SO_4^{2-}$$

For simplicity, we omit the aluminum hydrolysis species (and some of the Fe(III) hydrolysis species) since these are not important at low pH. As in the previous example, we again use the tableau more to account for the stoichiometry of the weathering reactions than to calculate equilibria with solid phases. Although equilibrium calculations may be applied to the dissolution of some minerals such as gibbsite, such calculations are chemically meaningless (though mathematically feasible) for the oxidation reactions. As we shall see in Chapter 7, the value of the equilibrium constants are so large that the reactions effectively proceed to completion in the presence of oxygen. With this caveat in mind, we may compose Tableau 5.6 using the solutes listed in Table 5.7 (i.e., the products of the weathering reactions above) and $Fe(OH)_3(s)$ as components.

The advantage of setting up the tableau in this form is that of all the solutes listed in Table 5.7, only H^+ and the iron species appear in the $TOTH$ equation.

$$TOTH = [H^+] - [OH^-] + 2[Fe^{2+}] + 3[Fe^{3+}] + 2[FeOH^{2+}] - [HSiO_3^-]$$
$$= 4[\text{pyrite}]_T - 3[\text{gibbsite}]_T - 2[\text{enstatite}]_T - 8[\text{anorthite}]_T \qquad (95)$$

TABLEAU 5.6

	H^+	Ca^{2+}	Mg^{2+}	Al^{3+}	SO_4^{2-}	Zn^{2+}	Cu^{2+}	H_2SiO_3	$Fe(OH)_3(s)$	O_2	$\log K$
H^+	1										
OH^-	-1										-14
Ca^{2+}		1									
Mg^{2+}			1								
Al^{3+}				1							
SO_4^{2-}					1						
Zn^{2+}						1					
Cu^{2+}							1				
Fe^{2+}	2								1	$-1/4$	
Fe^{3+}	3								1		3.2
$FeOH^{2+}$	2								1		1.0
H_2SiO_3								1			
$HSiO_3^-$	-1							1			-9.86
$[Pyrite]_T$	4				2				1	$-15/4$	$10^{-2.52}$
$[ZnS]_T$					1	1				-2	$10^{-3.58}$
$[CuS]_T$					1		1			-2	$10^{-3.59}$
$[Gibbsite]_T$	-3			1							$10^{-3.47}$
$[Enstatite]_T$	-2		1					1			$10^{-3.18}$
$[Anorthite]_T$	-8	1		2				2			$10^{-3.28}$

The amount of a particular mineral that has reacted can be obtained from the TOT equation for the appropriate cation and the solute concentrations of Table 5.7. For example,

$$TOTCa = [Ca^{2+}] = [Anorthite]_T = 10^{-3.28} \tag{96}$$

Thus the total concentrations noted in the final column of Tableau 5.6 are based on the measured concentrations of Table 5.7. Note that the negative concentration for $[gibbsite]_T$ corresponds to gibbsite precipitation; the measured dissolved aluminum concentration is *less* than that contributed by anorthite dissolution. The relative contributions of sulfate from dissolution of pyrite (FeS_2) and of other sulfide minerals $(ZnS$ or $CuS)$ may be distinguished by comparing the $TOTSO_4$, $TOTZn$, and $TOTCu$ equations, that is,

$$TOTSO_4 = [SO_4^{2-}] = 2[pyrite]_T + [ZnS]_T + [CuS]_T = 10^{-2.18} \tag{97}$$

$$TOTZn = [Zn^{2+}] = [ZnS]_T = 10^{-3.58} \tag{98}$$

$$TOTCu = [Cu^{2+}] = [CuS]_T = 10^{-3.59} \tag{99}$$

Rearranging these equations, we obtain

$$[pyrite]_T = \tfrac{1}{2}([SO_4^{2-}] - [Zn^{2+}] - [Cu^{2+}]) = 3.04 \times 10^{-3} = 10^{-2.52} \tag{100}$$

With similar manipulations, we may obtain a numerical value for the $TOTH$ equation (neglecting $[HSiO_3^-]$):

$$TOTH = [H^+] - [OH^-] - 2[Fe^{2+}] + 3[Fe^{3+}] + 2[FeOH^{2+}]$$
$$= 4[pyrite]_T - 3[gibbsite]_T - 2[enstatite]_T - 8[anorthite]_T$$
$$= 4 \times 10^{-2.52} + 3 \times 10^{-3.47} - 2 \times 10^{-3.18} - 8 \times 10^{-3.28} = 10^{-2.11} \tag{101}$$

Then

$$[H^+] = 10^{-2.11} + 2[Fe^{2+}] - 3[Fe^{3+}] - 2[FeOH^{2+}] \tag{102}$$

If we assume that the dissolved Fe(III) is Fe^{3+}, then

$$[H^+] = 10^{-2.11} + 2 \times 10^{-3.91} - 3 \times 10^{-2.69} = 10^{-2.73}$$

Since $[FeOH^{2+}]$ is not negligible at this pH, we may, for an exact solution, express $[Fe^{3+}]$ and $[FeOH^{2+}]$ in terms of $[Fe(III)]_T$ and $[H^+]$, thus,

$$[Fe^{3+}] = \frac{[Fe(III)]_T}{1 + 10^{-2.2}[H^+]^{-1}} \tag{103}$$

and

$$[FeOH^{2+}] = \frac{10^{-2.2}[Fe(III)_T][H^+]^{-1}}{1 + 10^{-2.2}[H^+]^{-1}} \tag{104}$$

By substituting these expressions into Equation 102, we obtain a quadratic equation in $[H^+]$ with the solution

$$pH = 2.49$$

in reasonable agreement with the observed pH of 2.8 at this site.

We should note here that the persistence of Fe(II) in the acid mine drainage water is due to the kinetics of oxidation which are slow at low pH unless the reaction is catalyzed by autotrophic bacteria. Also, the reported dissolved Fe(III) concentration is very high (even for pH $\simeq$ 2–3). Precipitation of iron hydroxides, resulting in formation of orange coatings on the stream bed, is commonly observed in areas affected by acid mine drainage.[12]

6. THE CONTROL OF ALKALINITY IN THE OCEANS

The question of the alkalinity of seawater can be addressed on two different time and space scales. First, considering geological times—over which the oceans are well mixed—one may ask what geochemical processes control the concentration of the major ions in seawater and thus the alkalinity of the oceans. Second, on a much finer scale, one may want to examine the relatively small spatial and temporal variations in seawater alkalinity and to understand the causes of such variations.

6.1 Seawater Alkalinity over Geological Times

If we compare the major ion composition of the average river to that of seawater (Table 5.8), it is quite striking that seawater is not simply concentrated river water. The relative concentrations of the major cations are proportionally much lower in seawater than in rivers compared, for example, to chloride concentrations. As noted by Sillén,[13] we are very close to the equivalence point of the acidimetric titration of the oceans. What then are the oceanic mechanisms that remove cations in seawater more effectively than they remove anions? What are the alkalinity sinks that maintain the alkalinity of the ocean at its present value?

Chemical oceanography texts of yesteryear, which relied on experimental information before the discovery of the thermal, chemical, and biological activity associated with underwater volcanism along midocean ridges, had few known alkalinity sinks available to them.[14] They listed only three kinds of sedimentary material that seem to be removed to ocean sediments at rates in any way

TABLE 5.8   Major Ion Composition of Average River[a] and Seawater[b]

	Average River (mM)	Average Seawater (mM)
HCO_3^-	0.86	2.38
SO_4^{2-}	0.069	28.2
Cl^-	0.16	545.0
Ca^{2+}	0.33	10.2
Mg^{2+}	0.15	53.2
Na^+	0.23	468.0
K^+	0.03	10.2

[a] Data from Berner and Berner (1987)[2]. Note that the reported concentrations exclude pollution.
[b] Data from Holland (1978).[1]

comparable to river inputs of major ions:

1. The suspended load of rivers (physical weathering) that is deposited essentially unaltered at the bottom of the oceans (and is referred to as "detrital" material by oceanographers to confuse biologists).
2. Organic matter from the "soft parts" of sea-living organisms.
3. Calcium carbonate (calcite and aragonite) and silica (opal) precipitated as the "hard parts" of plants and animals (and thus also referred to as "organic" sediments by oceanographers to confuse chemists).

Of these various sedimentary deposits, only calcium carbonate represents an alkalinity sink. It was then widely believed that the other major cations, Mg^{2+}, Na^+, and K^+, were removed by some as yet unidentified "reverse weathering" reactions. The formation of authigenic alumino-silicates in ocean sediments was generally hypothesized as a major sink of cations and alkalinity, although there was really no experimental evidence for it.

On the basis of a rough mass balance, taking into account the principal sinks now documented for the major ions in seawater, we can calculate that calcium carbonate precipitation may in fact be the only important alkalinity sink in the ocean. As shown in Table 5.9, the other major ions appear to be simply exchanged for calcium in a variety of processes. (Note that Table 5.9, much like Table 5.5, is partly tautological since some numbers are chosen specifically to balance out elemental fluxes, and electroneutrality guarantees that everything works out in the end.)

An input–output balance for the major elements in seawater starts with an estimate of riverine inputs (Table 5.9, column 2). The chloride cycle is then closed by ascribing it entirely to aerosol formation and dissolution and—over geological times—to evaporite formation and weathering. The other elements

TABLE 5.9 Long Term Input–Output Balance for Major Seawater Ions and Alkalinity[a]

	Ocean Inventory (10^{18} mol)	River Input[b]	Atmospheric–Evaporite Cycling[c]	Ion Exchange[d]	Hydrothermal Activity[e]	Pyrite Burial or Other[f]	Carbonate Deposition[g]
Cl^-	710	+6.1	(−6.1)		(−0.9)		
Na^+	608	+8.5	−5.7	−1.9	(−3.1)		
Mg^{2+}	69	+5.2	−0.3	−1.2	(−1.7)		−0.6
SO_4^{2-}	37	+3.2	−0.3			−1.2	
K^+	13	+1.2	−0.06	−0.4	(−0.6)	−0.1	
Ca^{2+}	13	+12.5	−0.06	+2.6	(+2.0)		−17
Alk	3.1	+31.9	−0.06	+0.5	−0.4	+2.4	−35

[a] Data from McDuff and Morel (1980)[15] and Berner and Berner (1987).[2]

[b] Concentrations from Berner and Berner (1987);[2] discharge 3.74×10^{16} L y^{-1} (world average).

[c] Includes pore water burial.

[d] Values from McDuff and Morel (1980).[15]

[e] Includes both high- and low-temperature basalt alternation.

[f] For K, fixation on deltaic clays.

[g] Values from Berner and Berner (1987).[2]

are included in this cycle in proportions typical of sea salt aerosols and evaporite rocks. The ion exchange process taking place when suspended river material (detrital clays) contacts seawater is then estimated on the basis of laboratory experiments. Its net effect is to exchange Na^+ for Ca^{2+} which, together with some removal through seawater–volcanic reactions, roughly balances out the sodium fluxes. Sulfate is removed in part through biogenic pyrite formation and burial. This flux is estimated from the pyrite S/organic C ratio in anoxic marine sediments and the organic C removal rate. The balance of sulfate removal is attributed to seawater–basalt reactions, which occur as seafloor spreading exposes fresh basalt. High-temperature seawater-basalt reactions occur at active spreading centers at ridge crests; basalt reacts with seawater at lower temperatures on the ridge flanks. Such reactions are also the major sink for magnesium in the ocean. The hot water from the hydrothermal "vents" is effectively stripped of magnesium. (The results for Na^+ and Cl^- are variable.) For potassium, the major removal mechanisms have not been established. High-temperature seawater–basalt reactions are a source of K^+. The output flux (here chosen for mass balance) results from some combination of low temperature basalt alteration, reverse weathering, and fixation on deep-sea sediments. As noted above, Table 5.9, column 5 includes both high- and low-temperature basalt alteration. This distinction is significant because of the temperature-dependence of the rates of chemical reactions, for example, the reduction of sulfate by basaltic Fe(II). Quantification of element fluxes due to hydrothermal activity is complicated by insufficient information on relative contributions of high- and low-temperature processes to both heat and chemical fluxes. Although the values listed in Table 5.9 are consistent with the current understanding of hydrothermal processes, they were chosen to balance inputs and outputs of elements and should be regarded with due caution.[16-18]

Even with these caveats, it is clear that the net effects of the processes described above on alkalinity are fairly negligible. With an estimate for calcium removal by carbonate deposition based on analysis of ocean sediments, the calcium budget (for the last 25 million years) is essentially in balance. Thus alkalinity is also balanced ipso facto. The only other oceanic processes affecting alkalinity, biogenic pyrite removal, and basalt alteration, support only a small alkalinity flux compared with the major input from rivers and output by carbonate deposition.

According to this simple model of ocean chemistry, the concentrations of chloride and sodium in seawater are controlled by a balance between the rate of evaporite formation and its weathering. This is mostly dependent on the chance geological formation of closed basins and is probably reasonably constant on a scale of millions of years. Sulfate, potassium, and magnesium are controlled by the balance of weathering and hydrothermal activities (and, for sulfate, biogenic pyrite formation). Both weathering and seawater–basalt reactions may be dependent on tectonic activity, and their relative intensities may be less variable than their absolute intensities. Finally, calcium and carbonate are controlled by calcium carbonate precipitation, the partial pressure of CO_2 in the atmosphere (with the assumption that P_{CO_2} is controlled by other processes over geological time), and the constraint of mass balance. The resulting carbonate–pH model of the ocean then consists of three governing equations:

1. Charge balance. Neglecting weak acids and placing on the right-hand side the ions controlled by input–output processes:

$$2[Ca^{2+}] \simeq [Cl^-] + 2[SO_4^{2-}] - [Na^+] - [K^+] - 2[Mg^{2+}] \tag{105}$$

where the present-day $[Ca^{2+}] \simeq 10^{-2} M$.

2. Solubility of $CaCO_3(s)$. Using an apparent constant that takes into account ionic strength effects, complex formation, and pressure and temperature effects (e.g., $pK_{a2} = 9.3$ for carbonate; the solubility constant is actually chosen to be equal to the reaction quotient corresponding to present-day ocean composition):

$$\frac{[Ca^{2+}][HCO_3^-]}{[H^+]} = K_1 = 10^{3.5} \tag{106}$$

3. Solubility of $CO_2(g)$. Again using an apparent constant:

$$[HCO_3^-][H^+] = K_2 P_{CO_2} = 10^{-7.4} P_{CO_2} (\simeq 10^{-10.9} \text{ now}) \tag{107}$$

Algebraic manipulation of these three equations yields

$$pH \simeq 5.45 - \tfrac{1}{2}\log(P_{CO_2}) - \tfrac{1}{2}\log[Ca^{2+}] = 8.2 \tag{108}$$

$$Alk \simeq [HCO_3^-] + 2[CO_3^{2-}] \tag{109}$$

therefore

$$Alk \simeq 10^{-1.95} \left[\frac{P_{CO_2}}{[Ca^{2+}]}\right]^{1/2} + \frac{10^{-5.5}}{[Ca^{2+}]} \simeq 2.3 \times 10^{-3} \text{ eq L}^{-1} \tag{110}$$

With such a model one can make a simple study of the long-term stability of the acid–base chemistry of the oceans with respect to possible historical or eventual geological events such as changes in geothermal activity, variations in weathering rate, or increases in atmospheric CO_2.

6.2   Local Variations

Although on a large time and space scale the ocean might be considered globally at equilibrium with $CaCO_3(s)$, measurements in the water column show that $CaCO_3(s)$ is in fact supersaturated in surface waters and undersaturated at depth (Figure 5.10). This is largely due to the concretive activities of plants and animals

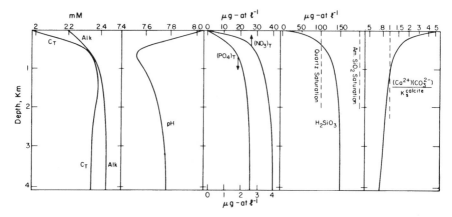

Figure 5.10 Typical vertical profiles for total carbonate, alkalinity, pH, phosphate, nitrate, and silicate concentrations, and calcium carbonate saturation in the Pacific Ocean. These idealized profiles[14] show the depletion of the principal algal nutrients nitrogen, phosphorus, and silicon in surface waters and the resulting alkalinity decrease and pH increase (Example 13). Note the pH minimum at the point where C_T and alkalinity are approximately equal. Calcite, which is supersaturated in surface waters, becomes undersaturated at depth due in part to the decrease in pH and the effect of pressure on the solubility of $CaCO_3(s)$ (see Figure 2.6).

in the euphotic zone. Coccolithophorids and foraminifera precipitate calcite and aragonite; diatoms and radiolarians precipitate opal (to much below its saturation). As both the hard and the soft parts of such organisms settle to deepwater and the organic matter is oxidized, the solids dissolve progressively. This is because both $CaCO_3$ and SiO_2 are undersaturated in the deep ocean, as shown in Figure 5.10. (Recall from Chapter 2 that the solubility product of $CaCO_3$ increases markedly with depth.) Obviously, these processes must have an effect on the alkalinity of the water. Let us calculate in a simple example the magnitude of such an effect and the relative importance of the various processes.

Example 13. Removal of Alkalinity by Biological Activity in Surface Water*
Consider a volume of Pacific deepwater upwelled into the euphotic zone of the productive equatorial region. The relevant initial conditions of the deepwater are

$$C_T^0 = 2.35 \times 10^{-3}\,M$$

$$Alk^0 = 2.45 \times 10^{-3}\,eq\,L^{-1}$$

$$[PO_4]_T^0 = 2.5 \times 10^{-6}\,M$$

*The conditions for this example were provided by R. Collier.

$$[NO_3]_T^0 = 4.0 \times 10^{-5} \, M$$

$$[H_2SiO_3]_T^0 = 1.5 \times 10^{-4} \, M$$

$$[Ca]_T^0 = 1.030 \times 10^{-2} \, M$$

Consider now that the productivity of the water may be attributed partly to diatoms and partly to coccolithophorids in such a way that roughly 0.5 mol of SiO_2 and 0.6 mol of $CaCO_3$ are precipitated for every mole of carbon fixed photosynthetically. The overall reaction of formation for living matter can thus be written in this particular case (using the principal components at the CO_2 equivalence point):

$$H_2PO_4^- + 16NO_3^- + 166H_2CO_3^* + 50H_2SiO_3 + 60\,Ca^{2+} \rightarrow$$

$$C_{106}H_{263}O_{110}N_{16}P_1 + 50SiO_2 + 60CaCO_3 + 103H^+ + 34H_2O + 138O_2$$

$$\text{(soft parts)} \qquad\qquad \text{(hard parts)}$$

As the reaction proceeds to the right, the alkalinity of the water decreases by 103 equivalents per mole carbon fixed (120 eq from $CaCO_3$ precipitation; -17 eq from photosynthetic nitrate and phosphate uptake). Considering that the productivity is limited by the nutrient supply, we obtain the change in composition (note that N and P are present in the desired 16:1 ratio):

$$\Delta[PO_4]_T = -2.5 \times 10^{-6}$$

$$\Delta[NO_3]_T = -4.0 \times 10^{-5}$$

$$\Delta C_T = -4.0 \times 10^{-5} \times \tfrac{166}{16} = -4.15 \times 10^{-4}$$

$$\Delta \text{Alk} = -4.0 \times 10^{-5} \times \tfrac{103}{16} = -2.57 \times 10^{-4}$$

$$\Delta[Ca]_T = -4.0 \times 10^{-5} \times \tfrac{60}{16} = -1.50 \times 10^{-4}$$

$$\Delta[H_2SiO_3]_T = -4.0 \times 10^{-5} \times \tfrac{50}{16} = -1.25 \times 10^{-4}$$

therefore

$$[PO_4]_T = 0$$

$$[NO_3]_T = 0$$

$$C_T = 1.94 \times 10^{-3} \, M$$

$$\text{Alk} = 2.19 \times 10^{-3} \, \text{eq L}^{-1}$$

$$[Ca]_T = 1.015 \times 10^{-2} \, M$$

$$[H_2SiO_3]_T = 2.5 \times 10^{-5} \, M$$

This example illustrates a few salient points with regard to the effect of biological activity on alkalinity in the oceans. First it is important to underscore that most of the alkalinity changes are due to $CaCO_3$ precipitation (and redissolution) with only a minor effect due to NO_3^- uptake (and regeneration) and no effect at all because of SiO_2 precipitation. Alkalinity variations are thus highly dependent on the kind of organism present in the surface water, the principal parameter being the ratio of carbon precipitated as $CaCO_3$ to the carbon fixed photosynthetically in organic matter. (Note that another factor in alkalinity variations implicit to this discussion — and always important in ocean chemistry — is the advection of water masses, since adjacent water layers may have different origins and exhibit different chemistry.)

All the changes in concentrations calculated in this example are measurable and rather typical of the differences between deep and surface Pacific water (original vs. final composition). A detailed explanation of profiles such as those of Figure 5.10 is complicated by the fact that suspended particulate nitrogen, organic carbon, and inorganic carbon are regenerated at different rates in the water column. For example, the organic carbon to nitrogen ratio in particles increases with depth (e.g., from 6 to 8), particularly in oligotrophic areas, demonstrating a faster regeneration of nitrogen than of carbon in the upper layers. However, the nitrogen released to the upper part of the water column, mostly as ammonia and urea, is reutilized rapidly by the biota with a zero net effect on alkalinity. Although the total net carbon fixation can thus be in excess of the original nitrate concentration compared to the Redfield ratio, the alkalinity variations are still stoichiometrically linked to the changes in calcium and nitrate concentrations. In all cases a major factor in determining the shape of profiles such as those in Figure 5.10 is the kinetics of calcium carbonate dissolution in the water column, a subject of great complexity and active research.

7. KINETICS OF PRECIPITATION AND DISSOLUTION

As noted previously, many of the precipitation–dissolution reactions that take place in aquatic systems are not rapid compared to the hydraulic residence time. A kinetic description of the processes would then be desirable. The physical, chemical, and biological processes that control the kinetics of solid precipitation and dissolution in natural waters are highly complex, intertwined, and frequently poorly understood. We may, however, develop some understanding of dissolution and precipitation kinetics in natural waters by elucidating the controlling mechanisms for these processes in simpler systems. Here we examine the dependence of reaction rates on the concentrations of the reacting species, on the concentrations of species that enhance mineral dissolution and on the nature, reactivity, and surface area of the reacting solid. Then observations of weathering rates in the field may be used to determine whether mechanistic models of dissolution (based on laboratory studies) are applicable under field conditions. The last example in the chapter illustrates the incorporation of dissolution kinetics into a pseudoequilibrium model for alumino-silicate weathering (as a reprise of Example 11).

7.1   The Solid–Solution Interface at the Molecular Level

Aside from the nucleation process (i.e., the initial formation of solid nuclei), the precipitation and dissolution of a solid phase are reasonably symmetrical processes. As depicted in Figure 5.11, four steps can be distinguished in the attachment and detachment of solutes to and from a solid surface: (1) diffusion in the solution boundary layer adjacent to the surface (three-dimensional or bulk diffusion), (2) adsorption–desorption reaction with the solid surface (3) migration on the surface to or from a step edge (two-dimensional or surface diffusion), and (4) migration along a step edge to or from a kink (one-dimensional or edge diffusion). Any of these steps, alone or in combinations, may limit the kinetics of precipitation or dissolution. When Step 1 only is limiting, the kinetics are said to be diffusion controlled; when Steps 2, 3, and/or 4 only are limiting, the kinetics are said to be controlled by a surface reaction. If, for simplicity, the precipitation–dissolution reaction is taken to involve only one solute with an equilibrium concentration, C_s, and a bulk concentration, C, Figure 5.12 illustrates the difference between these two extreme cases. If the overall reaction

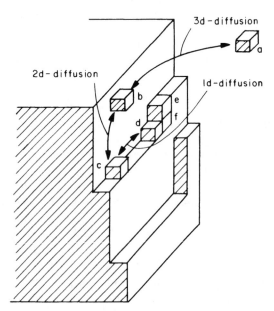

Figure 5.11   The various steps in the attachment and detachment of an ion or molecule to and from a solid lattice. During precipitation the ion or molecule becomes increasingly stable (lower free energy) as it becomes embedded deeper in the solid: a < b < c < d < e < f. Adapted from Nancollas and Reddy.[19]

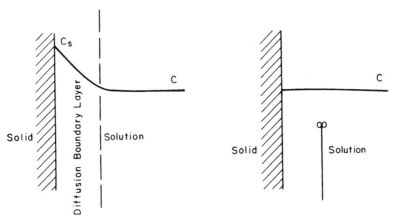

Figure 5.12   Control of dissolution kinetics by diffusion or surface reactions. If the kinetics are controlled by diffusion, there is a concentration gradient in a fluid layer adjacent to the solid: the thickness of that diffusion layer and the rate of dissolution are dependent on the mixing rate. At a sufficiently high mixing rate the diffusion gradient is effectively eliminated; the dissolution becomes controlled by a surface reaction independent of the mixing rate. A symmetrical situation could be depicted for the kinetics of precipitation.

is entirely controlled by the kinetics of the surface reaction (with comparatively rapid transport of solute to and from the surface), then the solute concentration is uniform in the solution and equal to C up to the solid-solution interface. Conversely, if transport is limiting, the solute concentration near the solid surface varies from the equilibrium value, C_s, at the surface, to the bulk concentration, C, far from the surface. Transport cannot, however, be limiting close to equilibrium (i.e., saturation) since at equilibrium there is no net flux of dissolved species at the solid-solution interface.[20]

It is rather intuitive that, in a saturated solution, molecules at the surface of a solid are less stable when they are in contact with more water and more stable when they are embedded deeper in the solid. For example, in the diagram of Figure 5.11 the surface molecules are increasingly bonded in the order b, c, d, e, and f. Thus both the thermodynamic stability and kinetic reactivity of a surface are highly dependent upon its geometry at the molecular level as well as on its chemical nature. A very smooth surface, without "kinks" or "holes," is less susceptible to dissolution than one that is chemically similar but has a more rugged aspect.[21,22] The thermodynamics and the kinetics of new hole formation on a smooth surface are in fact rather similar to that of nuclei formation in a homogeneous solution. The contribution of reaction at a particular site type, such as edges, kinks, or dislocations, to the overall dissolution reaction reflects both the reactivity (related to the instability) and the density of that site (i.e., its proportion of the total sites). Thus dissolution may not occur chiefly at the highest-energy sites but at more abundant, though lower-energy, sites. This explanation is consistent with the observation that dislocation density has little or no effect on the dissolution rates of quartz and calcite.[23,24] In natural waters, foreign ions or organic matter that adsorb preferentially at discontinuities on the surface may significantly retard both dissolution and precipitation. For example, the dissolution of calcite in artificial seawater[25] is inhibited by phosphate and its precipitation by Mg^{2+}.

7.2   Effects of Particle Size

Particle size influences both the thermodynamics and kinetics of dissolution and precipitation. These effects are due to the contribution of the surface energy to the free energy of the solid phase. Owing to their large surface to volume ratios, and hence their large surface (interfacial) free energies, small solids (< 1 μm) are inherently less stable than large solids. The calculated increase in solubility (decrease in pK_s) with decreasing particle size is shown in Figure 5.13 for goethite and hematite. The calculated decrease in pK_s ($-\delta pK_s$) depends on the specific interfacial free energy (or interfacial tension) of the solid. This effect is thus more important for goethite and hematite than for many other minerals, which tend to have lower specific interfacial free energies.[21] The greater total surface

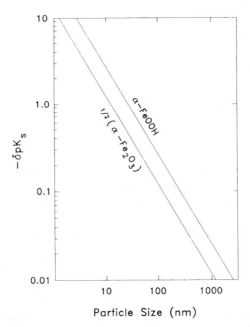

Figure 5.13   Effect of particle size on solubility for goethite (α-FeOOH) and hematite (1/2 α-Fe$_2$O$_3$)]. Solubility decreases as indicated by the decrease in $pK_s(-\delta pK_s)$, with increasing particle size. From Langmuir and Whitemore 1971.[26]

free energy of small particles is also the basis of the phenomenon of Ostwald ripening, a recrystallization process in which small particles are replaced by larger particles. Thus, in a closed system, Ostwald ripening results in an increase in mean particle size, broadening of the particle size distribution, and a decrease in the number of particles. Observed particle size distributions of clay minerals in shales have been attributed to Ostwald ripening.[27-29]

The effect of the surface energy of small particles on solubility is intimately related to the kinetics of precipitation and dissolution. The effects of particle size on reaction rates may, however, be difficult to distinguish from effects of the heterogeneity of the surface (i.e., the relative density of surface sites of varying reactivity).

7.3 Surface- versus Transport-Controlled Dissolution

Laboratory experiments on mineral dissolution have generally shown surface reactions, rather than transport, to be rate-limiting. Several types of observations support this conclusion. The hydrodynamics of the system, as determined by the stirring rate, have no effect on the rate of reaction when the stirring exceeds some minimum value. If transport were rate-limiting, a decrease in the thickness of the diffusion boundary layer due to increased stirring would increase the dissolution rate. Activation energies for dissolution, determined from the dependence of reaction rates on temperature (see Chapter 3), are significantly higher than expected for diffusion-limited reactions.[30] The presence of etch pits on the surfaces of crystals that have undergone some dissolution is not consistent with transport control, which should produce a more uniform dissolution over the surface. And, as discussed below, the dependence of the dissolution rate both on time and on the concentration of reactive species in solution supports surface rather than transport control. In some of the earlier work on dissolution kinetics, the observed time dependence of dissolution rates was attributed to transport control. These observations, however, were an artifact of sample preparation.[31] Thus, except for some extremely soluble minerals (e.g., calcite at low pH[25]) the model of surface-controlled dissolution can be applied to dissolution reactions under laboratory conditions.

Some evidence, particularly the morphology of naturally weathered mineral grains, suggests that dissolution processes may also be surface controlled under field conditions (see also Section 7.7). Transport-controlled dissolution may be important in soils, however, if unsaturated flow through soil macropores limits wetting of mineral surfaces and solute transport.[32]

7.4 Kinetics of Transport-Controlled Dissolution

The case of the diffusion control of precipitation–dissolution reactions is formally very similar to the one we have examined for the transfer of gases at the air–water interface (see Section 9 of Chapter 4). Molecular diffusion of solutes occurs across a boundary layer whose thickness is defined by the physics of the system (turbulence and particle size). The possible reactions among solutes in the boundary layer must be considered since the net transport of a particular solute to and from a solid surface may be effectively "enhanced" by reactions of other solutes, as demonstrated for the kinetics of CO_2 exchange at the air–water interface. For example, it may be necessary to consider transport of a total constituent, such as C_T, rather than of an individual species, such as CO_3^{2-}, since the corresponding acid–base reactions are fast.

Usually, a steady state is assumed and the possible effects of electrical forces on the transport of ions are not accounted for; in some instances, these effects could be important, as most solids in natural waters exhibit a sizable surface charge. The flux of a solute is then taken to be simply proportional to its concentration gradient, leading to an expression of the form:

$$\frac{dC}{dt} = E(C_{sat} - C) \tag{111}$$

where dC/dt is the time rate of change of the concentration C in the bulk solution due to the precipitation-dissolution process (other processes may be taking place as well); C_{sat} is the concentration of the solute at equilibrium with the solid; and E is an exchange coefficient that includes the physical characteristics of the solid (size, surface area, and porosity in the case of compacted sediments), the solid concentration, and the diffusion coefficient of the solute. Although theoretical expressions of E can be developed for simple situations, it is generally taken as an empirical parameter, with dimensions of inverse time. The relative importance of transport-control vs. surface-control in natural weathering reactions depends strongly on the hydrodynamic regime.

7.5 Kinetics of Surface-Controlled Dissolution

In the surface-controlled dissolution model[33-35] dissolution is treated as a two-step process. The first step involves a rapid, reversible sorption of reactive chemical species, such as protons, ligands, or reductants, from solution onto the surface. Coordination of reactants results in destabilization of metal centers at the surface of the crystalline lattice. For example, surface protonation polarizes and weakens metal–oxygen bonds in an oxide mineral. Formation of a surface complex by a reductant and electron transfer to the coordinated surface metal results in a less stable (and thus more easily detached) reduced metal center at the surface. The second step results in detachment of a metal from the surface of the crystalline lattice. Although both of these steps certainly consist of a series of elementary reactions, the rate law for surface-controlled dissolution is based on the assumption that the first step is fast and the second step rate-limiting. Rapid regeneration of the surface and reequilibration of the reactive surface species is assumed. The surface-controlled dissolution mechanisms are illustrated schematically for proton-promoted and ligand-promoted dissolution in Figure 5.14.

Under appropriate conditions, the surface-controlled dissolution model pre-

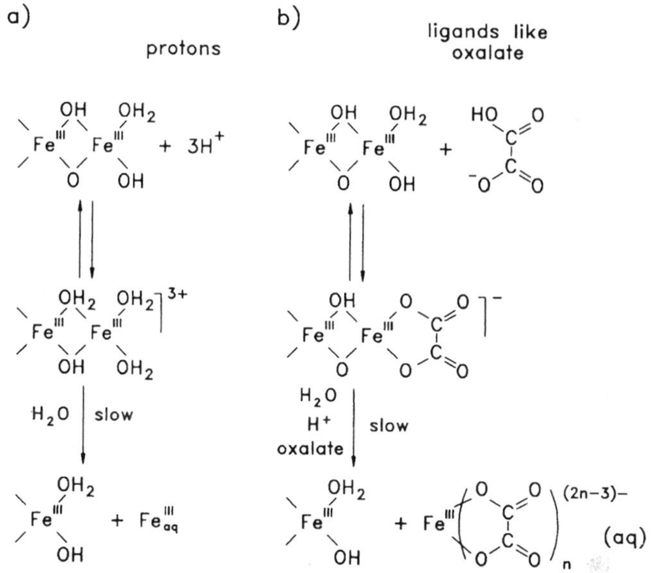

a)

protons

b)

ligands like
oxalate

Figure 5.14 Schematic representation of iron oxide dissolution. (*a*) Proton-promoted dissolution: protonation of surface hydroxyl groups enhances detachment of the surface Fe(III). (*b*) Ligand-promoted dissolution: formation of the surface complex (shown for the ligand oxalate) enhances detachment of the surface Fe(III).[33]

dicts steady-state dissolution, that is, a constant dissolution rate over time, where the pseudo zero-order dissolution rate is proportional to the concentration of the reactive surface species. Such a rate law would be valid under conditions that are far from equilibrium (so that back reactions can be neglected) and for which neither the surface area of the solid nor the dissolved concentration of the reactive species decreases significantly during the course of the dissolution. Steady-state dissolution of δ-Al_2O_3 is illustrated in Figure 5.15.

If, for a series of dissolution experiments, the dissolved concentration of the reactive species (e.g., pH in Figure 5.15) is varied, the dissolution rate also changes. Figure 5.16*a* shows the dependence of the dissolution rate of albite on solution pH. This dependence arises from the equilibrium between dissolved and sorbed chemical species, a topic that will be discussed in Chapter 8. For our purposes here, it is sufficient to say that, within a limited range in solute concentration, the surface concentration of a particular species will exhibit a fractional-order dependence on the dissolved concentration. The fractional-order dependence of the dissolution rate is then consistent with the surface-controlled mechanism for dissolution:

$$\equiv M-OH + H^+ \underset{}{\overset{fast}{\rightleftharpoons}} \equiv M-OH_2^+$$

$$\equiv M-OH_2^+ \xrightarrow{slow} M_{aq}^{n+} + \text{regenerated surface} \equiv M-OH$$

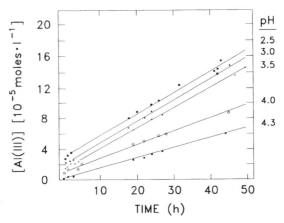

Figure 5.15   Concentration of dissolved Al over time during proton-promoted dissolution of aluminum oxide. Constant dissolution rates are consistent with control by a surface reaction. From Furrer and Stumm 1986.[33]

where $\equiv M-OH$ and $\equiv M-OH_2^+$ are surface species. The predicted rate law is

$$\text{Rate} = d[M_{aq}^{n+}]/dt = k[\equiv M-OH_2^+] \qquad (112)$$

The dissolution rate is often normalized to the surface area of the solid; then, if the surface concentration is expressed in moles per square meter, the rate constant, k, has units of reciprocal time. In a general form, this mechanism can also be applied to ligand-promoted, reductive, or oxidative dissolution reactions where, in each case, the rate is dependent on the surface concentration of the reactive species. In the case of albite dissolution (Figure 5.16a, b), the dependence of the dissolution rate on pH parallels the dependence of the surface concentration of H^+ (in the acidic region) or OH^- (in the basic region) on pH; this parallel indicates that the dissolution rate is directly related to the surface concentration of the reactive species. The dependence of the proton-promoted dissolution rate on the surface proton concentration, C_H^s, may not be simply first-order. Higher-order dependence, where

$$\text{rate} \propto (C_H^s)^i$$

has been observed for the proton-promoted dissolution of BeO ($i = 2$) and Al_2O_3 ($i = 3$). Some rates for proton-promoted mineral dissolution are given in Table 5.10. For a series of related minerals, dissolution rate constants may, in some cases, be estimated from a linear free energy relationship (LFER) with the corresponding thermodynamic solubility constants as seen in Chapter 3 (Figure 3.18).

One possible limitation on the applicability of the surface-controlled dissolu-

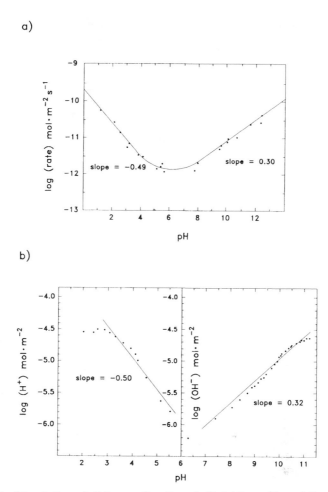

Figure 5.16 Dissolution of albite as a function of pH. (*a*) Logarithm of the dissolution rate as a function of pH. The dissolution rate decreases with increasing pH in the acidic region and increases with increasing pH in the basic region. This behavior may be explained by the dependence of the dissolution rate on surface concentrations of H^+ and OH^- as shown in (*b*). From Blum and Lasaga 1988.[36]

tion model involves the formation of altered surface layers. Formation of surface leached layers has been observed in the dissolution of silicate minerals. The initial mineral dissolution is *incongruent* (i.e., the stoichiometry of ions released into solution does not correspond to the mineral stoichiometry) as base cations (Ca^{2+}, Mg^{2+}, Na^+, and K^+) are depleted preferentially to silicon and aluminum. Continued growth of such leached layers, however, has not been observed and diffusion of ions through relatively thin leached layers (less than a few nanometers) is not thought to be rate limiting. Altered surface layers may also

TABLE 5.10 Dissolution Rates (at pH 5) for Silicate Minerals and Reaction Order in $[H^+]^a$

	Log Rate $(\text{mol SiO}_2\,\text{m}^{-2}\,\text{s}^{-1})$	Reaction Order
Framework Silicates		
Quartz	$-13.39,^b -11.85^c$	0
Opal	$-12.14,^b -10.92^c$	
Albite	$-11.32, -11.37^d$	0.49
Orthoclase	-11.77^d	
Oligoclase	-11.59^d	
Anorthite	-11.49	0.54
Nepheline	-8.55	1.0
Sheet Silicates		
Kaolinite	-12.55	
Muscovite	-12.70	
Chain Silicates		
Enstatite	-10.0	0.6
Diopside	-9.85	0.5
Tremolite	-10.47	
Augite	-10.60	0.7
Wollastonite	-7.70	0.7
Ring and Orthosilicates		
Olivine	-9.49	0.6

^{a}For detailed references see (for rates) Stumm and Wieland (1990)[37] and (for reaction orders in acid solutions) Schott (1990).[24]
^{b}In pure water.
^{c}In 0.2 M NaCl.
$^d P_{CO_2} = 1$ atm.

result from redox reactions at mineral surfaces. Dissolution of reduced iron minerals under oxic conditions has been shown to produce Fe(III)-containing surface layers. Under such conditions, accumulation of Fe(III) oxides on the surface of pyrite and of Fe^{3+} in a silicate matrix on the surface of Fe(II) silicates has been observed.[38,39] In this case, slow diffusion through the precipitated coating could control the rate of dissolution.

7.6 Precipitation Kinetics

The initial formation of solid nuclei by precipitation in a saturated solution (homogeneous nucleation) is a very complicated process, usually involving the formation of polymeric species. The interfacial free energy of small particles, mentioned above, results in an energy barrier to nucleation. The activation energy for formation of crystallization nuclei is related to the specific interfacial free energy, σ. For spherical particles,

$$\Delta G^{\ddagger} = \frac{16\pi\sigma^3 V^2}{3(\mathbf{k}T \ln \Omega)^2} \qquad (113)$$

where V is the volume of a formula unit of the mineral and Ω is the saturation quotient.[8,40] Experimentally, the kinetics of such homogeneous precipitation reactions are often difficult to reproduce, resulting in variable degrees of super-saturation before nucleation is observed.

Fortunately, in natural waters there are usually plenty of suspended particles to serve as nuclei for precipitating substances (heterogeneous nucleation). Thus the focus of this elementary discussion of precipitation kinetics in aquatic systems is on crystal growth.[28] Lest we should forget, however, natural waters are not sterile, and, as noted in Example 13, organisms are responsible for carrying out important precipitation–dissolution reactions. The kinetics of these biologically mediated reactions are often only indirectly controlled by the chemical composition of the water. Indeed, diatoms can precipitate opal even in undersaturated waters. Thus, the rate of SiO_2 or $CaCO_3$ precipitation may depend on the nitrate or phosphate concentrations that limit the growth of the precipitating organisms.

7.6.1 Kinetics of Crystal Growth Crystal growth, that is, precipitation on preformed nuclei, is the reverse of dissolution. Detailed balancing (introduced in Chapter 3) may be applied to precipitation and dissolution kinetics. Dissolution and precipitation of calcite have been studied at high pH and low P_{CO_2}.[41] Under these conditions, for the reaction

$$Ca^{2+} + CO_3^{2-} \underset{k_d}{\overset{k_p}{\rightleftharpoons}} CaCO_3(s)$$

the dissolution rate (normalized to surface area) is constant:

$$r_d = k_d \qquad (114)$$

The precipitation rate depends on the concentrations of the ions

$$r_p = k_p[Ca^{2+}][CO_3^{2-}] \qquad (115)$$

as shown in Figure 5.17. The calculated precipitation rates in this figure were obtained from the difference between the dissolution rate determined under conditions far from equilibrium (where precipitation is negligible) and the net dissolution rates observed near equilibrium

$$r_p = r_d - r_d^{net} \qquad (116)$$

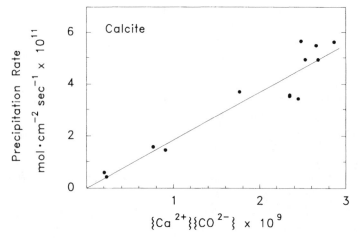

Figure 5.17 Rate of calcite precipitation as a function of the ion activity product $\{Ca^{2+}\}\{CO_3^{2-}\}$ determined during experiments in which net dissolution of calcite occurs. The precipitation rate is calculated from the difference between the net dissolution rate observed close to equilibrium and the maximum dissolution rate observed in very undersaturated solutions. From Chou et al. 1989.[41]

Good agreement is observed between the ratio of the rate constants for dissolution and precipitation (k_d and k_p) and the solubility product of calcite (where k_p corresponds to the slope of the line in Figure 5.17) in accordance with detailed balancing:

$$K_s = \frac{k_d}{k_p} \tag{117}$$

By applying detailed balancing, we obtain the result that the net precipitation rate of calcite is a function of the supersaturation of the solution. For

$$r_p^{\text{net}} = r_p - r_d = k_p[Ca^{2+}][CO_3^{2-}] - k_d \tag{118}$$

we obtain, by substituting for k_d (from Equation 117),

$$r_p^{\text{net}} = k_p K_s \left[\frac{[Ca^{2+}][CO_3^{2-}]}{K_s} - 1 \right] \tag{119}$$

or

$$r_p^{\text{net}} = k_p K_s (\Omega - 1) \tag{120}$$

Thus, net precipitation occurs when the solution is supersaturated, $\Omega > 1$, and net negative precipitation (i.e., net dissolution) when the solution is undersaturated, $\Omega < 1$.

The general result is that the net rates of dissolution and precipitation can be related as follows:

$$r_d^{net} = r_d - r_p = k_d(1 - \Omega) = -r_p^{net} = k_p K_s(1 - \Omega) \tag{121}$$

Note that a similar, though *not* identical, empirical expression

$$r = k(1 - \Omega)^n \tag{122}$$

is often used to describe observed precipitation–dissolution kinetics. The saturation quotient Ω does not appear in the rate law for the surface-controlled dissolution model as described in Section 7.5 where the model is applied *only* far from equilibrium (r_p negligible).

A more theoretical derivation of net crystal growth rates also gives the same dependence between the precipitation rate and the supersaturation (for highly supersaturated solutions). In the Burton–Cabrera–Frank (BCF) theory, the precipitation rate is considered to be limited by the rate of surface diffusion of ions adsorbed on the flat of the solid surface (Figure 5.11 type b) to "step" or "kink" sites. The "step" sites are formed by the intersection of crystal dislocations with the surface. The "step" sites propagate in a spiral during crystal growth with the overall result that, at high supersaturation, the precipitation rate is directly proportional to the supersaturation.[22,40]

In natural waters, however, many precipitation reactions are biologically mediated. In such cases, biological processes dictate the type and morphology of the mineral precipitated.[42] For example, calcite is precipitated in seawater by coccolithophorids (plants) and foraminifera (animals) and aragonite by pteropods (animals). Clearly, the rates of biological mineral precipitation may be quite different from the rates of abiotic, chemical precipitation.

7.7 Rates of Chemical Weathering

A first question in estimating rates of chemical weathering in nature is whether dissolution rates obtained in laboratory experiments for oxide and silicate minerals can be extrapolated to the field. As mentioned in Chapter 3, such an extrapolation can only be valid if the same mechanism is operative under both laboratory and field conditions. Laboratory weathering experiments on natural soils support this extrapolation.[32] Although initial release of cations is dominated by rapid ion-exchange reactions and dissolution of amorphous mineral phases, the long-term, steady dissolution observed for all ions is consistent with the model of surface-controlled dissolution.

The importance of surface control in natural weathering reactions is indicated by the inverse correlation between concentration of dissolved constituents in river waters and runoff (Figure 5.18). This correlation suggests that, above some

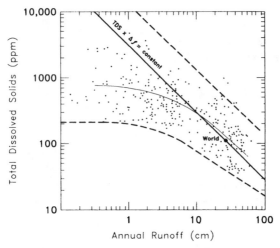

Figure 5.18   Total dissolved solids in river waters of the United States as a function of annual runoff. The dissolved load of rivers is a result of mineral dissolution. The observed relationship of dissolved solids concentration and runoff indicates that, at high flow, mineral dissolution rates are not transport limited, whereas at low flow, transport limitation appears to be important. From Holland 1978.[1]

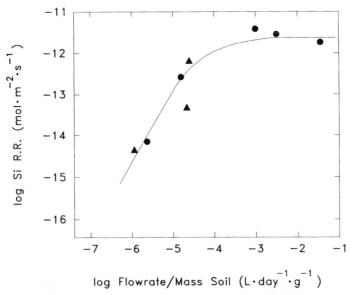

Figure 5.19   Weathering rate (silica release rate) as a function of the ratio of flow rate to mass. Maximum weathering rates are obtained in laboratory studies. The plateau in the weathering rates indicates surface-controlled dissolution. Decreasing weathering rates with lower flow rate:mass ratios are observed in field studies and are indicative of transport limitation. From Schnoor 1990.[32]

minimum value, increasing flow rates do not result in increasing rates of weathering; that is, weathering is not transport controlled except at very low flow. In studies of individual watersheds, the dependence of weathering rates on mass-normalized flowrate (Figure 5.19) indicates a shift from hydrologic control at low flow to surface control at high flow; weathering rates at high flow are comparable to laboratory experiments. It should be noted that these calculations depend on assumptions of the effective water depth in soil and of the wetted surface area and abundance of reactive minerals. These parameters are not easily estimated. Transport control may also influence the rate of mineral dissolution in lake or ocean sediments.

We may now attempt to predict the rates of chemical weathering in natural systems by using rate laws for surface-controlled dissolution obtained from laboratory experiments. Again, we must be cautious in extrapolating rate constants obtained in laboratory experiments to field conditions. In general, field-estimated weathering rates tend to be lower than those observed under laboratory conditions.[32]

Example 14. Kinetics of Chemical Weathering In Example 11, we examined the effect of gibbsite dissolution and precipitation and anorthite weathering on streamwater acidity. We did not apply equilibrium conditions in this example. We asked how much anorthite must be weathered to achieve a certain pH, rather than what the equilibrium pH would be for water in contact with some mineral phases. The concentrations of dissolved species as a function of time can also be examined if we can estimate the rate of anorthite weathering under natural conditions.

The weathering rate (determined as the silica release rate) for anorthite at pH 5 is given in Table 5.10 as

$$R_{Si} = 10^{-11.49} \, \text{mol SiO}_2 \cdot \text{m}^{-2} \, \text{s}^{-1} \qquad (123)$$

and the dependence of the dissolution rate on pH, in this pH range, by

$$R_{Si} \propto [H^+]^{0.54} \qquad (124)$$

Thus we may express R_{Si} as an explicit function of $[H^+]$ where

$$R_{Si} = 10^{-11.49} = k_{Si}(10^{-5})^{0.54} \qquad (125)$$

and

$$k_{Si} = 10^{-8.79} \, \text{mol SiO}_2 \cdot \text{m}^{-2} \, M^{-0.54} \, \text{s}^{-1} \qquad (126)$$

If we assume congruent dissolution of anorthite, then Ca^{2+} will be released in 1:2 stoichiometry and, converting to a slightly more convenient unit of time, we obtain

$$k_{Ca} = 10^{-4.155} \, \text{mol} \, Ca \cdot m^{-2} \, M^{-0.54} \, d^{-1} \qquad (127)$$

then the change in $[Ca^{2+}]$ over time is a function of the rate of Ca release (per m^2 of anorthite) and the mineral surface area subject to weathering (A_{surf}), that is,

$$\frac{\Delta[Ca^{2+}]}{\Delta t} = R_{Ca} \cdot A_{surf} = k_{Ca}[H^+]^{0.54} A_{surf} \qquad (128)$$

By assuming a value of $10 \, m^2 L^{-1}$ for A_{surf}, we may examine the effect of anorthite weathering on streamwater composition over time. From Example 11 (Tableau 5.5), we have the $TOTH$ equation (neglecting OH^-, $HSiO_3^-$, and $Al(OH)_4^-$):

$$TOTH = [H^+] + 3[Al^{3+}] + 2[AlOH^{2+}] + [Al(OH)_2^+] + 2[Ca^{2+}] = 10^{-4.13} \qquad (129)$$

where the numerical value for the $TOTH$ equation is the acidity of the rainwater (at pH = 4.13). Again assuming pseudoequilibrium between dissolved Al species and gibbsite, we calculate the concentrations of all dissolved Al species as a function of $[H^+]$ and obtain (as in Example 11) an expression relating $[Ca^{2+}]$ and $[H^+]$:

$$[Ca^{2+}] = \tfrac{1}{2}(10^{-4.13} - (3)10^{8.5}[H^+]^3 - (2)10^{3.5}[H^+]^2 - 10^{-0.8}[H^+]) \qquad (130)$$

From anorthite weathering, we can also write $[Ca^{2+}]$ as a function of time (based on Equation 128):

$$[Ca^{2+}]_{t+\Delta t} = [Ca^{2+}]_t + \Delta[Ca^{2+}] = [Ca^{2+}]_t + (k_{Ca}[H^+]^{0.54} A_{surf})\Delta t \qquad (131)$$

where the subscript $t + \Delta t$ refers to a time later (by the increment Δt) than time t.

At $t = 0$, no anorthite weathering has occurred, that is, $[Ca^{2+}]_{t=0} = 0$, and (pre)equilibrium with gibbsite results in the streamwater composition given in the first part of Example 11, that is,

$$pH = 4.5$$

$$[Al]_{diss} = 18.5 \, \mu M$$

Then for each time increment Δt, we may calculate $[Ca^{2+}]_{t+\Delta t}$ (from Equation 131) based on the (pH-dependent) anorthite weathering rate and solve explicitly for $[H^+]$ (using Equation 130) as a function of time. This calculation, however, is rather inconvenient and it is equally correct, though not as satisfying logically,

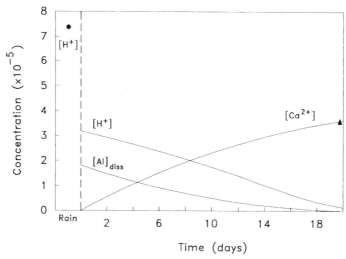

Figure 5.20 Kinetics of chemical weathering. Calculated concentrations of dissolved aluminum, calcium, and protons with time during weathering of anorthite. Rapid equilibrium with gibbsite is assumed. The initial dissolution of gibbsite results in the rapid decrease in $[H^+]$ from its value in acid rain (shown by ●) and release into the solution. Weathering of anorthite results in further pH increase and Ca^{2+} release. The value of $[Ca^{2+}]$ calculated in Example 11 is shown by the ▲. Dissolved aluminum decreases as gibbsite precipitates. The plateau observed in $[Ca^2]$ is a function of the decrease in the dissolution rate with increasing pH; equilibrium with anorthite is not attained over this time.

to calculate the time required to achieve various pH values. For an increment $\Delta[H^+]$, we may solve easily for the corresponding $\Delta[Ca^{2+}]$. Then the Δt required for this extent of reaction (based on Equation 131) is

$$\Delta t = \frac{[Ca^{2+}]_{t+\Delta t} - [Ca^{2+}]_t}{k_{Ca}[H^+]_t^{0.54} A_{surf}} \qquad (132)$$

The results of this calculation are shown in Figure 5.20. These calculations are, of course, only valid below neutral pH and far from equilibrium with anorthite (because the species OH^-, $HSiO_3^-$, and $Al(OH)_4^-$ and any reprecipitation of anorthite have been neglected). A more complete kinetic model could include additional species and reactions.

From Figure 5.20 we may see that, initially, the neutralization of $[H^+]$ by anorthite weathering is countered by the reprecipitation of gibbsite. This effect is less important at higher pH because of the lower dissolved aluminum concentrations and the shift in aluminum speciation toward more hydrolyzed species.

An important consideration in applying such models is the sensitivity of the calculation to the surface area parameter. The weathering rate (and thus the time dependence of water composition) is directly dependent on surface area. If the calculation given above is repeated with different values of A_{surf}, the time axis is compressed or expanded, but the shape of the curves obtained for concentration of dissolved species vs. time does not change. The surface area of weatherable minerals in the field is, unfortunately, not easily estimated. With the value chosen for A_{surf} in this case, almost 20 days of anorthite weathering are required to increase the pH and Ca concentrations to the values calculated in Example 11 (i.e., corresponding to the lower elevation streamwater).

Finally, we should recognize that the plateau in $[Ca^{2+}]$ in Figure 5.20 is a kinetic rather than an equilibrium effect; it results from the decreasing anorthite weathering rate at higher pH rather than from equilibrium with the mineral.

REFERENCES

1. H. D. Holland, *The Chemistry of the Atmosphere and Oceans*, Wiley, New York, 1978.

2. E. K. Berner and R. A. Berner, *The Global Water Cycle*, Prentice-Hall, New Jersey, 1987.

3. R. M. Garrels and F. T. MacKenzie, *Evolution of Sedimentary Rocks*, Norton, New York, 1971.

4. H. C. Helgeson, J. M. Delany, H. W. Nesbitt, and D. K. Bind, *Am. J. Sci.*, **278** (1978).

5. R. M. Smith and A. E. Martell, *Critical Stability Constants: Inorganic Complexes*, Vol. 4, Plenum, New York, 1976.

6. C. F. Baes, Jr. and R. E. Mesmer, *The Hydrolysis of Cations*, Wiley, New York, 1976.

7. R. A. Robie and D. R. Waldbaum, *US Geol. Survey Bull.*, **1259**, 1968.

8. W. Stumm and J. J. Morgan, *Aquatic Chemistry*, 2nd ed., Wiley-Interscience, New York, 1981.

9. J. W. Gibbs, *The Collected Works*, Yale University Press, New Haven, CN, 1948.

10. N. M. Johnson, C. T. Driscoll, J. S. Eaton, G. E. Likens, and W. H. McDowell, *Geochim. Cosmochim. Acta*, **45**, 1421 (1981).

11. R. Giovanoli, J. L. Schnoor, L. Sigg, W. Stumm, and J. Zobrist, *Clay Clay Min.*, **36**, 521 (1988).

12. L. H. Filipek, D. K. Nordstrom, and W. H. Flicklin *Environ. Sci. Technol.*, **21**, 388 (1987).

13. L. G. Sillén, *Science*, **165**, 1189 (1967).

14. W. S. Broeker, *Chemical Oceanography*, Harcourt, Brace, Jovanovich, New York, 1980.

15. R. E. McDuff and F. M. M. Morel, *Environ. Sci. Technol.*, **14**, 1182 (1980).

16. R. E. McDuff and J. M. Edmond, *Earth Planet. Sci. Lett.*, **57**, 117 (1982).

17. K. L. von Damm, J. M. Edmond, B. Grant, and C. I. Measures, *Geochim. Cosmochim. Acta*, **49**, 2197 (1985).

18. F. Albarede and A. Michard, *Chem. Geol.*, **57**, 1 (1986).

19. G. H. Nancollas and M. M. Reddy, in *Aqueous Environmental Chemistry of Metals*, Vol. 5, A. J. Rubin, Ed., Wiley, New York, 1974.

20. R. Wollast, in *Aquatic Chemical Kinetics*, W. Stumm, Ed., Wiley-Interscience, New York, 1990 pp. 431–445.

21. R. A. Berner, *Early Diagenesis*, Princeton University Press, Princeton, New Jersey, 1980.

22. A. E. Blum and A. C. Lasaga, in *Aquatic Surface Chemistry*, W. Stumm, Ed. Wiley-Interscience, New York, 1987.

23. A. Blum, R. A. Yund, and A. C. Lasaga, *Geochim. Cosmochim. Acta*, **54**, 283 (1990).

24. J. Schott, in *Aquatic Chemical Kinetics*, W. Stumm, Ed., Wiley-Interscience, New York, 1990.

25. R. A. Berner and J. W. Morse, *Am. J. Sci.*, **274**, 108 (1974).

26. D. Langmuir and D. O. Whittemore, in *Nonequilibrium Systems in Natural Water Chemistry*, J. D. Hem, Ed., ACS Advances in Chemistry Series No. 106, American Chemical Society, Washington, DC, 1971, pp. 209–234.

27. D. D. Eberl, J. Srodon, M. Kralik, B. E. Taylor, Z. E. Peterman, *Science*, **248**, 474 (1990).

28. C. I. Steefel and P. van Cappellen, *Geochim. Cosmochim. Acta*, **54**, 2657 (1990).

29. J. W. Morse and W. H. Casey, *Am. J. Sci.*, **288**, 537 (1988).

30. A. C. Lasaga, in *Kinetics of Geochemical Processes*, Vol. 8, A. C. Lasaga and R. J. Kirkpatrick, Eds., Mineralogical Society of America, Washington, DC, 1981, pp. 1–68.

31. G. R. Holdren and R. A. Berner, *Geochim. Cosmochim. Acta*, **43**, 1161 (1979).

32. J. L. Schnoor, in *Aquatic Chemical Kinetics*, W. Stumm, Ed., Wiley-Interscience, New York, 1990, pp. 475–504.

33. G. Furrer and W. Stumm *Geochim. Cosmochim. Acta*, **50**, 1847 (1986).

34. E. Wieland, B. Wehril, and W. Stumm, *Geochim. Cosmochim. Acta*, **52**, 1969 (1988).

35. W. Stumm and R. Wollast, *Rev. Geophysics*, **28**, 53 (1990).

36. A. Blum and A. C. Lasaga, *Nature*, **331**, 431 (1988).

37. W. Stumm and E. Wieland, in *Aquatic Chemical Kinetics*, W. Stumm, Ed., Wiley-Interscience, New York, 1990.

38. R. V. Nicholson, R. W. Gilham, and E. J. Reardon, *Geochim. Cosmochim. Acta*, **54**, 395 (1990).

39. J. Schott and R. A. Berner, *Geochim. Cosmochim. Acta*, **47**, 2233 (1983).

40. M. Ohara and R. C. Reid, *Modeling Crystal Growth from Solution*, Prentice-Hall, New Jersey, 1973.

41. L. Chou, R. M. Garrels, and R. Wollast, *Chem. Geol.*, **78**, 269 (1989).

42. S. Mann, *Nature*, **332**, 119 (1988).

PROBLEMS

5.1   Given $P_{CO_2} = 10^{-3.5}$ atm, $P_{H_2S} = 10^{-6}$ atm, and $Cd_T = 10^{-8}$ M, what is the principal stable form of cadmium at pH = 7? In what pH range(s) do any two or three solids coexist? (Use solubility products from Table 6.3.)

5.2   Calculate the free cadmium ion activity $[Cd^{2+}]$, given $[Ca^{2+}] = 10^{-2}$ M, pH = 8.5, $Cd_T = 10^{-5}\,M$, and assuming precipitation of carbonates. (Use solubility products from Table 6.3.)

5.3 What is the pH of a solution saturated with $CaCO_3(s)$ containing $Ca_T = 10^{-4}\,M$ and $C_T = 10^{-2}\,M$ in solution?

5.4 **a.** For a solution saturated with calcite and having $C_T = 10^{-3}\,M$, calculate the maximum $[Ca^{2+}]$, as controlled by calcite solubility, as a function of pH. (Hint: Use the expression for $[CO_3^{2-}]$ as a function C_T and $[H^+]$.)

b. For a solution saturated with calcite and having $Alk = 10^{-3}\,eq\,L^{-1}$, calculate the maximum $[Ca^{2+}]$, as controlled by calcite solubility, as a function of pH. (Hint: Use the expression for $[CO_3^{2-}]$ as a function of Alk and $[H^+]$.)

5.5 At what Fe(II) concentration would $FeCO_3(s)$ precipitate in groundwater at pH = 7.5, at equilibrium with $P_{CO_2} = 10^{-1.5}\,atm$?

5.6 The predominant forms of heavy metals (Zn, Pb, Fe, Cu, Ni, etc.) in oxic sediments are the carbonates and the oxide or hydroxide forms. An important consideration in modeling the sediment is the stable chemical form of the metal in question. Consider the reaction

$$Pb(OH)_2(s) + CO_2(aq) = PbCO_3(s) + H_2O$$

ΔG° for the reaction is $= -25\,kJ\,mol^{-1}$.

a. At what partial pressure of CO_2 will $Pb(OH)_2(s)$ and $PbCO_3(s)$ be at equilibrium?

b. If the system is at equilibrium with the atmosphere, $P_{CO_2} = 10^{-3.5}\,atm$, calculate ΔG for the reaction. What is the stable form of lead in the sediments?

5.7 Given $[CO_3]_T = 10^{-2}\,M$, $[S\text{-}II]_T = 10^{-7}\,M$, $Cu_T = 10^{-8}\,M$, what are the stable forms of copper as a function of pH? Make an appropriate diagram. Do some of the solids coexist at any pH?

5.8 **a.** Draw a $\log C$–pH diagram for the aluminum species (i) at equilibrium with $Al(OH)_3(s)$ and, (ii) at equilibrium with kaolinite and am·$SiO_2(s)$.

b. For the recipe $[AlCl_3]_T = 10^{-3}\,M$, what are the pH and the composition of the system, assuming equilibrium with $Al(OH)_3(s)$? Assuming equilibrium with kaolinite and am·$SiO_2(s)$?

c. In each case for what concentrations (minimum and/or maximum) is an addition of $[AlCl_3]_T$ entirely soluble?

5.9 Following the algebraic and graphical method given in the chapter, study the solubility of Fe(II) in a system consisting of

	Case 1	Case 2	Case 3
$[Fe \cdot II]_T =$	10^{-4}	10^{-4}	10^{-4}
$[CO_3]_T =$	10^{-3}	10^{-5}	$10^{-3.8}$

5.10   **a.** Calculate the composition of a groundwater in equilibrium with $P_{CO_2} = 10^{-1.5}$ atm, calcite, albite, and kaolinite.

b. As a function of C_T (total dissolved carbon), calculate the composition of this water as it flows out (no more rock to dissolve but precipitation can occur) and equilibrates with the atmosphere ($P_{CO_2} = 10^{-3.5}$ atm).

5.11   Following Example 10, calculate the pH and Alk of water at equilibrium with Na-montmorillonite and kaolinite given the recipe

$P_{CO_2} = 10^{-2.5}$ atm
Na-montmorillonite: $Na_2O \cdot 7Al_2O_3 \cdot 22SiO_2 \cdot 6H_2O(s)$
Kaolinite: $Al_2O_3 \cdot 2SiO_2 \cdot 2H_2O(s)$

5.12   Use seawater apparent constants (see Chapter 6), consider seawater to be at equilibrium with $CaCO_3(s)$ and the calcium concentration to be fixed (independently) at $[Ca^{2+}] = 10^{-2} M$.

a. Derive a plot of pH versus C_T.

b. Does photosynthesis, independent of hard part formation, tend to precipitate or dissolve $CaCO_3(s)$? (Must Ca^{2+} be added or withdrawn from the system to maintain equilibrium?)

c. How does the alkalinity of the water vary with C_T?

d. Discuss the issue of the stability (homeostasis) of the carbon system with respect to $[CO_2]_T$ changes.

5.13   Consider a system made up of $Al(OH)_3$ and H_2SiO_3 in distilled water (no carbonate), total concentrations $TOTAl$ and $TOTSi$, and dissolved concentrations, Al_T and Si_T.

a. Supposing the pH to be buffered at 7.0, develop a $TOTAl$ versus $TOTSi$ phase diagram. Use a range of 10^{-10}–1 M for both components. Indicate clearly the regions of coexistence of solids.

b. Supposing the pH to be unbuffered, calculate the pH as a function of Al_T and Si_T. How is the phase diagram modified?

c. What is the equilibrium composition of the solution when $TOTAl = 0.10 M$ and $TOTSi = 0.11, 0.101, 0.10$, and $0.09 M$? At high concentrations (say $TOTAl \simeq TOTSi \simeq 0.1 M$), how much in excess of $TOTAl$ should $TOTSi$ be, or vice versa, for $SiO_2(s)$ or $Al(OH)_3(s)$ to precipitate?

d. For each representative total concentration indicate on the diagram the final solution composition (i.e., draw arrows from $TOTAl$, $TOTSi$ to Al_T, Si_T coordinates).

5.14   This problem deals with the characteristics of rain and groundwater (resulting from equilibrium of rain with rock minerals) in polluted and unpolluted regions. Although the questions are obviously related, most of them can be answered independently.

a. What is the pH of the rain in an unpolluted region if it can be considered simply as distilled water in equilibrium with the atmosphere ($P_{CO_2} = 10^{-3.5}$ atm)? What is its buffer capacity?

b. In this case, what are the pH and alkalinity of the groundwater ($P_{CO_2} = 10^{-1.5}$ atm) in a calcite region? In a dolomite region? In an albite region?

c. In a polluted region the rain is actually acidic, pH = 3.7, as a result of equimolar concentrations of nitric and sulfuric acids. What is the mineral acidity of the polluted rainwater? What is its buffer capacity?

d. In the case of acidic precipitation, what are the pH and the alkalinity of the groundwater ($P_{CO_2} = 10^{-1.5}$ atm) in a calcite region? In a dolomite region? In an albite region?

CHAPTER 6

COMPLEXATION

In Chapters 4 and 5 we examined chemical processes that control the geochemical cycles of major elements. The dissolution and dissociation of weakly acidic gases and the weathering and sedimentation of rocks control the gross chemical composition of natural waters. Complex formation—defined here loosely as the reversible reaction of two dissolved species to form a third one—plays a relatively minor role in these major element cycles. Nonetheless, as we shall see, the solubility of a solid such as $CaCO_3(s)$ is markedly increased by complexation reactions in seawater compared to freshwater.

For many trace elements, however, the situation is quite different. Their global geochemical cycles may also be controlled by precipitation and dissolution of minerals, but their chemistry in the water column is dominated by complexation, biological uptake, and sorption on suspended solids. All three of these processes, which are particularly important in surface waters where biological activity is most intense, are controlled by the same coordinative mechanisms, all obeying the principles expounded in this chapter. We shall briefly discuss biological uptake at the end of this chapter and leave a detailed study of sorption reactions for Chapter 8.

The topic of aquatic complexation is in a paradoxical state. We know a great deal about coordination chemistry and we can make many chemical models that predict the existence of complexes in natural waters. Yet it is so difficult to analyze for individual chemical species under the conditions prevailing in aquatic systems where the concentrations are very low and the constituents very many, that we can rarely demonstrate the existence of these complexes unambiguously. The elucidation of the chemical speciation of trace elements in natural waters is probably the greatest remaining challenge to analytical chemists; the objective is to demonstrate and quantify the existence of fractions of chemical constituents as picomolar concentrations of perhaps ephemeral species. Thus, seemingly simple questions are yet unanswered: what are the principal dissolved species of the major and minor components of natural waters? What are the concentrations or the activities of free ions in comparison to the total (analytical) concentrations of chemical constituents?

Besides irrepressible chemical curiosity, a major motivation for asking these questions arises from our interest in the aquatic biota. Chemical constituents of natural waters affect the biota as essential nutrients and as potential toxicants. It is now firmly established that these interactions are directly dependent on the chemical speciation of the constituents. For example, the availability and toxicity of many trace metals to planktonic microorganisms are determined by their free concentrations rather than their total concentrations and are thus decreased by complexation. Trace metal availability and toxicity are not just matters of academic exercises or pollution control; they are thought to be natural controlling environmental factors for aquatic ecosystems. We need to understand the speciation of elements in water if we want to understand the interactions of aquatic organisms with their external milieu. Although the role of complexation in natural waters may be more subtle than geochemists typically care about or than analysts can presently measure, it is a dominant aspect of aquatic chemistry for biologists. Thus, much of this chapter will focus on "biologically interesting" elements such as Fe, Mn, Zn, Co, Ni, Cu, and Cd.

In keeping with the thermodynamic view that we have taken so far, we say little in this chapter regarding the fundamental theories of coordination chemistry. Questions relating to the nature of coordination bonds or comparisons among the coordinative properties of elements and compounds are addressed only briefly. For the most part, we describe chemical species by their stoichiometries and their free energies. Electronic interactions in metal–ligand complexes are discussed primarily to provide some basis for macroscopic observations of the structure and reactivity of these species.

Chemical complexes in natural waters can be conveniently classified into three groups: ion pairs of major constituents, inorganic complexes of trace elements, and organic complexes. These are discussed briefly at the beginning of the chapter where necessary definitions are provided. From simple mole balance considerations, the speciation of abundant aquatic constituents such as Ca^{2+}, Na^+, Cl^-, and HCO_3^- cannot be affected by complexation with those present in trace amounts. Although the nature of the coordination processes may not be very different, the topics of inorganic complexation of major and of trace constituents are thus effectively separate and are presented in consecutive sections. Owing to our relative ignorance of the nature and properties of the dissolved organic matter in natural waters, the question of organic complexation must be addressed largely at a theoretical level. To make the presentation more tangible the scant available data are thus generalized liberally and perhaps speculatively. We close the chapter with the subjects of complexation kinetics and biological uptake, the former serving as a conceptual framework for the latter.

1. AQUEOUS COMPLEXES

Consider a metal ion such as Cu^{2+} in water. Although we commonly talk of the "free cupric ion" in solution, this is really a misnomer, for the metal ion is actually associated with its surrounding water molecules. Inorganic chemists distinguish four solvent regions around a metal ion[1] (see Figure 6.1):

1. A primary solvation shell in which the water molecules are considered chemically bound to the ion. In the case of copper (and many other metals) there are six such water molecules, leading to the more refined chemical symbolism $Cu(H_2O)_6^{2+}$ instead of Cu^{2+} for the hydrated cupric ion.
2. A secondary solvation shell in which the water molecules are ordered by the electrostatic influence of the ion. The volume of this shell increases with the charge of the ion and is inversely related to its size.
3. A transition region separating the hydrated metal ion from the bulk

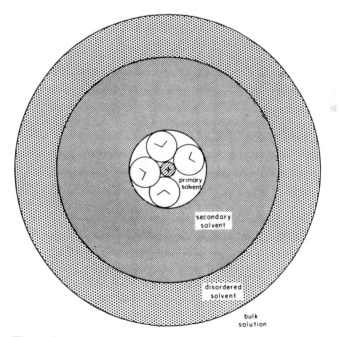

Figure 6.1   The various solvent regions around a metal ion. Adapted from Burgess, 1978.[1]

solution. In this region the water molecules are less ordered than in either the solvation shell or the bulk solution.

4. The bulk solution where the presence of the metal ion is not felt.

From a thermodynamic point of view, the energy of hydration of an ion is critical to its very existence. Ionization of a metal or separation of cations and anions from a crystal lattice are energetically very unfavorable processes. For example, it takes some $2700 \, kJ \, mol^{-1}$ to ionize Cu to Cu^{2+} in the gas phase. The presence of the cupric ion in solution is thus made possible only by a considerable energy of solvation which exceeds the unfavorable ionization energy. Of course, the energetics of the metal–solvent interactions do not normally concern us when we study the thermodynamics of complex formation in aqueous solution and consider only the differences among the free energies of aquated species. However, the large energy of solvation is important for the kinetics of complex formation which are often controlled by the rate of dissociation of the coordinated water molecules, an energetically unfavorable process. Free metal ions in solutions are really aquo complexes, the water itself is a *ligand* that binds metals, and every complexation reaction in water is effectively a *ligand-exchange* reaction.

Other ligands that can replace water molecules around the *central atom* (metal ion) are chemical species that have a nonbonding pair of electrons to share with the metal. These include simple anions such as the halides Cl^-, F^-, Br^-, I^-, more complex inorganic compounds such as NO_3^-, CO_3^{2-}, SO_4^{2-}, NH_3, S^{2-}, PO_4^{3-}, SO_3^{2-}, CN^-, and a great variety of organic molecules with suitable functional groups, usually containing oxygen, nitrogen, or sulfur atoms as purveyors of electron pairs (e.g., $R-COO^-$; $R-OH$; $R-NH_2$; $R-SH$). For the sake of generality, it is convenient to consider H^+ as a metal and OH^- as a ligand, thus including all acid–base reactions as a subset of coordination reactions.

The reaction of a metal with a ligand can be of an electrostatic or covalent nature or both. When it is primarily electrostatic and the reactants retain some water of hydration between them, the product is called an *ion pair* or an *outer sphere complex*. In natural waters this type of interaction is particularly important among the major ions in high ionic stength media such as seawater. When the reaction of a metal with a ligand involves coordination at several positions, one speaks of *chelation* (in Greek, *chelos* = crab—has two binding claws). Such a reaction requires the combination of a metal with a *coordination number* greater than 1 (i.e., more than one site for coordination), and a *multidentate* ligand, an organic compound with several reactive functional groups. These organic compounds are called *chelators, chelating agents,* or *complexing agents*. Among metals, only H^+ has a coordination number of 1; most metals of interest in aquatic chemistry have a coordination number of 6 and thus form octahedral complexes[2] (see Table 6.1). In natural waters, *polynuclear complexes*, those involving more than one metal, are probably rare except for polymeric hydroxide species. In this chapter, we use the word complex somewhat loosely

TABLE 6.1 Common Coordination Numbers and Geometries for Some Metals

Metal Species	CN[a]	Geometry[a]	Example
Li(I)	4	Tetrahedral	$Li(H_2O)_4^+$
Cr(II)	6	Octahedral	$Cr(H_2O)_6^{2+}$
Cr(III)	6	Octahedral	$Cr(H_2O)_6^{3+}$
Cr(VI)	4	Tetrahedral	CrO_4^{2-}
Mn(II)	6	Octahedral	$Mn(H_2O)_6^{2+}$
Mn(III)	6	Octahedral	$Mn(oxalate)_3^{3-}$
Mn(IV)	6	Octahedral	$MnCl_6^{2-}$
Fe(II)	6	Octahedral	$Fe(H_2O)_6^{2+}$
Fe(III)	6	Octahedral	$Fe(H_2O)_6^{3+}$
	(4)	(Tetrahedral)	$FeCl_4^-$
Co(II)	6	Octahedral	$Co(H_2O)_6^{2+}$
	4	Tetrahedral	$CoCl_4^{2-}$
Co(III)	6	Octahedral	$Co(NH_3)_6^{3+}$
Ni(II)	6	Octahedral	$Ni(H_2O)_6^{2+}$
	4	Square planar	$Ni(CN)_4^{2-}$
	(4)	(Tetrahedral)	$NiCl_4^{2-}$
Cu(I)	4	Tetrahedral	$Cu(CN)_4^{3-}$
	(2)	(Linear)	$CuCl_2^-$
Cu(II)	6	Octahedral (distorted)	$Cu(H_2O)_6^{2+}$
	(4)	(Square planar)	$CuCl_4^{2-}$

[a]Less common CN and geometries indicated by parentheses. Based on Cotton and Wilkinson (1972).[2]

to cover all dissolved species resulting from the metal–ligand combinations illustrated in Figure 6.2, including ion pairs and chelates.

Our primary interest is to use thermodynamic data describing metal–ligand interactions, specifically equilibrium constants, to predict the distribution of metal species and the reactivity of metals in natural waters. For this purpose, it is not strictly necessary to understand the factors determining the magnitude of the equilibrium constants for a particular complex or the details of the interaction between metals and ligands. Nonetheless, a brief discussion of the fundamentals of metal–ligand interactions may lend some insight into the chemical characteristics (particularly the stability and reactivity) of metal complexes. We thus now examine the relative importance of enthalpic and entropic factors in determining the thermodynamics of metal–ligand interactions and the effects of electronic interactions and overlap of metal and ligand electron orbitals on the structure and reactivity of metal complexes.

1.1 Thermodynamics of Complex Formation

The changes in standard free energy, ΔG°, upon formation of metal complexes and the corresponding stability constants, $K = \exp(-\Delta G^\circ/RT)$, are due, in large

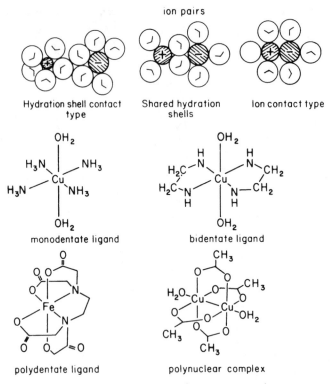

Figure 6.2 Various types of aqueous complexes.

part, to entropic factors. In some cases, for example the hydrolysis of Ca^{2+} or Fe^{3+}, the enthalpy change is actually unfavorable and the reaction proceeds solely because of the favorable change in entropy[3] (Table 6.2). As already mentioned, formation of a complex between a metal ion in solution and a ligand requires displacement of one or more coordinated water molecules from the initial aquo complex. The release of coordinated waters from the relatively constrained metal complex results in a significant increase in entropy. This may be compared to the entropy gained when ice melts as the water molecules are freed from the rigid structure of the solid. The increase in entropy associated with complex formation depends on the strength of the initial interaction between the metal and the coordinated water—the stronger the initial association, the greater the entropy increase as the water is displaced. Decreasing entropic effects would be expected with decreasing charge on the central metal ion or with the stepwise addition of subsequent ligands (cf. Table 6.2). The increased stability of multidentate ligands over monodentate ligands with the same donor atoms (the chelate effect) is also due to entropic factors[2]. This effect may be illustrated by comparing the stability of nickel complexes with NH_3 and ethylenediamine

TABLE 6.2 Enthalpy and Entropy of Complex Formation[a]

Ligand = OH⁻ Metal	ΔH° (kJ·mol⁻¹)	ΔS° (J·mol⁻¹ K⁻¹)	$\log K$	Ionic Strength
Ca^{2+}	8.4	51.5	1.2	0
	5.0	43.5	1.4	0
Fe^{3+}	5.0	209	11.7	0

Ligand = malonate complex	ΔH° (kJ·mol⁻¹)	ΔS° (J·mol⁻¹ K⁻¹)	$\log K$	Ionic Strength
FeL/Fe.L	11.3	181.4	7.5	1
FeL₂/FeL.L	3.1	117	5.54	1
FeL₃/FeL₂.L	−4.6	53	3.56	1

[a]Data for 25°C from Christensen and Izatt (1983).[3]

(en = $H_2NCH_2CH_2NH_2$). In the reaction

$$Ni(NH_3)_6^{2+} + 3(en) = Ni(en)_3^{2+} + 6NH_3$$

the chelate complex is favored ($\Delta G^\circ = -67 \text{ kJ mol}^{-1}$) predominantly by entropy ($-T\Delta S^\circ = -55 \text{ kJ mol}^{-1}$). The chelate effect has also been rationalized in terms of the increased probability of binding additional donor atoms of a multidentate ligand, subsequent to attachment of the first donor atom, as compared to the stepwise binding of monodentate ligands.

These thermodynamic arguments, however, cannot be used to rationalize the effects of specific ligands on the characteristics of metal–organic complexes, particularly the spectroscopic characteristics of the complexes, their coordination number and geometry, or the stabilization of different oxidation states of the metal. To understand these aspects of coordination chemistry, we must consider the electronic structure of metals and the electronic interactions between metals and ligands.

1.2 Electronic Configurations of Metals

The chemical and physical properties of the metallic elements depend on their electron configurations, that is, the occupancy of their electron orbitals. Most of the elements (roughly three-fourths) are metals, characterized by their luster, high electrical conductivity (decreasing with increasing temperature), high thermal conductivity, and mechanical properties such as strength and ductility. The metallic elements fall into several groups: *main group* elements, whose outer electron (or *valence*) shells consist solely of s and p electrons; the *transition* elements, which have, either as the neutral atom or as an important ion or both, an incomplete set of d electrons; and the *post-transition* elements, Zn, Cd, and Hg. (Note: The designations s, p, and d refer to electron orbitals of different shape and symmetry. Superscripts, e.g., s^2, are used to denote the number of electrons in an orbital or occupancy. The maximum occupancy is 2 for s orbitals, 6 for p orbitals, and 10 for d orbitals; the higher orbitals are not of much concern for our purposes.)

Metal chemistry is strongly influenced by the ease with which one or more electrons can be removed from the neutral metal atom. For example, the alkali metals (Li, Na, K, etc.) have only a single electron outside a noble-gas core (corresponding to the very stable electron configuration of the inert or noble gases, e.g. $1s^2$ for He, $1s^2 2s^2 p^6$ for Ne, etc.). Thus, of all the elements, the alkali metals have the lowest first ionization energies, defined as the energies required to remove an electron from the metals as isolated gaseous atoms. These metals react violently with water, the resulting alkali hydroxides are strong bases, and the chemistry of the alkali metals is essentially that of their $+1$ ions. It is a common feature of the main group and post-transition metals that they occur as ionic species only in a single oxidation state.

In contrast, most transition metals exhibit variable valence, such as $+2$ and $+3$ for Fe or $+2$, $+3$, $+4$, $+6$ and $+7$ for Mn. Transition metal ions with partially filled d shells are strongly influenced by their surroundings, particularly by their coordinated ligands, because the d orbitals project well out to the periphery of the ions. (Note: We do not consider here the lanthanide or actinide elements, which have partially filled f orbitals, although these elements are formally included among the transition elements.)

1.3 Electronic Interactions in Metal Complexes

The specific influence of ligands on transition metals may be treated in two ways. The first and simpler treatment, ligand field theory (LFT, also called crystal field theory), considers the perturbation of the atomic orbitals of the metal, specifically the d orbitals, by the electrostatic effect of the ligands. The second, molecular orbital theory (MOT), allows for overlap between the metal and ligand orbitals.

In LFT the approach of the ligands, considered as point charges, toward the metal affects the stability of the metal d orbitals because of electrostatic repulsion. As a result, some ions with partially filled d orbitals are energetically more (or less) stable than they would be if the d orbitals were unperturbed. This difference in stability is the ligand field stabilization energy (LFSE). Specifically, for octahedral complexes, the stability of the d orbitals oriented along axes (i.e., $d_{x^2-y^2}$ and d_{z^2}) is decreased and the stability of the d orbitals oriented off the axes (d_{xy}, d_{xz}, d_{yz}) is increased due to the approach of the ligands along the axes (Figure 6.3). An energy diagram (Figure 6.4a) then shows the five degenerate levels of the d orbitals split into two higher-energy levels (with the symmetry designation e_g) and three lower-energy levels (with the symmetry designation t_{2g}). Since the geometry of the complex, which is most often either octahedral or tetrahedral, affects the interaction of the coordinating ligands with the metal d orbitals, the different contributions of LFSE partly determine which geometries are favored for various metal complexes (cf. Table 6.1).

a)

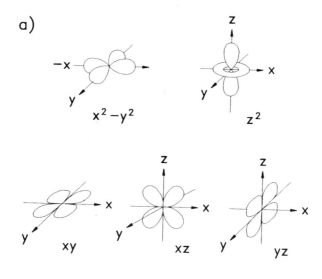

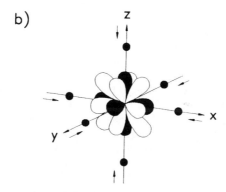

Figure 6.3   Representations of the five d orbitals (xy, xz, yz, z^2, and $x^2 - y^2$). As shown in a, two of the orbitals (z^2 and $x^2 - y^2$) are oriented along the axes and the other three orbitals (xy, xz, and yz) are oriented off the axes. Because of these different orientations, the interactions of the d orbitals with the ligands approaching along the axes, shown in b for an octahedral complex, destabilizes the z^2 and $x^2 - y^2$ orbitals and stabilizes the xy, xz, and yz orbitals. Adapted from Huheey (1983).[4]

The contribution of LFSE to hydration energies of the $+2$ ions in the first transition series, corresponding to the reaction

$$M^{2+}(g) + \infty H_2O = M(H_2O)_6^{2+}(aq)$$

can be seen in Figure 6.5. The ions $Ca^{2+}(d^0)$, $Mn^{2+}(d^5)$, and $Zn^{2+}(d^{10})$ have

a)

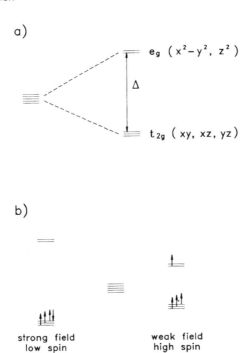

b)

strong field
low spin

weak field
high spin

Figure 6.4 (a) Splitting of the energy of the five d orbitals as a result of interactions with ligands in an octahedral complex. The energies of the z^2 and $x^2 - y^2$ orbitals (with symmetry designation e_g) are increased and the energies of the xy, xz, and yz orbitals (with symmetry designation t_{2g}) are decreased. The energy gap between the e_g and t_{2g} orbitals is indicated by Δ. (b) The occupancy of the orbitals, shown for metals with four d electrons, depends on the nature of the ligand. Interactions with strong field ligands produce a large energy gap (Δ) and favor the low spin configuration. With all four d electrons in the three t_{2g} orbitals, two of the electrons must have opposite spin (indicated by the directions of the arrows). In contrast, weak field ligands produce only a small energy gap and favor the high spin configuration in which all four electrons have the same spin.

no LFSE and their hydration energies lie along a smooth curve. The hydration energies of the other ions lie above this curve; the difference corresponds to the spectroscopically determined LFSE. Although the values of LFSE for divalent ions are on the order of $100 \, kJ \, mol^{-1}$ and thus comparable to the energies of many chemical changes, they are small (ca. 5–10%) compared to the total binding energies involved in metal complexation.[2] Thus, the LFSE is mostly important in explaining the difference in reactivity of various metal complexes.

Spectroscopic evidence, such as the absorbance of ultraviolet or visible light by metal complexes, demonstrates the effects of specific ligands on the metal d orbitals. Differences in the spectra of metal complexes indicate a consistent pattern in the extent to which metal orbitals are perturbed by ligands; this

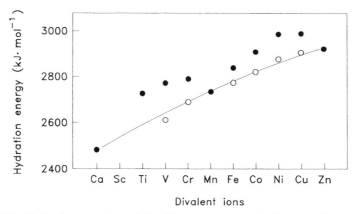

Figure 6.5 Hydration energies of the divalent cations including the first row of the transition metals. Solid symbols indicate measured hydration energies. Open symbols are corrected for the contributions of LFSE (ligand field stabilization energies). With this correction, hydration energies increase smoothly across the row. From Cotton and Wilkinson, 1972.[2]

ranking of ligands is termed the "spectrochemical series." Strong field ligands generate a large energy gap (Δ) between the t_{2g} and e_g levels and weak field ligands a small energy gap. In the case of a d^4 metal (e.g., Cr^{2+}), there can be either one electron in the higher-energy e_g level or a pair of electrons in the lower-energy t_{2g}. Then, the electron configuration (see Figure 6.4b) will depend on how the energy gap (Δ) between the t_{2g} and e_g levels compares with the spin-pairing energy, which arises from the inherent repulsion between the electrons within one orbital. Thus the ligands in a metal complex affect the distribution of electrons among the d orbitals, which influences both the chemical reactivity and magnetic and spectroscopic properties of the complexes.

Some metal–ligand interactions, however, cannot be explained simply in terms of electrostatics. In MOT overlap between metal and ligand orbitals is considered explicitly. For our purposes, the important difference between MOT and LFT is seen if the ligand has either extra lone pairs of electrons (π donor ligands) or empty orbitals of energy comparable to the metal d orbital energies (π acceptor ligands). Then interaction of the ligand p or d orbitals and the metal d orbitals (see Figure 6.6) results in formation of π bonds. In metal complexes with π donor ligands, there is increased electron density at the metal center and with π acceptor ligands, decreased electron density at the metal center. These variations in electron density influence the relative stability of metal oxidation states in the complexes. Figure 6.7 shows that Fe(II) is stabilized by π acceptor ligands and Fe(III) is stabilized by π donor ligands.[5]

Although necessarily brief, this discussion of the fundamentals of metal–ligand interactions provides some context for the discussion of (equilibrium) metal speciation in natural waters. Complexation kinetics have also been

a)

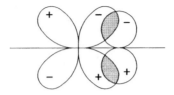

b)

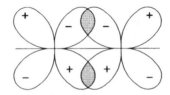

Figure 6.6 Overlap between metal and ligand orbitals that results in π bonds between the metal and ligand described by molecular orbital theory (MOT). (*a*) Overlap between metal d orbital and ligand p orbital. (*b*) Overlap between metal and ligand d orbitals.

interpreted, to some extent, on the basis of electronic interactions in metal complexes, particularly on the basis of the changes in LFSE in the transition state of the reaction. This is discussed in more detail in Section 5.

1.4 Prediction of Metal Speciation

Metal–ligand interactions are a particularly convenient framework within which to organize the reactions of interest in natural waters, including acid–base reactions, complexation, and solid formation. Equilibrium constants and the stoichiometric coefficients for reactions among some 20 metals and 32 ligands are presented, in a compact way, in Table 6.3. The first 15 ligands represent practically all reactive inorganic constituents commonly encountered in natural waters; they are roughly arranged in order of decreasing abundance. The other 17 ligands are chosen as representative compounds containing the types of functionalities present in aquatic organic matter, all of them being actually present as individual species at trace concentrations in natural waters. The metals are also roughly organized in order of decreasing abundance, the list of trace metals being biased toward the more reactive elements.

The logs of equilibrium constants given in Table 6.3 correspond to a general

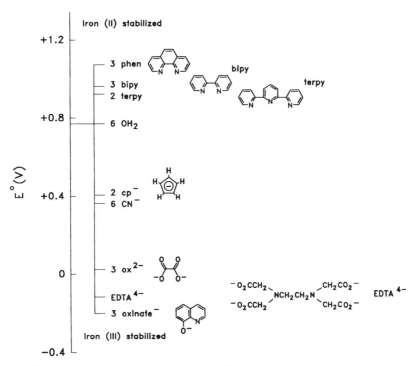

Figure 6.7 Standard reduction potentials for iron complexes. Iron(II) is stabilized by π acceptor ligands such as phenanthroline (phen) and iron(III) is stabilized by π donor ligands such as EDTA. Adapted from Burgess (1988).[5]

complexation or precipitation reaction of the type

$$mM + lL + hH^+ = M_mL_lH_h$$

$$\beta_{mlh} = \frac{[M_mL_lH_h]}{[M]^m[L]^l[H^+]^h}$$

For any given complex or solid, several different reactions of formation or dissociation can be written making it necessary to specify the reaction considered when giving an equilibrium constant. This is achieved implicitly by standardizing the notation for the constants themselves as shown in Table 6.4.

An important caveat must accompany any compilation of thermodynamic data such as that of Table 6.3. Although the choice of constants has been made with some care, there might still be some glaring errors or omissions. The original references have been examined only in a few instances, and previous compilations, particularly those of Smith and Martell, and Martell and Smith,[6] have been relied upon extensively. A major issue is that of data consistency: for example, a complex formation constant reported by one author may have been calculated

TABLE 6.3 Stability Constants for Formation of Complexes and Solids from Metals and Ligands[a]

	OH^-	CO_3^{2-}	SO_4^{2-}	Cl^-	Br^-	F^-	NH_3	$B(OH)_4$
H^+	$HL\cdot w$ 14.00	HL 10.33 H_2L 16.68 $H_2L\cdot g$ 18.14	HL 1.99			HL 3.2	HL 9.24 $L\cdot g$ −1.8	HL 9.24 HL_3 10.4 H_2L_3 20.4 H_2L_4 21.0 H_4L_5 38.8
Na^+		NaL 1.27 $NaHL$ 10.08	NaL 1.06					
K^+			KL 0.96					
Ca^{2+}	CaL 1.15 $CaL_2\cdot s$ 5.19	CaL 3.2 $CaHL$ 11.59 $CaL\cdot s$ 8.22 $CaL\cdot s$ 8.35	CaL 2.31 $CaL\cdot s$ 4.62			CaL 1.1 $CaL_2\cdot s$ 10.4		
Mg^{2+}	MgL 2.56 Mg_4L_4 16.28 $MgL_2\cdot s$ 11.16	MgL 3.4 $MgHL$ 11.49 $MgL\cdot s$ 4.54 $MgL\cdot s$ 7.45	MgL 2.36			MgL 1.8 $MgL_2\cdot s$ 8.2		
Sr^{2+}		$SrL\cdot s$ 9.0	SrL 2.6 $SrL\cdot s$ 6.5			$SrL_2\cdot s$ 8.5		
Ba^{2+}		BaL 2.8 $BaL\cdot s$ 8.3	BaL 2.7 $BaL\cdot s$ 10.0			$BaL_2\cdot s$ 5.8		
Cr^{3+}	CrL 10.0 CrL_2 18.3 CrL_3 24.0 CrL_4 28.6 Cr_3L_4 47.8 $CrL_3\cdot s$ 30.0		CrL 3.0	CrL	.23	CrL 5.2 CrL_2 9.2 CrL_3 12.0		

332

Ion						
Al^{3+}	AlL 9.0 AlL_2 18.7 AlL_3 27.0 AlL_4 33.0 Al_3L_4 42.1 $AlL_3 \cdot s$ 33.5					AlL 7.0 AlL_2 12.6 AlL_3 16.7 AlL_4 19.1
Fe^{3+}	FeL 11.8 FeL_2 22.3 FeL_4 34.4 Fe_2L_2 25.0 $FeL_3 \cdot s$ 42.7 $FeL_3 \cdot s$ 38.8		FeL 4.0 FeL_2 5.4	FeL 1.5 FeL_2 2.1	FeL 0.6	FeL 6.0 FeL_2 10.6 FeL_3 13.7
Mn^{2+}	MnL 3.4 MnL_2 5.8 MnL_3 7.2 MnL_4 7.7 $MnL_2 \cdot s$ 12.8	$MnHL$ 12.1 $MnL \cdot s$ 9.3	MnL 2.3	MnL 0.6	MnL 1.3	MnL 1.0 MnL_2 1.5
Fe^{2+}	FeL 4.5 FeL_2 7.4 FeL_3 11.0 $FeL_2 \cdot s$ 15.1	$FeL \cdot s$ 10.7	FeL 2.2		FeL 1.4	
Co^{2+}	CoL 4.3 CoL_2 9.2 CoL_3 10.5 $CoL_2 \cdot s$ 15.7	$CoL \cdot s$ 10.0	CoL 2.4	CoL 0.5	CoL 1.0	CoL 2.0 CoL_2 3.5 CoL_3 4.4 CoL_4 5.0
Ni^{2+}	NiL 4.1 NiL_2 9.0 NiL_3 12.0 $NiL_2 \cdot s$ 17.2	$NiL \cdot s$ 6.9	NiL 2.3	NiL 0.6	NiL 1.1	NiL 2.7 NiL_2 4.9 NiL_3 6.6 NiL_4 7.7 NiL_5 8.3

TABLE 6.3 (Continued)

	OH⁻		CO₃²⁻		SO₄²⁻		Cl⁻		Br⁻		F⁻		NH₃		B(OH)₄	
Cu^{2+}	CuL	6.3	CuL	6.7	CuL	2.4	CuL	0.5			CuL	1.5	CuL	4.0		
	CuL₂	11.8	CuL₂	10.2	Cu₄(OH)₆L·s	68.6							CuL₂	7.5		
	CuL₄	16.4	CuL·s	9.6									CuL₃	10.3		
	Cu₂L₂	17.7	Cu₂(OH)₂L·s	33.8									CuL₄	11.8		
	CuL₂·s	19.3	Cu₃(OH)₂L₂·s	46.0												
	CuL₂·s	20.4														
Zn^{2+}	ZnL	5.0	ZnL·s	10.0	ZnL	2.1	ZnL	0.4			ZnL	1.2	ZnL	2.2		
	ZnL₂	11.1			ZnL₂	3.1	ZnL₂	0.2					ZnL₂	4.5		
	ZnL₃	13.6					ZnL₃	0.5					ZnL₃	6.9		
	ZnL₄	14.8					Zn₂(OH)₃L·s	26.8					ZnL₄	8.9		
	ZnL₂·s	15.5														
	ZnL₂·s	16.8														
Pb^{2+}	PbL	6.3	PbL·s	13.1	PbL	2.8	PbL	1.6	PbL	1.8	PbL	2.0				
	PbL₂	10.9			PbL·s	7.8	PbL₂	1.8	PbL₂	2.6	PbL₂	3.4				
	PbL₃	13.9					PbL₃	1.7	PbL₃	3.0	PbL₂·s	7.4				
	PbL₂·s	15.3					PbL₄	1.4	PbL₂·s	5.7						
							PbL₂·s	4.8								
Hg^{2+}	HgL	10.6	HgL·s	16.1	HgL	2.5	HgL	7.2	HgL	9.6	HgL	1.6	HgL	8.8		
	HgL₂	21.8			HgL₂	3.6	HgL₂	14.0	HgL₂	18.0			HgL₂	17.4		
	HgL₃	20.9					HgL₃	15.1	HgL₃	20.3			HgL₃	18.4		
	HgL₂·s	25.4					HgL₄	15.4	HgL₄	21.6			HgL₄	19.1		
							HgOHL	18.1	HgL₂·s	19.8						
Cd^{2+}	CdL	3.9	CdL·s	13.7	CdL	2.3	CdL	2.0	CdL	2.1	CdL	1.0	CdL	2.6		
	CdL₂	7.6			CdL₂	3.2	CdL₂	2.6	CdL₂	3.0	CdL₂	1.4	CdL₂	4.6		
	CdL₂·s	14.3			CdL₃	2.7	CdL₃	2.4					CdL₃	5.9		
							CdL₄	1.7					CdL₄	6.7		
Ag^+	AgL	2.0	Ag₂L·s	11.1	AgL	1.3	AgL	3.3	AgL	4.7	AgL	0.4	AgL	3.3	AgL	0.6
	AgL₂	4.0			Ag₂L·s	4.8	AgL₂	5.3	AgL₂	6.9			AgL₂	7.2	AgHL₂·s	22.9
	AgL·s	7.7					AgL₃	6.4	AgL₃	8.7						
							AgL·s	9.7	AgL₄	9.0						
									AgL·s	12.3						

334

	SiO$_3^{2-}$	S^{2-}	S$_2$O$_3^{2-}$	PO$_4^{3-}$	P$_2$O$_7^{4-}$	P$_3$O$_{10}^{5-}$	CN$^-$
H$^+$	HL 13.1 H$_2$L 23.0 H$_2$L$_2$ 26.6 H$_4$L$_4$ 55.9 H$_6$L$_4$ 78.2 H$_2$L·s 25.7	HL 13.9 H$_2$L 20.9 H$_2$L·g 21.9	HL 1.6 H$_2$L 2.2	HL 12.35 H$_2$L 19.55 H$_3$L 21.70	HL 9.4 H$_2$L 16.1 H$_3$L 18.3 H$_4$L 19.7	HL 9.3 H$_2$L 18.8 H$_3$L 21.3 H$_4$L 22.3	HL 9.2
Na$^+$			NaL 0.5	NaHL 13.5	NaL 2.3 Na$_2$L 4.2 NaHL 10.8	NaL 2.7 NaHL 11.6	
K$^+$			KL 1.0	KHL 13.4	KL 2.1	KL 2.8	
Ca^{2+}	CaL 4.2 CaHL 14.1 CaH$_2$L$_2$ 29.9		CaL 2.0	CaL 6.5 CaHL 15.1 CaH$_2$L 21.0 CaHL·s 19.0	CaL 6.8 CaHL 13.4 CaOHL 8.9 Ca$_2$L·s 14.7	CaL 8.1 CaHL 14.1 CaOHL 10.4	
Mg^{2+}	MgL 5.3 MgHL 14.3 MgH$_2$L$_2$ 30.8		MgL 1.8	MgL 4.8 MgHL 15.3 MgH$_2$L 20.0 Mg$_3$L$_2$·s 25.2 MgHL·s 18.2	MgL 7.2 MgHL 14.1 MgOHL 9.3	MgL 8.6 MgHL 14.5 MgOHL 11.0	
Sr^{2+}			SrL 2.0	SrL 5.5 SrHL 14.5 SrH$_2$L 20.3 SrHL·s 19.3	SrL 5.4 SrOHL 7.7 Sr$_2$L·s 12.9	SrL 7.2 SrHL 13.6 SrOHL 9.3	
Ba^{2+}			BaL 2.3 BaL·s 4.8	BaHL·s 19.8		BaL 6.3 BaHL 12.9 Ba$_2$L·s 16.1	
Cr^{3+}							
Al^{3+}							

TABLE 6.3 (Continued)

	SiO$_3^{2-}$	S^{2-}	S$_2$O$_3^{2-}$	PO$_4^{3-}$	P$_2$O$_7^{4-}$	P$_3$O$_{10}^{5-}$	CN$^-$
Fe^{3+}	FeHL 22.7		FeL 3.3	FeHL 22.5 FeH$_2$L 23.9 FeL·s 26.4			FeL$_6$ 43.6
Mn^{2+}		MnL·s 10.5 13.5	MnL 2.0			MnL 9.9 MnHL 14.8	
Fe^{2+}		FeL·s 18.1		FeHL 16.0 FeH$_2$L 22.3 Fe$_3$L$_2$·s 36.0			FeL$_6$ 35.4
Co^{2+}		CoL·s 21.3 CoL·s 25.6	CoL 2.1	CoHL 15.5	CoL 7.9 CoHL 14.1	CoL 9.7 CoHL 14.8	
Ni^{2+}		NiL·s 19.4 NiL·s 24.9 NiL·s 26.6	NiL 2.1	NiHL 15.4	NiL 7.7 NiHL 14.4	NiL 9.5 NiHL 14.7	NiL 7.3 NiL$_4$ 30.2 NiH$_2$L$_4$ 40.8 NiHL$_4$ 36.1
Cu^{2+}		CuL·s 36.1		CuHL 16.5 CuH$_2$L 21.3	CuL 9.8 CuHL 15.5 CuL$_2$ 12.5 CuH$_2$L 19.2	CuL 11.1 CuHL 15.5	CuL$_2$ 16.3 CuL$_3$ 21.6 CuL$_4$ 23.1
Zn^{2+}		ZnL 16.6 ZnL·s 24.7	ZnL 2.4 ZnL$_2$ 2.5 ZnL$_3$ 3.3 Zn$_2$L$_2$ 7.0	ZnHL 15.7 ZnH$_2$L 21.2 Zn$_3$L$_2$ 35.3	ZnL 8.7 ZnL$_2$ 11.0 ZnOHL 13.1	ZnL 10.3 ZnHL 14.9 ZnOHL 13.6	ZnL 5.7 ZnL$_2$ 11.1 ZnL$_3$ 16.1 ZnL$_4$ 19.6 ZnL$_2$·s 15.9
Pb^{2+}		PbL·s 27.5	PbL 3.0 PbL$_2$ 5.5 PbL$_3$ 6.2 PbL$_4$ 7.3	PbHL 15.5 PbH$_2$L 21.1 Pb$_3$L$_2$·s 43.5 PbHL·s 23.8	PbL 9.5 PbL$_2$ 10.2		

Hg²⁺

HgL	7.9	HgL₂	29.2	HgOHL	18.6	HgL	17.0
HgL₂	14.3	HgL₃	30.6			HgL₂	32.8
HgOHL	18.5					HgL₃	36.3
HgL·s	52.7					HgL₄	39.0
HgL·s	53.3					HgOHL	29.6

Cd²⁺

CdL	19.5	CdL	3.9	CdL	8.7	CdL	9.8	CdL	6.0
CdHL	22.1	CdL₂	6.3	CdOHL	11.8	CdHL	14.6	CdL₂	11.1
CdH₂L₂	43.2	CdL₃	6.4			CdOHL	12.6	CdL₃	15.7
CdH₃L₃	59.0	CdL₄	8.2					CdL₄	17.9
CdH₄L₄	75.1	Cd₂L₂	12.3						
CdL·s	27.0								

Ag⁺

AgL	19.2	AgL	8.8	Ag₃L·s	17.6	AgL₂	20.5
AgHL	27.7	AgL₂	13.7			AgL₃	21.4
AgHL₂	35.8	AgL₃	14.2			AgOHL	13.2
AgH₂L₂	45.7	Ag₂L₄	26.3			AgL·s	15.7
Ag₂L·s	50.1	Ag₃L₅	39.8				
		Ag₆L₈	78.6				

TABLE 6.3 (Continued)

	Ethylene-diamine	NTA	EDTA	CDTA	IDA	Picolinate	Cysteine	Desferri-ferrioxamine B
H^+	HL 9.93 H_2L 16.78	HL 10.33 H_2L 13.27 H_3L 14.92 H_4L 16.02	HL 11.12 H_2L 17.8 H_3L 21.04 H_4L 23.76 H_5L 24.76	HL 13.28 H_2L 20.0 H_3L 23.98 H_4L 26.62 H_5L 28.34	HL 9.73 H_2L 12.63 H_3L 14.51	HL 5.39 H_2L 6.40	HL 10.77 H_2L 19.13 H_3L 20.84	HL 10.1 H_2L 19.4 H_3L 27.8
Na^+		NaL	NaL 1.9		NaL 0.8			
K^+			KL 1.7					
Ca^{2+}		CaL 7.6	CaL 12.4 $CaHL$ 16.0	CaL 15.0	CaL 3.5	CaL 2.2 CaL_2 3.8		CaL 3.5
Mg^{2+}	MgL 0.4	MgL 6.5	MgL 10.6 $MgHL$ 15.1	MgL 12.8	MgL 3.8	MgL 2.6 MgL_2 4.0		MgL 5.2
Sr^{2+}		SrL 6.3	SrL 10.5 $SrHL$ 14.9	SrL 12.4	SrL 3.1	SrL 1.8 SrL_2 3.0		SrL 3.1
Ba^{2+}		BaL 5.9	BaL 9.6 $BaHL$ 14.6	BaL 10.5 $BaHL$ 17.8	Ba 2.5	BaL 1.6		
Cr^{3+}			CrL 26.0 $CrHL$ 28.2 $CrOHL$ 32.2		CrL 12.2 CrL 23.2			
Al^{3+}		AlL 13.4 $AlOHL$ 22.1	AlL 18.9 $AlHL$ 21.6 $AlOHL$ 26.6 $Al(OH)_2L$ 30.0	AlL 22.1 $AlHL$ 24.3 $AlOHL$ 28.1	AlL 9.9 AlL_2 17.5			
Fe^{3+}		FeL 17.9 FeL_2 26.3	FeL 27.7 $FeHL$ 29.2 $FeOHL$ 33.8 $Fe(OH)_2L$ 37.7	FeL 32.6 $FeOHL$ 36.5	FeL 12.5	FeL_2 13.9 $FeOHL_2$ 24.9		FeL 31.9 $FeHL$ 32.6

338

	C1	C2	C3	C4	C5	C6	C7	C8
Mn^+	MnL 2.8 MnL_2 3.7 MnL_3 5.8	MnL 8.7 MnL_2 11.6	MnL 15.6 $MnHL$ 19.1	MnL 19.2 $MnHL$ 22.4		MnL 4.0 MnL_2 7.1 MnL_3 8.8	MnL 5.6	
Fe^{2+}	FeL 4.3 FeL_2 7.7 FeL_3 9.7	FeL 9.6 FeL_2 13.6 $FeOHL$ 12.6	FeL 16.1 $FeHL$ 19.3 $FeOHL$ 20.4 $Fe(OH)_2L$ 23.7	FeL 20.8 $FeHL$ 23.9	FeL 6.7 FeL_2 11.0	FeL 5.3 FeL_2 9.7 FeL_3 13.0		$FeHL$ 18.7 FeH_2L 21.0
Co^{2+}	CoL 6.0 CoL_2 10.8 CoL_3 14.1	CoL 11.7 CoL_2 15.0 $CoOHL$ 14.5	CoL 18.1 $CoHL$ 21.5	CoL 21.4 $CoHL$ 24.7	CoL 7.9 CoL_2 13.2	CoL 6.4 CoL_2 11.3 CoL_3 14.8		CoL 11.2 $CoHL$ 18.0 $CoHL$ 23.6
Ni^{2+}	NiL 7.4 NiL_2 13.6 NiL_3 17.9	NiL 12.8 NiL_2 17.0 $NiOHL$ 15.5	NiL 20.4 $NiHL$ 24.0 $NiOHL$ 21.8	NiL 22.1 $NiHL$ 25.4	NiL 9.1 NiL_2 15.7	NiL 7.2 NiL_2 12.5 NiL_3 17.9	NiL 10.7 NiL_2 20.9	NiL 11.8 $NiHL$ 18.3 NiH_2L 23.8
Cu^{2+}	CuL 10.5 CuL_2 19.6 $CuOHL$ 11.8	CuL 14.2 CuL_2 18.1 $CuOHL$ 18.6	CuL 20.5 $CuHL$ 23.9 $CuOHL$ 22.6	CuL 23.7 $CuHL$ 27.3	CuL 11.5 CuL_2 17.6	CuL 8.4 CuL_2 15.6	$Cu(II) \rightarrow Cu(I)$	CuL 15.0 $CuHL$ 24.1 CuH_2L 27.0
Zn^{2+}	ZnL 5.7 ZnL_2 10.6 ZnL_3 13.9	ZnL 12.0 ZnL_2 14.9 $ZnOHL$ 15.5	ZnL 18.3 $ZnHL$ 21.7 $ZnOHL$ 19.9	ZnL 21.1 $ZnHL$ 24.4	ZnL 8.2 ZnL_2 13.5	ZnL 5.7 ZnL_2 10.3 ZnL_3 13.6	ZnL 10.1 ZnL 19.1 $ZnHL$ 16.4	ZnL 11.0 $ZnHL$ 17.5 ZnH_2L 22.9
Pb^{2+}	PbL 7.0 PbL_2 8.5	PbL 12.6	PbL 19.8 $PbHL$ 23.0	PbL 22.1 $PbHL$ 25.3	PbL 8.3	PbL 5.0 PbL_2 8.6	PbL 12.5	
Hg^{2+}	HgL 14.3 HgL_2 23.2 $HgOHL$ 24.2 $HgHL_2$ 28.0	HgL 15.9	HgL 23.5 $HgHL$ 27.0 $HgOHL$ 27.7	HgL 26.8 $HgHL$ 30.3 $HgOHL$ 29.7	HgL 11.7	HgL 8.1 HgL_2 16.2	HgL 15.3	
Cd^{2+}	CdL 5.4 CdL_2 9.9 CdL_3 11.7	CdL 11.1 CdL_2 15.1 $CdOHL$ 13.4	CdL 18.2 $CdHL$ 21.5	CdL 21.7 $CdHL$ 25.1	CdL 6.6 CdL_2 11.1	CdL 5.0 CdL_2 8.3 CdL_3 11.4		CdL 8.8 $CdHL$ 16.2 CdH_2L 22.7
Ag^+	AgL 4.7 AgL_2 7.7 $AgHL$ 11.9	AgL 5.8	AgL 8.2 $AgHL$ 14.9	AgL 9.9		AgL 3.6 AgL_2 6.1		

TABLE 6.3 (Continued)

	Glycine	Glutamate	Acetate	Glycolate	Citrate	Malonate	Salicylate	Phthalate
H^+	HL 9.78 H_2L 12.13	HL 9.95 H_2L 14.47 H_3L 16.70	HL 4.76	HL 3.83	HL 6.40 H_2L 11.16 H_3L 14.29	HL 5.70 H_2L 8.55	HL 13.74 H_2L 16.71	HL 5.51 H_2L 8.36
Na^+					NaL 1.4 KL 1.3	NaL 0.7		NaL 0.7
K^+								
Ca^{2+}	CaL 1.4	CaL 2.1	CaL 1.2	CaL 1.6	CaL 4.7 CaHL 9.5 CaH_2L 12.3	CaL 2.4 CaHL 6.6	CaL 0.4	CaL 2.4
Mg^{2+}	MgL 2.7	MgL 2.8	MgL 1.3	MgL 1.3	MgL 4.7 MgHL 9.2	MgL 2.9 MgHL 7.1		
Sr^{2+}	SrL 0.9	SrL 2.3	SrL 1.1	SrL 1.2	SrL 4.1	SrL 2.1 SrHL 6.5		
Ba^{2+}	BaL 0.8	BaL 2.2	BaL 1.1	BaL 1.1	BaL 4.1 BaHL 9.0 BaH_2L 12.4	BaL 2.1	BaL 0.2	BaL 2.3
Cr^{3+}			CrL 5.4 CrL_2 8.4 CrL_3 11.2			CrL 9.6		
Al^{3+}			AlL 2.4				AlL 14.2 AlL_2 25.1 AlL_3 31.1	AlL 5.0 AlL_2 8.7
Fe^{3+}	FeL 10.8	FeL 13.8	FeL 4.0 FeL_2 7.6 FeL_3 9.6	FeL 3.7 FeOHL 19.6 $FeOHL_2$ 22.3 $FeOHL_3$ 23.8	FeL 13.5 $Fe_2(OH)_2L_2$ 56.3	FeL 9.3	FeL 17.6 FeL_2 28.6 FeL_3 36.2	

Ion	Col 1	Col 2	Col 3	Col 4	Col 5	Col 6	Col 7	Col 8
Mn^{2+}	MnL 3.2		MnL 1.4	MnL 1.6	MnL 5.5, MnHL 9.4	MnL 3.3	MnL 6.8, MnL_2 10.7	MnL 2.7
Fe^{2+}	FeL 4.3	FeL 4.6	FeL 1.4	FeL 1.9	FeL 5.7, FeHL 9.9		FeL 7.4, FeL_2 12.1	
Co^{2+}	CoL 5.1, CoL_2 9.0, CoL_3 11.6	CoL 5.4, CoL_2 8.7	CoL 1.5	CoL 2.0, CoL_2 3.0	CoL 6.3, CoHL 10.3, CoH_2L 12.9	CoL 3.7, CoL_2 5.1, CoHL 7.0	CoL 7.5, CoL_2 12.3	CoL 2.8, CoHL 7.2
Ni^{2+}	NiL 6.2, NiL_2 11.1, NiL_3 14.2	NiL 6.5, NiL_2 10.6	NiL 1.4	NiL 2.3, NiL_2 3.4, NiL_3 3.7	NiL 6.7, NiHL 10.5, NiH_2L 12.9	NiL 4.1, NiL_2 5.8, NiHL 7.2	NiL 7.8, NiL_2 12.6	NiL 3.0, NiHL 6.6
Cu^{2+}	CuL 8.6, CuL_2 15.6	CuL 8.8, CuL_2 15.0	CuL 2.2, CuL_2 3.6	CuL 2.9, CuL_2 4.7, CuL_3 4.7	CuL 7.2, CuHL 10.7, CuH_2L 13.8, CuOHL 16.4, Cu_2L_2 16.3	CuL 5.7, CuL_2 8.2, CuHL 8.3	CuL 11.5, CuL_2 19.3	CuL 4.0, CuHL 7.1
Zn^{2+}	ZnL 5.4, ZnL_2 9.8, ZnL_3 12.3	ZnL 5.8, ZnL_2 9.5, ZnL_3 9.8	ZnL 1.6, ZnL_2 1.8	ZnL 2.4, ZnL_2 3.6, ZnL_3 3.9	ZnL 6.1, ZnL_2 6.8, ZnHL 10.3, ZnH_2L 13.3	ZnL 3.8, ZnL_2 5.4, ZnHL 7.1	ZnL 7.7	ZnL 2.9, ZnL_2 4.2
Pb^{2+}	PbL 5.5, PbL_2 8.9		PbL 2.7, PbL_2 4.1	PbL 2.5, PbL_2 3.7, PbL_3 3.6	PbL 5.4, PbL_2 8.1, PbHL 10.2, PbH_2L 13.1	Pb 4.0, PbL_2 4.5		
Hg^{2+}	HgL 10.9, HgL_2 20.1		HgL 6.1, HgL_2 10.1, HgL_3 14.1, HgL_4 17.6		HgL 12.2			

TABLE 6.3 (*Continued*)

	Glycine	Glutamate	Acetate	Glycolate	Citrate	Malonate	Salicylate	Phthalate
Cd^{2+}	CdL 4.7	CdL 4.8	CdL 1.9	CdL 1.9	CdL 5.0	CdL 3.2	CdL 6.4	CdL 3.4
	CdL_2 8.4		CdL_2 3.2	CdL_2 2.7	CdL_2 7.2	CdL_2 4.0		
	CdL_3 10.7				CdHL 9.5	CdHL 6.9		
					CdH_2L 12.6			
Ag^+	AgL 3.5		AgL 0.7	AgL 0.4				
	AgL_2 6.9		AgL_2 0.6	AgL_2 0.5				

^{a}*Note*: Constants are given as logarithms of the overall formation constants, β, for complexes and as logarithms of the overall precipitation constants for solids, at zero ionic strength and 25°C [From Smith and Martell 1975, 1976 and Martell and Smith 1974, 1977.[6] Exceptions are major ion interaction constants (Na^+, K^+, Ca^{2+}, Mg^{2+}, CO_3^{2-}, SO_4^{2-}, Cl^-) taken from Whitfield;[7] hydrolysis (OH^-) and some carbonate (CO_3^{2-}) formation constants taken from Baes and Mesmer;[8] Cu^{2+}—CO_3^{2-} complexes constants taken from Sunda and Hanson;[9] the ZnS(aq) constant recalculated from the data of Sainte Marie et al.[10]; the $MnCO_3$(s) constant taken from Morgan.[11]] When necessary, constants have been extrapolated to $I = 0\ M$ using the following values of ($-\log$ of) activity coefficients (applied to all ions including H^+ and tri- and tetravalent ions):

I	z	1	2	3	4
0.1 M,	2 M	0.11	0.44	0.99	1.76
0.3 M		0.13	0.52	1.17	2.08
0.5 M		0.15	0.60	1.35	2.40
1.0 M		0.14	0.56	1.26	2.24
3.0 M		0.07	0.28	0.63	1.12
4.0 M		0.03	0.12	0.27	0.48

TABLE 6.4   Formulation of Stability Constant[a]

Mononuclear Complexes

Addition of ligand

$$M \xrightarrow[K_1]{L} ML \xrightarrow[K_2]{L} ML_2 \cdots \xrightarrow[K_i]{L} ML_i \cdots \xrightarrow[K_n]{L} ML_n$$

$$\xrightarrow{\quad \beta_2 \quad}$$
$$\xrightarrow{\quad \beta_i \quad}$$
$$\xrightarrow{\quad \beta_n \quad}$$

$$K_i = \frac{[ML_i]}{[ML_{(i-1)}][L]}$$

$$\beta_i = \frac{[ML_i]}{[M][L]^i}$$

Addition of protonated ligands

$$M \xrightarrow[*K_1]{HL} ML \xrightarrow[*K_2]{HL} ML_2 \cdots \xrightarrow[*K_i]{HL} ML_i \cdots \xrightarrow{HL} ML_n$$

$$\xrightarrow{\quad *\beta_2 \quad}$$
$$\xrightarrow{\quad *\beta_i \quad}$$
$$\xrightarrow{\quad *\beta_n \quad}$$

$$*K_i = \frac{[ML_i][H^+]}{[ML_{(i-1)}][HL]}$$

$$*\beta_i = \frac{[ML_i][H^+]^i}{[M][HL]^i}$$

Polynuclear Complexes

In β_{nm} and $*\beta_{nm}$ the subscripts n and m denote the composition of the complex M_mL_n formed. [If $m = 1$, the second subscript ($= 1$) is omitted.]

$$\beta_{nm} = \frac{[M_mL_n]}{[M]^m[L]^n}$$

$$*\beta_{nm} = \frac{[M_mL_n][H^+]^n}{[M]^m[HL]^n}$$

Source: Adapted from Stumm and Morgan (1981).[12]

[a]*Note*: The notation given above is the same as that used by Sillén and Martell (1964, 1971).[13] In the text the notation β_1 or K_1 is used indifferently to indicate the constants of formation of M_1L_1 complexes. β' or K' indicate mixed acidity constants expressed as a function of the activity of H^+ and of the concentrations of other reactants and products.

on the basis of a solubility constant that is not the same as that chosen in the compilation. For precise calculations it is essential to examine the original literature and crosscheck the methods for estimating the constants from experimental data. Note also that the absence of a reported constant for a metal–ligand combination in Table 6.3 does not rule out the existence of a complex. It is in fact the value of such a table to make the missing information particularly visible. (The price paid for such advantage is that mixed complexes, such as HgClGly, cannot be accommodated in the format of the table.)

Some simple generalities about coordination chemistry can be deduced by inspection of Table 6.3. Consider first the sulfate column. The similarities among stability constants for metals of like charge is striking (e.g., $\log K = 2.2–2.8$ for divalent metal ions). This is a reflection of the principally electrostatic binding of the sulfate ion pair complexes (see Section 5.1). Similarly low equilibrium constants for complex formation, correlated with ionic charge, are also observed for carbonate (and bicarbonate) and halides (Cl^-, Br^-, and F^-).

A dominant feature of Table 6.3 is the high degree of similarity among organic ligands having the same donor atoms in their relative affinities for various metals. The absolute values of the stability constants may be different from one ligand to another but their relative values from metal to metal are highly correlated. A good example is the complex stability sequence of the transition metals, $Mn^{2+} < Fe^{2+} < Co^{2+} < Ni^{2+} < Cu^{2+} > Zn^{2+}$, well known as the Irving–Williams series.[14] This empirical observation is related to both the increase in effective nuclear charge with atomic number (which is due to imperfect shielding by the electrons) and the LFSE effects described previously. On the basis of such empirical observations, the metals can be organized into various groups exhibiting similar coordinative properties. Table 6.3 has been organized to highlight some of these correlations; the grouping of the metals reflects their coordination properties which can be explained by the configurations of their electron shells.[15]

Most of the organic ligands in Table 6.3 coordinate metals through oxygen donor atoms. A marked difference in the affinity of ligands for various metals may be noted when ligands with O, N, and S donor atoms are compared. Both metals and ligands may be categorized in terms of their polarizability. Interactions between "soft" (easily polarizable) donors and acceptors or between "hard" (usually compact) donors and acceptors are more favorable than hard–soft interactions. The empirically observed order of ligand affinities for "hard" metals (typically, alkali and alkaline earth metal ions and smaller, highly charged ions such as Fe^{3+}, Co^{3+}, or Al^{3+}) and for "soft" metals (typically, heavier transition metal ions such as Hg^{2+}) may be summarized as follows:[15]

strongest		→	weakest	complexes with "hard" metals
F^-	Cl^-	Br^-	I^-	
R_2O	R_2S	R_2Se	R_2Te	
R_3N	R_3P	R_3As	R_3Sb	
weakest	←		strongest	complexes with "soft" metals

where R indicates that the heteroatoms (O, N, S, P, Te, Se, As, and Sb) are bonded to organic carbon. Although useful, this classification is entirely empirical.

It is apparent that trivalent metals are typically more reactive than divalent ones, but this is largely offset by the greater insolubility of their corresponding oxides and hydroxides. For example, many ligands have a relatively high affinity for Fe^{3+}, but the very high stability of the ferric hydroxides, including the solid $Fe(OH)_3$, keeps the free ferric ion activity so low that the extent of Fe complexation is limited. As a result, Cu^{2+} and Hg^{2+}, the most reactive of the divalent metals, are most apt to form complexes in natural waters. Copper complexation is one of the major foci of experimental studies of trace metal speciation in aquatic systems and will serve as one of our principal examples throughout this chapter.

2. ION ASSOCIATION AMONG MAJOR AQUATIC CONSTITUENTS

If two constituents are present in widely differing concentrations, say, by a factor of 100 or more, the less abundant constituent can affect the activity of the other only negligibly through complex formation. To study the complexation of the major constituents of natural waters, it is then sufficient to consider the components that account for 99% or so of the dissolved solids. For most natural waters, including seawater, these components are the metals Na^+, Ca^{2+}, Mg^{2+}, K^+ and the ligands Cl^-, SO_4^{2-}, and CO_3^{2-}. (H^+ and OH^- are included implicitly and so is HCO_3^- as a "complex" of H^+ and CO_3^{2-}.) Our objective is to describe quantitatively to what degree these seven major aquatic constituents are bound to each other as complexes. For the purpose of comparison, let us consider a freshwater and a seawater model, both at pH = 8.1.

Example 1. Ion Association in Freshwater; pH = 8.1*

$$TOTNa = 2.8 \times 10^{-4} M = 10^{-3.55} \qquad TOTCl = 2.0 \times 10^{-4} M = 10^{-3.70}$$
$$TOTCa = 3.7 \times 10^{-4} M = 10^{-3.43} \qquad TOTSO_4 = 1.0 \times 10^{-4} M = 10^{-4.00}$$
$$TOTMg = 1.6 \times 10^{-4} M = 10^{-3.80} \qquad TOTCO_3 = 1.0 \times 10^{-3} M = 10^{-3.00}$$
$$TOTK = 6.0 \times 10^{-5} M = 10^{-4.22}$$

Example 2. Ion Association in Seawater; pH = 8.1*

$$TOTNa = 4.68 \times 10^{-1} M = 10^{-0.33} \qquad TOTCl = 5.45 \times 10^{-1} M = 10^{-0.26}$$
$$TOTCa = 1.02 \times 10^{-2} M = 10^{-1.99} \qquad TOTSO_4 = 2.82 \times 10^{-2} M = 10^{-1.55}$$
$$TOTMg = 5.32 \times 10^{-2} M = 10^{-1.27} \qquad TOTCO_3 = 2.38 \times 10^{-3} M = 10^{-2.62}$$
$$TOTK = 1.02 \times 10^{-2} M = 10^{-1.99}$$

The upper left corner of Table 6.3 provides the necessary list of species and thermodynamic constants and the equilibrium problems can be formulated with

---

*Total concentrations from Holland (1978).[16]

TABLE 6.5 Free Single Ion Activity Coefficients Used in the Ion Association Model for Seawater

Ion	γ	$-\log \gamma$
H^+	0.95	0.02
Na^+	0.71	0.15
K^+	0.63	0.20
All other $+1$ ions	0.68	0.17
Ca^{2+}	0.26	0.59
Mg^{2+}	0.29	0.54
OH^-	0.65	0.19
Cl^-	0.63	0.20
All other -1 ions	0.68	0.17
SO_4^{2-}	0.17	0.77
CO_3^{2-}	0.20	0.70
Uncharged species	1.13	-0.05

Source: After Whitfield (1974).[7]

appropriate mole balance and mass law equations. The complexation effects that we wish to study are reasonably subtle, however, since the formation constants of the complexes are only on the order of $10^0 - 10^2$. To achieve precision in the seawater model, it is imperative that proper ionic strength corrections be made. In this spirit of precision, we choose from the literature activity coefficients that are considered most appropriate for seawater[7] (see Table 6.5) rather than applying a general empirical expression such as the Davies equation. Concentration equilibrium constants are then readily calculated and choosing H^+, Na^+, Ca^{2+}, Mg^{2+}, K^+, Cl^-, SO_4^{2-}, and HCO_3^- as components yields Tableau 6.1 in which the constants are expressed for the formation of the species from the components.

Hand calculations corresponding to our two examples are a bit tedious and require some iterative solution scheme. For example, as shown in Tables 6.6 and 6.7 one can (1) start the calculation by assuming the principal component concentrations to be equal to the total concentrations, (2) calculate each of the species from the corresponding mass law, and (3) compare the sum of species in each mole balance equation to the imposed total concentration and increase or decrease accordingly the concentration for the next iteration. The calculation converges in two iterations for the freshwater model, and in four iterations for the seawater model. The test of convergence is of course the satisfaction of the mole balance equations.

In the freshwater example, the complexes are of little importance for the speciation of the major constituents. Complexation is most significant for sulfate which is calculated to be 10% bound to calcium and magnesium. By contrast, in the seawater example the effect of association among major ions is quite important; major fractions of both sulfate (60%) and carbonate (34%) are complexed by the metals, and some 10% of the calcium and magnesium bound to the ligands. The results of Example 2 are by and large comparable to those of the historical

TABLEAU 6.1

	H^+	Na^+	K^+	Ca^{2+}	Mg^{2+}	HCO_3^-	SO_4^{2-}	Cl^-	Freshwater	Seawater
H^+	1									+0.02
OH^-	−1								−14.0	−13.81
CO_3^{2-}	−1					1			−10.33	−9.80
HCO_3^-						1				
H_2CO_3	1					1			+6.35	+6.13
SO_4^{2-}							1			
HSO_4^-	1						1			
Cl^-								1	+1.99	+1.39
Na^+		1								
$NaCO_3^-$	−1	1				1			−9.06	−9.21
$NaHCO_3$		1				1			−0.25	−0.62
$NaSO_4^-$		1					1		+1.06	+0.31
K^+			1							
KSO_4^-			1				1		+0.96	+0.16
Ca^{2+}				1						
$CaOH^+$	−1			1					−12.85	−13.27
$CaCO_3$	−1			1		1			−7.13	−7.94
$CaHCO_3^+$				1		1			+1.26	+0.67
$CaSO_4$				1			1		+2.31	+0.90
Mg^{2+}					1					
$MgOH^+$	−1				1				−11.44	−11.81
$MgCO_3$	−1				1	1			−6.93	−7.69
$MgHCO_3^+$					1	1			+1.16	+0.62
$MgSO_4$					1		1		+2.36	+1.00

TABLE 6.6 Calculation of Major Ion Interactions in Freshwater

	Iteration		Metal	Ligand
	1st	2nd	(%)	(%)
H^+	8.1	8.1		
Na^+	3.55	3.55	100	
K^+	4.22	4.22	100	
Ca^{2+}	3.43	3.45	95	
Mg^{2+}	3.80	3.82	95	
HCO_3^-	3.00	3.02		96
SO_4^{2-}	4.00	4.05		90
Cl^-	3.70	3.70		100
OH^-	5.9	5.9		
CO_3^{2-}	5.23	5.25		1
H_2CO_3	4.75	4.77		2
HSO_4^-	10.11	10.16		
$NaCO_3^-$	7.51	7.53		
$NaHCO_3$	6.80	6.82		
$NaSO_4^-$	6.49	6.54		
KSO_4^-	7.26	7.31		
$CaOH^+$	8.18	8.20		
$CaCO_3$	5.46	5.50 ⎤	3	1
$CaHCO_3^+$	5.17	5.21 ⎦		
$CaSO_4$	5.12	5.19	2	6
$MgOH^+$	7.14	7.16		
$MgCO_3$	5.63	5.67 ⎤	3	
$MgHCO_3^+$	5.64	5.68 ⎦		
$MgSO_4$	5.44	5.51	2	3
ΣNa	$10^{-3.55}$	$10^{-3.55}$		
ΣK	$10^{-4.22}$	$10^{-4.22}$		
ΣCa	$10^{-3.41}$	$10^{-3.43}$		
ΣMg	$10^{-3.78}$	$10^{-3.80}$		
ΣHCO_3	$10^{-2.98}$	$10^{-3.00}$		
ΣSO_4	$10^{-3.95}$	$10^{-4.00}$		
ΣCl	$10^{-3.7}$	$10^{-3.7}$		

Garrels and Thompson[17] model for seawater, and of course similar to those of Whitfield[7] since most of the same constants and activity coefficients have been selected. The differences that exist among these various calculations point out the difficulty in estimating stability constants and activity coefficients in a system as complex as seawater.

In all these traditional models of ion interactions in seawater, chloride complexes are considered unimportant. However, there is evidence that species such

TABLE 6.7   Calculation of Major Ion Interactions in Seawater

	Iteration				Metal (%)	Ligand (%)
	1st	2nd	3rd	4th		
H^+	8.1	8.1	8.1	8.1		
Na^+	0.33	0.35	0.34	0.34	98	
K^+	1.99	2.01	2.00	2.00	98	
Ca^{2+}	1.99	2.08	2.03	2.03	91	
Mg^{2+}	1.27	1.38	1.32	1.32	89	
HCO_3^-	2.62	2.83	2.80	2.81		64
SO_4^{2-}	1.55	1.96	1.93	1.95		40
Cl^-	0.26	0.26	0.26	0.26		100
OH^-	5.71	5.71	5.71			
CO_3^{2-}	4.32	4.53	4.50	4.51		1
H_2CO_3	4.59	4.80	4.77	4.78		1
HSO_4^-	8.26	8.67	8.64	8.66		
$NaCO_3^-$	4.06	4.29	4.25	4.26		2
$NaHCO_3$	3.57	3.80	3.76	3.77		7
$NaSO_4^-$	1.57	2.00	1.96	1.98	2	37
KSO_4^-	3.38	3.81	3.78	3.80	2	1
$CaOH^+$	7.16	7.25	7.20	7.20		
$CaCO_3$	4.45	4.75	4.67	4.68		1
$CaHCO_3^+$	3.94	4.24	4.16	4.17	1	3
$CaSO_4$	2.64	3.14	3.06	3.08	8	3
$MgOH^+$	4.98	5.09	5.03	5.03		
$MgCO_3$	3.48	3.80	3.71	3.72		8
$MgHCO_3^+$	3.27	3.59	3.50	3.51	1	13
$MgSO_4$	1.82	2.34	2.25	2.27	10	19
ΣNa	$10^{-0.31}$	$10^{-0.34}$	$10^{-0.33}$	$10^{-0.33}$		
ΣK	$10^{-1.97}$	$10^{-2.00}$	$10^{-1.99}$	$10^{-1.99}$		
ΣCa	$10^{-1.90}$	$10^{-2.04}$	$10^{-1.99}$	$10^{-1.99}$		
ΣMg	$10^{-1.16}$	$10^{-1.33}$	$10^{-1.27}$	$10^{-1.27}$		
ΣHCO_3	$10^{-2.41}$	$10^{-2.65}$	$10^{-2.61}$	$10^{-2.62}$		
ΣSO_4	$10^{-1.14}$	$10^{-1.58}$	$10^{-1.53}$	$10^{-1.55}$		
ΣCl	$10^{-0.26}$	$10^{-0.26}$	$10^{-0.26}$	$10^{-0.26}$		

as $NaCl$, KCl, $MgCl^+$, and $CaCl^+$ may represent a significant fraction (13, 17, 43, 47%, respectively) of the total metal concentrations.[18] The recalculation of the ion-pairing model with the additional chloride constants is straightforward. However, the extension of these results to trace elements (see Section 3) would require a reinterpretation of the original experimental coordination data with equilibrium constants that are consistent with the new ion-pairing model.

As discussed in Chapter 2, interactions among ions in complex systems span the whole spectrum from unspecific long-range electrostatic interactions (of the type accounted for by the Debye–Hückel theory) to specific complex formation. In the *ion-association model* of seawater (Example 2), the mutual interactions among major ions are divided into these two types, utilizing simultaneously single ion activity coefficients and formation constants for the ion pair complexes. Such division between ideal and nonideal interactions among ions is largely arbitrary, particularly in the case of ion pairs. As is the case for nonideal effects, the major part of the energy for ion pair formation is electrostatic and may be considered "long range" since the primary solvation shell is thought to be intact. Still, we have chosen to consider that these long-range electrostatic interactions, but not others, result in the formation of independent chemical entities. To paraphrase Horne,[19] we have chosen to bless some liaisons into marriages and considered the others to be outside the ideality laws.

A different approach is clearly possible, and all electrostatic interactions among ions, including what we have taken to be ion pair formation, can be considered as nonideal interactions. In such a case, a *total activity coefficient*, γ_i^T, is defined for each ion as the ratio of the free ion activity to the *total* ion concentration:

$$\gamma_i^T = \frac{\{C_i\}}{[C_i]_T} \tag{1}$$

For example, according to Example 2, the total activity coefficients for the major seawater ions are

$$\gamma_{Na}^T = \frac{\{Na^+\}}{Na_T} = \gamma_{Na}\frac{[Na^+]}{Na_T} = 0.71 \times 0.98 = 0.70 \tag{2}$$

$$\gamma_K^T = \frac{\{K^+\}}{K_T} = \gamma_K\frac{[K^+]}{K_T} = 0.63 \times 0.98 = 0.62 \tag{3}$$

$$\gamma_{Ca}^T = \frac{\{Ca^{2+}\}}{Ca_T} = \gamma_{Ca}\frac{[Ca^{2+}]}{Ca_T} = 0.26 \times 0.91 = 0.24 \tag{4}$$

$$\gamma_{Mg}^T = \frac{\{Mg^{2+}\}}{Mg_T} = \gamma_{Mg}\frac{[Mg^{2+}]}{Mg_T} = 0.29 \times 0.89 = 0.26 \tag{5}$$

$$\gamma_{SO_4}^T = \frac{\{SO_4^{2-}\}}{[SO_4]_T} = \gamma_{SO_4}\frac{[SO_4^{2-}]}{[SO_4]_T} = 0.17 \times 0.4 = 0.068 \tag{6}$$

$$\gamma_{Cl}^T = \frac{\{Cl^-\}}{Cl_T} = \gamma_{Cl}\frac{[Cl^-]}{Cl_T} = 0.63 \times 1.0 = 0.63 \tag{7}$$

For the carbonate system, the coefficients are defined for each acid–base species:

$$\gamma_{H_2CO_3}^T = \frac{\{H_2CO_3^*\}}{[H_2CO_3]_T} = \gamma_{H_2CO_3}\frac{[H_2CO_3^*]}{[H_2CO_3]_T} = 1.13 \tag{8}$$

$$\gamma_{HCO_3}^T = \frac{\{HCO_3^-\}}{[HCO_3]_T} = \gamma_{HCO_3}\frac{[HCO_3^-]}{[HCO_3^-]+[NaHCO_3]+[CaHCO_3^+]+[MgHCO_3^+]}$$
$$= 0.68 \times 0.74 = 0.50 \tag{9}$$

$$\gamma_{CO_3}^T = \frac{\{CO_3^{2-}\}}{[CO_3]_T} = \gamma_{CO_3}\frac{[CO_3^{2-}]}{[CO_3^{2-}]+[NaCO_3^-]+[CaCO_3]+[MgCO_3]}$$
$$= 0.20 \times 0.10 = 0.020 \tag{10}$$

Such total activity coefficients are of course dependent on the ionic composition of the system and they must be defined anew for each different system. It is important to realize that total activity coefficients rather than ion pair formation constants are the quantities most directly amenable to experimental determination. Our calculation of total activity coefficients on the basis of the ion pair model is essentially a reversal of the process by which the parameters of the ion pair model are estimated in the first place.

In the same way that concentration equilibrium constants can be defined for any ionic strength by including activity coefficients into thermodynamic constants, *apparent equilibrium constants* can be defined for a particular solution by incorporation of total activity coefficients. For example, for the carbonate system in seawater, the first and second apparent mixed acidity constants are defined as

$$(K_1^{app})' = \frac{\{H^+\}[HCO_3]_T}{[H_2CO_3]_T} = K_1'\frac{\gamma_{H_2CO_3}^T}{\gamma_{HCO_3}^T} = 10^{-6.00} \tag{11}$$

$$(K_2^{app})' = \frac{\{H^+\}[CO_3]_T}{[HCO_3]_T} = K_2'\frac{\gamma_{HCO_3}^T}{\gamma_{CO_3}^T} = 10^{-8.93} \tag{12}$$

[These apparent constants vary somewhat depending on the chosen ion pair formation model. For example, Stumm and Morgan[20] report: $(pK_1^{app})' = 6.035$; $(pK_2^{app})' = 9.09$.] In the same way, we obtain the apparent solubility product of calcite:

$$K_s^{app} = [Ca^{2+}]_T[CO_3^{2-}]_T = \frac{K_s}{\gamma_{Ca}^T\gamma_{CO_3}^T} = 10^{-6.03} \tag{13}$$

Much of the interest in quantifying the extent of major ion association in seawater centers on its effects on the carbonate system and particularly calcium carbonate solubility. [Note that $CaCO_3(s)$ has not been considered in Example 2, thus allowing for calcite supersaturation in the seawater.]

In reference to the ion pair model of Example 2, apparent acidity and solubility constants effectively lump together the various ion pairs for each carbonate species, much in the way that H_2CO_3 and $CO_2(aq)$ are lumped together to define $H_2CO_3^*$. This is a very convenient convention to study carbonate chemistry in seawater: by simple redefinition of the acidity constants, $pK_{a1} = 6.0$, $pK_{a2} = 8.9$, we can consider seawater as a typical CO_2, C_T, Alk, pH system, ignoring the complications of major ion interactions. Calculations with apparent constants for the carbonate system in seawater and complete ion-pairing calculations are equivalent only because the major cations are little affected by ion pairing. Otherwise, the free concentrations of the metals and hence their effects on the carbonate speciation would be affected by the pH through carbonate complexation and the apparent constants would vary with pH.

The two approaches—total activity coefficients or ion pair complexes—provide equally valid thermodynamic representations of ion interactions in water. Whether or not we wish to consider major ion interactions as actual complex formation is more a matter of philosophy and terminology than a matter of fact. In the end, it is really a matter of practicality. If one is considering systems with highly variable ionic strength and composition, then it seems more convenient to use a universal description of ion association according to the ion pairing approach. (For concentrated brines, the ion pairing model is insufficient and other ion interaction models including first and second order effects among all major ions must be considered; see Chapter 2.) On the contrary, if one is dealing exclusively with seawater composition, then the total activity coefficient approach becomes very efficient. The whole major seawater ion model of Tableau 6.1 can then be reduced to 12 lines, eliminating all ion pairs and including the proper total activity coefficients in the last column.

3. INORGANIC COMPLEXATION OF TRACE ELEMENTS

Unlike the question of interactions among major ions, in which everything depends on everything, the question of the inorganic speciation of minor constituents of natural waters can be treated one element at a time, independently of the other constituents. This is because the important complexes are formed with constituents in large excess whose free concentrations (activities) are unaffected by complexation with a trace element.

To focus this discussion, let us consider a divalent trace metal M^{2+}. The inorganic ligands that may form important complexes with this metal in natural waters are relatively few: OH^-, Cl^-, SO_4^{2-}, CO_3^{2-}, S^{2-} (HCO_3^- and HS^- are included implicitly). Other reactive inorganic ligands (e.g., F^-, Br^-, NH_3, PO_4^{3-}, CN^-) are usually present at concentrations too low to result in any significant metal binding.

Consider the following complex formation reactions with ligands A^- and B^-:

$$M^{2+} + mA^- = MA_m^{(m-2)-}; \qquad \beta_m \qquad (14)$$

$$M^{2+} + nB^- = MB_n^{(n-2)-}; \qquad \beta_n \qquad (15)$$

Introducing the mass laws for Reactions 14 and 15 into the mole balance equation for M^{2+} allows the separation of the metal and ligand concentrations into the two factors of a product:

$$TOTM = [M^{2+}] + [MA_m^{(m-2)-}] + [MB_n^{(n-2)-}] + \cdots \qquad (16)$$

$$TOTM = [M^{2+}](1 + \beta_m[A^-]^m + \beta_n[B^-]^n + \cdots) \qquad (17)$$

Such separation is possible as long as there are no polynuclear species, which would introduce higher exponents for $[M^{2+}]$. To resolve the question of metal speciation it is then sufficient to evaluate the term in brackets which depends exclusively on the ligand concentrations.

Since the inorganic ligands are normally in large excess of the trace metals that they bind, the problem of evaluating the free ligand concentrations is independent of the nature and concentration of the metals. Only two considerations then enter in the calculation: (1) at high ionic strength the possible effect of major ion interactions on the ligand speciation as discussed in Section 2, and (2) the acid–base chemistry of weak acid ligands as a function of pH. The use of apparent constants accounting for all ion-pairing effects allows the resolution of the acid–base speciation independently of the major ion interactions. For example, using the ionization fraction formalism for carbonate in seawater one can write

$$[CO_3^{2-}] = \alpha_2 C_T$$

in which α_2 is calculated on the basis of the appropriate apparent acidity constants.

In general, Equation 17 can thus be rewritten:

$$TOTM = [M^{2+}](1 + \beta_m(\alpha_A)^m(A_T)^m + \beta_n(\alpha_B)^n(B_T)^n + \cdots) \qquad (18)$$

where the coefficients α are uniquely determined for a given pH and a given extent of major ion association.

In most cases, only a few of the terms inside the parentheses are important, most typically only one:

if $1 \gg \beta_m(\alpha_A)^m(A_T)^m, \beta_n(\alpha_B)^n(B_T)^n, \ldots$
then M^{2+} is the dominant species;
if $\beta_m(\alpha_A)^m(A_T)^m \gg 1, \beta_n(\alpha_B)^n(B_T)^n, \ldots$
then $MA_m^{(m-2)-}$ is the dominant species, and so on.

Species that are not dominant, but may be significant in the speciation of M^{2+}, are those for which the expression in parentheses is not too small compared to the largest one (say $> 1\%$). Since L_T is an upper limit on $[L]$, the coefficients α are always smaller than 1; thus ligands for which $\beta_n(L_T)^n$ is less than 1 can never form dominant complexes.

The methodology to calculate the inorganic complexation of a trace metal is thus remarkably simple:

1. List all the species considered.
2. Calculate the free ligand concentrations on the basis of some ion interaction model and of the acid–base speciation of the weak acids—most simply by using apparent acidity constants that include the effects of major ion interactions.
3. Calculate the various terms in brackets and retain only those that are significant.
4. Obtain the distribution of the trace metal among its major inorganic species by division.

Example 3. Inorganic Speciation of Copper and Cadmium in Freshwater   Let us consider the freshwater system of Example 1 and the following data from Table 6.3:

$CuOH^+$; $\log \beta_1 = 6.3$ $CuCO_3$; $\log \beta_1 = 6.7$ $CuSO_4$; $\log \beta_1 = 2.4$
$Cu(OH)_2$; $\log \beta_2 = 11.8$ $Cu(CO_3)_2^{2-}$; $\log \beta_2 = 10.2$ $CuCl^+$; $\log \beta_1 = 0.5$

$CdOH^+$; $\log \beta_1 = 3.9$ $Cd(SO_4)$; $\log \beta_1 = 2.3$ $CdCl^+$; $\log \beta_1 = 2.0$
$Cd(OH)_2$; $\log \beta_2 = 7.6$ $Cd(SO_4)_2^{2-}$; $\log \beta_2 = 3.2$ $CdCl_2$; $\log \beta_2 = 2.6$
                                  $Cd(SO_4)_3^{4-}$; $\log \beta_3 = 2.7$ $CdCl_3^-$; $\log \beta_3 = 2.4$
                                                                          $CdCl_4^{2-}$; $\log \beta_4 = 1.7$

From the calculations of Example 1, we already know the free ligand concentrations:

$$[OH^-] = 10^{-5.9}$$

$$[CO_3^{2-}] = 10^{-5.3}$$

$$[SO_4^{2-}] = 10^{-4.0}$$

$$[Cl^-] = 10^{-3.7}$$

A rapid examination of the thermodynamic data shows the predominant complexes to be the following.

For copper:

$$CuOH^+: \quad \beta_1[OH^-] = 10^{+0.4} \tag{19}$$

$$Cu(OH)_2: \quad \beta_2[OH^-]^2 = 10^{0.0} \tag{20}$$

$$CuCO_3: \qquad \beta_1[CO_3^{2-}] = 10^{1.4} \qquad (21)$$

$$Cu(CO_3)_2^{2-}: \quad \beta_2[CO_3^{2-}]^2 = 10^{-0.4} \qquad (22)$$

For cadmium:

$$CdOH^+: \beta_1[OH^-] = 10^{-2.0} \qquad (23)$$

$$CdSO_4: \beta_1[SO_4^{2-}] = 10^{-1.7} = 2 \times 10^{-2.0} \qquad (24)$$

$$CdCl^+: \quad \beta_1[Cl^-] \quad = 10^{-1.7} = 2 \times 10^{-2.0} \qquad (25)$$

Other complexes are negligible. The copper is thus mostly present as carbonate and hydroxide complexes and only about 3% as the free (hydrated) cupric ion:

$$TOTCu = [Cu^{2+}](1 + 10^{0.4} + 10^{0.0} + 10^{1.4} + 10^{-0.4})$$

$$= [Cu^{2+}] \times 10^{1.5} = [Cu^{2+}] \times 30 = Cu_T \qquad (26)$$

therefore

$$\frac{[Cu^{2+}]}{Cu_T} = \frac{1}{30} = 3\%$$

$$\frac{[CuOH^+]}{Cu_T} = \frac{10^{0.4}}{30} = 8\%$$

$$\frac{[Cu(OH)_2]}{Cu_T} = \frac{10^{0.0}}{30} = 3\%$$

$$\frac{[CuCO_3]}{Cu_T} = \frac{10^{1.4}}{30} = 84\%$$

$$\frac{[Cu(CO_3)_2^{2-}]}{Cu_T} = \frac{10^{-0.4}}{30} = 1\%$$

On the other hand, only about 5% of the cadmium is complexed, and the major species is, by far, the free cadmium ion, Cd^{2+}:

$$TOTCd = [Cd^{2+}](1 + 10^{-2} + 2 \times 10^{-2} + 2 \times 10^{-2})$$

$$= [Cd^{2+}] \times 1.05 = Cd_T \qquad (27)$$

therefore

$$\frac{[Cd^{2+}]}{Cd_T} = \frac{1}{1.05} = 95\%$$

$$\frac{[\text{CdOH}^+]}{\text{Cd}_T} = \frac{0.01}{1.05} = 1\%$$

$$\frac{[\text{CdSO}_4]}{\text{Cd}_T} = \frac{0.02}{1.05} = 2\%$$

$$\frac{[\text{CdCl}^+]}{\text{Cd}_T} = \frac{0.02}{1.05} = 2\%$$

Example 4. *Inorganic Speciation of Copper and Cadmium in Seawater* Consider the seawater model of Example 2; the free ligand concentrations are

$$[\text{OH}^-] = 10^{-5.7}$$
$$[\text{CO}_3^{2-}] = 10^{-4.5}$$
$$[\text{SO}_4^{2-}] = 10^{-2.0}$$
$$[\text{Cl}^-] = 10^{-0.26}$$

Extrapolation of the previous copper and cadmium constants to $I = 0.5\,M$ according to the Davies equation yields

CuOH$^+$; $\log\beta_1 = 5.7$ CuCO$_3$; $\log\beta_1 = 5.5$ CuSO$_4$; $\log\beta_1 = 1.2$
Cu(OH)$_2$; $\log\beta_2 = 10.9$ Cu(CO$_3$)$_2^{2-}$; $\log\beta_2 = 9.0$ CuCl$^+$; $\log\beta_1 = -0.2$

CdOH$^+$; $\log\beta_1 = 3.3$ CdSO$_4$; $\log\beta_1 = 1.1$ CdCl$^+$; $\log\beta_1 = 1.4$
Cd(OH)$_2$; $\log\beta_2 = 6.7$ Cd(SO$_4$)$_2^{2-}$; $\log\beta_2 = 2.0$ CdCl$_2$; $\log\beta_2 = 1.7$
                   Cd(SO$_4$)$_3^{4-}$; $\log\beta_3 = 2.7$ CdCl$_3^-$; $\log\beta_3 = 1.5$
                                            CdCl$_4^{2-}$; $\log\beta_4 = 1.1$

For copper, the hydroxide and carbonate complexes are still the most important and the speciation of copper is changed from that of Example 3, mostly because of ionic strength effects:

CuOH$^+$: $\beta_1[\text{OH}^-] = 10^{0.0}$ (28)

Cu(OH)$_2$: $\beta_2[\text{OH}^-]^2 = 10^{-0.5}$ (29)

CuCO$_3$: $\beta_1[\text{CO}_3^{2-}] = 10^{1.0}$ (30)

Cu(CO$_3$)$_2^{2-}$: $\beta_2[\text{CO}_3^{2-}]^2 = 10^{0.0}$ (31)

CuSO$_4$: $\beta_1[\text{SO}_4^{2-}] = 10^{-0.8}$ (32)

CuCl$^+$: $\beta_1[\text{Cl}^-] \cong 10^{-0.5}$ (33)

Copper is now about 7% in the free ionic form:

$$TOTCu = [Cu^{2+}](1 + 10^{0.0} + 10^{-0.5} + 10^{1.0} + 10^{0.0} + 10^{-0.8} + 10^{-0.5})$$

$$= Cu_T$$

$$\frac{[Cu^{2+}]}{Cu_T} = \frac{1}{13.8} = 10^{-1.14} = 0.07 \tag{34}$$

For cadmium, the chloride complexes are now obviously the most important:

$$CdCl^+: \quad \beta_1[Cl^-] = 10^{1.14} \tag{35}$$

$$CdCl_2: \quad \beta_2[Cl^-]^2 = 10^{1.18} \tag{36}$$

$$CdCl_3^{2-}: \quad \beta_3[Cl^-]^3 = 10^{0.72} \tag{37}$$

$$CdCl_4^{2-}: \quad \beta_4[Cl^-]^4 = 10^{0.06} \tag{38}$$

Since the constants are not very precisely known, we can consider that cadmium is present in seawater predominantly as chloro complexes and that the free cadmium ion is roughly 3% of the total metal:

$$TOTCd = [Cd^{2+}](1 + 10^{1.14} + 10^{1.18} + 10^{0.72} + 10^{0.06}) = Cd_T \tag{39}$$

therefore

$$\frac{[Cd^{2+}]}{Cd_T} = 10^{-1.56} = 0.03$$

Considering the results of Examples 3 and 4, we note that the speciation of metals that form important carbonate or hydroxide complexes (e.g., Cu^{2+}) varies strongly with pH. For metals that form important chloride or sulfate complexes (e.g., Cd^{2+}, Hg^{2+}) in seawater, their speciation in estuaries as freshwater mixes with seawater provides an interesting, though somewhat tedious, case study of inorganic complexation, as illustrated in Figure 6.8.

Hydroxide complexes, which may be considered to be produced by the dissociation of the weakly acidic hydrated metal ions, are of course important for many metals in natural waters, as illustrated in our study of ferric hydroxide precipitation (see Example 1 in Chapter 5). Many metal hydroxides form polymers [e.g., $Fe_n(OH)_n^{2n+}$] on the way to precipitation as hydrous oxide solids. These polynuclear complexes are poorly studied and often metastable. Note also that the presence of stoichiometric coefficients greater than 1 for the metal complicates the speciation calculations. At the limit there is a continuum between large metal hydroxide polymers and small suspended colloids of hydrous oxides.

From examination of the thermodynamic data in Table 6.3 it appears that apart from the hydroxides there are relatively few inorganic complexes of trace metals that are expected to be dominant in oxic waters. The major exceptions

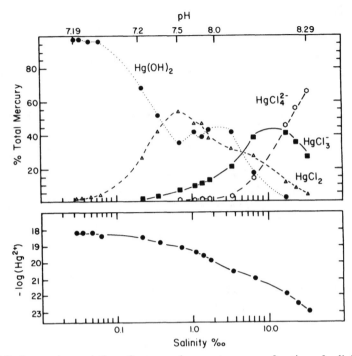

Figure 6.8   Inorganic speciation of mercury in an estuary as a function of salinity. This example has been calculated by considering the mixture of a freshwater ([Cl$^-$]) = 5×10^{-4} M; Alk = 1.3×10^{-4} M; pH = 7.2) with seawater (salinity = 36‰, pH = 8.20) and maintaining the total mercury concentration at 1 nM. Note the complicated changes in the inorganic mercuric species [($\bullet$)Hg(OH)$_2$, (Δ)HgCl$_2$, ($\blacksquare$)HgCl$_3^-$, ($\bigcirc$)HgCl$_4^{2-}$] due to concomitant variations in pH (see top scale) and major ion (chiefly Cl$^-$) concentrations.

are the carbonate complex of copper in sufficiently alkaline systems and the chloride complexes of cadmium, silver, and mercury in the presence of high chlorinity. Generalizing to all species that account for a few percent of the total metal concentrations, Table 6.8 shows—for the trace metals of Table 6.3—the inorganic species that are expected to be important in natural waters.

In anoxic waters the formation of bisulfide (HS$^-$), thiosulfate (S$_2$O$_3^{2-}$), and polysulfide (S$_n^{2-}$) complexes is important in maintaining a fraction of some trace metals in solution despite the low solubility of metal sulfides.[21,22]

4. ORGANIC COMPLEXATION

The concentration of dissolved organic matter in natural waters is typically in the range 1–100 mg C L^{-1} (dissolved organic carbon, DOC). Typical average values in freshwater are 1 mg L^{-1} in ground- and rainwater, 2–10 mg L^{-1} in lakes

TABLE 6.8 Predominant Inorganic Species for Selected Trace Metals in Aquatic Systems

Ba
Ba^{2+}
$BaSO_4(s)$, $BaCO_3(s)$

Cr
$Cr(OH)_2^+$, $Cr(OH)_3$, $Cr(OH)_4^-$, $HCrO_4^-$, CrO_4^{2-}
$Cr(OH)_3(s)$

Al
$Al(OH)_3$, $Al(OH)_4^-$, AlF^{2+}, AlF_2^+
$Al(OH)_3(s)$, $Al_2O_3(s)$, $Al_2Si_2O_5(OH)_4(s)$, Al-silicates

Fe
Fe^{2+}, $FeCl^+$, $FeSO_4$, $Fe(OH)_2^+$, $Fe(OH)_4^-$
$FeS(s)$, $FeS_2(s)$, $FeCO_3(s)$, $Fe(OH)_3(s)$, $Fe_2O_3(s)$, $Fe_3O_4(s)$, $FePO_4(s)$,
$Fe_3(PO_4)_2(s)$, Fe-silicates

Mn
Mn^{2+}, $MnCl^+$
$MnS(s)$, $MnCO_3(s)$, $Mn(OH)_2(s)$, $MnO_2(s)$

Co
Co^{2+}, $CoCl^+$, $CoSO_4$
$CoS(s)$, $Co(OH)_2(s)$, $CoCO_3(s)$, $Co(OH)_3(s)$

Ni
Ni^{2+}, $NiCl^+$, $NiSO_4$
$NiS(s)$, $Ni(OH)_2(s)$

Cu
Cu^{2+}, $CuCO_3$, $CuOH^+$
$CuS(s)$, $CuFeS_2(s)$, $Cu_2CO_3(OH)_2(s)$, $Cu_3(CO_3)_2(OH)_2(s)$, $Cu(OH)_2(s)$, $CuO(s)$

Zn
Zn^{2+}, $ZnCl^+$, $ZnSO_4$, $ZnOH^+$, $ZnCO_3$, ZnS
$ZnS(s)$, $ZnCO_3(s)$, $ZnSiO_3(s)$

Pb
Pb^{2+}, $PbCl^+$, $PbCl_2$, $PbCl_3^-$, $PbOH^+$, $PbCO_3$
$PbS(s)$, $PbCO_3(s)$, $Pb(OH)_2(s)$, PbO_2

Hg
Hg^{2+}, $HgCl^+$, $HgCl_2$, $HgCl_3^-$, $HgOHCl$, $Hg(OH)_2$, HgS_2^{2-}, $HgOHS^-$
$HgS(s)$, $Hg(liq)$, $Hg(OH)_2(s)$

Cd
Cd^{2+}, $CdCl^+$, $CdCl_2$, $CdCl_3^-$, $CdOH^+$, CdS, $CdHS^+$, $Cd(HS)_2$, $Cd(HS)_3^-$, $Cd(HS)_4^{2-}$
$Cd(s)$, $CdCO_3(s)$, $Cd(OH)_2(s)$

Ag
Ag^+, $AgCl$, $AgCl_2^-$, AgS^-, $AgHS$, $AgHS_2^{2-}$, $Ag(HS)_2^-$
$Ag_2S(s)$, $AgCl(s)$, $AgBr(s)$

and rivers, 10–50 mg L^{-1} in bogs and marshes. In the ocean, DOC varies from 1 mg C L^{-1} in deep waters to as much as 10 times this value in highly productive surface waters. Recent determinations of DOC by high temperature catalytic oxidation have given values two to three times those of earlier methods.[23] These controversial data exhibit strong vertical DOC gradients whose correlation with oxygen utilization suggests significant respiration of DOC in deep waters.[24]

The chemical nature of dissolved organic matter is poorly characterized. For example, in attempts to fractionate organic compounds in seawater, 50–90% of the DOC could not be identified and is thus broadly classified as humic and fulvic material[25,26] (Table 6.9). The term *humic acids* is normally used to indicate the fraction (ca. 10%) of DOC that precipitates at very low pH. The term *fulvic acids*

TABLE 6.9 Composition of Dissolved Organic Matter in Surface and Deep Ocean Water[a]

Compound Class[b] (Units)	Surface Ocean Concentration	Surface Ocean DOC-C[c] (%)	Deep Ocean Concentration	Deep Ocean DOC-C[c] (%)
DOC (μg/L)	400–2500		400–1600	
TFAA (nM)	50–500	0.4–1.8	25–40	0.3–0.4
THAA (nM)	50–1600	0.4–5.9	50–200	0.6–2.4
Hexosamines (nM)	10–30	0.1–0.2		
Urea (nM)	30–1700	0.05–1.4	<70	<0.17
TFMS (nM)	90–1700	1.0–8.7	51–120	0.8–1.8
THMS (nM)	600–4700	6.5–24	1000–1400	15–21
Combined uronic acids (nM)	75–260	0.8–1.3		
THFA (nM)	19–190	0.5–2.5	37–80	1.4–3.0
Chlorophyll (ng L^{-1})	10–2000	0.001–0.1	<100	<0.02
Indoles (μg L^{-1})	1	0.1		
Glycollic acid (nM)	260–520	0.9		
Phenols (μg L^{-1})	1–2	0.2		
Nucleic acids (μg L^{-1})	13–80	1.8–5.0		
Sterols (μg L^{-1})	0.2–0.4	0.02		
Hydrocarbons (μg L^{-1})	0.3–30	0.04–2.0	0–4	0–0.8
Vitamin B12 (ng L^{-1})	0.1–6.0		1	
Thiamine (ng L^{-1})	8.0–100			
Biotin (ng L^{-1})	1.6–4.6			
Unidentified		46–87[d]		70–82[d]

[a] Adapted from Buffle (1988).[25]
[b] TFAA—total free amino acids; THAA—total hydrolyzable amino acids; TFMS—total free monosaccharides; THMS—total hydrolyzable monosaccharides; THFA—total hydrolyzable fatty acids.
[c] Percentage contribution of carbon in the compound class to DOC estimated based on surface DOC = 0.7–1.5 mg L^{-1}, mean deep DOC = 0.5 mg L^{-1}.
[d] Estimated by difference based on other values in this table.

refers to acid-soluble compounds, usually of lower molecular weight (ca. 40% of DOC). Both of these fractions are generally defined operationally by a method of extraction, most often as what is retained on a particular type of chromatographic column. Thus humic and fulvic acids are those compounds that are sufficiently hydrophobic to be retained on acrylic-ester resins (XAD-8). A more hydrophilic fraction that passes through such columns can be extracted on others (e.g., XAD-4). This fraction (ca. 30% of DOC) has been less extensively studied than the more hydrophobic one but appears to have generally the same coordination properties. Here we shall use the general terms humate, humic acid, or humic compounds to designate all fractions: humic, fulvic, and hydrophilic acids.

One could undoubtedly identify a myriad of other trace organic compounds from natural and pollution sources in all natural waters, but the compounds in Table 6.9 may in fact account for the bulk of the DOC in the samples. A majority of the organic matter is put into the ill-defined class of humic acids because it possesses the inherent complexity and structural variability characteristic of this class, probably not because we ignore the existence of some well-defined substance that accounts for a sizeable fraction of the dissolved organic pool.

There is a major difference between the geochemical behavior of the humic organic fraction and the rest. Relatively small and reactive molecules such as sugars, amino acids, urea, or phenols provide readily available energy or nitrogen sources to aquatic microorganisms and are rapidly degraded. Their concentrations are maintained in the water column by an equally rapid production rate from microorganisms. One should visualize this fraction of the total DOC as turning over very rapidly—on a time scale of minutes to days. The rapid turnover time of this material makes it analytically elusive; some researchers believe that compounds such as carbohydrates may represent a sizeable fraction of the DOC.[27,28] The polymeric compounds that compose the humic fraction are refractory (a good definition for dissolved humic material is the acidic fraction of the DOC that is resistant to oxidation), and they are eliminated, at least in part, by eventual incorporation into the sediments rather than by degradation. Although recent work indicates some turnover of higher molecular weight DOC by microbial respiration[23,24] or by photochemical degradation,[29,30] still the residence time of this DOC fraction in the water column is on a time scale of weeks to thousands of years.

There is an ongoing scientific debate over the role that natural organic compounds play in complexing metal ions in aquatic systems.[31–33] An imposing body of experimental evidence supports the argument that some metal ions are largely, if not mostly, bound to organic ligands. Determinations of trace metal speciation in natural waters are based on the nonreactivity of the metals in chemical or biological assays; this nonreactivity is generally attributed to metal complexation by dissolved organic ligands. Thus the analytical method employed must be responsive only to some definable fraction of the total (dissolved) metal, ideally the free metal ion alone or with its inorganic complexes. For example, the dependence of copper toxicity on the free cupric ion concentration (discussed

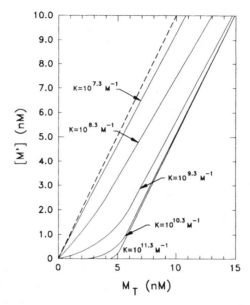

Figure 6.9 Voltammetric titration of a 5-nM ligand of variable metal binding strength. Only the inorganic form of the metal, M', is assumed to be measured by reaction at the electrode surface. The effective complexation constant K is defined as $K = [MY]/[M'] [Y_T - [MY]]$.

in Section 6) has been used as the basis for bioassays. In these assays, however, copper must be added to the water sample in order to observe any deleterious effect on the organisms. In almost all cases, such a metal titration is necessary because of the detection limit of the analytical method. Estimations of the ambient free metal ion concentrations are then obtained by modeling the metal titration and extrapolating to ambient total metal concentrations as discussed in Sidebar 6.1.

According to such estimations, the extent of metal complexation in natural waters can be close to 100% for reactive metals such as copper. Recently, significant complexation of lead, mercury, and even the less reactive metals zinc and cadmium has also been reported (see references from Table 6.10). Table 6.10 lists some of the reported values for trace metal complexation in both fresh- and seawater. This table is meant to provide an introduction to the substantial literature on metal complexation rather than a comprehensive review.

It can also be argued that very few important metal complexes have been directly demonstrated to exist and that, in evaluating reported metal complexation, certain limitations and assumptions of the analytical methods must be considered. The necessity for data modeling and extrapolating in the case of strongly complexed metals has already been mentioned. The attribution of the nonreactivity of metals strictly to organic complexation (or organometallic compounds) neg-

SIDEBAR 6.1

Electrochemical Measurements of Metal Complexation

Although bioassays have been used with some success and chromatographic techniques now show promise, the overwhelming majority of the analytical data on metal complexation in natural waters has been obtained with electrochemical methods. These electrochemical methods fall into two categories: potentiometric and voltammetric. Potentiometric methods involve the use of ion-sensitive electrodes, which exhibit a logarithmic response to free metal ion activity. An example of potentiometric data on copper complexation by humic substances is shown later (Figure 6.18). Although ion-sensitive electrodes have excellent sensitivity for free metal ion activities, the measurements require fairly high *total* metal concentrations ($\geqslant 10^{-7} M$), usually in excess of those present in natural waters. In addition, for some metals, measurements cannot be made in high-chloride media, such as seawater.[50]

In voltammetric methods, the current generated when metals are oxidized or reduced at an electrode surface is measured. In direct application, the measured signal is linearly related to the concentration of labile metal—which may include weak organic complexes as well as inorganic complexes of the metal. The sensitivity of such direct measurements is usually insufficient for environmental applications. Thus voltammetric techniques applied in natural waters include an electrochemical preconcentration step, which dramatically increases sensitivity. For example, a widely used preconcentration technique consists of plating the metal by reduction at a hanging mercury electrode before measuring it in an oxidative "stripping" step (reverse polarography). Recent applications of such methods in their multifarious variations have allowed determination of labile metals at concentrations down to $0.2 \, nM$.[44]

In some cases this sensitivity is good enough to obtain direct measurement of labile metal concentrations in natural water. In many cases, however, the reactive metal concentrations are still too low and a titration of the water sample with added metal is then performed. The shape of such a titration curve in the presence of a ligand at $5 \times 10^{-9} M$ concentration is shown in Figure 6.9. This figure assumes that the added metal has reached equilibrium with the ligand before measurement and that the metal–ligand complex is strictly unreactive at the electrode surface. Neither assumption is necessarily always correct (see Section 5). As seen in Figure 6.9, the ligand concentration L_T can in principle be obtained by extrapolation of the linear portion of the graph and the ligand's strength, K,

(continued)

deduced from its curvature (various linearizations of the graph can be used to simplify the data reduction). Thus in a simple system the inorganic (viz. "reactive") metal concentration can be calculated from knowledge of L_T, K, and M_T. In complex systems containing several ligands of varying concentrations and strengths, the data-fitting problem is much more complicated. In all cases the validity of the calculated inorganic metal concentration depends on the assumption that the ligand controlling it is the one characterized by the titration. For example one must assume that the unmeasurable leftmost point in Figure 6.9 is determined by the 5-nM ligand, not by another much stronger 1-nM ligand.

TABLE 6.10 Extent of Organic Complexation of Metals in Surface Waters

Metal	Percent Complexed	Water Type	Method	Reference
Cu	100	Oceanic	Electrochemical	34
	> 99	Coastal	Electrochemical	35
	> 95		Electrochemical	36
	> 99.9		Ligand competition	37
	> 99.9		Ligand competition	38
	> 99		Bioassay	35, 39
Zn	> 98	Oceanic	Electrochemical	40
	> 95		Electrochemical	41
	60–95	Estuarine	Electrochemical	42
	14	Riverine	Sorption	43
Pb	50–70	Oceanic	Electrochemical	44
Hg	2–89[a]	Freshwaters	Atomic fluorescence	45
Al	≃ 40	Freshwaters	Ion-exchange chromatography	46
Cd	0	Oceanic	Electrochemical	47
	0		Radiotracer addition and chromatography	48
	70		Electrochemical	49
Mn	0	Oceanic	Radiotracer addition and chromatography	48

[a] For Hg, "organic complexation" includes organomercury compounds (cf. Section 4.5).

lects the possible importance of colloidal species. Usually, the "dissolved" metal fraction is operationally defined by filtration or other physical methods of separation that do not exclude colloids. More importantly, equilibrium between (dissolved) metal and ligand species on the analytical time scale of titration experiments is usually assumed. Possible artifacts arising from this assumption are discussed in Section 5 on complexation kinetics. In addition, competition among metals has been considered only in a few cases.[37,44]

The greatest difficulty in predicting trace metal speciation from field measurements stems from a lack of information about naturally occurring complexing agents. Modeling requires information on ligand concentrations and stability constants, yet field measurements of trace metal speciation provide these parameters only indirectly. Direct information on the distribution and structures of naturally occurring metal complexes is a crucial link between measuring and modeling trace metal speciation in natural waters.

In the absence of such information, we may gain some quantitative understanding of the effects of organic complexation on trace metals in natural waters by making equilibrium calculations of model systems. As we shall see, the identifiable fraction of marine DOC cannot contribute much to metal complexation. Even the amino acids, which have the most interesting complexing properties, are not present in sufficient concentrations to affect metal speciation in most natural waters. We therefore focus on the following three categories of aquatic organic ligands:

1. Well-defined, strong complexing agents, such as EDTA and NTA, that are introduced into the aquatic environment through human activities.

2. Strong ligands produced by the planktonic biota. These may be present at only trace concentrations but exhibit much higher affinities for metals than the metabolites listed in Table 6.9. The information concerning these ligands originates mostly from culture studies in which high concentrations can be obtained. Extrapolation of these results to natural waters is of course problematic.

3. Humic substances isolated from a variety of aquatic systems and whose coordination properties have been characterized on concentrated samples by an assortment of methods.

Let us then examine each of these categories and obtain some quantitative estimates of organic complexation, recognizing that the question must ultimately be resolved through direct analysis, not model calculations.

4.1 Trace Metal Complexation by Strong (Anthropogenic) Chelating Agents

Metal complexation by strong artificial chelating agents such as NTA $[N(CH_2COO^-)_3]$ or EDTA (see Figure 6.7) is probably the best studied topic in aquatic coordination chemistry, and the effects of these chelating agents in complex mixtures have been thoroughly analyzed. EDTA has been and continues to be used for a wide range of industrial, pharmaceutical, and agricultural applications.[51,52] Because of its widespread use and limited biodegradation, EDTA has been found in groundwaters, sewage effluents, freshwaters, including drinking water, and estuarine waters.[53-56] Of particular concern is the mobilization of trace metals in groundwater due to the presence of EDTA. The use of EDTA in decontamination of nuclear facilities has been implicated in radionuclide migration in groundwater.[57,58]

The effect of EDTA on trace metal speciation is examined in the following example. According to the data of Table 6.3, copper is one of the metals expected to be most complexed by EDTA. For comparison purposes let us consider both copper and cadmium.

Example 5. Complexation of Copper and Cadmium by EDTA in Seawater
Measured concentrations of EDTA in the Mississippi River are in the range of $1-10\,nM$, attaining the upper limit of the range downstream of major urban inputs. In the United State municipal sewage contains typically $100-500\,nM$. (In countries where EDTA is used to replace phosphates in some cleaning products, the concentration is higher.) Thus, $10^{-7}\,M$ EDTA is a reasonable upper limit to the possible ligand concentration in a harbor or estuary. To the recipe of our seawater model (Example 2), let us then add typical surface seawater concentrations of copper ($10^{-8.5}\,M$) and cadmium ($10^{-9}\,M$) and $10^{-7}\,M$ EDTA:

$$Y_T = 10^{-7}\,M$$
$$Cu_T = 10^{-8.5}\,M$$
$$Cd_T = 10^{-9}\,M$$

Recall that at pH = 8.1,

$$[OH^-] = 10^{-5.7}$$
$$[CO_3^{2-}] = 10^{-4.5}$$
$$[SO_4^{2-}] = 10^{-2.0}$$
$$[Cl^-] = 10^{-0.26}$$
$$[Ca^{2+}] = 10^{-2.0}$$
$$[Mg^{2+}] = 10^{-1.3}$$

The relevant EDTA constants are obtained from Table 6.3 after appropriate ionic strength corrections*:

$HY^{3-}; \log\beta_1' = 10.1$ $CaY^{2-}; \log\beta_1 = 10.0$ $CuY^{2-}; \log\beta_1 = 18.1$

$H_2Y^{2-}; \log\beta_2' = 16.0$ $MgY^{2-}; \log\beta_1 = 8.2$ $CdY^{2-}; \log\beta_1 = 15.8$

---

*Ionic strength corrections according to the Davies equation have dubious validity for ions of charge 3 and 4. However, as seen in the example and as is discussed below, the results of the calculations are controlled by the ratios K_{Cu}/K_{Ca} and K_{Cd}/K_{Ca}, which are unaffected by the ionic strength corrections.

To simplify the problem, let us recall the results of Example 4:

$$\Sigma \text{ inorganic Cu species} = [Cu'] = 10^{1.14}\,[Cu^{2+}]$$

$$\Sigma \text{ inorganic Cd species} = [Cd'] = 10^{1.56}\,[Cd^{2+}]$$

The mole balance equations for copper and cadmium thus simplify to:

$$TOTCu = [Cu'] + [CuY^{2-}] = [Cu^{2+}](10^{1.14} + 10^{18.1}[Y^{4-}])$$

$$= Cu_T = 10^{-8.5} \tag{40}$$

$$TOTCd = [Cd'] + [CdY^{2-}] = [Cd^{2+}](10^{1.56} + 10^{15.8}[Y^{4-}])$$

$$= Cd_T = 10^{-9.0} \tag{41}$$

The free ligand concentration is obtained from the EDTA mole balance equation:

$$TOTY = [HY^{3-}] + [H_2Y^{2-}] + [CaY^{2-}] + [MgY^{2-}] + [CuY^{2-}] + [CdY^{2-}]$$

$$= Y_T = 10^{-7} \tag{42}$$

The last two terms can obviously be neglected since Cu_T and $Cd_T \ll Y_T$:

$$TOTY = [Y^{4-}](10^{10.1}[H^+] + 10^{16.0}[H^+]^2 + 10^{10.0}[Ca^{2+}] + 10^{8.2}[Mg^{2+}])$$

$$= 10^{-7} \tag{43}$$

$$TOTY = [Y^{4-}](10^{2.0} + 10^{-0.2} + 10^{8.0} + 10^{6.9}) = 10^{-7} \tag{44}$$

Note that the first two terms are also negligible and that EDTA is predominantly in the calcium form. As a result,

$$[Y^{4-}] = 10^{-8.0} \times 10^{-7} = 10^{-15.0} \tag{45}$$

Substituting into the copper and cadmium mole balances,

$$TOTCu = [Cu^{2+}](10^{1.14} + 10^{3.1}) = 10^{-8.5} \tag{46}$$

$$TOTCd = [Cd^{2+}](10^{1.56} + 10^{0.8}) = 10^{-9.0} \tag{47}$$

The concentrations of free, total inorganic, and organically-complexed metals are:

$$[Cu^{2+}] = 10^{-11.6}; \quad [Cu'] = 10^{-10.5}; \quad [CuY^{2-}] = 10^{-8.5}$$

$$[Cd^{2+}] = 10^{-10.6}; \quad [Cd'] = 10^{-9.1}; \quad [CdY^{2-}] = 10^{-9.9}$$

The copper is thus totally bound to EDTA, but only some 12% of the cadmium is in the organic complex form. In this example, there is no competition between the two trace metals for complexation with the organic ligand since EDTA is present in large excess.

$$* \quad * \quad *$$

Example 5 illustrates several important points concerning organic complexation of trace metals in natural waters. First and foremost, weak complexing agents, such as amino acids or other identifiable organic compounds listed in Table 6.9, cannot, at their ambient concentrations in natural waters, affect metal speciation. Even the strong ligand EDTA at the relatively high concentration of $10^{-7} M$ can only slightly affect the speciation of cadmium in seawater. If the affinity and concentration of the ligand were somewhat smaller, say, a factor of 10 each, even copper, one of the most reactive trace metals, would be present largely as inorganic species. For organic complexation to be important in the speciation of a trace metal, it is thus necessary that the ligand exhibit a very high specificity for the metal (see next example) or that the ligand concentration and effective binding constant be quite high.

In addition to higher organic ligand concentrations and lesser competition from inorganic ligands, a reason to expect organic complexation of trace metals to be more important in freshwater than in seawater is the dominant role of the major divalent cations, Ca^{2+} and Mg^{2+}, on the organic ligand speciation in seawater. As seen in Example 5, while copper is effectively 100% bound with EDTA, EDTA itself is 100% bound with calcium, which is itself 100% free. This is possible of course because of the wide differences in total concentrations: $Cu_T \ll Y_T \ll Ca_T$. The result is a large decrease in the apparent binding strength of EDTA for the metals:

$$[CuY^{2-}] = K_{Cu}[Cu^{2+}][Y^{4-}] \tag{48}$$

$$[CuY^{2-}] = \frac{K_{Cu}}{K_{Ca}[Ca^{2+}]}[Cu^{2+}]Y_T \tag{49}$$

therefore

$$K_{Cu}^{app} = \frac{K_{Cu}}{K_{Ca}[Ca^{2+}]} = 10^{-8} K_{Cu} \tag{50}$$

In other words, the complexation of copper by EDTA is controlled by the equilibrium of the reaction:

$$CaY^{2-} + Cu^{2+} = CuY^{2-} + Ca^{2+}; \quad K = \frac{K_{Cu}}{K_{Ca}} \tag{51}$$

which is pushed to the left by a large calcium concentration. Although sometimes the magnesium rather than the calcium complex is dominant, this result is applicable to many organic ligands in seawater, and it affects the complexation of all trace metals. In freshwaters, where the calcium and magnesium concentrations are 1–3 orders of magnitude smaller, the effective binding constants are increased accordingly. In some cases, depending on the particular ligand and the pH, the ligand speciation may in fact depend directly on its acid–base chemistry. This is often the case for weak ligands. Even for EDTA the calcium and magnesium complexes become unimportant in soft, acidic waters (e.g., where $pH = 4$, $[Ca^{2+}] = 10^{-4}$, $I = 0 M$):

$$TOTY = [Y^{4-}](10^{11.1}[H^+] + 10^{17.8}[H^+]^2 + 10^{21.0}[H^+]^3 + 10^{12.4}[Ca^{2+}])$$

$$(52)$$

The ligand speciation is then dominated by the species H_2Y^{2-} and H_3Y^-. At low ligand concentrations, however, when the metal–ligand concentration ratio exceeds unity for some metals, the ligand speciation is dominated by the corresponding complexes. Figure 6.10 illustrates what happens to EDTA and to various trace metals when a freshwater is titrated with EDTA. In this example EDTA at very low concentrations is present chiefly in the nickel complex. When nickel itself becomes titrated as the EDTA concentration exceeds that of nickel, the copper and zinc complexes become prevalent. Finally, when those two

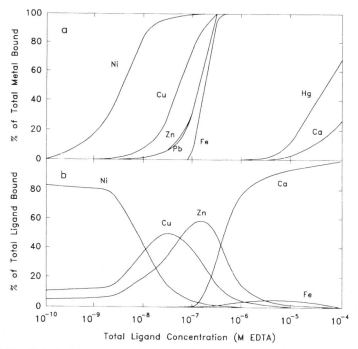

Figure 6.10   Calculated titration of a model freshwater with EDTA. (*a*) Speciation of the metals; (*b*) EDTA speciation. The pH is fixed at 8.10 and the total concentrations are as follows: $Ca_T = 3.7 \times 10^{-4} M$; $Mg_T = 1.6 \times 10^{-4} M$; $K_T = 6.0 \times 10^{-5} M$; $Na_T = 2.8 \times 10^{-4} M$; $Fe(III)_T = 5.0 \times 10^{-7} M$; $Cu_T = 5 \times 10^{-8} M$; $Hg_T = 10^{-9} M$; $Zn_T = 1.5 \times 10^{-7} M$; $Ni_T = 5 \times 10^{-9} M$; $Pb_T = 10^{-9} M$; $C_T = 10^{-3} M$; $[SO_4]_T = 10^{-4} M$; $Cl_T = 2 \times 10^{-4} M$.

metals are themselves titrated $(Y_T > Cu_T + Zn_T = 50 + 150 \, nM)$, EDTA becomes controlled by its calcium complex. Besides Ni, Cu, and Zn, the other three metals included in this model calculation, Hg, Pb, and Fe, also become bound to EDTA although they never account for a sizeable fraction of the ligand. Lead and zinc are titrated simultaneously. Lead, in contrast to zinc, does not affect the ligand speciation because of its low total concentration; zinc is present in 150-fold excess of lead. Iron, which is controlled by its amorphous hydroxide precipitate, becomes organically bound only when the free EDTA increases markedly upon titration of the zinc. Mercury is the last of the trace metals to be titrated.

4.2   Trace Metal Complexation by Strong Biogenic Chelating Agents

Practically all aquatic microorganisms release metabolites into their growth medium, including a variety of compounds whose affinity for metals ranges from that of simple amino acids up to that of fairly strong chelating agents. But, as argued before, unless these compounds have unusual metal binding strength or specificity, they are unlikely to affect metal speciation in natural waters. Aquatic microorganisms, however, synthesize two families of strong metal complexing agents: transport ligands to acquire essential trace metals and detoxifying (or buffering) ligands to defend themselves against the toxicity of others.

Transport Ligands. For the most part we do not know the nature of the metal transport ligands in phytoplankton and bacteria. But we know that they are normally membrane-bound and exhibit high affinity and specificity for the target metals. (This is clearly a necessity to acquire trace metals from oligotrophic waters.) In the case of iron, three types of transport ligands may be important. The catechol and hydroxamate siderophores of terrestrial bacteria (Figure 6.11) are also produced by heterotrophic marine bacteria, cyanobacteria (blue green algae), fungi, and marine dinoflagellate phytoplankton.[59-63] In the organisms that have been studied so far, siderophores are produced under iron-limited growth conditions and then released to the medium. (Such release may not be typical of organisms living in dilute media such as the high seas.) The iron–siderophore complexes are then taken up through an active (i.e., ATP-requiring) process involving specific membrane proteins.[64,65] The detailed structures and the exact properties of several of these siderophores have been characterized and one of them, desferriferrioxamine B, is included in Table 6.3. Siderophores have been isolated from soils[66] and chemical and biological assays have indicated their presence in algal mats[67] and in lakewater during a blue-green algal bloom.[68] It has been suggested that siderophore production by blue-green algae may suppress the growth of competing species, which are unable to utilize the iron–siderophore complex for growth, thus fostering the bloom of blue-green algae.[68] In the next example, we use the well-characterized trihydroxamate siderophore desferriferrioxamine B as a model compound to

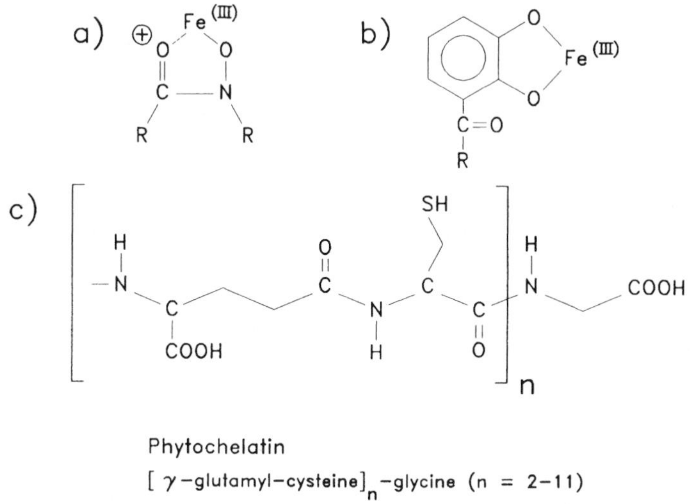

Phytochelatin

[γ -glutamyl-cysteine]$_n$ -glycine (n = 2-11)

Figure 6.11 Complexing functionalities of some biogenic ligands: (*a*) hydroxamate siderophores; (*b*) catechol siderophores; (*c*) phytochelatins. Only partial structures are shown for the siderophores. The side-chain moieties (indicated by R-) are, in many siderophores, quite substantial and often include additional complexing functional groups of the types illustrated. For example, desferriferrioxamine B, the siderophore that appears in Table 6.3, is a hexadentate ligand with three of the hydroxamate functional groups shown in *a*. For detailed structures see Neilands, 1981.[59]

examine the situation created by the presence of complexing agents with very high affinity and specificity for a particular metal.

Example 6. Complexation of Iron(III) and Copper by a Trihydroxamate Siderophore in Freshwater Consider the effect of adding desferriferrioxamine B to the freshwater system of Example 1 containing $10^{-6} M$ Fe(III) and $10^{-7} M$ Cu. (Recall that at pH = 8.1, $[OH^-] = 10^{-5.9}$, $[CO_3^{2-}] = 10^{-5.3}$, $[SO_4^{2-}] = 10^{-4.0}$, $[Cl^-] = 10^{-3.2}$, $[Ca^{2+}] = 10^{-3.5}$, and $[Mg^{2+}] = 10^{-4.1}$.) Desferriferrioxamine B is a hexadentate ligand with three hydroxamates and a terminal amine which is protonated somewhere below pH = 11. We thus use the symbol Y^{2-} to indicate the ligand with all hydroxamates deprotonated. The relevant organic complexation constants are given in Table 6.3:

$HY^-; \log \beta_1 = 10.1$ $CaY^{2-}; \log \beta_1 = 3.5$ $FeY^+; \log \beta_1 = 31.9$

$H_2Y; \log \beta_2 = 19.4$ $MgY^{2-}; \log \beta_1 = 5.2$ $FeHY^{2+}; \log \beta_{11} = 32.6$

$H_3Y^+; \log \beta_3 = 27.8$ $CuY^{2-}; \log \beta_1 = 15.0$

                                                            $CuHY^+; \log \beta_{11} = 24.1$

                                                            $CuH_2Y^{2+}; \log \beta_{12} = 27.0$

At a pH of 8.1, in the absence of chelating agents, iron is precipitated as $Fe(OH)_3(s)$ (see Example 1 in Chapter 5), and the free ferric ion concentration is given by

$$[Fe^{3+}] = 10^{3.2}[H^+]^3 = 10^{-21.1} \tag{53}$$

The free cupric ion concentration in the absence of a chelator has been calculated as 3% ($10^{-1.5}$) of the total inorganic copper in Example 3:

$$[Cu^{2+}] = 10^{-1.5}[Cu'] = 10^{-8.5}$$

The mole balance equation for desferriferrioxamine at low ligand concentration, low enough not to affect $[Fe^{3+}]$ or $[Cu^{2+}]$, can then be simplified:

$$
\begin{aligned}
TOTY = [Y^{2-}]&(1 + 10^{10.1}[H^+] + 10^{19.4}[H^+]^2 + 10^{27.8}[H^+]^3 \\
&+ 10^{3.5}[Ca^{2+}] + 10^{5.2}[Mg^{2+}] + 10^{31.9}[Fe^{3+}] \\
&+ 10^{32.6}[H^+][Fe^{3+}] + 10^{15}[Cu^{2+}] + 10^{24.1}[H^+][Cu^{2+}] \\
&+ 10^{27}[H^+]^2[Cu^{2+}]) = Y_T
\end{aligned}
\tag{54}
$$

All terms are negligible compared to the FeY^+ concentration:

$$TOTY = [FeY^+] = 10^{31.9}[Fe^{3+}][Y^{2-}] = 10^{10.8}[Y^{2-}] = Y_T \tag{55}$$

Given the enormous affinity of the organic ligand for the ferric ion, as long as $Y_T < Fe_T$, the speciation problem is simple: the ligand is entirely bound to iron; the free feric ion is controlled by precipitation of $Fe(OH)_3(s)$; and the copper remains chiefly in the copper carbonate complex ($K_{CO_3}[CO_3^{2-}]$ $= 10^{1.4}$).

As soon as the hydroxamate is in excess of Fe_T, however, it completely dissolves the iron, and the high affinity of the ligand for copper becomes important for copper speciation. Consider, for example, $Y_T = 10^{-5.8}\,M$. In addition to FeY^+, the major species at pH $= 8.1$ are clearly H_2Y and H_3Y^+:

$$TOTY = [H_2Y] + [H_3Y^+] + [FeY^+] = Y_T = 10^{-5.8} \tag{56}$$

At this point a reasonable assumption to be verified subsequently is that all the iron is bound to the organic ligand:

$$[FeY^+] = Fe_T = 10^{-6} \tag{57}$$

therefore

$$[H_2Y] + [H_3Y^+] = Y_T - Fe_T = 10^{-6.23} \tag{58}$$

and

$$[Y^{2-}](10^{19.4}[H^+]^2 + 10^{27.8}[H^+]^3) = 10^{-6.23}$$

$$[Y^{2-}](10^{3.2} + 10^{3.5}) = 10^{-6.23} \tag{59}$$

$$[Y^{2-}] = 10^{-9.9}$$

We can verify that the iron is indeed dissolved:

$$[FeY^+] = 10^{31.9}[Y^{2-}][Fe^{3+}] = Fe_T = 10^{-6} \tag{60}$$

therefore

$$[Fe^{3+}] = 10^{-28.0} < 10^{-21.1}$$

The mole balance for copper is now written:

$$TOTCu = [Cu'] + [CuY] + [CuHY^+] = Cu_T = 10^{-7} \tag{61}$$

$$TOTCu = [Cu^{2+}][10^{1.5} + 10^{15.0}[Y^{2-}] + 10^{24.1}[H^+][Y^{2-}]] = 10^{-7}$$

$$TOTCu = [CuHY^+] = 10^{6.1}[Cu^{2+}] = 10^{-7} \tag{62}$$

therefore

$$[Cu^{2+}] = 10^{-13.1}$$

The copper becomes immediately complexed by the hydroxamate, and its free concentration is lowered by a factor of approximately one million. Thus, siderophores can alleviate copper toxicity to algae. Note, however, that this is possible only if the organic ligand is in excess of the iron, a situation that may not be commonly encountered in natural waters. In seawater, as illustrated in Example 5, the major cations would affect (slightly) the complexing ability of the organic ligand toward copper since MgY is one of the major ligand species in seawater (once the iron is entirely complexed). In the most probable situation where iron siderophores are present at very low concentrations, lower than that of trace metals, their expected effect on metal speciation is a simple stoichiometric binding of Fe(III).

$$* \qquad * \qquad *$$

Detoxifying Ligands. The other major family of strong biogenic chelating agents, the metal detoxifying molecules, are synthesized by bacteria, plants, and animals in response to metal toxicity. They consist of cysteine-rich polypeptides and include two major types, the phytochelatins of plants and the metallothioneins of animals. Phytochelatins are small polypeptides (Figure 6.11) produced enzymatically by all plants.[69] In microscopic algae they seem to play a metal buffering role even when the algae are not under metal stress.[70] Metallothioneins are small proteins (mw $\simeq$ 6000–7000) with high cysteine content and no aromatic amino acids.[71] They bind several metal ions per molecule and are ubiquitous throughout the animal kingdom. Both phytochelatins and metallothioneins are highly soluble and may thus represent a sizeable source of

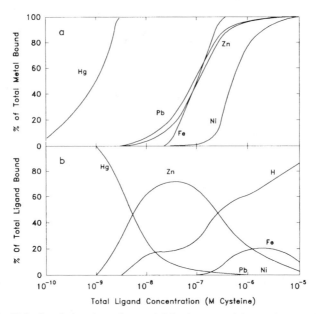

Figure 6.12 Calculated titration of a model freshwater with cysteine. (*a*) Speciation of the metals. (*b*) Cysteine speciation. Same conditions as Figure 6.10 except that Cu(II), which is reduced by cysteine, has been omitted.

strong, unspecific metal complexing agents in water upon cell lysis. Cysteine is included in Table 6.3 to exemplify lower limits on the binding strength of these compounds.

Because detoxifying ligands, like EDTA, are strong unspecific chelators, their effect on metal speciation is very similar to that discussed in Section 4.1. The major difference stems from the different relative metal affinities of the "softer" S in the thiols of the phytochelatins and metallothioneins compared to the "harder" O in the carboxylate of EDTA. For example, Figure 6.12 is the result of complexation calculations identical to those of Figure 6.10, in which EDTA has been replaced by cysteine as a model thiol chelator. Owing to its very high affinity for the thiol group, Hg^{2+} is readily titrated by cysteine. However, because (in this example!) the more abundant zinc controls the cysteine speciation at low ligand concentrations (compared to the less abundant nickel in the case of EDTA), the other trace metals never influence the ligand speciation. Nickel, whose affinity for thiols is relatively poor, is the last metal titrated. [Note that Cu(II) which is reduced to Cu(I) in the presence of cysteine is not included in the calculations.]

*   *   *

In the preceding discussion, the problem of trace metal complexation has been kept simple by considering only one organic ligand at a time. The problem appears much more complicated when one has to consider several metals and several ligands together. Determining the extent of competition among metals for the same ligand and among ligands for the same metal in a complex mixture seems formidable, and computer programs have been used extensively for this purpose.

In fact, examination of the results of such computer calculations shows that the problems are often not as difficult as they may appear at first and that competition among metals for the same ligand seldom occurs. A few rules of thumb can be given:

1. Like the hydrogen ion (H^+), major cations (Ca^{2+}, Mg^{2+}) do not really "compete" with trace metals for strong ligands, they merely decrease the ligand's apparent affinities for the metals (recall the $K_{Cu}/K_{Ca}[Ca^{2+}]$ factor in Example 5).

2. Trace metals do not compete with each other for the same ligand as long as the ligand is in large excess of the metals ($L_T > M_T$'s). This is the situation most often considered.

Table 6.11 presents the results of a computer calculation for a model freshwater system containing citrate, glycine, cysteine, and NTA, all in excess of all the trace metals but iron. Except for mercury, which has a very high affinity for cysteine, and iron, which is partly present as the hydroxide solid, the reactive trace metals are found chiefly as NTA complexes. Most of the NTA itself is present as the calcium complex. A series of hand calculations such as those presented in Examples 5 and 6 would have yielded the same result ponderously but easily.

4.3 Complexation by Humic Compounds

As mentioned previously, a major fraction of the dissolved organic matter in natural waters is composed of refractory humic compounds, polymers of variable composition. Although their overall structure, particularly their three-dimensional conformation, may be complicated and ill-defined, humic compounds appear to contain reasonably simple and consistent functional groups for coordination: carboxyl, alcohol, and phenol.[72,73] Humic substances have substantially different chemistry in the open ocean and in fresh or coastal waters. In the open ocean where organic matter is almost entirely autochthonous (formed in situ), humic compounds are formed by condensation, polymerization, and partial oxidation of smaller molecules such as triglycerides, sugars, or amino acids, and they exhibit little aromatic character.[74] This is illustrated in Figure 6.13, which shows a hypothetical pathway of formation of marine humates from polyunsaturated triglycerides, leading to crosslinked aliphatic acids.[75] Near land, by contrast, the DOC is largely allochthonous (imported from foreign sources) and derived from the decaying material of higher plants, particularly

TABLE 6.11 Results of an Equilibrium Computation for a Freshwater System Containing Four Organic Ligands[a]

	Total Conc	Free Conc	CO_3	SO_4	Cl	F	NH_3	PO_4	SiO_2	CIT	GLY	NTA	CYST	OH
	→(Total)	→(Free)	3.00	4.50	3.50	5.50	5.50	5.00	4.00	5.00	5.00	5.00	5.00	—
			6.15	4.56	3.50	5.51	7.50	11.84	12.51	17.00	7.84	9.97	10.51	7.00
Ca	3.00	**3.02**	4.82	5.54	—	7.56	10.62	5.51s	—	5.88	9.19	5.18	—	8.28
Mg	3.50	**3.51**	5.30	5.93	—	7.35	10.91	7.70	—	7.17	7.58	6.67	—	7.77
Na	3.50	**3.50**	8.58	7.49	—	—	—	—	—	—	—	11.26	10.08	—
Fe(III)	5.00	17.80	—	18.56	20.10	17.47	—	16.53	14.14	7.59	14.94	6.66	5.75	8.70 **5.10s**
Mn(II)	6.50	**6.55**	7.86	9.07	9.08	—	13.36	9.64	—	9.61	10.82	7.71	12.62	9.72
Cu(II)	6.00	10.15	9.86	12.68	13.18	14.49	11.86	12.85	—	7.65	9.32	**6.01**	—	11.12
Ba	7.00	**7.00**	10.76	—	—	—	—	—	—	27.06	13.67	10.86	—	12.87
Cd	6.00	7.56	—	10.08	9.19	12.10	12.56	15.88	—	9.41	11.12	**6.01**	8.82	9.62
Zn	7.00	9.05	—	11.57	13.18	13.39	14.36	12.65	—	13.25	11.72	**7.01**	10.63	11.62
Ni	6.50	9.24	—	11.76	12.37	13.78	14.05	—	—	10.38	10.91	**6.50**	—	11.01
Hg	9.00	32.91	—	35.34	26.10	36.95	30.40	—	—	—	28.76	27.33	**9.00**	24.18
Pb	7.00	10.34	9.35	12.46	12.27	—	—	—	—	11.56	12.21	**7.10**	7.70	10.50
Co(II)	7.50	9.55	—	11.87	12.68	—	15.05	—	—	27.20	12.41	**7.51**	10.12	11.81
Ag	9.00	**9.13**	—	12.72	9.59	—	13.43	—	6.03s	17.85	13.13	—	—	13.83
Al	5.00	13.95	—	15.70	—	12.16	—	5.48	—	**5.07**	—	8.03	—	9.18
H	7.00	7.00	3.01	9.49	—	9.58	5.50	6.07	4.01	7.32	5.00	6.66	5.19	—

Source: Morel et al. (1973).[33]

[a]*Note:* CIT = citrate, GLY = Glycine, NTA = nitrilotriacetate, CYST = cysteine. Numbers are negative logarithms of molar concentrations in solution. The presence of a solid is indicated by s. A dash signifies the absence of a computable species. Bold-face numbers indicate the major species or solid for each metal.

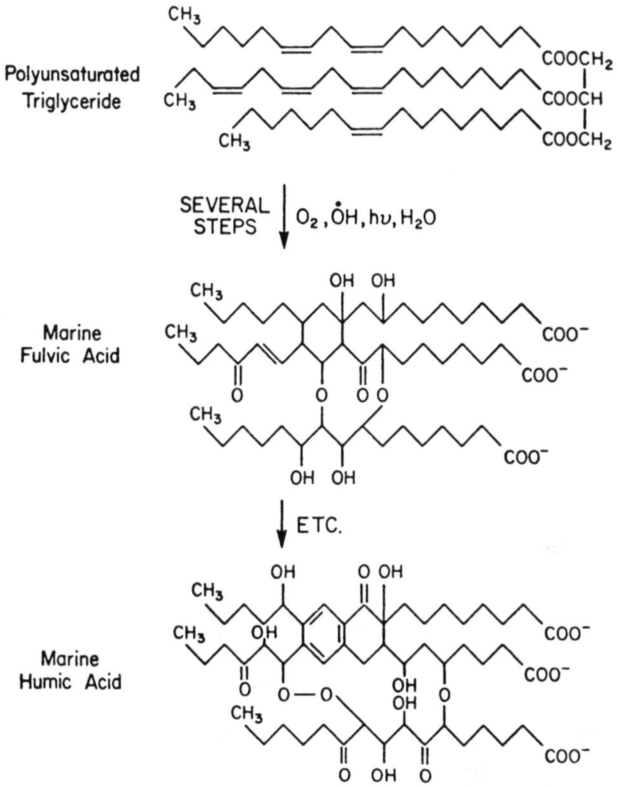

Figure 6.13 A possible pathway for the formation of marine humic acids from a triglyceride. From Harvey et al., 1983.[75]

lignins[73,76] (Figure 6.14). The resulting terrigenous humic compounds retain a relatively high degree of aromaticity from their precursors[77] (Figure 6.15). Thus, carboxylic and alcoholic functional groups are characteristic of all humates, but phenolic groups are found predominantly in the humic compounds near land.

The total number of titratable acid groups in humic substances is typically in the range of 10 to 20 meq $(gC)^{-1}$. Terrigenous humic compounds are usually close to 50% carbon by weight. Using calorimetric techniques it has been estimated that 60–90% of these acid groups are carboxyl groups and the rest phenolic groups.[78] Chelation of carboxyl and phenolic groups in close proximity to each other is generally considered to be the major mode of metal complexation by such humates. Thus, compounds such as malonate, phthalate, salicylate, or catechol serve as convenient model compounds for the coordinative properties of humic acids (Figure 6.16).

Convenient though they may be for didactic or illustrative purposes, model compounds do not provide an accurate image of the complications involved in

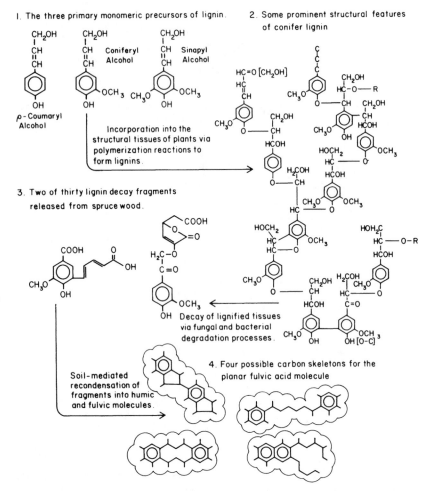

1. The three primary monomeric precursors of lignin.

2. Some prominent structural features of conifer lignin

3. Two of thirty lignin decay fragments released from spruce wood.

Incorporation into the structural tissues of plants via polymerization reactions to form lignins.

Decay of lignified tissues via fungal and bacterial degradation processes.

Soil-mediated recondensation of fragments into humic and fulvic molecules.

4. Four possible carbon skeletons for the planar fulvic acid molecule

Figure 6.14 A possible schematic pathway for the formation of terrigenous humic acids from lignins. Based on Aiken et al. (1985)[73] and Crawford (1981).[76]

studying the coordinative properties of humic substances. Reduction of data from typical acid–base titrations or metal-coordination experiments with humates usually do not yield a simple set of discrete constants; the titration curves are "smeared" and the resulting constants vary continuously with the experimental conditions. There are three principal causes for this confusing state of affairs:

1. As a result of chemical and steric differences in their neighboring groups, coordination sites on humates may well have a continuous range of affinities for metals including H^+.

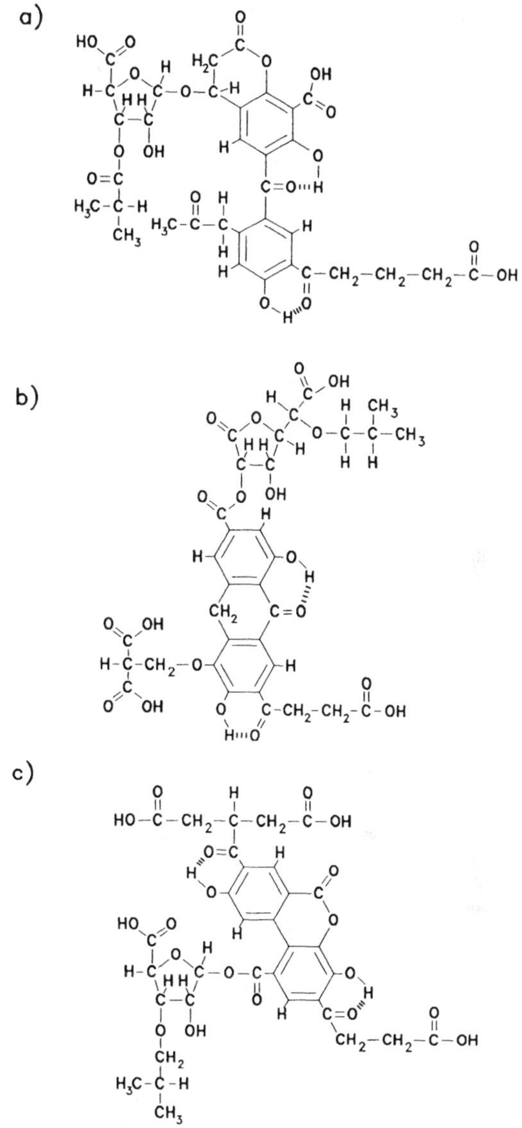

Figure 6.15 Three proposed structures for terrigenous humic acids. After Leenheer in reference 72.

2. Conformational changes resulting from electrostatic interactions among the various functional groups on a single molecule may make the coordination properties of the molecule highly dependent on the extent of cation binding and on the ionic strength of the solution.

3. Electrostatic attraction and repulsion from neighboring ionized groups

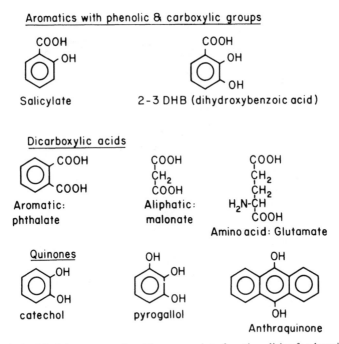

Figure 6.16 Model compounds with appropriate functionalities for humic acids.

may affect markedly the metal affinity of a given group, thus making the free energy of complexation, and hence the complexation constant, dependent on the extent of cation complexation of the ligand, particularly its protonation (and on ionic strength).

Three different approaches have been taken to model the complicated co-ordination chemistry of such complex mixtures of complex molecules. The easiest and most common is to fit titrimetric data with an ad hoc set of well-defined discrete ligands. Though simple, this approach has the fundamental weakness of being strictly an empirical data fitting procedure without power of prediction or extrapolation. This weakness is not ameliorated by the second approach in which humates are modeled as a mixture of ligands with a continuous (Gaussian) distribution of acidity or metal binding constants. Here we take a third approach and assume that the coordination properties of humates can be described by a superposition of the individual affinities of a few well-defined ligands and the coulombic attraction of neighboring groups. The idea is to capture in such a model enough of the fundamental physical–chemical processes controlling humate chemistry to obtain true predictive capability.

Polyelectrolyte effects. As noted in Chapter 2, coulombic forces are effective over relatively large distances and affect the free energies of ions even in dilute solutions. When a cation reacts with an acidic functional group of a molecule with multiple functional groups (a "polyanion" such as a humic acid or a protein), the chemical free energy of interaction is augmented by the long-range coulombic attraction emanating from all the neighboring, nonreacting, negative sites. This so-called polyelectrolyte effect which can strengthen greatly the binding of metal ions by polyanions has been well studied in physical-biochemistry. Its role in the chemistry of humic compounds is beginning to be understood. A simple quantitative treatment of coulombic effects in polyion binding is obtained by the same blending of thermodynamics and electrostatics and the very same starting equation involved in the Debye–Hückel theory.[79]

The free energy of reaction between a metal M^{n+} and the functional group L^{m-}

$$M^{n+} + L^{m-} \rightleftharpoons LM^{(n-m)+}$$

is the sum of the *chemical free energy* due to the reaction with the functional group itself and the *coulombic free energy* due to electrostatic interactions between M^{n+} and all the charges on the polyanion.

$$\Delta G^\circ = \Delta G^\circ_{chem} + \Delta G^\circ_{coul} \tag{63}$$

In the Debye–Hückel theory, the coulombic term was calculated as the work necessary to charge the molecule from zero to $-Ze$ ($= ZF$ per mole) in an ionic atmosphere. In the case of a polyelectrolyte, the coulombic work involved in reacting with M^{n+} can be taken as the product of the charge increase because of the reaction ($= nF$) and an average potential (Ψ_0) at the surface of the polyelectrolyte.

$$\Delta G^\circ_{coul} = nF\Psi_0 \tag{64}$$

(Recall that F is the Faraday constant, 96,500 C).

In terms of equilibrium constants,

$$K = e^{-\Delta G^\circ/RT} = K_{chem} \cdot K_{coul} \tag{65}$$

where

$$K_{chem} = e^{-\Delta G^\circ_{chem}/RT} \tag{66}$$

$$K_{coul} = e^{-\Delta G^\circ_{coul}/RT} = e^{-nF\Psi_0/RT} = P^n \tag{67}$$

The convenient parameter P is the inverse of the exponential of the nondimensional average surface potential, Ψ_0.

The issue then is to calculate what the average electrical potential, Ψ_0, at the binding site might be. For this purpose, the polyion is represented as an impenetrable sphere of uniform surface charge density. The distribution of

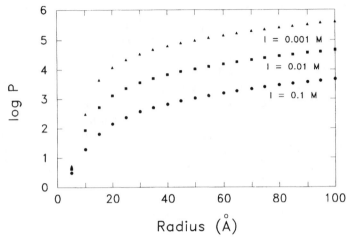

Figure 6.17 Coulombic factor ($\log P = -2.3\,F\Psi_0/RT$) for an impenetrable sphere as a function of its radius and the ionic strength. The total charge is assumed to be proportional to R^3 so that the surface charge density increases proportionally to R. A charge density of $9.2\,\mu C\,cm^{-2}$ was chosen for $R = 7.7\,\text{Å}$. After Bartschat et al., 1992.[79]

co- and counterions from the surrounding electrolyte is represented as a continuously varying charge in a medium of uniform dielectric constant.

As discussed in Chapter 2 for the Debye–Hückel theory, the Poisson–Boltzmann equation then describes the relationship between the variations in potential and the concentration of co- and counterions in the medium. With the appropriate boundary conditions (see Sidebar 6.2), we can solve this equation to obtain the potential Ψ as a function of distance from the spherical polyelectrolyte. More importantly, we also obtain the value of the surface potential Ψ_0 as a function of the surface charge density, the radius (R) of the polyelectrolyte, and the concentration of co- and counterions in solution (Figure 6.17). As may be expected, these surface potentials approach the Debye–Hückel values at small radii and the values for infinite flat plates (see Gouy–Chapman theory in Chapter 8) at large radii.

To calculate the extent of binding of cations by polyanions requires an iterative procedure since the equilibrium constant, $K = K_{chem} \cdot K_{coul}$, depends on the surface charge of the polyanion, which itself depends on the extent of binding of cations, including protons. A general approach to such calculations will be presented in Chapter 8 dealing with binding at the surface of solids. In the following section we describe a simple model of humates, discuss the general features of their acid–base and complexometric titrations (from experimental data and corresponding model calculations), and study how to calculate by hand the extent of metal complexation by humates in the simple and typical case of low metal to ligand ratios.

SIDEBAR 6.2

Solutions to the Poisson–Boltzmann Equation
II. COULOMBIC EFFECTS ON OLIGOELECTROLYTES

Our humic acid model is the same as the Debye–Hückel model for small ions introduced in Chapter 2 and is based on the spherically symmetrical form of the Poisson–Boltzmann equation:

$$\frac{d^2\Psi}{dr^2} + \frac{2}{r}\frac{d\Psi}{dr} = \frac{-1}{\varepsilon\varepsilon_0}\sum Z_i F(S_i)_0 e^{-Z_i F\Psi/RT} \tag{i}$$

There are two important differences, however.

First, if the model is to be applicable to oligoelectrolytes (which have a larger charge number and a larger surface charge density than small ions), we cannot assume that the potentials are small enough to justify the linearization step used in the Debye–Hückel theory. We instead have to solve equation (i) numerically. The result is Ψ as a function of r (the distance from the center of the sphere) for a given bulk ionic composition, molecular radius R, and charge on the molecule (Q^-).[79]

Second, we make an approximation which spares us from having to integrate our numerical solution of Ψ over the charge of the molecule as we did in Chapter 1.

Consider a reaction between a polyion of charge Q^- and a small ion of charge Z^+:

$$A^{Q^-} + M^{Z^+} \rightarrow A-M^{Z-Q}$$

The apparent (conditional) equilibrium constant for this reaction is

$$K_{app} = \frac{[A-M^{Z-Q}]}{[A^{Q^-}][M^{Z^+}]} = \frac{\{A-M^{Z-Q}\}}{\{A^{Q^-}\}\{M^{Z^+}\}}\frac{\gamma_Q\gamma_Z}{\gamma_{Z-Q}} = K_{int}\gamma_Z\frac{\gamma_Q}{\gamma_{Z-Q}} \tag{ii}$$

Thus, it is the ratio of polyion activity coefficients γ_Q/γ_{Z-Q} and the simple Debye–Hückel activity coefficient for the small ion that will appear as the corrections to K_{int} in the equilibrium expression. One can define this ratio as an effective activity coefficient:

$$\ln\gamma_{eff} = \ln\frac{\gamma_Q}{\gamma_{Z-Q}} = \ln\gamma_Q - \ln\gamma_{Z-Q}$$

(*continued*)

$$= \frac{-1}{kT} \int_{-Qe}^{(Z-Q)e} \Psi_0(q)dq \qquad \text{(iii)}$$

If Q is much larger than Z, $\Psi_0(q)$ remains approximately constant over the range of the integral, so that

$$\ln \gamma_{eff} = \frac{-Ze}{kT} \Psi_0(Q) \qquad \text{(iv)}$$

The "local" activity of a small ion at the surface of the oligoelectrolyte is then given by the "bulk" activity multiplied by a Boltzmann factor,

$$P = \exp\left(-\frac{Ze}{kT} \Psi_0 \right)$$

given in Figure 6.17. A physical interpretation of the Boltzmann factor is that it represents the work required to bring the small ion from an infinite distance to the surface of the oligoelectrolyte. This interpretation is often used to derive the expression for γ_Z, but it is only strictly correct when the oligoelectrolyte does not come close to being neutralized by the reaction with the small ion. In reality, it is not a "local" activity coefficient which we use to correct the equilibrium constants, but a ΔG_{coul}, and if the charge on the oligoelectrolyte changes enough so that the ionic atmosphere also rearranges itself significantly, the former is not a good approximation of the latter.

A Simple Model Humate. A simple model of a humic acid is presented in Table 6.12. It consists of a mixture of three ligands, acetate, malonate, and catechol, distributed among two size groups, 75% in molecules of molecular weight 700 daltons ($R = 7.7$ Å) and 25% in molecules of molecular weight 5000 daltons ($R = 15$ Å). Such relatively low molecular weight compounds may be termed "oligoelectrolytes" to distinguish them from true polyelectrolytes such as proteins. The types and concentrations of the ligands have been chosen to be consistent with available structural data and to approximate titrimetric data with H^+ and Cu^{2+}. The total concentration of carboxylic acid groups [$6 \, \text{meq} \, g^{-1} \simeq 12 \, \text{meq} \, (gC)^{-1}$] is in the middle range of reported values. As discussed above, the molecules are considered as impenetrable spheres with their ionizable functional groups producing a uniform charge distribution on the surface—the maximum charge is $4.3 \, \text{eq} \, \text{mol}^{-1}$ for the small size range and $32 \, \text{eq} \, \text{mol}^{-1}$ for the large size range.[79]

Acid–Base Properties of Humates. In Chapter 4 we showed a typical acid-base

TABLE 6.12 A Model Humate

Two size groups

$R = 7.7$ Å $(Z = 4.3$ eq mol^{-1}; mw $\simeq 700)$
75% of all functional groups

$R = 15$ Å $(Z = 32$ eq mol^{-1}; mw $\simeq 5000)$
25% of all functional groups

Three ligands

"Acetate"	4×10^{-3} mol g^{-1}	HL_0	pK_a	$= 3.8$
"Malonate"	10^{-3} mol g^{-1}	H_2L_1	pK_{a1}	$= 4.5$
		HL_1^-	pK_{a2}	$= 4.5$
		CuL_1	pK_{Cu}	$= 5.1$
"Catechol"	5×10^{-4} mol g^{-1}	H_2L_2	pK_{a1}	$= 9.4$
		HL_2^-	pK_{a2}	$= 12.6$
		CuL_2	pK_{Cu}	$= 14.7$

titration of a humic acid sample (Figure 4.13) and the influence of ionic strength. The oligoelectrolyte model described above accounts for the observed ionic strength effects (though it has not been designed to fit the data of Chapter 4 and may, in particular, lack sufficiently strong acid groups). Both the distribution of the pK_a's and the coulombic polyelectrolyte effect contribute to the smeared aspect of the titration curve in which no clear inflection point can be distinguished. The importance of the polyelectrolyte effect which is due to a gradual decrease in the coulombic attraction of protons can be quantified by expressing the mass law for the deprotonation of an acid site.

$$HL = H^+ + L^-$$

$$\frac{[H^+][L^-]}{[HL]} = K_a \cdot P^{-1} \tag{68}$$

where K_a is the intrinsic (chemical) acidity constant of the site and P^{-1} $(= e^{F\Psi_0/RT})$ is the coulombic term. The increase in the value of P with increasing pH—hence the increase in the apparent pK of the acid site: $pK_a^{app} = pK_a + \log P$—is given in Table 6.13 for each of the two sizes of the model humate molecules. Because the smaller humates account for 75% of the acid groups, they dominate the acid–base titrations. The magnitude of the ionic strength effects is thus chiefly a reflection of the dependence of P on I as given in Table 6.13.

Metal Binding by Humates. Much of the experimental data on metal binding by humates concerns copper. The effects of pH and ionic strength on copper

TABLE 6.13 Calculated Values of the Electrostatic Factor P as a Function of pH and Ionic Strength[a]

pH	$I = 0.01\ M$		$I = 0.1\ M$	
	$\log P_S$	$\log P_L$	$\log P_S$	$\log P_L$
2.7	0.061	0.158	0.045	0.105
2.8	0.074	0.185	0.054	0.124
2.9	0.089	0.215	0.065	0.147
3.0	0.105	0.248	0.078	0.172
3.1	0.124	0.284	0.093	0.199
3.2	0.146	0.323	0.110	0.230
3.3	0.169	0.364	0.128	0.262
3.4	0.195	0.408	0.149	0.297
3.5	0.223	0.454	0.171	0.335
3.6	0.252	0.502	0.195	0.374
3.7	0.284	0.551	0.221	0.415
3.8	0.317	0.603	0.248	0.458
3.9	0.352	0.656	0.276	0.502
4.0	0.387	0.710	0.305	0.547
4.1	0.424	0.766	0.335	0.594
4.2	0.462	0.823	0.365	0.641
4.3	0.501	0.880	0.397	0.689
4.4	0.540	0.939	0.428	0.737
4.5	0.580	0.998	0.460	0.785
4.6	0.620	1.057	0.492	0.834
4.7	0.661	1.117	0.523	0.883
4.8	0.701	1.178	0.555	0.931
4.9	0.742	1.238	0.587	0.980
5.0	0.782	1.298	0.617	1.028
5.1	0.823	1.359	0.648	1.075
5.2	0.863	1.419	0.677	1.122
5.3	0.902	1.479	0.705	1.168
5.4	0.941	1.539	0.731	1.214
5.5	0.979	1.598	0.755	1.259
5.6	1.016	1.656	0.778	1.302
5.7	1.051	1.714	0.798	1.345
5.8	1.084	1.772	0.815	1.386
5.9	1.115	1.828	0.831	1.425
6.0	1.144	1.884	0.844	1.463
6.1	1.171	1.939	0.854	1.499
6.2	1.194	1.993	0.863	1.533
6.3	1.215	2.045	0.871	1.564
6.4	1.233	2.097	0.877	1.593
6.5	1.248	2.147	0.881	1.618
6.6	1.261	2.196	0.885	1.641
6.7	1.271	2.243	0.888	1.661
6.8	1.280	2.288	0.891	1.678

(*continued*)

TABLE 6.13   (*Continued*)

pH	I = 0.01 M		I = 0.1 M	
	$\log P_S$	$\log P_L$	$\log P_S$	$\log P_L$
6.9	1.287	2.331	0.893	1.692
7.0	1.293	2.372	0.894	1.704
7.2	1.302	2.446	0.897	1.722
7.4	1.307	2.508	0.898	1.734
7.6	1.311	2.557	0.899	1.741
7.8	1.313	2.593	0.900	1.746
8.0	1.314	2.619	0.900	1.749
8.2	1.315	2.637	0.901	1.751
8.4	1.316	2.648	0.901	1.752
8.6	1.317	2.655	0.902	1.753
8.8	1.317	2.660	0.903	1.754
9.0	1.318	2.663	0.904	1.754

[a]The fulvic acid parameter of Bartschat et al. (1992)[79] were used to obtain values of P for the small (P_S) and large (P_L) size classes.

titration of humic acids are shown in Figure 6.18. The oligoelectrolyte model described in Table 6.12 accounts reasonably well for these effects, particularly at low total copper concentrations, in the range of Cu_T/L_T representative of some natural waters. As one expects, pH has a dramatic effect on the extent of binding as is the case for any weak acid ligand whose apparent affinity for a

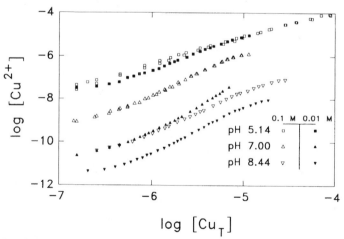

Figure 6.18   Copper-sensitive electrode titrations of humic acids at various pH's and ionic strengths. Note the characteristic shapes of such potentiometric titration curves (contrast these log–log graphs to those in Figure 6.9). Data from Cabaniss as reported in Bartschat et al., 1992.[79]

metal increases with pH. This effect is exacerbated by the increase in coulombic attraction of the metal ion to the binding sites as the humates become deprotonated as can be seen by writing the mass law equation for copper binding:

$$Cu^{2+} + L^{2-} = CuL$$

$$\frac{[CuL]}{[Cu^{2+}][L^{2-}]} = K_{Cu} \cdot P^2 \tag{69}$$

where K_{Cu} is the intrinsic (chemical) copper-binding constant of the site. As pH increases, the coulombic term P increases and so does the apparent binding constant.

Both the high apparent affinity of the humates for copper and its large (and pH-dependent) sensitivity to ionic strength are due to polyelectrolyte effects. At low Cu_T/L_T ratios, the larger humates, whose copper affinities are most enhanced by coulombic attraction, account for the bulk of copper binding. Because the coulombic term P for these molecules (at high pH) is inversely proportional to I (see Figure 6.17), the effective affinity for copper binding increases dramatically at low ionic strength. The actual dependency of the free cupric ion concentration, $[Cu^{2+}]$, on I (at a given pH, L_T, and Cu_T) depends on the stoichiometry of H^+ exchange at a specific binding site. If $H_\alpha L^{n-}$ is the dominant protonated form of the site, the reaction with copper is written:

$$Cu^{2+} + H_\alpha L^{n-} \rightleftharpoons CuL^{(n+\alpha-2)-} + \alpha H^+$$

$$K_{coul} = P^{2-\alpha} \propto I^{2-\alpha} \tag{70}$$

Thus the effect of I on K_{coul} may lead to a dependency of the apparent copper affinity on I that is anywhere between null ($\alpha = 2$) or up to a square function ($\alpha = 0$). Though it is complicated by the multiplicity of binding sites, Figure 6.18 illustrates that the maximum ionic effect ($\alpha = 0$) is observed at pH = 7.

Example 7. Cu Complexation by Humates in Freshwater To get a better quantitative understanding of these questions, let us reconsider the issue of copper speciation in a freshwater at pH = 8.1 that we now assume to contain $10 \, mg \, L^{-1}$ of humic acids. We have already calculated the ratio of free to inorganic copper in Example 3 and found $[Cu^{2+}]/[Cu'] = 10^{-1.5}$. Following the usual methodology, we can write the mole balance for copper as

$$TOTCu = [Cu'] + [CuL_{0S}] + [CuL_{0L}] + [CuL_{1S}] + [CuL_{1L}]$$
$$+ [CuL_{2S}] + [CuL_{2L}] = Cu_T \tag{71}$$

In our model humic acid, the acetate-type ligand (L_0) is much weaker than the other two, and despite its somewhat higher concentration it can have no sizable effect on copper speciation at the low metal to ligand ratio of this example. We thus neglect the species CuL_{0S} and CuL_{0L} and limit our calculation to the other two ligands, malonate (L_1) and catechol (L_2). As in all the previous examples, we first need to calculate the speciation of these ligands. Ignoring the possible complexation by Ca^{2+} and Mg^{2+} we can write the mole balance for the malonate-type ligands in the small humates, L_{1S}, as

$$TOTL_{1S} = [L_{1S}^{2-}] + [HL_{1S}^-] + [H_2L_{1S}] + [CuL_{1S}] = [L_{1S}]_T$$
$$= 0.75 \times 10^{-3}\,\text{mol g}^{-1} \times 10^{-2}\,\text{g L}^{-1} = 10^{-5.12}\,M \tag{72}$$

Because copper is assumed to be in trace quantities, it has a negligible effect on the ligands. We can thus ignore $[CuL_{1S}]$ in the mole balance. We can also calculate the coulombic factor P a priori for a given pH and ionic strength. According to Table 6.13, at a pH = 8.1 and $I = 0.01\,M$ the relevant value of P for small humates is

$$\log P_S = 1.32$$

The mass laws for the acid–base reactions of the malonate-type ligands in the small humates can then be written

$$\frac{[L_{1S}^{2-}]}{[HL_{1S}^-]} = \frac{K_{a2}}{[H^+] \cdot P} = \frac{10^{-4.5}}{10^{-8.1} \cdot 10^{+1.3}} \gg 1$$

$$\frac{[HL_{1S}^-]}{[H_2L_{1S}]} = \frac{K_{a1}}{[H^+] \cdot P} = \frac{10^{-4.5}}{10^{-8.1} \cdot 10^{+1.3}} \gg 1$$

Thus, the completely deprotonated ligand is the dominant species:

$$TOTL_{1S} \simeq [L_{1S}^{2-}] = [L_{1S}]_T = 10^{-5.12} \tag{73}$$

Similar calculations show that L_{1L}^{2-} is also the dominant species for malonate-type ligands in the larger humates [despite the larger shift in apparent pK_a ($= pK_a + \log P = pK_a + 2.63$) due to coulombic effects] and that H_2L_{2S} and H_2L_{2L} are the dominant species for the catechol-type ligands, in both size classes:

$$[L_{1L}^{2-}] = [L_{1L}]_T = 0.25 \times 10^{-5.0} = 10^{-5.60} \tag{74}$$

$$[H_2L_{2S}] = [H_2L_{2S}]_T = 0.75 \times 10^{-5.3} = 10^{-5.42} \tag{75}$$

$$[H_2L_{2L}] = [H_2L_{2L}]_T = 0.25 \times 10^{-5.3} = 10^{-5.90} \tag{76}$$

The mole balance equation for copper can then be written:

$$TOTCu = Cu_T = [Cu^{2+}]\left[10^{1.5} + 10^{5.1} \times P_S^2 \times 10^{-5.12} + 10^{5.1} \times P_L^2 \times 10^{-5.60} \right.$$

$$+ 10^{14.7}P_S^2 \times \frac{10^{-9.4}}{P_S[H^+]} \times \frac{10^{-12.6}}{P_S[H^+]} \times 10^{-5.42}$$

$$\left. + 10^{14.7} \times P_L^2 \times \frac{10^{-9.4}}{P_L[H^+]} \times \frac{10^{-12.6}}{P_L[H^+]} \times 10^{-5.90} \right] \qquad (77)$$

Introducing $P_S = 10^{1.32}$, $P_L = 10^{2.63}$ (Table 6.13), and $[H^+] = 10^{-8.1}$:

$$TOTCu = [Cu^{2+}][10^{1.5} + 10^{2.62} + 10^{4.76} + 10^{3.48} + 10^{3.00}] \qquad (78)$$

Thus the copper is 99.9% bound to humates, the free cupric ion is decreased more than three orders of magnitude compared to the purely inorganic case

$$\frac{[Cu^{2+}]}{Cu_T} = 10^{-4.79}$$

and the malonate-type ligand in the large humic compounds, L_{1L}, accounts for 93% of the copper in solution.

*   *   *

Example 7 provides an extreme case of metal complexation by humates since the humic acid content of the system is relatively large, the ionic strength is low, and the pH is high. The apparent complexation constants would decrease by more than one order of magnitude per pH unit at this ionic strength (see Figure 6.18). Nonetheless, it is clear that humic compounds may control the speciation of some trace metals in waters of high humic content. According to Example 7, some four orders of magnitude decrease in the product $K_{Cu}L_{1L}$ is necessary for the metal–humate complex to be unimportant. If the model ligands malonate and catechol are truly representative of the relative metal affinities of all complexing sites on humates, metals such as Cd^{2+}, Ni^{2+}, or Zn^{2+} should have humate affinities roughly 10 times less than that of copper. Depending on the pH and the humic content of the water, the complexation of such metals by humic material may thus vary from 0% (pH < 7, DOC < 1 mg L^{-1}) to 100% in situations such as that of Example 7.

An important difference between the oligoelectrolyte theory and the Debye–Hückel theory is the role that divalent background electrolyte ions play in determining the charge-potential relationship. In the linear approximation of the Debye–Hückel theory, the effects of all the ions in the background electrolyte are accounted for by one parameter, the ionic strength of the solution.

In the oligoelectrolyte theory, the nonlinearity of the equations results in a proportionally much greater effect of di- and trivalent ions on the surface potential. The relative role of each counterion in neutralizing the surface potential can be calculated by comparing the relative *surface concentrations* of metal ions of charge number z^+:

$$[M^{z+}]_{surface} = P^z \cdot [M^{z+}]_{bulk} \tag{79}$$

For example, in a medium of 0.1 M ionic strength, a concentration of 1 mM Ca^{2+} will dominate the coulombic interactions of moderate size polyacids.[79] Thus, the divalent cations Ca^{2+} and Mg^{2+}, which are abundant in natural waters, can diminish the tendency of humates to complex trace metals both by competing for specific functional groups and by decreasing (shielding) the coulombic attraction of neighboring ionized groups (the oligoelectrolyte effect).

4.4 Conditional Stability Constants

In earlier sections of this chapter, we have introduced various apparent or effective stability constants for complexation reactions. Often researchers who deal with metal complexation in natural waters find it convenient to describe the thermodynamics of the reactions with conditional stability constants that are applied to some analytical fraction of the reactants and products. Thus these constants are only valid under a specified set of conditions. To avoid confusion, let us briefly review the different types of conditional constants and provide a systematic terminology for them.

Consider the complexation reaction

$$M^{2+} + L^- = ML^+$$

The thermodynamic constant for this reaction is written

$$K = \frac{\{ML^+\}}{\{M^{2+}\}\{L^-\}} \qquad \text{(thermodynamic constant)} \tag{80}$$

In Chapter 2 we defined the corresponding concentration constant that we are using throughout the text:

$$^cK = \frac{[ML^+]}{[M^{2+}][L^-]} \qquad \text{(concentration constant)} \tag{81}$$

This is a conditional constant valid for a particular ionic strength, and it includes implicitly all unspecific long-range (mostly electrostatic) interactions among ions, namely, activity coefficients.

Going one step further, in Section 2 of this chapter we defined apparent constants for carbonate species in seawater, including in the constants all the specific and unspecific interactions among the major ions:

$$K^{app} = \frac{[\sum ML^+]}{[M^{2+}][\sum L^-]} \quad \text{(apparent constant)} \quad (82)$$

The symbol $\sum$ indicates that all complexes formed with the major metals, $Na^+, K^+, Ca^{2+}, Mg^{2+}$, are included in the formulation. The domain of validity of such apparent constants is of course restricted to solutions with similar major ion composition. For dilute solutions such as freshwaters, $^cK = K^{app}$.

To study organic complexation of metals it is convenient to introduce yet another type of conditional constant, an effective constant that includes the acid–base speciation of the ligand ($[\sum H_x L] = [L] + [HL] + [H_2 L] + [H_3 L] + \cdots$):

$$K^{eff} = \frac{[ML]}{[M^{2+}][\sum H_x L]} \quad \text{(effective constant)} \quad (83)$$

Such a constant is of course highly pH dependent and is obtained simply by introducing the ligand ionization fractions in the mass law expression. It can be generalized slightly by considering also all the various acid–base species of the complex (e.g., $[\sum ML] = [ML^+] + [MHL^{2+}] + [MOHL] + \cdots$):

$$K^{eff} = \frac{[\sum ML]}{[M^{2+}][\sum H_x L]} \quad (84)$$

Over a limited pH range the $[H^+]$ dependency of such a constant can often be made explicit:

$$K^{eff} = K_x^{eff}[H^+]^x \quad (85)$$

where K_x^{eff} is constant over some pH range; see Section 4.1. In our study of inorganic speciation, we have used effective constants as the product $\beta\alpha^m$ (see Section 3).

We can combine the notion of apparent constant with that of effective constant to obtain an overall conditional constant that includes the interactions of the ligand with all major cations, H^+, Ca^{2+}, Mg^{2+}:

$$K^{cond} = \frac{[\sum ML]}{[M^{2+}] \times L_T} \quad (86)$$

In cases where the calcium or magnesium complexes of the ligand are dominant, $K^{cond} = K^{app}$; in cases where the acid–base species dominate the ligand speciation, $K^{cond} = K^{eff}$.

In general, the inorganic metal speciation is *not* included in the expressions of conditional constants and the free metal ion concentration $[M^{2+}]$ appears in all the foregoing mass law expressions. In situations where the complex is a sizeable fraction of the ligand, only the unbound ligand should appear in the denominator of the mass law expression:

$$K^{cond} = \frac{[\sum ML]}{[M^{2+}][L_T - \sum ML]} \tag{87}$$

Conditional constants are the most directly attainable quantities in complexation experiments where the measured concentrations are simply the fractions of bound and free metal:

$$K^{cond} = \frac{[M \cdot bound]}{[M \cdot free][L_T - M \cdot bound]} \tag{88}$$

Although such constants may have little thermodynamic meaning and only a limited range of applicability, they are very convenient to use. In addition, for humates whose thermodynamic constants are difficult to define, they provide a direct link between experimental data and calculations, without an intervening and often misunderstood extrapolation procedure. When using conditional constants one should always remember the particular conditions of applicability and be wary that different authors may include implicitly different factors in their constants (e.g., acid–base chemistry or major ion complexation). As noted earlier, internal consistency in thermodynamic data is as important as the absolute values of the individual constants. Often, to increase the range of applicability of conditional constants, particularly for humates, a pH dependency is given explicitly:

$$K^{cond} = K_x^{cond}[H^+]^x = \frac{[H^+]^x \times [\sum ML]}{[M^{2+}][L_T - \sum ML]} \tag{89}$$

In this case, the exponent x reflects not only the acid–base chemistry of the ligand and the complex over a certain pH range, but also the pH dependency of the electrostatic effects on the metal affinity for the ligand; that is, the value of x can be interpreted as the average number of protons released per bound metal ion, or it can be a more generalized fitting parameter.

4.5 Organometallic Compounds

There is one class of coordination compounds that cannot be addressed satisfactorily within the framework of complexation equilibria. These are the organometallic compounds, in which the ligand bonds to the metal through carbon atoms (excluding the inorganic ligand CN^-). Although organometallic compounds may be formed with almost any metal, only a few metals, notably mercury and tin, form organometallic compounds that are stable in water. The metalloids arsenic and selenium also form compounds that are chemically similar to organometallic compounds and are stable in water. Other organometallic compounds are hydrolyzed by reactions such as

$$RM + H_2O \longrightarrow RH + M^+ + OH^-$$

The "ligand" in an organometallic compound is a carbanion R^-, that is, an organic anion with a carbon atom bearing the negative charge. These species are extremely reactive and organometallic compounds have a strong covalent character. Dissociation of an organometallic compound, as in the complexation equilibrium

$$RM = R^- + M^+$$

would result in immediate reaction of the carbanion with water (or protons).

Concern over environmental contamination by organomercury and organotin compounds arises because of their toxicity and because they tend to become concentrated in the biota. These compounds are directly introduced into the environment through their use as biocides.[80,81] Organometallic species are also produced by enzymatic methylation of cationic metals or by their reaction with biogenic methylating agents, such as methyl iodide.[82,83] Organometallic

TABLE 6.14   Occurrence of Organometallic Compounds in the Environment

Element	Compound	Concentration (M)	Sampling Site	Ref.
Hg	Monomethylmercury	$< 0.5–2.8 \times 10^{-13}$	Equatorial Pacific (subthermocline)	84
		$< 0.02–3.2 \times 10^{-12}$	Freshwaters	85
		8.0×10^{-14}	Coastal seawater	85
	Dimethylmercury	$0.3–6.7 \times 10^{-13}$	Equatorial Pacific	84
Sn	Monobutyltin	up to 1.4×10^{-8}	Freshwaters	86
	Dibutyltin	up to 7.8×10^{-9}		86
	Tributyltin	up to 1.4×10^{-8}		86
	"organotin"[a]	$(2.8–6.7) \times 10^{-10}$	Rivers	87
		$(0.25–4.7) \times 10^{-9}$	Harbors	87
		$(1.3–7.6) \times 10^{-9}$		88
As	Methylarsonate	4×10^{-10}	Baltic Sea	89
		$(< 0.1–8.6) \times 10^{-9}$	Lakes	90
	Dimethylarsenate	$(0.77–6.15) \times 10^{-9}$	Baltic Sea	89
		$(0.4–33) \times 10^{-9}$	Lakes	90

[a]Includes tri- and dibutylated species.

compounds are widely distributed in the aquatic environment, both freshwater and marine (see Table 6.14).

The occurrence and reactions of organometallic compounds can be quite significant in the biogeochemical cycles and mass balances of some metals in natural waters. An important feature of many organometallic compounds in this regard is their volatility. Aquatic systems may be a source of Hg, Sn, As, and Se to the atmosphere through the volatilization of methylated species.[83,91,92] Reactions involving organometallic compounds, however, may be better addressed in the context of reactions of organic rather than inorganic species.[93]

5. COMPLEXATION KINETICS

The preceding discussion of complexation reactions is based on equilibrium arguments. All reactions that are considered to take place in the system are also considered to have reached equilibrium. In many cases, complexation of metals by ligands is indeed rapid relative to other processes of interest, and equilibrium models then provide a good representation of complexation phenomena in natural waters. Here, however, we wish to consider the general question of whether complexation kinetics, that is, the rates of formation or dissociation of metal–organic complexes, may affect the overall rates of biogeochemical processes or the analytical determination of metal speciation in natural waters.[94]

The ambient trace metal speciation in a natural water is, as we have seen, determined by the mixture of metals and complexing agents present and also by the pH, ionic strength, and major ion composition. Changes in one or more of these parameters may result from the physical mixing that occurs at stream confluences, upwelling of deep ocean water, or input of sewage effluents to surface waters. Such changes may also be effected by biogeochemical processes, such as uptake of metals or release of complexing agents by microorganisms, changes in pH of surface waters due to photosynthesis, or photochemical reactions. Clearly one question concerning complexation kinetics is how quickly the metal and ligand species distributions change in response to such perturbations.

This question may be particularly important in analytical determinations of trace metal speciation where the initial composition of the water sample is often significantly perturbed by the additions of either metals or ligands or both (see Table 6.10 for references) and the equilibration time between the perturbation and the analytical measurement tends to be short. In metal titrations, the equilibration time between metal additions and the analytical measurement can affect the measured concentration of inorganic (labile) metal[44,95] (see Figure 6.19). As discussed below, the stronger ligands may, in some cases, react more slowly with added metals than weaker ligands; this effect depends on the initial ligand speciation. When the sample is titrated with a metal, the final

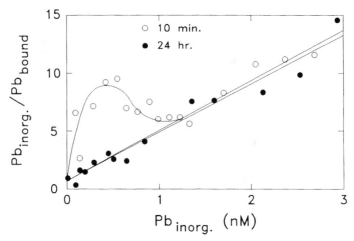

Figure 6.19 Ratios of inorganic-to-bound lead as a function of inorganic lead observed in a titration experiment after 10 min (open circles) or 24 h (closed circles) equilibration time. Because equilibration of added lead is incomplete after 10 min, increasing lead complexation (decreasing ratio of inorganic-to-bound lead) is observed over time. From Capodaglio et al. (1990).[44]

composition of the sample (e.g., the total metal concentration at each step of the titration) uniquely determines the equilibrium metal and ligand speciation but *not* the path to equilibrium. Since the fastest reactions do not necessarily result in formation of the thermodynamically favored complexes, attainment of equilibrium speciation may be slow on an analytical time scale.

When perturbation of metal and ligand speciation occurs as a result of biogeochemical processes, such as biological metal uptake, sorption, or chemical or photochemical transformation, the rates of these processes may be linked to the rates of complexation reactions. For example, as we shall see in the next section, if a metal occurs predominantly as an organic complex and the intact complex is not transported into the cells, then the kinetics of complex dissociation may, under some conditions, limit the rate at which the metal can be accumulated by microorganisms.

To evaluate the importance of complexation kinetics in natural or laboratory systems, it is necessary to consider the pathways both for the complexation reactions and for any co-occurring biogeochemical processes. From the reaction pathways, the intermediate species whose low or transient concentrations determine the rates of reactions may be identified. In the following discussion, we examine the effects of initial speciation and reaction pathways on the kinetics of copper complexation by humic acid and EDTA.

Humic acids exhibit a natural fluorescence which is decreased (quenched) when the humic acid is associated with Cu(II) ions.[96] As shown in Figure 6.20,

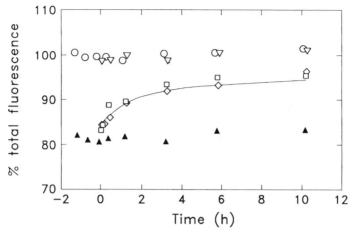

Figure 6.20   Reaction of copper with a mixture of EDTA and humic acid. Percent fluorescence of the humic acid is shown over time after addition of copper at $t = 0$ to the ligand mixture in the absence of calcium (∇) or with 0.01 M calcium ($\square$, $\diamond$). For reference, the percent fluorescence over time of the ligand mixture in the absence of copper ($\bigcirc$) and of humic acid and copper in the absence of EDTA ($\blacktriangle$) with 0.01 M Ca are shown. (——) calculated values obtained by assuming pseudoequilibrium between inorganic copper and humate binding sites and reaction of inorganic copper with CaEDTA as described in Section 5.3. From Hering and Morel (1989).[97]

the quenching of humate fluorescence (i.e., formation of humate–Cu complexes) when Cu(II) is added to a mixture of humic acid and EDTA is strongly dependent on the calcium concentration.[79] In the absence of calcium, no decrease in fluorescence is observed, indicating rapid formation of the thermodynamically favored CuEDTA species. With $10^{-2}\,M$ calcium, however, the initial decrease in fluorescence indicates formation of Cu–humate complexes; the slow recovery of the fluorescence over time corresponds to exchange of copper between the humic acid and EDTA.

We have already seen in Example 5 that calcium can have little effect on the equilibrium copper speciation in these systems. Why then does it have such a dramatic effect on the rate at which the equilibrium speciation is attained?

The presence or absence of calcium determines the initial EDTA speciation. Without calcium, *complex formation* between the protonated EDTA species and the added copper is rapid. At high calcium concentrations, however, reaction of the added copper with the dominant CaEDTA complex involves a slower *metal-exchange* reaction and formation of Cu–humate complexes is kinetically favored. Subsequent reequilibration requires a *double-exchange* reaction, that is, both a *ligand exchange* in which EDTA replaces the humate in the copper complex and a *metal exchange* in which copper replaces calcium in the EDTA complex. Then to understand these experimental observations, we must understand the mechanisms and rates of the reactions, complex formation and dissociation, as well as metal, ligand, and double exchange.

5.1 Kinetics of Complex Formation and Dissociation

Observed rates of complex formation are generally consistent with a mechanism in which formation of an outer-sphere complex between the metal and ligand, with the stability constant K_{os}, is followed by a rate-limiting loss of water from the inner coordination sphere of the metal, with the water loss rate constant k_{-w}:

$$M(H_2O)_6^{2+} + L^{n-} \xrightleftharpoons{K_{os}} M(H_2O)_6,L^{(2-n)+}$$

$$M(H_2O)_6,L^{(2-n)+} \xrightarrow{k_{-w}} M(H_2O)_5 L^{(2-n)+} + H_2O$$

Then the rate constant for complex formation, k_f, can be written (omitting coordinated waters) as:

$$\frac{d[ML^{(2-n)+}]}{dt} = k_f[M^{2+}][L^{n-}] \tag{90}$$

where

$$k_f = K_{os}k_{-w} \tag{91}$$

The stability constant for the outer-sphere complex, K_{os}, is primarily dependent on the charges of the reacting species and the ionic strength of the medium. This term can be calculated as shown in Sidebar 6.3. Experimentally determined values of K_{os} (which may be obtained for relatively slow-reacting species) are in agreement with calculated values. For $3+$ ions, observed values of K_{os} range from approximately $10 M^{-1}$ for complexes with $1-$ ions to approximately 3×10^3 for complexes with $3-$ ions.[98] The rate constant for water-loss, k_{-w}, is characteristic of the reacting metal (Table 6.15). For many metals, the magnitude of k_{-w} can be explained by the electrostatic interaction between the metal cation and the coordinated water molecules, which is related to the charge and, inversely, to the size of the metal cation as shown in Figure 6.21. The particular inertness of some metals, such as Cr^{3+}, Rh^{3+}, Co^{3+}, and Ni^{2+}, relative to other metals has been attributed to unfavorable changes in ligand field stabilization energy (LFSE) during complex formation (c.f. Section 1.2). A ligand field activation energy (LFAE) can be derived by comparing the LFSE of the initial 6-coordinate octahedral complex with the LFSE of the 5-coordinate square pyramidal complex that results from dissociation of a coordinated water molecule.[4] This change in LFSE is most unfavorable for $d^3(Cr^{3+})$, $d^8(Ni^{2+})$, and (strong field) $d^6(Rh^{3+})$ configurations.

The rate of water loss can be significantly accelerated by metal hydrolysis; this effect is especially important for slow-reacting metals[101] such as Cr(III),

SIDEBAR 6.3

Stability Constants for Outer-Sphere Complexes (K_{os})

A theoretical expression for K_{os}, derived from statistical considerations and including both the electrostatic interaction of the ions and the effect of ionic strength on those interactions,[98,99] is given by

$$K_{os} = \frac{4000\pi \mathcal{N} a^3}{3} \exp\left[\frac{-z_M z_L e^2}{4\pi\varepsilon_0\varepsilon \mathbf{k}\, Ta}\right]\exp\left[\frac{z_M z_L e^2\kappa}{4\pi_0\varepsilon \mathbf{k} T(1+\kappa a)}\right] \qquad (i)$$

where κ is the Debye–Hückel ion atmosphere parameter[100]

$$\kappa^2 = \frac{2000 e^2 \mathcal{N} I}{\varepsilon_0\varepsilon \mathbf{k} T} \qquad (ii)$$

These equations are written for SI units with the constants e (elementary charge) in C, $\mathbf{k}$ (Boltzmann constant) in $J \cdot K^{-1}$, $\mathcal{N}$ (Avogadro constant) in mol^{-1}, ε_0 (vacuum permittivity) $8.854 \times 10^{-12}\, J^{-1} C^2 m^{-1}$, and T (absolute temperature) in K. The relative permittivity of the medium, ε, is 78.54 for water at 25°C. The parameter a is the distance of closest approach of the ions (usually 5×10^{-10} m). In the table below, values of K_{os} are calculated for different values of the product of the charges on the ions ($z_M z_L$) for ionic strengths (I) of both 0 and 0.1.

$z_M z_L$	$\log K_{os}$		$z_M z_L$	$\log K_{os}$	
	$I=0$	$I=0.1$		$I=0$	$I=0.1$
0	−0.50	−0.50			
−1	0.12	−0.93	+1	−1.12	−0.91
−2	0.74	0.32	+2	−1.74	−1.32
−3	1.36	0.72	+3	−2.36	−1.73
−4	1.98	1.13	+4	−2.98	−2.13
−6	3.22	1.95	+6	−4.22	−2.95
−8	4.46	2.76	+8	−5.46	−3.76
−9	5.08	3.17	+9	−6.08	−4.17
−12	6.94	4.39	+12	−7.94	−5.40

TABLE 6.15   Rate Constants for Water Exchange[a]

Metal Ion	k_{-w} (s^{-1})	Range of Reported Values	Ref.
Pb^{2+}	7×10^9		
Hg^{2+}	2×10^9		
Cu^{2+}	1×10^9	$(0.2-5) \times 10^9$	
Ca^{2+}	6×10^8	$(>0.5-9) \times 10^8$	
Cd^{2+}	3×10^8	$(>1-5) \times 10^8$	
La^{3+}	1×10^8	$(0.8-2) \times 10^8$	
Zn^{2+}	7×10^7	$(3-50) \times 10^7$	
Mn^{2+}	3×10^7	$(0.4-5) \times 10^7$	
Fe^{2+}	4×10^6	$(1-4) \times 10^6$	101
Co^{2+}	2×10^6	$(0.2-2) \times 10^6$	
Mg^{2+}	3×10^5	$(1-5) \times 10^5$	
Ni^{2+}	3×10^4	$(1-4) \times 10^4$	
Fe^{3+}	2×10^2	$(2-200) \times 10^2$	
Ga^{3+}	8×10^2	$(8-20) \times 10^2$	101
Al^{3+}	1	$0.2-16$	
Cr^{3+}	5×10^{-7}	$(4-6) \times 10^{-7}$	
Rh^{3+}	3×10^{-8}		
Fe^{3+}	2×10^2		101
$FeOH^{2+}$	1×10^5		101
$Fe(OH)_2^+$	10^7 (est)		102
$Fe(OH)_4^-$	10^9 (est)		102

[a] Unless otherwise noted, reference is to Margerum et al. (1978).[98] Values for k_{-w} assigned based on most current reference, most reliable method, or as average for narrow ranges of reported values.

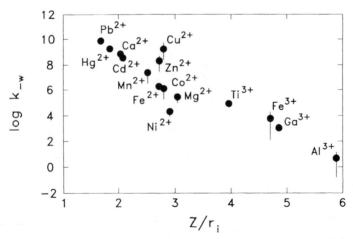

Figure 6.21   Water-loss rate constants for metal cations as a function of the ratio of the charge to the ionic radius of the metal ion. Rate constants from Margerum et al. (1978).[98] Adapted from Crumbliss and Garrison (1988).[101]

Al(III), and Fe(III) (see Table 6.15). Complexation by other inorganic ligands (e.g, Cl^-, NH_3, CO_3^{2-}) does not generally influence the rate of water loss from the metal as strongly as complexation by hydroxide. In practice, it may often be convenient to define an apparent rate constant for complex formation (k'_f) that includes the contribution of various inorganic metal species, thus

$$k'_f[M'] = k_{f0}[M^{2+}] + k_{f1}[MOH^+] + k_{f2}[MCl^+] + k_{f3}[MCO_3^0] + \cdots$$
(92)

where $[M']$ is the concentration of all inorganic metal species. This apparent rate constant must, of course, vary with pH and ionic strength.

The above mechanism for complex formation also holds for the reaction of metals with multidentate ligands if the rate-limiting step for complexation is the formation of the first metal–ligand bond. The effects of ligand structure, other than charge, are unfortunately difficult to generalize. Ligand protonation decreases the rate of complex formation often more than can be accounted for based on the change in ligand charge (see Margerum et al.[98] for further discussion).

Rate constants for complex dissociation can be related to formation rate constants through detailed balancing:

$$M + L \underset{k_b}{\overset{k_f}{\rightleftharpoons}} ML$$

$$k_b = k_f/K \tag{93}$$

If the formation rate constants for a reaction of a single metal with a series of ligands show little variation (being, as discussed above, mostly dependent on k_{-w}), then the dissociation rate constants should show an inverse correlation with the stability constant K. We have already seen an example of this correlation in Chapter 3 (Figure 3.16).

5.2 Kinetics of Metal- and Ligand-Exchange Reactions

Both metal- and ligand-exchange reactions involve three reacting species. For reaction of an initial complex ML with an incoming metal or ligand X, the rate of the overall reaction

$$ML + X \longrightarrow \begin{cases} XL + M \ (\text{metal exchange}) \\ \text{or} \\ MX + L \ (\text{ligand exchange}) \end{cases}$$

can be described by the rate expression

$$\frac{-d[ML]}{dt} = k_{ex}[ML][X] \tag{94}$$

This overall exchange reaction can proceed through two general pathways: a *disjunctive pathway*, involving complete dissociation of the initial complex, and an *adjunctive* pathway, involving direct attack of the incoming species on the initial complex.[94,103] For the metal-exchange reaction $ML + M^* \rightarrow M^*L + M$, these may be written:

Disjunctive mechanism	Adjunctive mechanism
$ML \rightleftharpoons M + L$	$M^* + ML \rightleftharpoons M^*LM$
$M^* + L \longrightarrow M^*L$	$M^*LM \longrightarrow M^*L + M$

where charges and protonated ligand species are omitted for simplicity. Analogous pathways can be written for ligand-exchange reactions. The rate law for the overall reaction then reflects the contributions of both of these pathways. When pseudoequilibrium can be assumed for the first step in the disjunctive pathway (which will usually be valid when the outgoing metal, M is in large excess), the overall rate of metal exchange is

$$\frac{-d[ML]}{dt} = \frac{-d[M^*]}{dt} = \frac{d[M^*L]}{dt} = k_{dis}\frac{[ML][M^*]}{[M]} + k_{adj}[ML][M^*]$$

$$= k_{ex}[ML][M^*] \qquad (95)$$

where k_{dis} is the rate constant for the disjunctive pathway and k_{adj} is the rate constant for the adjunctive pathway. These *stoichiometric* mechanisms can then

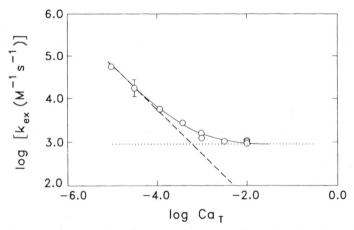

Figure 6.22   Rate constants for the metal-exchange reaction of Cu(II) with CaEDTA at pH = 8.2 as a function of (excess) calcium for rate = $-d[Cu]/dt = k_{ex} [Cu] [CaEDTA]$, where $k_{ex} = (k_{dis}/[Ca] + k_{adj})$. Data ($\bigcirc$) are shown with empirically derived contributions of the adjunctive pathway ($\cdots$) and disjunctive pathway ($---$) to the observed rate constant (———). Adapted from Hering and Morel (1988).[103]

be distinguished by the dependence of the observed rate constant k_{ex} on the concentration of the outgoing metal M:

$$k_{ex} = \frac{k_{dis}}{[M]} + k_{adj} \qquad (96)$$

The variation in k_{ex} as a function of [M] is shown in Figure 6.22 for the reaction of Cu(II) with CaEDTA. The kinetic inhibition of formation of the CuEDTA complex at high calcium concentration is part of the explanation for the data shown in Figure 6.20.

The rate constants k_{dis} and k_{adj} are dependent on the ligand structure and, for k_{dis}, on pH. The disjunctive rate constant is determined by the rate constant for reaction of the incoming metal with the steady-state concentration of the free or protonated ligand intermediate. It is thus inversely proportional to the conditional equilibrium constant for the initial metal complex. The adjunctive rate constant is more dependent on ligand structure since either formation or dissociation of the intermediate dinuclear complex M*LM can be rate limiting. When the incoming metal M* is a transition metal and the outgoing metal M an alkaline earth metal, the formation of the dinuclear complex is more likely to be limiting. The adjunctive rate constant can be rationalized on the basis of the rate stability of the (outgoing) metal complex with a ligand fragment (corresponding to a partially dissociated complex), and the rate of reaction of

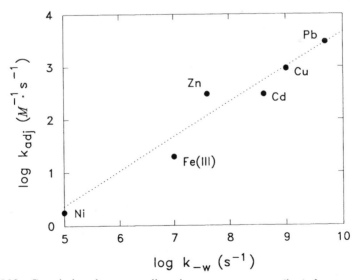

Figure 6.23 Correlation between adjunctive rate constants (k_{adj}) for reactions of CaEDTA with various metals in seawater and the water-loss rate constants of the metals (k_{-w}). From Hudson (1989).[104]

the incoming metal with this partially dissociated complex.[98,103] On this basis a correlation between the adjunctive rate constant for reactions of a specific complex with various metals and their water loss rate constants is expected.[104] Such a correlation for the reaction of CaEDTA is shown in Figure 6.23.

5.3 Kinetics of Double-Exchange Reactions

The overall double-exchange reaction

$$ML + M^*L^* \longrightarrow ML^* + M^*L$$

involves four reacting species in simultaneous metal- and ligand-exchange reactions. The overall reaction may be either "ligand-initiated"

$$ML \rightleftharpoons M + L$$
$$L + M^*L^* \longrightarrow M^*L + L^*$$

or "metal-initiated"

$$ML \rightleftharpoons M + L$$
$$M + M^*L^* \longrightarrow ML^* + M^*$$

Both of these pathways are disjunctive with respect to the complex ML and may be either adjunctive or disjunctive with respect to the reaction of the intermediate free M or L with the complex M*L*. The complete dissociation of both initial complexes is an unlikely pathway unless both are weak, as may be the case for alkaline earth metal complexes. Double-exchange reactions of transition metals often involve coordination chain mechanisms (e.g., further reaction of L*, produced by a "ligand-initiated" pathway, with initial complex ML) and the overall rates may be strongly dependent on trace concentrations of reactants that promote or terminate coordination chain mechanisms.[98]

* * *

Now we can understand why calcium has such a dramatic effect on the overall rate of formation of CuEDTA species when a competing ligand (i.e., humic acid) is present as shown in Figure 6.20. At high calcium concentrations, reaction of CaEDTA with added copper by a disjunctive pathway is not kinetically favored and reaction by an adjunctive pathway is slow compared to the formation of Cu–humate complexes. Thus because EDTA speciation in seawater is dominated by Ca, EDTA is effectively a much slower-reacting ligand than, in this case, humic acid. Once the Cu–humate complexes are formed, equilibrium can only be attained through slow double-exchange reactions. Note that this corresponds to the conditions chosen arbitrarily in the illustration of pseudoequilibrium in Chapter 3 where the stronger ligand was assigned a slower rate constant for complex formation than the weaker ligand. The experimental data in Figure 6.20 can be modeled by assuming a "metal-initiated" double-exchange in which inorganic copper (in pseudoequilibrium with humate binding sites) reacts with CaEDTA. In this model, the formation of CuEDTA is taken to proceed via the same pathways, in which CaEDTA reacts with inorganic copper, whether or not humates and Cu–humate complexes are present. Then, since copper complexation by humates markedly decreases the inorganic copper concentration $[Cu']$, the rate of formation of CuEDTA

$$\frac{-d[\text{CuEDTA}]}{dt} = \left[\frac{k_{\text{diss}}}{[\text{Ca}^{2+}]} + k_{\text{adj}} \right] [\text{Cu}'][\text{CaEDTA}] \tag{97}$$

is slow in the presence of humates.

This example illustrates the somewhat counterintuitive point that the availability of multiple reaction pathways can, in some instances, decreases the rate at which the equilibrium distribution of ligand and metal species is attained. The most facile complexation reactions may not be the ones that lead most directly to the ultimate equilibrium distribution. Although the equilibrium speciation is determined by the lowest free energy of the system, the predominant path to this lowest energy state may not be the most direct one. As we have seen, slow complexation kinetics are most likely to be observed in systems containing mixtures of strong and weak ligands and high concentrations of alkaline earth metals or mixtures of competing transition metals.

The extrapolation of rates and mechanisms of complexation reactions to natural waters is hampered because of limited knowledge of the structure and reactivity of naturally occurring ligands. Certain principles, however, should still hold. The initial speciation of both metals and ligands in a system subject to some perturbation can profoundly influence the reaction pathways through which equilibrium is reestablished. In this way, species that have only little effect on the final equilibrium speciation can have a dramatic effect on the rate at which that equilibrium is attained. In estimating the rates of complexation reactions in natural waters and their effects on the overall rates of biogeochemical processes, the relative and absolute concentrations of all reacting species and the most probable reaction pathways under the natural conditions of interest must be considered.

6. TRACE METALS AND MICROORGANISMS

As mentioned early in this chapter, the complexation of trace metals in natural waters modulates their effects on the aquatic biota. The growth of all organisms is dependent on the acquisition of the proper quantities of many trace elements. Some trace metals such as iron, zinc, manganese, and copper are required for growth, but too high a concentration of these metals or others produce toxic

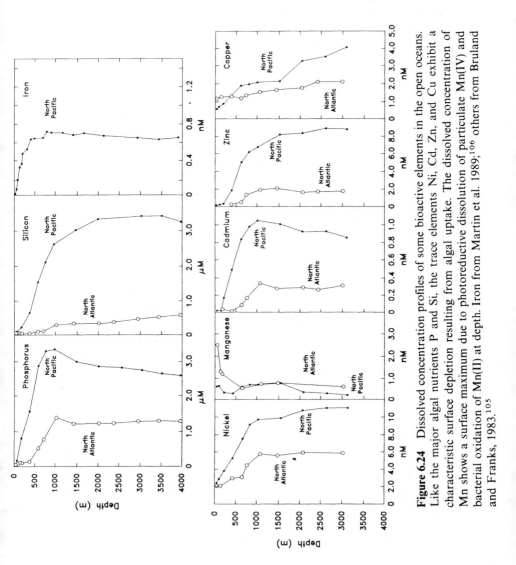

Figure 6.24 Dissolved concentration profiles of some bioactive elements in the open oceans. Like the major algal nutrients P and Si, the trace elements Ni, Cd, Zn, and Cu exhibit a characteristic surface depletion resulting from algal uptake. The dissolved concentration of Mn shows a surface maximum due to photoreductive dissolution of particulate Mn(IV) and bacterial oxidation of Mn(II) at depth. Iron from Martin et al. 1989;[106] others from Bruland and Franks, 1983.[105]

effects. Whereas higher organisms maintain a finely poised *internal milieu* for the proper operation of the various physiological tasks of the individual cells, aquatic microorganisms, particularly unicellular organisms such as phytoplankton and bacteria, must function in an *external milieu* whose chemistry is governed both by abiotic geochemical processes and by the action of the organisms themselves. Biological uptake of essential trace elements from the photic zone of lakes and oceans can be clearly seen in characteristic surface depletion.[105,106] As seen in Figure 6.24 both major nutrients N, P, and Si, and the trace nutrients Fe, Zn, Ni, Cu, and Cd are scavenged from oceanic surface waters. In addition to depleting surface concentrations through uptake, microorganisms can also influence trace metal chemistry by releasing complexing agents (see Section 4.2) and by inducing redox transformation (see Chapter 7). In this section we examine the kinetics of trace metal uptake by microorganisms, a process that is chiefly one of coordination chemistry and which controls both essential trace element availability and trace element toxicity.

Trace Element Uptake. Some compounds, including some trace metal complexes, are sufficiently lipid soluble that they can diffuse at significant rates through the lipid-bilayer membrane that is the major barrier between the inside of a unicellular organism and its external milieu. Passive diffusion governs the uptake of the dichloro-mercuric complex, $HgCl_2$, which is very toxic to aquatic microorganisms. In most cases, however, trace metal ions and their complexes have very low solubility in lipids and consequently very slow diffusion rates through cell membrances. Uptake of such trace metals is then catalyzed by a two-step process consisting first of chemical binding to a membrane uptake site, followed by transfer through the cell membrane and delivery to the inside of the cell. The steady-state kinetic description of such an uptake process is known as the Michaelis–Menten equation, as discussed in Chapter 3:

$$\rho = \frac{[M]}{K_M + [M]} \rho_{max} \tag{98}$$

TABLE 6.16   Michaelis–Menten Uptake Kinetics

Reaction	$M + L \underset{k_{-L}}{\overset{k_L}{\rightleftharpoons}} ML \xrightarrow{k_{in}} M_{cell}$
Uptake rate	$\rho = \dfrac{\rho_{max}[M]}{K_M + [M]}$
Maximum uptake rate	$\rho_{max} = L_T k_{in}$
Half-saturation constant	$K_M = \dfrac{k_{-L} + k_{in}}{k_L}$

The maximum uptake rate ρ_{max} (mol L^{-1} cell^{-1} s^{-1}) and the half saturation constant K_M(mol L^{-1}) are related to the concentration of uptake sites and the reaction rate constants as shown in Table 6.16. At low concentrations of the metal, [M] < K_M, the uptake rate is linearly related to [M]. At high concentrations, [M] > K_M, when most of the uptake ligands are bound to M, the uptake rate becomes saturated ($\rho = \rho_{max}$) and independent of [M]. Under normal circumstances, trace metal uptake ligands are far from saturated in natural waters: most of the ligands are free and available to react with the rare metals in the water. This is the case we shall focus on in the following discussion. The uptake rate is then a linear function of the metal concentration:

$$\rho = \frac{[M]}{K_M} \rho_{max} \tag{99}$$

Importance of Trace Metal Speciation. Coordination with transport sites on the surface of microorganisms is only one of the possible reactions for a trace metal in a natural water. The diagram of Figure 6.25 exemplifies a situation in which a coordination reaction with an organic chelating agent Y in excess dominates the speciation of a trace metal M which is taken up by cells via a cellular ligand L.[107] Although we shall consider only two dissolved species, M and MY, the following discussion applies to any set of species, dissolved or precipitated. In particular, M may represent the sum of all inorganic dissolved species of a given metal (denoted M') and MY may be a solid phase, such as an oxide or a hydroxide rather than an organic chelate.

The concentration of M, which depends on the relative rates of formation and dissociation of the competing complexes MY in solution and ML on the cells' surfaces, will reach a steady state after some characteristic time as discussed in Chapter 3 (see Figure 3.8).

Three possible limiting cases are shown on Figure 6.25: (A) The reactions with the ligands L and Y may be fast enough that the trace metal is in a condition of pseudoequilibrium in the medium; (B) the rate of binding to the surface ligand L may limit the metal uptake rate; (C) the rate of dissociation of the complex MY may limit the metal uptake rate.

A. Consider first a surface binding reaction that is reversible and reaches a pseudoequilibrium with M ($k_{in} \ll k_{-L}$, thus $k_L/k_{-L} = K_M$, the half saturation constant). This may be the case for a necessary metal that is relatively abundant or for a toxic metal transported adventitiously by the cells. At low cell concentrations the concentration of cellular ligands L is also low and the total concentration of M is determined by its equilibrium with Y. As depicted in Figure 6.25, both coordination reactions are then at pseudoequilibrium and the uptake rate, which is determined by the free metal concentration [M], is slow compared to all complexation and dissociation reactions.

B. The second limiting case is obtained for metal uptake systems that are largely irreversible ($k_{in} \gg k_{-L}$) as is observed for the transport of essential

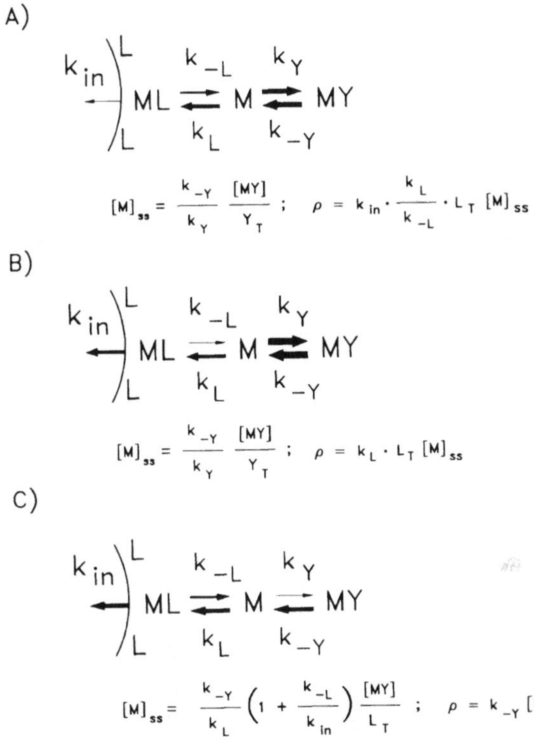

A)

$$[M]_{ss} = \frac{k_{-Y}}{k_Y} \frac{[MY]}{Y_T} \quad ; \quad \rho = k_{in} \cdot \frac{k_L}{k_{-L}} \cdot L_T [M]_{ss}$$

B)

$$[M]_{ss} = \frac{k_{-Y}}{k_Y} \frac{[MY]}{Y_T} \quad ; \quad \rho = k_L \cdot L_T [M]_{ss}$$

C)

$$[M]_{ss} = \frac{k_{-Y}}{k_L}\left(1 + \frac{k_{-L}}{k_{in}}\right)\frac{[MY]}{L_T} \quad ; \quad \rho = k_{-Y}[MY]$$

Figure 6.25 Limiting cases for the kinetics of biological uptake of a trace metal M in the presence of a chelator Y. (*a*) Pseudo-equilibrium between M and Y and between M and the surface uptake ligand L. (*b*) Pseudo-equilibrium between M and Y (fast kinetics) and quasi-irreversible uptake of M (slow). (*c*) Steady state between dissociation of the complex MY and quasi-irreversible uptake of M. The magnitudes of the fluxes are indicated by the widths of the arrows.

elements present as low concentrations in the medium. At low cell concentrations, the free metal concentration is again determined by pseudoequilibrium with the dominant ligand Y (compare A and B). However, the cell surface complex ML is then controlled by a steady state between binding and internalization, not pseudoequilibrium, and the uptake rate is simply the rate of formation of the surface complex. This case is characteristic of limiting trace elements in aquatic systems.

C. The third limiting case is characteristic of systems with high cell (and total cellular ligands) concentrations or with slow-dissociating ligands, Y, in the medium. The rate of uptake then becomes limited by the rate of dissociation of the complex MY. The free metal concentration decreases along with the uptake rate of individual cells as the concentrations of cells and total cellular ligands increase (the concentration of L per cell being constant). If the rate of dissociation from the surface ligand is much faster than the rate of internalization, the cellular ligand and the metal remain at pseudoequilibrium as in the first case discussed above.

The Importance of Free Metal Concentrations. The conditions illustrated by cases A and B are those most relevant to natural waters where the ambient cell concentrations are usually modest and to laboratory cultures where the trace metals are extensively buffered by chelation. In both of those cases, the free concentration of the metal [M] determines its rate of uptake.

In case A this is true regardless of reaction mechanisms since pseudoequilibrium is achieved among all parties. The uptake rate is proportional to the concentration of the surface complex [ML], which depends on the free metal ion concentration [M]; [M] itself can be calculated from its equilibrium with the ligand Y ("thermodynamic control"). Thus the dependence of the biological response—the uptake rate or, as a result of limitation or toxicity, the growth rate—on the free metal ion concentration [M] does not signify that the free metal ion is the only species available (or toxic) to the organism. In an equilibrium system, the extent of all complexation reactions, including those that control the uptake rate, are determined by the free metal ion concentration regardless of reaction mechanisms. The free metal ion concentration merely provides a thermodynamic measure of the reactivity of the metal. The same result would obtain even if the chelated metal MY were "available"; that is, if a fast (adjunctive) metal exchange reaction occurred between MY and L.[108]

In contrast, in case B the importance of the free metal concentration results from the reasonable assumption that the exchange of the metal between the chelator Y and the uptake ligand L proceeds via dissociation of the complex MY (disjunctive process). Then it is the steady state concentration of the fast reacting metal species that determines the reaction rate with the transport ligands and, hence the uptake rate ("kinetic control"). Typically all the inorganic metal species may be considered to react rapidly and the relevant concentration is then their sum, denoted [M']. As seen in Section 5.8, an effective reaction rate constant, k'_L, can be determined for M' at given pH and major ion concentrations.

Because in most experiments the total metal and the total ligand concentrations are varied but the pH and major ion concentrations are kept constant, [M'] is proportional to [M]. Thus one cannot easily distinguish between "thermodynamic control" and "kinetic control" of the uptake. Figure 6.26a illustrates how in the presence of various complexing agents, the uptake of iron by microscopic algae is indeed dependent on the free iron concentration. In this case, we know from short-term studies of the reaction kinetics that the uptake is kinetically controlled as shown in Figure 6.26b, which depicts the individual reaction rates of iron in a typical experiment.

Metal Competition and Toxicity. No ligand, not even an uptake ligand, can be totally specific for a trace metal. Thus one expects that the binding of another metal to the uptake ligand for an essential metal may competitively inhibit its

a)

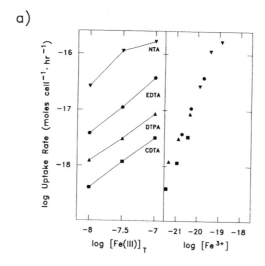

b)

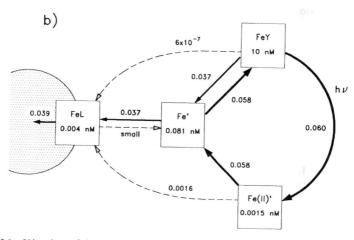

Figure 6.26 Kinetics of iron uptake in laboratory cultures of the marine diatom *Thalassosira weissflogii*. (*a*) Uptake rates at varying iron concentrations in the presence of 10^{-5} M of various chelating agents (1.4×10^7 cells L^{-1}). The uptake rate is a unique function of the free ferric ion concentration. Symbols indicate the organic ligand used to control iron speciation: (▼) NTA, (●) EDTA, (▲) DTPA, (■) CDTA. After Anderson and Morel, 1982.[108] (*b*) Measured and estimated rates of individual processes in an iron-limited culture containing 10^{-8} M iron and 10^{-5} M EDTA in the presence of 10^7 cells L^{-1} and with white-light illumination (ca. 100 μmol quanta $m^{-2} s^{-1}$). Both the thermal dissociation of FeEDTA and its photoreduction and reoxidation contribute to the formation of the dissolved inorganic Fe(III) pool which is responsible for the uptake. After Hudson and Morel, 1990.[107]

uptake. The extent of such inhibition can be simply calculated by considering the fraction of uptake ligand bound to the competing metal M* assumed to be at (pseudo)equilibrium with the ligand:

$$\text{percent of inhibition} = \frac{[M*L]}{L_T} \qquad (100)$$

If only a small fraction of the transport ligands is bound to the target metal M, this ratio can be written

$$\text{percent of inhibition} = \frac{K*[M*]}{1 + K*[M*]} \qquad (101)$$

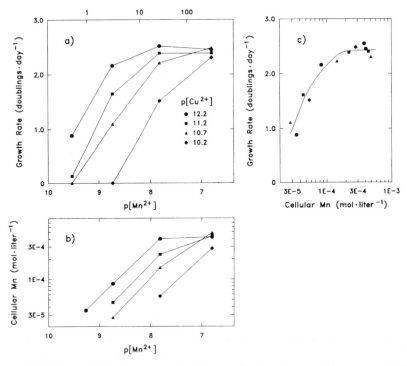

Figure 6.27 The relation between growth rate, free manganese ion concentration, cellular manganese concentration, and free cupric ion concentration in manganese-limited cultures of *Thalassosira pseudonana*. Increasing $[Mn^{2+}]$ or decreasing $[Cu^{2+}]$ increases the growth rate (A). In either case this is a reflection of an increasing intracellular concentration of the limiting nutrient manganese (B). The growth rate is a unique function of the intercellular manganese concentration (C), demonstrating that the toxicity of copper is the result of its inhibition of manganese uptake. After Sunda and Huntsman.[109]

Thus the uptake inhibition—and hence the inhibition of growth rate if M is limiting—is a function of the free concentration of the competing metal [M*] and of its affinity for the transport site.

An example of such competitive inhibition of manganese uptake by copper in a marine phytoplankton is illustrated in Figure 6.27. Under manganese-limited conditions, increasing copper decreases the growth rate. This is because the growth rate depends solely on the manganese content of the cells and as the copper concentration is increased, the cellular manganese is decreased. Additional manganese in the medium reverses the toxicity of copper.[109] Such antagonism between copper and manganese is thought to cause the low productivity of newly upwelled waters which contain low dissolved manganese and, presumably, low concentrations of chelating agents to reduce the inorganic concentration and hence the toxicity of copper.[110]

In the example of Figure 6.27, at a given manganese concentration, the growth rate is only a function of $[Cu^{2+}]$. This situation is a rather general one and may result from competitive inhibition of uptake of an essential metal (Mn) or from competitive inhibition of its assimilation or utilization. Thus, through proper calibration, the growth rate (or other appropriate measure of toxicity) of a microorganism can be used as a measure of the free concentration of toxic metals. This is the basis of the bioassay techniques that have been utilized to quantify metal chelation in natural waters.

Kinetic Control of Growth Rate. At steady state, the cellular concentration of a limiting element M is determined by a balance between the uptake rate and the rate of cell division:

$$\rho = \mu Q \tag{102}$$

where μ is the specific growth rate (in inverse time) and Q the cellular concentration of M (the "quota" expressed in mol cell^{-1}).

As illustrated above for iron, the acquisition of limiting trace metals is usually determined by a kinetically controlled uptake process involving uptake ligands in large excess (Case B in Figure 6.25). The uptake rate is then simply the rate of reaction of the metal with the uptake ligands:

$$\rho = k'_L [M'] L_T \tag{103}$$

Thus from Equations 102 and 103 we obtain the growth rate as a simple expression of chemical kinetics:

$$\mu = \frac{1}{Q} k'_L [M'] L_T \tag{104}$$

As discussed in Section 5.1, the effective complexation rate constant k'_L (applicable

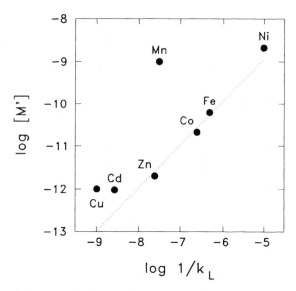

Figure 6.28. Apparent relationship between the concentration of inorganic metals in surface ocean water and their complexation rate constants. (Adapted from Morel et al., 1991[111] as corrected in Hudson and Morel, 1992.[112])

to the total inorganic metal concentration [M']) is characteristic of the metal and fixed for a given pH and ionic medium. For any organism, the cellular parameters Q and L_T are adjustable within limits: for a given volume (or cellular mass) there is a minimum requirement of the essential metal M; for a given surface area there is a maximum concentration of uptake ligands that can be fit in the membrane without compromising its integrity. When these physiological limits are approached, Equation 104 provides an expression of the maximum growth rate attainable for a given inorganic concentration [M'] of the limiting trace metal. Equation 104 also shows how low the inorganic concentration of a trace element may get as a result of uptake by an organism growing at a given rate.[111,112]

If we postulate that for all essential trace elements there is a constant relationship between the maximum number of uptake ligands and the minimum cellular quota, then Equation 104 predicts an inverse relationship between [M'] and k'_L. Though the available data on inorganic trace metal concentrations in seawater are very few and still questionable, they seem indeed to follow such inverse dependence on complexation rates (Figure 6.28). If this is correct it would support the contention that many elements are, on average, limiting the growth of microorganisms in the sea. Complexation kinetics may be one of the keys to marine ecology.

REFERENCES

1. J. Burgess, *Metal Ions in Solution*, Ellis Horwood-Wiley, New York, 1978.
2. F. A. Cotton and G. Wilkinson, *Advanced Inorganic Chemistry*, 3rd ed., Wiley Interscience, New York, 1972, 1145 pp.
3. J. J. Christensen and R. M. Izatt, *Handbook of Metal–Ligand Heats*, Marcel Dekker, New York, 1983, 783 pp.
4. J. E. Huheey, *Inorganic Chemistry*, 3rd ed., Harper and Row, New York, 1983, 936 pp.
5. J. Burgess, *Ions in Solution*, Ellis Horwood, Chichester, 1988, 191 pp.
6. A. E. Martell and R. M. Smith, *Critical Stability Constants*, Vol. 1, *Amino Acids*, Plenum, New York, 1974; R. M. Smith and A. E. Maltell, Vol. 2, *Amines*, Plenum, New York, 1975; A. E. Martell and R. M. Smith, Vol. 3, *Other Organic Ligands*, Plenum, New York, 1977; R. M. Smith and A. E. Martell, Vol. 4, *Inorganic Ligands*, Plenum, New York, 1976.
7. M. Whitfield, *Limnol. Oceanogr.*, **19**, 235 (1974).
8. C. F. Baes, Jr. and R. E. Mesmer, *The Hydrolysis of Cations*, Wiley, New York, 1976.
9. W. G. Sunda and P. J. Hanson, in *Chemical Modeling in Aqueous Systems*, E. A. Jenne, Ed., ACS Symposium Series 93, American Chemical Society, Washington, DC, 1979.
10. J. Sainte Marie, A. E. Torma, and A. O. Gübeli, *Can. J. Chem.*, **42**, 662 (1964).
11. J. J. Morgan, in *Principles and Application in Water Chemistry*, S. D. Faust and J. V. Hunter, Eds., Wiley, New York, 1967.
12. W. Stumm and J. J. Morgan, *Aquatic Chemistry*, 2nd ed., Wiley, New York, 1981, p. 332.
13. L. G. Sillén and A. E. Martell, *Stability Constants of Metal–Ion Complexes*, Special Publications Nos. 17 and 25, Chemical Society, London, 1964 and 1971.
14. H. Irving and R. J. P. Williams, *J. Chem. Soc.*, 3192 (1953).
15. F. A. Cotton and G. Wilkinson, *Basic Inorganic Chemistry*, Wiley, New York, 1976, 579 pp.
16. H. D. Holland, *The Chemistry of the Atmosphere and Oceans*, Wiley, New York, 1978.
17. R. M. Garrels and M. E. Thompson, *Am. J. Sci.*, **57**, 260 (1962).
18. K. S. Johnson and R. M. Pytkowicz, *Mar. Chem.*, **8**, 87 (1979).
19. R. A. Horne, *Marine Chemistry*, Wiley, New York, 1969.
20. W. Stumm and J. J. Morgan, *Aquatic Chemistry*, 2nd ed., Wiley, New York, 1981, p. 364.
21. J. Boulègue, J. P. Ciabrini, C. Fouillac, G. Michard, and G. Ouzounian, *Chem. Geol.*, **25**, 19 (1979).
22. J. Boulègue and G. Michard, in *Chemical Modeling in Aqueous Systms*, E. A. Jenne, Ed., ACS Symposium 93, American Chemical Society, Washington, DC, 1979.
23. Y. Sugimura and Y. Suzuki, *Mar. Chem.*, **24**, 105 (1988).
24. J. R. Toggweiler, *Nature*, **334** 468 (1988).

25. J. Buffle, *Complexation Reactions in Aquatic Systems*, Ellis Horwood, Chichester, 1988, 692 pp.

26. E. M. Thurman, Organic Geochemistry of Natural Waters, Martinus Nijhoff/Dr. W. Junk Publishers, Dordrecht, 1985, 497 pp.

27. C. M. Burney, K. M. Johnson, D. M. Lavoie, and McN. Sieburth, *Deep Sea Res.*, **26**, 1267 (1979).

28. R. Benner, J. D. Pakulski, M. McCarthy, J. I. Hedges, and P. G. Hatcher, *Science*, **255**, 1561 (1992).

29. R. J. Kieber, X. Zhou, and K. Mopper, *Limnol. Oceanogr.*, **37**, 1503 (1990).

30. K. Mopper and X. Zhou, *Science*, **250**, 661 (1990).

31. W. Stumm and P. A. Brauner, in *Chemical Oceanography*, 2nd ed., E. K. Duursma and R. Dawson, Eds., Academic, London, 1975.

32. R. F. C. Mantoura, in *Marine Organic Chemistry*, E. K. Duursma and R. Dawson, Eds., Elsevier, Amsterdam, 1981.

33. F. M. M. Morel, R. E. McDuff, and J. J. Morgan, in *Trace Metals and Metal–Organic Interactions in Natural Waters*, P. C. Singer, Ed., Ann Arbor Science, Ann Arbor, MI, 1973.

34. K. H. Coale and K. W. Bruland, *Limnol. Oceanogr.*, **33**, 1084 (1988).

35. J. G. Hering, W. G. Sunda, R. L. Ferguson, and F. M. M. Morel, *Mar. Chem.*, **20**, 299 (1987).

36. D. L. Huizenga and D. R. Kester, *Mar. Chem.*, **13**, 281 (1983).

37. W. G. Sunda and A. K. Hanson, *Limnol. Oceanogr.*, **32**, 537 (1987).

38. J. W. Moffett and R. G. Zika, *Mar. Chem.*, **21**, 301 (1987).

39. W. G. Sunda and R. L. Ferguson, in *Trace Metals in Sea Water*, C. S. Wong, E. Boyle, K. W. Bruland, J. D. Burton, and E. D. Goldberg, Eds., Plenum, New York, 1983.

40. K. W. Bruland, *Limnol. Oceanogr.*, **34**, 269 (1989).

41. J. R. Donat and K. W. Bruland, *Mar. Chem.*, **28**, 301 (1990).

42. C. M. G. van den Berg, A. G. A. Merks, and E. K. Duursma, *Est. Coast. Shelf Sci.*, **24**, 784 (1987).

43. C. M. G. van den Berg and S. Dharmvanij, *Limnol. Oceanogr.*, **29**, 1025 (1984).

44. G. Capodaglio, K. H. Coale, and K. W. Bruland, *Mar. Chem.*, **29**, 221 (1990).

45. G. A. Gill and K. W. Bruland, *Environ. Sci. Technol.*, **24**, 1392 (1990).

46. C. T. Driscoll, J. P. Baker, J. J. Bisogni, and C. L. Schofield, in *Acid Precipitation: Geological Aspécts*, O. Bricker, Ed., Butterworths, Boston, 1984.

47. J. C. Duinker and C. J. M. Kraemer, *Mar. Chem.*, **5**, 207 (1977).

48. W. G. Sunda, *Mar. Chem.*, **14**, 365 (1984).

49. K. W. Bruland, *Limnol. Oceanogr.*, **36**, 1555 (1991).

50. J. C. Westall, F. M. M. Morel, and D. N. Hume, *Anal. Chem.*, **51**, 1792 (1979).

51. J. L. Means, T. Kucak, and D. A. Crerrar, *Environ. Poll. (ser. B)*, **1**, (1980).

52. J. Gardiner, *Wat. Res.*, **10**, 507 (1976).

53. J. L. Means, D. A. Crerar, and J. O. Duguid, *Science*, **200**, 1477 (1978).

54. W. Giger, H. Ponusz, A. Alder, D. Baschnagel, D. Renggli, C. Schaffner, EAWAG Jahresbericht, EAWAG/ETH, Zurich, 1987.

55. A. C. Alder, H. Siegrist, W. Gujer, and W. Giger, *Water Res.*, **24**, 733 (1990).

56. H.-J. Brauch and S. Schullerer, *Vom Wasser*, **69**, 155 (1987).

57. J. M. Cleveland and T. F. Rees, *Science*, **212**, 1506 (1981).

58. R. W. D. Killey, J. O. McHugh, D. R. Champ, E. L. Cooper, and J. L. Young, *Environ. Sci. Technol.*, **18**, 148 (1984).

59. J. B. Neilands, *Ann. Rev. Biochem.*, **50**, 715 (1981).

60. B. F. Matzambe, G. Müller-Matzambe, and K. M. Raymond, *Phys. Bioinorg. Chem.*, Series 5, 3 (1989).

61. C. G. Trick, R. J. Andersen, A. Gillam, and P. J. Harrison, *Science*, **219**, 306 (1983).

62. C. M. Brown and C. G. Trick, *Arch. Microbiol.*, **157**, 349 (1992).

63. C. G. Trick and A. Kerry, *Curr. Microbiol.*, **24**, 241 (1992).

64. J. B. Neilands, *Ann. Rev. Microbiol.*, **36**, 285 (1982).

65. K. N. Raymond and C. J. Carrano, *Accts. Chem. Res.*, **12**, 183 (1979).

66. P. E. Powell, G. R. Cline, C. P. P. Reid, and P. J. Szaniszlo, *Nature*, **287**, 833 (1980).

67. M. Estep, J. E. Armstrong, and van Baalen, *App. Environ. Microbiol.*, **30**, 186 (1975).

68. T. P. Murphy, D. R. S. Lean, and C. Nalewajko, *Science*, **192**, 900 (1976).

69. E. Gill, E. L. Winnacker, and M. H. Zenk, *Science*, **30**, 674 (1985).

70. W. Gekeler, E. Grill, E.-L. Winnacker, and M. H. Zenk, *Arch. Microbiol.*, **150**, 197 (1988).

71. Y. Kojima and J. H. R. Kagi, *Metallothioneins TIBS*, **4**, (1978).

72. R. C. Averett, J. A. Leenheer, D. M. McKnight, and K. A. Thorn, Eds., *Humic Substances in the Sunwannee River, Georgia: Interactions, Properties and Proposed Structures.* U.S. Geological Survey Open File Report 87–557, 1989.

73. G. R. Aiken, D. M. McKnight, R. L. Wilson, and R. MacCarthy, Eds., *Humic Substances in the Soil, Sediment, and Water*, Wiley, New York, 1985.

74. D. H. Stuermer and G. R. Harvey, *Mar. Chem.*, **6**, 55 (1978).

75. G. R. Harvey, D. A. Boran, L. S. Chesal, and J. M. Tokar, *Mar. Chem.*, **12**, 119 (1983).

76. R. L. Crawford, *Lignin Biodegradation and Transformation*, Wiley, New York, 1981.

77. M. A. Wilson, P. F. Barron, and A. H. Gillam, *Geochim. Cosmochim. Acta*, **45**, 1743 (1981).

78. E. M. Perdue, *Geochim. Cosmochim. Acta*, **42**, 1351 (1978).

79. B. M. Bartschat, S. E. Cabaniss, and F. M. M. Morel, *Environ. Sci. Technol.*, **26**, 284 (1992).

80. T. C. Hutchinson and K. M. Meema, Eds., *Lead, Mercury, Cadmium, and Arsenic in the Environment, SCOPE 31*, John Wiley and Sons, Chichester, 1987, 360 pp.

81. K. Fent, *Mar. Environ. Res.*, **28**, 477 (1989).

82. C. C. Gilmour, J. H. Tuttle, and J. C. Means, in *Marine and Estuarine Geochemistry*, A. C. Sigleo and A. Hattori, Eds., Lewis, Chelsea, MI, 1985, pp. 239–258.

83. O. F. X. Donard, F. T. Short and J. H. Weber, *Can. J. Fish. Aq. Sci.*, **44**, 140 (1987).

84. R. P. Mason and W. F. Fitzgerald, *Nature*, **347**, 457 (1990).

85. N. Bloom, *Can. J. Fish. Aq. Sci.*, **46**, 1131 (1989).

86. R. J. Maguire, R. J. Tkacz, Y. K. Chan, G. A. Bengert, and P. T. S. Wong, *Chemosphere*, **15**, 253 (1986).

87. J. J. Cleary and A. R. D. Stebbing, *Mar. Poll. Bull.*, **18**, 238 (1987).

88. C. L. Alzieu, J. Sanjuan, J. P. Itreil, and M. Borel, *Mar. Poll. Bull.*, **17**, 494 (1986).

89. M. O. Andrea and P. N. Froelich, Jr., *Tellus*, **36B**, 101 (1984).

90. L. C. D. Anderson and K. W. Bruland, *Environ. Sci. Technol.*, **25**, 420 (1991).

91. J. P. Kim and W. F. Fitzgerald, *Science*, **231**, 1131 (1986).

92. J. T. Byrd and M. O. Andreae, *Science*, **218**, 565 (1982).

93. P. J. Craig, Ed., *Organometallic Compounds in the Environment*, Longman, Essex, 1986, 368 pp.

94. This discussion is based on J. G. Hering and F. M. M. Morel, in *Aquatic Chemical Kinetics*, W. Stumm, Ed., Wiley-Interscience, New York, 1990, pp. 145–171.

95. C. J. M. Kraemer, *Mar. Chem.*, **18**, 335 (1986).

96. S. E. Cabaniss and M. S. Shuman, *Anal. Chem.*, **60**, 2418 (1988).

97. J. G. Hering and F. M. M. Morel, *Geochim. Cosmochim. Acta*, **53**, 611 (1989).

98. D. W. Margerum, G. R. Cayley, D. C. Weatherburn, and G. K. Pagenkopf, in *Coordination Chemistry*, Vol. 2, A. E. Martell, Ed., ACS Monograph 174, Washington, DC, 1978, pp. 1–220.

99. J. N. Israelachvili, *Intermolecular and Surface Forces*, 2nd. ed., Academic Press, London, 1992, p. 33.

100. P. C. Hiemenz, *Principles of Colloid and Surface Chemistry*, 2nd ed., Dekker, New York, 1986, p. 691.

101. A. L. Crumbliss and J. M. Garrison, *Comments Inorg. Chem.*, **8**, 1 (1988).

102. W. Schneider, Personal Communication.

103. J. G. Hering and F. M. M. Morel, *Environ. Sci. Technol.*, **22**, 1469 (1988).

104. R. J. M. Hudson, "The Chemical Kinetics of Iron Uptake by Marine Phytoplankton," Ph.D. Thesis, Massachusetts Institute of Technology, Cambridge, MA, 1989.

105. K. W. Bruland and R. P. Franks, in *Trace Metals in Seawater*, C. S. Wong, E. Boyle, K. W. Bruland, J. D. Burton, and E. D. Goldberg, Eds., Plenum, New York, 1983.

106. J. H. Martin, R. M. Gordon, S. Fitzwater, and W. W. Broenkow, *Deep Sea Res.*, **36**, 649 (1989).

107. R. J. M. Hudson and F. M. M. Morel, *Limnol. Oceanogr.*, **35**, 1002 (1990).

108. M. A. Anderson and F. M. M. Morel, *Limnol. Oceanogr.*, **27**, 789 (1982).

109. W. G. Sunda and S. A. Huntsman, *Limnol. Oceanogr.*, **28**, 924 (1983).

110. W. G. Sunda, R. T. Barber, and S. A. Huntsman, *J. Mar. Res.*, **39**, 567 (1981).

111. F. M. M. Morel, R. J. M. Hudson, and N. M. Price, *Limnol. Oceanogr.*, **36**, 1742 (1991).

112. R. J. M. Hudson and F. M. M. Morel, *Deep Sea Res.*, (in press) (1992).

PROBLEMS

6.1 **a.** Using the same major ion composition as in Example 1, calculate the inorganic speciation of nickel and mercury in freshwater at pH = 8.1.

b. Maintaining C_T constant, calculate the effect of pH (from 5 to 12) on the inorganic speciation of nickel and mercury.

c. At what pH's would the carbonate or hydroxide solids precipitate given $[Ni]_T = [Hg]_T = 10^{-7} M$?

6.2 Consider the usual approximate stoichiometric reaction of photosynthesis: $CO_2 + H_2O \rightarrow \text{``CH}_2O\text{''} + O_2$. Note that this reaction changes both C_T and pH. Assuming the system to be closed to the atmosphere, discuss quantitatively the effect of an algal bloom on the formation of carbonate and bicarbonate trace metal complexes [e.g., $CuCO_3(aq)$ and $CuHCO_3^+$].

6.3 What is the effect of major ion associations in seawater on (a) pH (at equilibrium with the atmosphere), and (b) carbonate and bicarbonate trace metal complexes.

6.4 **a.** Redo Example 6 of Chapter 5 adding $Ca_T = 10^{-3} M$ and $[NTA]_T = 10^{-4} M$ to the recipe of the system and letting $CaCO_3(s)$ supersaturate.

b. How much NTA would have to be added at any pH to avoid precipitation of the iron $[Y_T = Y_T(pH)]$?

c. Redo parts a and b for $CaCO_3(s)$ precipitation.

6.5 Consider a seawater culture medium containing $[EDTA]_T = 5 \times 10^{-5} M$, $Fe(III)_T = 10^{-8} M$ and $Cd_T = 10^{-10} M$. ($I = 0.5 M$, $Ca_T = 10^{-2} M$, and $pH = 8.2$.)

a. Calculate the speciation and the free concentrations of EDTA, Fe, and Cd.

b. $4 \times 10^{-5} M$ $CdCl_2$ is added to the system. Calculate the new speciation and new free concentrations. What would the result have been if $4 \times 10^{-5} M$ CdEDTA had been added to the system?

c. The algae in the cultures (10^6 cells L^{-1}) have on their surface an iron-transport molecule characterized by $X_T = 10^{-16}$ mol cell^{-1}; $K_{Fe} = 10^{19}$; $K_{Cd} = 10^8$. How much Fe and Cd is bound to the cells before and after $CdCl_2$ addition? What is the net effect of the Cd addition on the Fe transport rate?

6.6 A lake with the simple composition

$$Na_T = Ca_T = Mg_T = K_T = Cl_T = [SO_4]_T = [CO_3]_T = 10^{-3} M$$

contains an organic complexing agent Y, characterized by

$$
\begin{array}{ll}
HY = H^+ + Y^- & pK_a = 6.0 \\
CuY^+ = Cu^{2+} + Y^- & pK = 7.0 \\
CaY^+ = Ca^{2+} + Y^- & pK = 2.0 \\
MgY^+ = Mg^{2+} + Y^- & pK = 2.0
\end{array}
$$

a. Consider no solid phase but all the relevant complexes. Calculate the copper speciation at pH $= 7.0$ ($Cu_T = 10^{-7} M$; $Y_T = 10^{-6} M$). Calculate the copper speciation as a function of pH.

b. Consider now that $Cu_T = 10^{-6} M$ and that copper carbonates and/or hydroxide may precipitate (same inorganic and organic species as in part a). Calculate the critical pH('s) of precipitation of the solid(s) in the absence of organic ligand. What solid(s) should form as pH varies from 6 to 12? Sketch a major Cu species diagram on a $\log Y_T$ versus pH graph.

CHAPTER 7

OXIDATION–REDUCTION

The geochemical cycles of elements are driven in part by oxidation–reduction reactions. Some mineral phases such as metal sulfides are dissolved through oxidation by oxygen, others such as iron oxides through reduction in anoxic environments; conversely, some solutes are ultimately precipitated as reduced sulfides or as oxides. Although significant, this auxiliary role of redox chemistry in the exogenic cycle of elements is not our major focus in this chapter. As unstable chemical entities fueled by a continual diet of decomposing compounds, sustained by a constant energy flux from oxidation reactions, we are motivated by more than pure academic curiosity to study redox chemistry; it is a simple, or not so simple, matter of life and death. Life is by nature a redox process and a majority of redox processes on the earth are life dependent. From our point of view the subject of the redox cycles of elements thus takes on an importance disproportionate to the relatively small elemental fluxes that are involved.

Fueling all the redox cycles on the earth's surface is one fundamental, thermodynamically unfavorable process, photosynthesis. Using solar energy through specialized pigments and organelles, plants reduce inorganic carbon to organic matter and produce oxygen, thus increasing the Gibbs free energy of the earth:

$$CO_2(g) + H_2O \longrightarrow \text{``}CH_2O\text{''} + O_2(g); \qquad \Delta G^\circ = +478 \, \text{kJ mol}^{-1} \qquad (1)$$

With very few exceptions, this energy capturing process drives all other redox reactions: organic matter is the ultimate reductant; oxygen, the ultimate oxidant.

Powered by photosynthesis, the cycle of organic carbon is the driving belt that propels the other elemental redox cycles. As a first approximation, we can thus view all molecules at the earth's surface as being subjected directly or indirectly to a reducing force (an electron supply) from organic carbon and to an oxidative force (an electron drain) from oxygen. The resulting redox species depend on a local balance between these opposing forces. Fortunately, this balance is rarely an equilibrium state. We are dependent for our survival on the sluggishness of redox reactions.

Besides carbon and oxygen, the elements most involved in redox cycles are those most abundant in living matter: nitrogen, sulfur, and some metals such as iron and manganese. (Phosphorus, which is also abundant in biological material, serves chiefly as a constituent of intracellular energy storage compounds, such as ATP and NADPH, but has no important redox chemistry of its own.) Because life is a matter of stoichiometry as well as one of energetics, the cycles of these elements are then tightly coupled in two different ways. First, the cycles are linked through the energetic relationships imposed by the thermodynamics of the redox reactions. For example, microorganisms must oxidize a certain quantity of organic carbon to obtain the energy necessary to reduce nitrogen. Second, in their ephemeral combination to form living matter, these elements are constrained to cycle within the rather strict stoichiometry of life, as expressed for example in the Redfield formula for aquatic biomass: $C_{106}H_{263}O_{110}N_{16}P_1$. These energetic and stoichiometric couplings are ultimately reflected in the chemical composition of the water itself.

Although equilibrium thermodynamics would seem to have little to do with such highly dynamic processes, it does in fact provide insight into the redox chemistry of aquatic systems in several ways. Continuing in the manner of the previous chapters, we can simply develop partial equilibrium models of natural waters, limiting them to include only the few redox couples that may reasonably be at equilibrium with each other. For example, this approach is helpful in studying the chemistry of trace metals in oxic and reduced systems. For those redox systems that are clearly in a state of disequilibrium, thermodynamics also allows us to calculate the energy required, at the minimum, to maintain that disequilibrium state. We can estimate how much net energy must be expended by plants to reduce carbon, or what is the maximum energy that can be derived by organisms from any particular oxidation reaction. On the basis of these energetic considerations, we can then decide what microbial redox processes are possible. In many cases we are also able to predict the sequence of redox processes in nature since, usually, microbes exploit first the most energetically favorable reactions. (This amounts to "quasi-kinetic" information on the basis of equilibrium thermodynamics.)

Although the global picture of photosynthesis driving all other redox reactions through production of reduced carbon and oxygen provides a useful conceptual image of the major elemental redox cycles, there are instances where oxidants or reductants originate from other processes. In surface waters a number of redox reactions are driven directly by solar energy through photochemical processes. An example is the photochemical reduction of ferric to ferrous iron in oxic waters. We are just beginning to quantify the importance of these photochemical reactions in the redox cycles of organic and inorganic compounds.

At the bottom of the oceans, a geochemically important set of redox reactions is driven by geothermal activity at oceanic ridges. Seawater constituents that circulate through ridge "vents" are mixed with reduced compounds from the earth's interior that do not ultimately owe their reduced condition to photosynthesis. These reduced compounds are important in the geochemical balance of many elements in the ocean. Note, however, that the corresponding redox processes—including the survival of the specialized ecosystems that inhabit the vents—are still dependent on photosynthesis since the oxidants that they utilize, O_2, SO_4^{2-}, Fe(III), are ultimately derived from photosynthetically produced oxygen.

The first section of this chapter is devoted to defining such fundamental concepts as half redox reactions, electron activities, and redox potentials, which are often a source of confusion. On the basis of these concepts, redox couples that are present in natural waters are then compared to characterize the redox status of the waters and to estimate the free energies that are involved in maintaining or exploiting existing disequilibria. Because the corresponding reactions are chiefly mediated by microorganisms, this is effectively a study of the energetics of aquatic microbial life. A series of examples of redox equilibrium calculations in complex systems is then presented, including a discussion of pe–pH diagrams. Finally, short introductions to aquatic photochemistry and redox kinetics in natural waters are provided.

1. DEFINITIONS, NOTATIONS, AND CONVENTIONS

1.1 The Electron as a Component

Historically, and etymologically, oxidation reactions are those that involve the combination of oxygen with another element or compound. In the general framework that we have developed so far, we then need one new component to account for these reactions, to introduce O_2 as a reactive aquatic species. If we consider our usual choice of components for aquatic systems, including H_2O and H^+, there are three simple and equivalent possible choices for this new component. The first two are rather obvious; oxygen itself (O_2) and hydrogen (H_2) both provide straightforward stoichiometric expressions for oxygen:

$$O_2 = (O_2)_1$$
$$O_2 = (H_2)_{-2}(H_2O)_2$$

The third choice is that of a component symbolizing electrical charge with no elemental significance. For consistency with what we know of the structure of matter, this last component is taken to have a negative unit charge, is symbolized by e^-, and is referred to as the electron. It provides a formula for oxygen according to

$$O_2 = (H^+)_{-4}(e^-)_{-4}(H_2O)_2$$

These three choices are strictly equivalent and each of them has in fact been made at different times in history and in various disciplines for the quantitative study of some particular family of redox reactions. Many textbooks still refer to hydrogen rather than electron transfer processes to describe biochemical redox reactions.

Following the modern custom, we choose to use here the electron as our basic redox component, although it may seem perhaps the least natural of the three possibilities. With this convention a reductant is a compound that reacts by releasing electrons (an electron donor), and an oxidant is a compound that takes up electrons (an electron acceptor).

Since we do not consider electrons to have an existence of their own (they have been introduced here as purely conceptual components, although free hydrated electrons may in fact have a half life approaching microseconds under favorable conditions[1]), our convention introduces a whole series of convenient but artificial concepts and quantities in the study of redox processes. In particular, we now have to define reactions involving electrons, called half redox reactions, and their associated thermodynamic constants.

1.2 Half Redox Reactions

Consider the formulae of oxygen and hydrogen as provided by our present choice of components:

$$O_2 = (H^+)_{-4}(e^-)_{-4}(H_2O)_2$$
$$H_2 = (H^+)_2(e^-)_2$$

According to our simple view of the world there are no free electrons in water, and the reactions corresponding to these formulae, the half redox reactions

$$2H_2O = O_2 + 4H^+ + 4e^- \tag{2}$$

$$2H^+ + 2e^- = H_2 \tag{3}$$

can have no tangible chemical reality of their own. The complete redox reaction, obtained by combining the two half reactions and eliminating the electrons,

$$2H_2O = O_2 + 2H_2 \tag{4}$$

can be studied experimentally to calculate a free energy change or an equilibrium constant, but the individual half redox reactions cannot. To obtain a thermodynamic description of half redox reactions we then have to decide arbitrarily on the standard free energy change (or the equilibrium constant) of one half redox reaction. By international convention, we take the half reaction between hydrogen ion and hydrogen gas to have a zero standard free energy change:

$$H^+ + e^- = \tfrac{1}{2}H_2(g); \qquad \Delta G^\circ = 0; \qquad K = 1 \tag{5}$$

The standard free energy change (and the equilibrium constant) of any other half redox reaction can then be calculated from the free energy change of complete redox reactions. Most simply, for complete redox reactions involving H^+ as an oxidant and H_2 as a reductant, the standard free energy change of the corresponding half redox reaction is equal to that of the complete reaction:

$$\left. \begin{aligned} Red + nH^+ &= Ox + \frac{n}{2}H_2 \\ Red &= ne^- + Ox \end{aligned} \right\} \text{ same } \Delta G^\circ; \text{ same } K \qquad \begin{aligned} (6) \\ (7) \end{aligned}$$

For example, the oxygen/water half redox reaction has the same standard free energy change and the same equilibrium constant as the reaction of formation of water from oxygen and hydrogen:

$$\tfrac{1}{2}O_2(g) + H_2 = H_2O; \qquad \Delta G^\circ = -236.6 \,\text{kJ mole}^{-1}; \qquad K = 10^{41.5} \tag{8}$$

$$\tfrac{1}{2}O_2(g) + 2e^- + 2H^+ = H_2O; \qquad \Delta G^\circ = -236.6 \,\text{kJ mole}^{-1}; \qquad K = 10^{41.5} \tag{9}$$

Even if the complete redox reaction with H^+ and H_2 is not conveniently studied experimentally, the thermodynamics of any half redox reaction can be indirectly related to the H^+/H_2 convention through any series of other complete redox reactions involving common half reactions. For example, the oxidation of glucose, $(CH_2O)_6$, by oxygen,

$$\tfrac{1}{6}\text{glucose} + O_2(g) = CO_2(g) + H_2O;$$
$$\Delta G^\circ = -477.7 \,\text{kJ mole}^{-1}; \qquad K = 10^{83.8} \tag{10}$$

provides the standard free energy change for the glucose/CO_2 half reaction by subtracting the O_2/H_2O half reaction, which is itself obtained from the reaction with hydrogen as shown above:

$$\tfrac{1}{6}\text{glucose} + H_2O = CO_2(g) + 4H^+ + 4e^-;$$
$$\Delta G^\circ = -4.5 \,\text{kJ mole}^{-1}; \qquad K = 10^{0.8} \tag{11}$$

1.3 Oxidation State

The "electron content" of an element in a given chemical species is given a formal and conventional value called the oxidation state of the element. It is symbolized by a Roman numeral with a sign in parentheses following the element: Fe($+$II), Fe($+$III), C(0), C($+$IV), S($-$II), and so on. (The plus sign is often omitted.) In complex molecules the assignment of the electrons to the correct element can be difficult and arbitrary. But, for most species of interest in aquatic chemistry, we can define oxidation states rather simply. Consider as a choice of components, H_2O, H^+, e^-, and the elements themselves, N, S, C, and Fe (but of course not O or H). The oxidation state of any element in a compound with oxygen, hydrogen, and electrons is then simply equal to the negative of the stoichiometric coefficient of the electron, normalized per atom of the element, in the formula of the compound. For a general chemical formula involving the element A,

$$A_a O_o H_h^{n-} = (A)_a (H_2O)_o (H^+)_{-2o+h} (e^-)_{-2o+h+n}$$

this translates into the equation

$$\text{oxidation state} = \frac{(2o - h - n)}{a} \tag{12}$$

Examples

$$N(0): N_2$$
$$N(-III): NH_3, NH_4^+$$
$$N(+V): NO_3^-$$
$$S(-II): H_2S, HS^-, S^{2-}$$
$$S(+II): S_2O_3^{2-}$$
$$S(+VI): SO_4^{2-}$$
$$C(-IV): CH_4$$
$$C(0): C, CH_2O$$
$$C(+IV): CO_2, HCO_3^-, CO_3^{2-}$$
$$Fe(+III): Fe^{3+}, Fe(OH)_3$$
$$Fe(+II): Fe^{2+}$$
$$Cr(+III): Cr^{3+}, Cr(OH)_3$$
$$Cr(+VI): CrO_4^{2-}, Cr_2O_7^{2-}$$

The oxidation state of elements is not changed by coordination reactions among metals and ligands. For example, in the ferrous carbonate complex $FeCO_3$, Fe has the oxidation state ($+$II) and C has the oxidation state ($+$IV). This leaves the oxidation state of elements in only a few compounds of interest in aquatic chemistry [e.g., cyanide CN^-: $N(-III)$; $C(+II)$] to be assigned by additional rules given in chemistry textbooks.

Any pair of species in which the same element occurs in different oxidation states constitutes a redox couple. A balanced half redox reaction can be written for any redox couple by: (1) balancing the stoichiometric coefficients of the element of interest; (2) adding H_2O to balance oxygen; (3) adding H^+ to balance hydrogen; (4) adding e^- to balance electrical charges. For example, the four steps to balance the $SO_4^{2-}/S_2O_3^{2-}$ reactions are as follows:

1. $2SO_4^{2-} = S_2O_3^{2-}$

2. $2SO_4^{2-} = S_2O_3^{2-} + 5H_2O$

3. $2SO_4^{2-} + 10H^+ = S_2O_3^{2-} + 5H_2O$

4. $2SO_4^{2-} + 10H^+ + 8e^- = S_2O_3^{2-} + 5H_2O$

1.4   Electron Activities and Redox Potentials

On the basis of the convention of an equilibrium constant of unity for the H^+/H_2 couple, we have now defined standard free energies and equilibrium constants for all half redox reactions:

$$Ox + ne^- = Red; \qquad \Delta G^\circ; \qquad K \qquad (7)$$

In so doing we have formally and mathematically defined the equilibrium activity of the electron for any redox couple:

$$\{e^-\} = \left[\frac{1}{K} \frac{[Red]}{[Ox]} \right]^{1/n} \qquad (13)$$

(The activities of the reductant and the oxidant are replaced here by their free concentrations for simplicity.) This electron activity is defined in any system where the free concentrations [Red] and [Ox] are defined, even though the electron is not considered to be an individual species and thus its concentration is certainly *not* defined. Electron activities are usually expressed on either of two scales, pe or E_H:

$$pe = -\log\{e^-\} = \frac{1}{n}\left[\log K - \log \frac{[Red]}{[Ox]} \right] \qquad (14)$$

$$E_H = \frac{2.3RT}{F} pe \qquad (15)$$

therefore

$$E_H = \frac{2.3RT}{nF}\left[\ln K - \ln \frac{[Red]}{[Ox]} \right] \qquad (16)$$

The parameter pe provides a nondimensional scale (like pH), while E_H, the redox potential, is measured in volts. [F is the Faraday constant (the electric charge of one mole of electrons = 96,500 coulombs), and $2.3\, RT/F$ has a value of 0.059 V at 25°C.]

One can gain insight into the reason why electron activities are logically expressed on an electric potential scale by considering energies. The progress to the right of the complete redox reaction

$$Ox_1 + Red_2 = Red_1 + Ox_2$$

corresponds to the transfer of n electrons from Red_2 to Ox_1. The energy (ΔG) necessary to transfer this electrical charge of $(-nF)$ coulombs per mole can be expressed as the product of the charge and a potential difference ($E_H^1 - E_H^2$):

$$\Delta G = (-nF)(E_H^1 - E_H^2) \tag{17}$$

This energy must also be equal to the free energy change of the reaction:

$$\Delta G = \Delta G° + RT \ln \frac{[Red_1][Ox_2]}{[Ox_1][Red_2]} \tag{18}$$

The standard free energy $\Delta G°$ can be expressed as a function of the equilibrium constants of the half reactions:

$$\Delta G° = -RT \ln K = -RT \ln \frac{K_1}{K_2} \tag{19}$$

$$\Delta G = RT \left[\ln K_2 - \ln \frac{[Red_2]}{[Ox_2]} \right] - RT \left[\ln K_1 - \ln \frac{[Red_1]}{[Ox_1]} \right] \tag{20}$$

A comparison of Equation 17 and 20 leads to the expression of the electrical potential E_H as given in Equation 16. (Note that Equation 17 also illustrates why the electron volt is a logical unit of energy for redox reaction: $1\,eV = 1.6 \times 10^{-19}\,J$.)

If we normalize the energy change to one electron transfer and introduce Equation 14 into Equation 20, we obtain the general expressions

$$\frac{\Delta G}{n} = 2.3RT(pe_2 - pe_1) \tag{21a}$$

$$= F(E_H^2 - E_H^1) \tag{21b}$$

The difference in electron activities between two half redox reactions is thus directly proportional to the free energy change of the complete redox reaction. In effect pe and E_H define two absolute energy scales (whose origins are fixed by the H^+/H_2 convention) and free energies of complete redox reactions are obtained by simple differences.

If we now consider the standard free energy changes of redox reactions (the free energy changes corresponding to unit activities of products and reactants; $\Delta G^\circ = -RT \ln K$), we can, for consistency, express the equilibrium constants of half redox reactions as pe° or standard potentials E_H°:

$$pe^\circ = \frac{F}{2.3RT} E_H^\circ$$

$$= \frac{1}{n} \log K$$

$$= -\frac{1}{n} \frac{\Delta G^\circ}{2.3RT} \tag{22}$$

At 25°C,

$$pe^\circ = 16.9 E_H^\circ \quad \text{(in V)} \tag{23a}$$

$$= \frac{1}{n} \log K \tag{23b}$$

$$= -\frac{0.175 \, \Delta G^\circ}{n} \quad \text{(in kJ mol}^{-1}) \tag{23c}$$

The coexistence of these four different scales (pe°, E_H°, K, ΔG°) to define the thermodynamic characteristics of redox reactions does much to confuse the subject. As shown in Table 7.1, which provides thermodynamic data for several redox reactions of interest in aquatic systems, we shall consistently use pe° and pe as our notation, remembering that pe° is the log of the equilibrium constant for the reaction written as a reduction involving one electron,

$$Ox + e^- = Red$$

and that the negative logarithm of the electron activity is then given by

$$pe = pe^\circ - \log \frac{[Red]}{[Ox]} \tag{24}$$

Equation 16, which defines an electrical potential corresponding to any half redox reaction, is known as the Nernst equation. This potential is in fact measurable compared to a reference potential defined by a hydrogen gas partial pressure of 1 atmosphere at equilibrium with a solution at pH = 0:

$$\{e^-\} = \frac{1}{1} \frac{P_{H_2}^{1/2}}{\{H^+\}} = 1; \quad E_H = 0 \tag{25}$$

TABLE 7.1   Half Redox Reactions

	$pe^\circ = \log K$
Hydrogen	
$H^+ + e^- = \frac{1}{2}H_2(g)$	0.0
Oxygen	
$\frac{1}{2}O_3(g) + H^+ + e^- = \frac{1}{2}O_2(g) + \frac{1}{2}H_2O$	$+35.1$
$\frac{1}{4}O_2(g) + H^+ + e^- = \frac{1}{2}H_2O$	$+20.75$
$\frac{1}{2}H_2O_2 + H^+ + e^- = H_2O$	$+30.0$
(Note also $HO_2^- + H^+ = H_2O$; $\log K = 11.6$)	
Nitrogen	
$NO_3^- + 2H^+ + e^- = \frac{1}{2}N_2O_4(g) + H_2O$	$+13.6$
(Note also $N_2O_4(g) = 2NO_2(g)$; $\log K = -0.47$)	
$\frac{1}{2}NO_3^- + H^+ + e^- = \frac{1}{2}NO_2^- + \frac{1}{2}H_2O$	$+14.15$
(Note also $NO_2^- + H^+ = HNO_2$; $\log K = 3.35$)	
$\frac{1}{3}NO_3^- + \frac{4}{3}H^+ + e^- = \frac{1}{3}NO(g) + \frac{2}{3}H_2O$	$+16.15$
$\frac{1}{4}NO_3^- + \frac{5}{4}H^+ + e^- = \frac{1}{8}N_2O(g) + \frac{5}{8}H_2O$	$+18.9$
$\frac{1}{5}NO_3^- + \frac{6}{5}H^+ + e^- = \frac{1}{10}N_2(g) + \frac{3}{5}H_2O$	$+21.05$
$\frac{1}{8}NO_3^- + \frac{5}{4}H^+ + e^- = \frac{1}{8}NH_4^+ + \frac{3}{8}H_2O$	$+14.9$
Sulfur	
$\frac{1}{2}SO_4^{2-} + H^+ + e^- = \frac{1}{2}SO_3^{2-} + \frac{1}{2}H_2O$	-1.65
(Note also $SO_3^{2-} + H^+ = HSO_3^-$; $\log K \simeq 7$)	
$\frac{1}{4}SO_4^{2-} + \frac{5}{4}H^+ + e^- = \frac{1}{8}S_2O_3^{2-} + \frac{5}{8}H_2O$	$+4.85$
$\frac{1}{6}SO_4^{2-} + \frac{4}{3}H^+ + e^- = \frac{1}{48}S_8^0(\text{s. ort.}) + \frac{2}{3}H_2O$	$+6.03$
(Note also S_8^0 (s. ort.) $= S_8^0$ (s. col.); $\log K = -0.6$)	
$\frac{3}{19}SO_4^{2-} + \frac{24}{19}H^+ + e^- = \frac{1}{38}S_6^{2-} + \frac{12}{19}H_2O$	$+5.40$
$\frac{5}{32}SO_4^{2-} + \frac{5}{4}H^+ + e^- = \frac{1}{32}S_5^{2-} + \frac{5}{8}H_2O$	$+5.88$
(Note also $S_5^{2-} + H^+ = HS_5^-$; $\log K = 6.1$ and	
$HS_5^- + H^+ = H_2S_5$; $\log K = 3.5$)	
$\frac{2}{13}SO_4^{2-} + \frac{16}{13}H^+ + e^- = \frac{1}{26}S_4^{2-} + \frac{8}{13}H_2O$	$+5.12$
(Note also $S_4^{2-} + H^+ = HS_4^-$; $\log K = 7.0$ and	
$HS_4^- + H^+ = H_2S_4$; $\log K = 3.8$)	
$\frac{1}{8}SO_4^{2-} + \frac{5}{4}H^+ + e^- = \frac{1}{8}H_2S(aq) + \frac{1}{2}H_2O$	$+5.13$
(Note also $H_2S(g) = H_2S(aq)$; $\log K_H = -1.0$, and other	
acid–base, coordination, and precipitation reactions)	
Carbon	

Inorganic

Carbon monoxide

$\frac{1}{2}CO_2(g) + H^+ + e^- = \frac{1}{2}CO(g) + \frac{1}{2}H_2O$	-1.74

Graphite

$\frac{1}{4}CO_2(g) + H^+ + e^- = \frac{1}{4}C(s) + \frac{1}{2}H_2O$	$+3.50$

TABLE 7.1   (*Continued*)

	$pe^o = \log K$

Organic—C.1

Formate$^-$
$$\tfrac{1}{2}CO_2(g) + \tfrac{1}{2}H^+ + e^- = \tfrac{1}{2}HCOO^-$$ -5.22

Formaldehyde
$$\tfrac{1}{4}CO_2(g) + H^+ + e^- = \tfrac{1}{4}HCHO(aq) + \tfrac{1}{4}H_2O$$ -1.20

Methanol
$$\tfrac{1}{6}CO_2(g) + H^+ + e^- = \tfrac{1}{6}CH_3OH(aq) + \tfrac{1}{6}H_2O$$ $+0.50$

Methane
$$\tfrac{1}{8}CO_2(g) + H^+ + e^- = \tfrac{1}{8}CH_4(g) + \tfrac{1}{4}H_2O$$ $+2.86$

Organic—C.2

Oxalate^{2-}
$$CO_2(g) + e^- = \tfrac{1}{2}(COO^-)_2$$ -10.7

Acetate
$$\tfrac{1}{4}CO_2(g) + \tfrac{7}{8}H^+ + e^- = \tfrac{1}{8}CH_3COO^- + \tfrac{1}{4}H_2O$$ $+1.27$

Acetaldehyde
$$\tfrac{1}{5}CO_2(g) + H^+ + e^- = \tfrac{1}{10}CH_3CHO(g) + \tfrac{3}{10}H_2O$$ $+0.99$

Ethanol
$$\tfrac{1}{6}CO_2(g) + H^+ + e^- = \tfrac{1}{12}CH_3CH_2OH(aq) + \tfrac{1}{4}H_2O$$ $+1.52$

Ethane
$$\tfrac{1}{7}CO_2(g) + H^+ + e^- = \tfrac{1}{14}C_2H_6(g) + \tfrac{2}{7}H_2O$$ $+2.41$

Ethylene
$$\tfrac{1}{6}CO_2(g) + H^+ + e^- = \tfrac{1}{12}C_2H_4(g) + \tfrac{1}{3}H_2O$$ $+1.34$

Acetylene
$$\tfrac{1}{5}CO_2(g) + H^+ + e^- = \tfrac{1}{10}C_2H_2(g) + \tfrac{2}{5}H_2O$$ -0.86

Organic—C.3

Pyruvate$^-$
$$\tfrac{3}{10}CO_2(g) + \tfrac{9}{10}H^+ + e^- = \tfrac{1}{10}CH_3COCOO^- + \tfrac{3}{10}H_2O$$ $+0.05$

Lactate$^-$
$$\tfrac{1}{4}CO_2(g) + \tfrac{11}{12}H^+ + e^- = \tfrac{1}{12}CH_3CHOHCOO^- + \tfrac{1}{4}H_2O$$ $+0.68$

Glycerol
$$\tfrac{3}{14}CO_2(g) + H^+ + e^- = \tfrac{1}{14}CH_2OHCHOHCH_2OH + \tfrac{3}{14}H_2O$$ $+0.21$

Alanine
$$\tfrac{1}{4}CO_2(g) + \tfrac{1}{12}NH_4^+ + \tfrac{11}{12}H^+ + e^- = \tfrac{1}{12}CH_3HCOO^-NH_3^+ + \tfrac{1}{3}H_2O$$ $+0.84$

Organic—C.4

Succinate^{2-}
$$\tfrac{2}{7}CO_2(g) + \tfrac{6}{7}H^+ + e^- = \tfrac{1}{14}(CH_2COO^-)_2 + \tfrac{2}{7}H_2O$$ $+0.77$

Organic—C.6

Glucose
$$\tfrac{1}{4}CO_2(g) + H^+ + e^- = \tfrac{1}{24}C_6H_{12}O_6 + \tfrac{1}{4}H_2O$$ -0.20

TABLE 7.1 *(Continued)*

	$pe^o = \log K$

Halogens

$\frac{1}{2}Cl_2(g) + e^- = Cl^-$	$+23.0$
$\frac{1}{2}HClO + \frac{1}{2}H^+ + e^- = \frac{1}{2}Cl^- + \frac{1}{2}H_2O$	$+25.3$
(Note also $HClO = H^+ + ClO^-$; $\log K = -7.50$)	
$\frac{1}{5}IO_3^- + \frac{6}{5}H^+ + e^- = \frac{1}{10}I_2(s) + \frac{3}{5}H_2O$	$+20.1$
$\frac{1}{2}I_2(s) + e^- = I^-$	$+9.05$

Trace Metals

Cr

$\frac{1}{3}HCrO_4^- + \frac{7}{3}H^+ + e^- = \frac{1}{3}Cr^{3+} + \frac{4}{3}H_2O$	$+20.2$
(Note also $HCrO_4^- = \frac{1}{2}Cr_2O_7^{2-} + \frac{1}{2}H_2O$, $\log K = -1.5$;	
$HCrO_4^- = H^+ + CrO_4^{2-}$, $\log K = -6.5$; and various Cr(III)	
precipitation and coordination reactions.)	

Mn

$\frac{1}{5}MnO_4^- + \frac{8}{5}H^+ + e^- = \frac{1}{5}Mn^{2+} + \frac{4}{5}H_2O$	$+25.5$
$\frac{1}{2}MnO_2(s) + 2H^+ + e^- = \frac{1}{2}Mn^{2+} + H_2O$	$+20.8$

Fe

$Fe^{3+} + e^- = Fe^{2+}$	$+13.0$
$\frac{1}{2}Fe^{2+} + e^- = \frac{1}{2}Fe(s)$	-7.5
$\frac{1}{2}Fe_3O_4(s) + 4H^+ + e^- = \frac{3}{2}Fe^{2+} + 2H_2O$	$+16.6$

Co

$Co(OH)_3(s) + 3H^+ + e^- = Co^{2+} + 3H_2O$	$+29.5$
$\frac{1}{2}Co_3O_4(s) + 4H^+ + e^- = \frac{3}{2}Co^{2+} + 2H_2O$	$+31.4$

Cu

$Cu^{2+} + e^- = Cu^+$	$+2.6$
$\frac{1}{2}Cu^{2+} + e^- = \frac{1}{2}Cu(s)$	$+5.7$

Se

$\frac{1}{2}SeO_4^{2-} + 2H^+ + e^- = \frac{1}{2}H_2SeO_3 + \frac{1}{2}H_2O$	$+19.4$
$\frac{1}{4}H_2SeO_3 + H^+ + e^- = \frac{1}{4}Se(s) + \frac{3}{4}H_2O$	$+12.5$
$\frac{1}{2}Se(s) + H^+ + e^- = \frac{1}{2}H_2Se$	-6.7
(Note also $H_2Se = H^+ + HSe^-$, $\log K = -3.9$;	
$H_2SeO_3 = H^+ + HSeO_3^-$, $\log K = -2.4$;	
$HSeO_3^- = H^+ + SeO_3^{2-}$, $\log K = -7.9$;	
$SeO_4^{2-} + H^+ = HSeO_4^-$, $\log K = +1.7$)	

Ag

$AgCl(s) + e^- = Ag(s) + Cl^-$	$+3.76$
$Ag^+ + e^- = Ag(s)$	$+13.5$

Hg

$\frac{1}{2}Hg^{2+} + e^- = \frac{1}{2}Hg(l)$	$+14.4$
$Hg^{2+} + e^- = \frac{1}{2}Hg_2^{2+}$	$+15.4$

Pb

$\frac{1}{2}PbO_2 + 2H^+ + e^- = \frac{1}{2}Pb^{2+} + H_2O$	$+24.6$
(Note many other reactions for Mn, Fe, Co, Cu, Se,	
Ag, Hg, and Pb)	

Source: From references 2–6.

This is achieved in electrochemical cells by using the standard hydrogen electrode as reference.

1.5 Similarities and Differences between Redox and Acid–Base Reactions

It is often noted that there is a parallel between acid–base and redox reactions; in one case H^+ is being exchanged, in the other e^-:

$$HA \quad = \quad H^+ \quad + \quad A^- \tag{26}$$
$$\text{acid} (H^+ \text{ donor}) \quad \text{proton} \quad \text{base} (H^+ \text{ acceptor})$$

$$Fe^{2+} \quad = \quad e^- \quad + \quad Fe^{3+} \tag{27}$$
$$\text{reductant} (e^- \text{ donor}) \quad \text{electron} \quad \text{oxidant} (e^- \text{ acceptor})$$

This formal similarity between acid–base and redox processes is conceptually helpful and serves as a thread throughout this chapter, but three major differences must be kept in mind:

1. H^+, in some hydrated form, exists as a species in water (at least in the sense of balancing electrical charges) but e^- does not.
2. Acid–base reactions in solution are typically fast, whereas redox reactions are typically slow and mediated by organisms. Total equilibrium for redox processes in natural water results in meaningless—and lifeless—models.
3. Unlike those for acid–base reactions, equilibrium constants for redox reactions are often extremely large or small. The free energies involved are much larger in redox than in acid–base processes and, at equilibrium, redox reactions proceed to completion in one direction or the other.

These differences between acid–base and redox processes are partly an artifact of the way we have introduced the corresponding concepts and definitions. For convenience we have systematically ignored water as a reactant and failed to differentiate between protons and electrons and their corresponding hydrated species. Reactions that produce or consume protons (per se) are half acid–base reactions, strictly similar to half redox reactions, and they necessitate another half acid–base reaction to yield a complete acid–base reaction. In aqueous systems this matching half acid–base reaction can often be $H_2O + H^+ = H_3O^+$, a reaction that is conceptually and mechanistically important even if in our convention it only depicts a strict identity. In this sense, the definition of pH, like that of pe, is independent of the existence of the species H^+ ($= H_3O^+$) in solution. The concept of pH, and its measurement, depend on the activity ratio for a given acid–base couple, not on the H^+ concentration; pH and pe are intensive thermodynamic quantities, not extensive ones. The energetic difference we have noted above (3), results in part from our focus on weak acid–base reactions and the fact that many redox reactions involve the transfer of several electrons, while acid–base reactions typically involve the transfer of only one or two protons.

2. COMPARISON AMONG REDOX COUPLES

2.1 pe's of Dominant Redox Couples

Since it is a measure of potential energy, the electron activity that can be calculated for any half redox reaction is a measure of the "reducing power" of the corresponding redox couple in the system of interest. Low pe's [high $\{e^-\}$] correspond to highly reducing species, and high pe's [low $\{e^-\}$] to oxidizing ones. Consider the complete redox reaction:

$$Ox_1 + Red_2 = Red_1 + Ox_2$$

As seen in Equation 21, a necessary and sufficient condition of equilibrium between these two redox couples is given by the equality of their electron activities:

$$\Delta G = 0 = 2.3RTn\,(pe_2 - pe_1) \tag{28}$$

$$pe_1 = pe_2 \tag{29}$$

Suppose that initially

$$pe_1 > pe_2$$

therefore

$$pe_1 - pe_2 = pe_1^\circ - pe_2^\circ - \log\frac{[Red_1][Ox_2]}{[Ox_1][Red_2]} > 0$$

To achieve equilibrium the logarithmic term must thus increase; its numerator must increase and its denominator must decrease. The complete redox reaction thus proceeds to the right: Ox_1 oxidizes Red_2 to Ox_2 and Red_2 reduces Ox_1 to Red_1. This may perhaps be seen more readily by considering that the free energy change ΔG of the overall reaction (see Equation 28) is negative. When two redox couples have different electron activities in a given system the one with the lower pe tends to reduce the other and vice versa.

If one redox couple is present in much larger concentrations than the other (e.g., $[Ox_1]$, $[Red_1] \gg [Ox_2]$, $[Red_2]$), the corresponding free concentrations of oxidant and reductant $[Ox_1]$ and $[Red_1]$ are unaffected by the advancement of the complete redox reaction toward equilibrium. The equilibrium electron activity is then effectively that of the corresponding dominant redox couple:

$$pe = pe_1 = pe_1^o - \log \frac{[Red_1]}{[Ox_1]}$$

Redox potentials (electron activities) corresponding to redox couples that are present in relatively large concentrations in natural waters are then particularly important. All other redox couples tend to achieve the same electron activity. The pe of the dominant redox couple is effectively characteristic of the aquatic system itself and is often (sloppily) referred to as the "pe of the system."

Example 1. pe of the Organic Matter/CO$_2$ Couple   In much of this chapter it is convenient to use some stoichiometric and thermodynamic description of organic matter as if it were a well defined unique compound. For this purpose we utilize the symbol "CH$_2$O" with the thermodynamic properties of $\frac{1}{6}$ glucose. Using this "average" organic compound permits us to define rough stoichiometric and energetic relationships. (There is, of course, no substitute for knowing the actual nature of reacting organic species in a particular situation.)

Consider water at neutral pH, at equilibrium with the atmosphere, and containing $10^{-5} M$ dissolved organic carbon:

$$\tfrac{1}{4}CO_2(g) + H^+ + e^- = \tfrac{1}{4}\text{"CH}_2O\text{"} + \tfrac{1}{4}H_2O; \qquad pe^o = -0.20$$

$$pe = pe^o - \log \frac{[CH_2O]^{1/4}}{P_{CO_2}^{1/4}[H^+]} = -6.83 \tag{30}$$

Note that this pe is rather insensitive to the actual concentration of organic carbon and the partial pressure of CO_2. However, the very low pe of the organic matter/CO_2 couple is not the dominant pe in natural waters because the oxidizing couples O_2/H_2O and SO_4^{2-}/S^{2-} are usually more abundant; yet it may be considered characteristic of intracellular compartments in living organisms. In aquatic systems, the biota provides strongly reducing microenvironments which are the loci of most of the reducing activity in the system.

Example 2. pe of Anoxic Waters: The Sulfate/Sulfide Couple   Anoxic waters are often characterized by the presence of sulfide. When oxygen is exhausted, organisms use sulfate as the electron acceptor to oxidize organic matter. As sulfate is usually abundant in natural waters, the pe of the sulfate/sulfide couple, S(VI)/S(−II), can be considered characteristic of anoxic systems. There are, of course, a number of exceptions to this and one may find anoxic systems dominated by N(V)/N(III), N(III)/N(−III), Fe(III)/Fe(II), and other couples.

Consider a system at pH = 8 where $[HS^-] = 10^{-5} M$ and $[SO_4^{2-}] = 10^{-3} M$:

$$\tfrac{1}{8}SO_4^{2-} + \tfrac{9}{8}H^+ + e^- = \tfrac{1}{8}HS^- + \tfrac{1}{2}H_2O; \qquad pe^o = 4.25$$

$$pe = pe^o - \log \frac{[HS^-]^{1/8}}{[SO_4^{2-}]^{1/8}[H^+]^{9/8}} = -4.5 \tag{31}$$

This pe is very insensitive to the actual concentrations of sulfate and sulfide: a decrease of sulfide concentration by four orders of magnitude would only result in a 0.5 increase in pe. The presence of any measurable sulfide in a natural water near neutral pH yields pe's in the range -3.5 to -5.5. Complications introduced by the presence of sulfur species with intermediate oxidation states (between $+\text{VI}$ and $-\text{II}$) are discussed in Section 4.

Example 3. pe of Oxic Waters: The Oxygen/Water Couple   The major oxidant in oxic systems is oxygen. A straightforward measure of its tendency to oxidize other compounds is given by the pe of the oxygen/water couple.

Consider at $\text{pH} = 7$ and $P_{O_2} = 10^{-0.7}$ atm

$$\tfrac{1}{4}O_2(g) + H^+ + e^- = \tfrac{1}{2}H_2O; \qquad pe^\circ = 20.75$$

$$pe = pe^\circ - \log \frac{1}{P_{O_2}^{1/4}[H^+]} = 13.58 \tag{32}$$

Once again, this pe is insensitive to the partial pressure of oxygen; pe's in the range 12 to 14 are characteristic of oxic natural waters.

Example 4. Redox Equilibrium of Iron in Oxic Waters   From the redox potential (pe) of the dominant redox couple in a given aquatic system, one may easily calculate the equilibrium redox speciation of minor species. What is the ratio of $[Fe^{2+}]$ to $[Fe^{3+}]$ at equilibrium with the oxygen of the atmosphere and water at neutral pH? The pe of the oxygen/water couple has been calculated in Example 3:

$$pe_1 = 13.58$$

$$Fe^{3+} + e^- = Fe^{2+}; \qquad pe^\circ = 13.0 \tag{33}$$

and

$$pe_2 = 13.0 - \log \frac{[Fe^{2+}]}{[Fe^{3+}]}$$

Equating the pe's,

$$\frac{[Fe^{2+}]}{[Fe^{3+}]} = 10^{-0.58} = 0.26 \tag{34}$$

Note that this calculation is simply an indirect way to express the mass law of the complete redox reaction $[Fe^{2+} + \tfrac{1}{4}O_2(g) + H^+ = Fe^{3+} + \tfrac{1}{2}H_2O]$ for specified concentrations of the reactants O_2 and H^+. Note also that it provides the equilibrium ratio of the free concentrations of the ferrous and ferric ions, not the ratio of total Fe(II) and Fe(III) concentrations. In Section 3.1 we shall see how more complicated redox equilibrium problems can be organized in the usual tableau form.

2.2 Redox Reactions as Irreversible Reactions

In Section 2.1 we noted that the pe of several redox couples is rather insensitive to the actual concentrations of the species involved (except H^+). For example, the presence of any sulfide in a sulfate-containing system leads to a very narrow range of redox potentials. Likewise, there can be no sulfide at equilibrium with sulfate in an oxygen dominated system:

$$pe = 13.58 \qquad \text{(Example 3)}$$

therefore

$$\log \frac{[HS^-]^{1/8}}{[SO_4^{2-}]^{1/8}[H^+]^{9/8}} = 4.25 - 13.58 = -9.33$$

$$[HS^-] = 10^{-74.6}[SO_4^{2-}][H^+]^9 \tag{35}$$

Choosing $[SO_4^{2-}] \simeq 10^{-2.4}\, M$ and pH $= 7$, we obtain the bisulfide concentration:

$$[HS^-] = 10^{-140}\, M \tag{36}$$

The value of $10^{-140}\, M$ is a truly meaningless concentration. Consider that it would correspond to only a single molecule in 10^{116} L, a volume greater than that of the world oceans (1.4×10^{21} L) by a factor of 10^{95}.

If we consider the general complete redox reaction, $Ox_1 + Red_2 = Red_1 + Ox_2$, its direction of progress depends in principle on the relative concentrations of the various species. However, if we consider reasonable limits on the values that these concentrations can actually take, the direction of progress of the reaction can be a foregone conclusion if the equilibrium constant is sufficiently large or small. This is the case for many redox reactions in natural systems. For example, the reaction of sulfide with oxygen discussed above,

$$HS^- + 2O_2(g) = SO_4^{2-} + H^+$$

has an equilibrium constant $K \simeq 10^{+132}$.

Consider another example, one that is sadly tangible to car owners in the Northeast United States:

$$2Fe^0 + \tfrac{3}{2}O_2(g) = Fe_2O_3(s); \qquad K = 10^{130} \tag{37}$$

Under any partial pressure of oxygen greater than 10^{-87} atm, elemental iron tends to be oxidized. Again, the partial pressure 10^{-130} atm is truly meaningless (it only makes conceptual sense) and correct thermodynamic statements can be made in an absolute qualitative way:

"Oxygen oxidizes elemental iron."
"Elemental iron is only stable in oxygen-free systems."
"Iron naturally rusts."

Chemically, we can symbolize the reaction as being irreversible:

$$2Fe^0 + \tfrac{3}{2}O_2(g) \rightarrow Fe_2O_3$$

This absolute approach to redox chemistry, which is typified by the question, "What can oxidize or reduce what?" with no regard for the composition of the system, is in some ways contrary to the law of mass action and to the spirit of equilibrium thermodynamics, but it clearly makes practical sense.

The approach can be systematized by comparing equilibrium constants (pe^o) of half redox reactions. The oxidant in reactions with large pe^o "can" oxidize a reductant at a lower pe^o and vice versa. Reactants or products that are involved in these reactions, but are not the oxidant/reductant couples of interest, can be fixed at some typical values such as $[H^+] = 10^{-7} M$ and $[HCO_3^-] = 10^{-3} M$. By including these values in the constants, which are then denoted pe_w^o, we can make more realistic comparisons among redox couples.

Consider, for example, the sulfate/sulfide couple:

$$\tfrac{1}{8}SO_4^{2-} + \tfrac{9}{8}H^+ + e^- = \tfrac{1}{8}HS^- + \tfrac{1}{2}H_2O; \qquad pe^o = \log K = 4.25$$

$$pe = pe^o - \log \frac{[HS^-]^{1/8}}{[SO_4^{2-}]^{1/8}[H^+]^{9/8}}$$

$$= 4.25 - \tfrac{1}{8}\log \frac{[HS^-]}{[SO_4^{2-}]} + \tfrac{9}{8}\log[H^+]$$

We obtain a good approximation (pe_w^o) of this pe for all reasonable values of the sulfide to sulfate ratio by ignoring the second term and setting $[H^+] = 10^{-7}$:

$$pe_w^o = 4.25 - (\tfrac{9}{8})7 = -3.63 \qquad (38)$$

(More precisely, at pH $= 7$, we could write the reaction with half of the sulfide as H_2S and half as HS^-.)

Of course, this introduction of pe_w^o as yet another thermodynamic scale (a fifth one) to describe redox reactions adds to the confusion. Yet it is quite practical for examining the possibility or impossibility of various redox processes.

Consider Table 7.2 which is organized according to pe_w^o's from strong

TABLE 7.2 Effective Equilibrium Constants of Aquatic Redox Couples

	pe^o	$pe^o_w (pH = 7)$
$\frac{1}{2}H_2O_2 + H^+ + e^- = H_2O$	$+30.0$	$+23.0$
$\frac{1}{4}O_2(g) + H^+ + e^- = \frac{1}{2}H_2O$	$+20.75$	$+13.75$
$\frac{1}{5}NO_3^- + \frac{6}{5}H^+ + e^- = \frac{1}{10}N_2(g) + \frac{3}{5}H_2O$	$+21.05$	$+12.65$
$\frac{1}{2}MnO_2(s) + 2H^+ + e^- = \frac{1}{2}Mn^{2+} + H_2O$	$+20.8$	$+9.8^a$
$\frac{1}{2}NO_3^- + H^+ + e^- = \frac{1}{2}NO_2^- + \frac{1}{2}H_2O$	$+14.15$	$+7.15$
$\frac{1}{8}NO_3^- + \frac{5}{4}H^+ + e^- = \frac{1}{8}NH_4^+ + \frac{3}{8}H_2O$	$+14.9$	$+6.15$
$\frac{1}{6}NO_2^- + \frac{4}{3}H^+ + e^- = \frac{1}{6}NH_4^+ + \frac{1}{3}H_2O$	$+15.2$	$+5.8$
$\frac{1}{2}O_2(g) + H^+ + e^- = \frac{1}{2}H_2O_2$	$+11.5$	$+4.5$
$Fe(OH)_3(am) + 3H^+ + e^- = Fe^{2+} + 3H_2O$	$+16.0$	$+1.0^a$
$\frac{1}{6}SO_4^{2-} + \frac{4}{3}H^+ + e^- = \frac{1}{48}S_8(col) + \frac{2}{3}H_2O$	$+5.9$	-3.4
$\frac{1}{8}SO_4^{2-} + \frac{5}{4}H^+ + e^- = \frac{1}{8}H_2S(g) + \frac{1}{2}H_2O$	$+5.25$	-3.5
$\frac{1}{8}SO_4^{2-} + \frac{9}{8}H^+ + e^- = \frac{1}{8}HS^- + \frac{1}{2}H_2O$	$+4.25$	-3.6
$\frac{1}{8}HCO_3^- + \frac{9}{8}H^+ + e^- = \frac{1}{8}CH_4(g) + \frac{3}{8}H_2O$	$+3.8$	-4.0
$\frac{1}{8}CO_2(g) + H^+ + e^- = \frac{1}{8}CH_4(g) + \frac{1}{4}H_2O$	$+2.9$	-4.1
$\frac{1}{16}S_8(col) + H^+ + e^- = \frac{1}{2}H_2S(g)$	$+3.2$	-3.8^b
$\frac{1}{16}S_8(col) + \frac{1}{2}H^+ + e^- = \frac{1}{2}HS^-$	-0.8	-4.3
$\frac{1}{6}N_2(g) + \frac{4}{3}H^+ + e^- = \frac{1}{3}NH_4^+$	$+4.65$	-4.7
$H^+ + e^- = \frac{1}{2}H_2(g)$	0	-7.0
$\frac{1}{4}HCO_3^- + \frac{5}{4}H^+ + e^- = \frac{1}{4}"CH_2O" + \frac{1}{2}H_2O$	$+1.8$	-7.0
$\frac{1}{4}CO_2(g) + H^+ + e^- = \frac{1}{4}"CH_2O" + \frac{1}{4}H_2O$	-0.2	-7.2
$\frac{1}{2}HCO_3^- + \frac{3}{2}H^+ + e^- = \frac{1}{2}CO(g) + H_2O$	$+2.2$	-8.3
$\frac{1}{2}CO_2(g) + H^+ + e^- = \frac{1}{2}CO(g) + \frac{1}{2}H_2O$	-1.7	-8.7

^{a} In the reaction of reductive dissolution of metal oxides, the concentrations of the dissolved metals (Mn^{2+} and Fe^{2+}) have been fixed at 1 μM to more accurately reflect their relative redox properties.
^{b} This reaction is listed out of order so as not to separate it from the reaction of formation of the bisulfide ion HS$^-$.

oxidants at the top to strong reductants at the bottom. Looking at such a table one can, for example, decide at a glance that sulfate "can" oxidize carbon ("CH_2O") to form carbon dioxide and sulfide, but that sulfate "cannot" oxidize nitrite to nitrate.

2.3 Energetics of Microbial Processes

Thermodynamic considerations can do more than just provide criteria for judging the possible reactions in chemical systems. The energies consumed or liberated by changes in chemical composition can actually be estimated. It is in fact remarkable how much insight into the workings of microbial communities, which mediate most redox reactions in aquatic systems, can be gained from purely energetic considerations, often with very limited knowledge of the actual biochemical mechanisms involved.

According to Equation 21 the energy liberated or consumed by any complete redox reaction involving n electrons is directly calculated from the pe^o's of the half reactions:

$$Ox_1 + Red_2 = Red_1 + Ox_2; \qquad \log K = n(pe_1^o - pe_2^o) \qquad (39)$$

therefore

$$\Delta G = -2.3RT\, n(pe_1 - pe_2) = -2.3RTn\left[(pe_1^o - pe_2^o) - \log \frac{[Ox_2][Red_1]}{[Ox_1][Red_2]} \right]$$
$$(40)$$

As discussed in Section 2.2, the concentration-dependent term is often small and can be ignored in such calculations or replaced by a constant term (e.g., pH = 7) included in the effective parameter pe_w^o. An approximate value, ΔG_w^o, of the energy liberated or consumed by a complete redox reaction is then obtained from a simple formula:

$$\Delta G_w^o \simeq -2.3RTn(pe_{w1}^o - pe_{w2}^o) \qquad (41)$$

therefore, at 25°C

$$\Delta G_w^o \simeq -5.70n(pe_{w1}^o - pe_{w2}^o)\, kJ\, mol^{-1} \qquad (42)$$

This last formula has been applied to some familiar microbial processes* listed in Table 7.3. (Table 7.2 provides the thermodynamic data used in the calculations.) The reactions of Table 7.3 correspond to very idealized processes and the calculated energies represent either the maximum energy available from each reaction or the minimum energy necessary to carry out the reaction. Inefficiencies in the energy transfer mechanisms and various energies necessary for maintenance of cellular integrity, reproduction, or ion transport are not accounted for. For example, about 1200 kJ of chemical energy are actually expended by plants to fix one mole of inorganic carbon (even more light energy must be absorbed) and only 170 kJ of useful energy can be obtained from oxidizing $\frac{1}{6}$ glucose to CO_2 with oxygen, rather than the standard value of 478 kJ.[7] Energetic efficiencies of the order of 30–40% are fairly typical of biochemical processes.

Microorganisms can be classified according to their principal energy source, their principal carbon source, and their electron source[8] (see Figure 7.1). All three can be separate, as is the case for plants which absorb light, fix CO_2, and oxidize water to oxygen; all three can be the same, as exemplified by many bacteria

---

*A majority of aquatic plants are phytoplankton, which can justifiably be classified as "microbes." Thus we include all photosynthetic reactions in our list of "microbial processes."

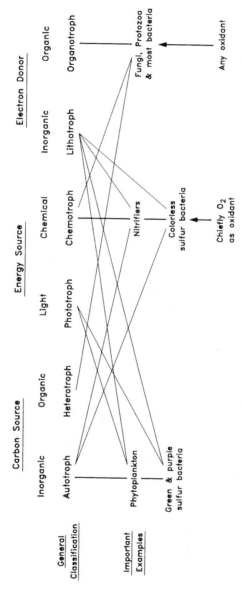

Figure 7.1 Metabolic classification of organisms.

TABLE 7.3 Energetics of Microbial Processes

Fixation of Elements (Organic Synthesis)

Carbon Fixation (Autotrophs)

A $\frac{1}{4}CO_2(g) + \frac{1}{4}H_2O = \frac{1}{4}``CH_2O'' + \frac{1}{4}O_2(g)$; $\Delta G_w^o = +119\,kJ\,mol^{-1}$

B $\frac{1}{4}CO_2(g) + \frac{1}{2}H_2S(g) = \frac{1}{4}``CH_2O'' + \frac{1}{16}S_8(col) + \frac{1}{4}H_2O$; $\Delta G_w^o = +19.4\,kJ\,mol^{-1}$

C $\frac{1}{4}CO_2(g) + \frac{1}{6}NH_4^+ + \frac{1}{12}H_2O = \frac{1}{4}``CH_2O'' + \frac{1}{6}NO_2^- + \frac{1}{3}H^+$; $\Delta G_w^o = +74.1\,kJ\,mol^{-1}$

Nitrogen Fixation (Nitrogen Fixers)

D $\frac{1}{6}N_2(g) + \frac{1}{3}H^+ + \frac{1}{4}``CH_2O'' + \frac{1}{4}H_2O = \frac{1}{3}NH_4^+ + \frac{1}{4}CO_2(g)$; $\Delta G_w^o = -14.3\,kJ\,mol^{-1}$

Nitrate Uptake and Reduction

E $\frac{1}{8}NO_3^- + \frac{1}{4}H^+ + \frac{1}{4}``CH_2O'' = \frac{1}{8}NH_4^+ + \frac{1}{4}CO_2(g) + \frac{1}{8}H_2O$; $\Delta G_w^o = -76.1\,kJ\,mol^{-1}$

Energy Sources

Light (Phototrophs; Mostly Autotrophs)

Energy per mole of photons of wave length $\lambda(nm) = 1.20 \times 10^5 \lambda^{-1}\,kJ\,mol^{-1}$

Oxidation of Reduced Inorganic Compounds (Chemolithotrophs, Mostly Autotrophs)

Nitrification

F $\frac{1}{6}NH_4^+ + \frac{1}{4}O_2(g) = \frac{1}{6}NO_2^- + \frac{1}{3}H^+ + \frac{1}{6}H_2O$; $\Delta G_w^o = -45.3\,kJ\,mol^{-1}$

G $\frac{1}{2}NO_2^- + \frac{1}{4}O_2(g) = \frac{1}{2}NO_3^-$; $\Delta G_w^o = -37.6\,kJ\,mol^{-1}$

Sulfide oxidation

H $\frac{1}{8}$"$H_2S(g) + \frac{1}{4}O_2(g) = \frac{1}{8}SO_4^{2-} + \frac{1}{4}H^+$; $\Delta G_w^o = -98.3\,\text{kJ}\,\text{mol}^{-1}$

Iron oxidation

I $Fe^{2+} + \frac{1}{4}O_2(g) + 5H_2O = Fe(OH)_3(s) + 2H^+$; $\Delta G_w^o = -106.8\,\text{kJ}\,\text{mol}^{-1}$

Hydrogen oxidation

J $\frac{1}{2}H_2(g) + \frac{1}{4}O_2(g) = \frac{1}{2}H_2O$; $\Delta G_w^o = -118\,\text{kJ}\,\text{mol}^{-1}$

Oxidation of Organic Compounds (Chemoorganotrophs, All Heterotrophs)

Aerobic respiration

K $\frac{1}{4}$"CH_2O" $+ \frac{1}{4}O_2(g) = \frac{1}{4}CO_2(g) + \frac{1}{4}H_2O$; $\Delta G_w^o = -119\,\text{kJ}\,\text{mol}^{-1}$

Denitrification

L $\frac{1}{4}$"CH_2O" $+ \frac{1}{5}NO_3^- + \frac{1}{5}H^+ = \frac{1}{4}CO_2(g) + \frac{1}{10}N_2(g) + \frac{7}{20}H_2O$; $\Delta G_w^o = -113\,\text{kJ}\,\text{mol}^{-1}$

Manganese reduction[a]

M $\frac{1}{4}$"CH_2O" $+ \frac{1}{2}MnO_2(s) + H^+ = \frac{1}{4}CO_2(g) + \frac{1}{2}Mn^{2+} + \frac{3}{4}H_2O$; $\Delta G_w^o = -96.9\,\text{kJ}\,\text{mol}^{-1}$

Iron reduction[a]

N $\frac{1}{4}$"CH_2O" $+ Fe(OH)_3(am) + 2H^+ = \frac{1}{4}CO_2(g) + Fe^{2+} + \frac{11}{4}H_2O$; $\Delta G_w^o = -46.7\,\text{kJ}\,\text{mol}^{-1}$

Sulfate reduction

O $\frac{1}{4}$"CH_2O" $+ \frac{1}{8}SO_4^{2-} + \frac{1}{8}H^+ = \frac{1}{4}CO_2(g) + \frac{1}{8}HS^- + \frac{1}{4}H_2O$; $\Delta G_w^o = -20.5\,\text{kJ}\,\text{mol}^{-1}$

Methane formation

P $\frac{1}{4}$"CH_2O" $= \frac{1}{8}CO_2(g) + \frac{1}{8}CH_4(g)$; $\Delta G_w^o = -17.7\,\text{kJ}\,\text{mol}^{-1}$

Hydrogen fermentation

Q $\frac{1}{4}$"CH_2O" $+ \frac{1}{4}H_2O = \frac{1}{4}CO_2(g) + \frac{1}{2}H_2(g)$; $\Delta G_w^o = -1.1\,\text{kJ}\,\text{mol}^{-1}$

[a]Note that the free energies of these reactions are calculated for $[Mn^{2+}] = [Fe^{2+}] = 1\,\mu M$.

that use organic compounds (such as glucose) as a source of energy, carbon, and electrons.

The first section of Table 7.3 shows some reactions by which inorganic species are incorporated into organic compounds. Carbon fixation can be achieved with a variety of electron donors, such as water, which is used by plants in aerobic systems, or reduced sulfur or nitrogen compounds, which are used by photosynthetic and chemosynthetic autotrophic bacteria. Incorporation of nitrogen into organic compounds necessitates reduction to the level of ammonia, followed by synthesis of amino acids, the latter reaction not being included in the table. Although organic carbon is shown as the electron donor (and the source of energy) for nitrogen reduction, in fact the process can be linked directly to light absorption, which then provides some of the necessary energy, while inorganic compounds, including H_2O, contribute some of the electrons for the overall reduction.

Organisms that use light as their energy source (phototrophs) also typically fix inorganic carbon (autotrophs). Phytoplankton and cyanobacteria (blue green algae), which live at the surface where light is plentiful, evolve oxygen from water as shown in Reaction A. Green and purple sulfur bacteria, which live in the bottom anaerobic zone of lakes, must make do with less light but they take advantage of reduced sulfur compounds to reduce CO_2 (Reaction B).

Bacteria that derive their energy from oxidizing reduced inorganic compounds (chemotrophs) are usually aerobes, and use oxygen to carry out the oxidation reaction. In some cases nitrate can be substituted as the oxidant. These bacteria are mostly autotrophs, fixing their own organic carbon from CO_2. For this purpose they use as electron donors the same inorganic compounds that serve as reductants for their energy source (Reactions F and C). Although few in number, these bacteria are important in the geochemical cycle of elements, particularly nitrogen and sulfur. Nitrifying bacteria oxidize ammonium to nitrite (F) and nitrate (G) leading to the regeneration of dinitrogen, $N_2(g)$, through denitrification (Reaction L). Chemosynthetic bacteria that oxidize sulfide with oxygen are called colorless sulfur bacteria to distinguish them from the photosynthetic green and purple sulfur bacteria. They can use a variety of reduced sulfur compounds (H_2S, $S_2O_3^{2-}$, S_8^0) and may produce a number of intermediate redox sulfur species. Other chemosynthetic bacteria oxidize iron, manganese, and hydrogen, although it is not certain that useful biochemical energy is obtained from Fe and Mn oxidation.

By far the largest number of microorganisms use organic compounds as their energy source (chemotrophs), as carbon source (heterotrophs), and as electron source (organotrophs). These organisms include fungi, protozoa, and most species of bacteria. Almost any oxidant (e.g., oxygen, nitrate, manganese, iron, or sulfate; Reactions K–O in Table 7.3) can be used for these "chemoorgano- trophs" to oxidize carbon. In fermentation processes (Reactions P and Q), organic carbon itself or water can serve as the oxidant. The oxidation of organic carbon rarely proceeds directly to the final end products shown in Table 7.3. In some cases the oxidant is reduced to an intermediate species such as N_2O or S_8^0. More importantly, intermediate organic oxidation products are usually formed and utilized by specialized microorganisms. For example, Table 7.4 presents complete redox reactions for three intermediate carbon compounds, acetate, propionate, and butyrate, which are formed and utilized by a consortium of bacteria during the anaerobic fermentation of glucose to methane.[9] Each of the reactions in the table is carried out by different bacterial species. Not all bacteria are so specialized, however. Some exhibit great versatility in both their organic substrate and their electron acceptors.

As seen in Table 7.4, the oxidation reactions of the various organic substrates involved in methane fermentation yield widely different free energies. The "average" organic formula "CH_2O," symbolizing $\frac{1}{6}$glucose, thus would not provide in this case an adequate representation of the thermodynamics of the process. This is true of all microbial processes; for example, the oxidation of glucose by any of the electron acceptors in Table 7.3 yields about 54 kJ more per carbon than a similar oxidation of acetate. This difference is of little significance (12%) when oxygen is the electron acceptor. However, when sulfate is the oxidant, this difference corresponds to a relatively large decrease (69%) in the energy made available to the organism.

This is shown for several organic compounds and several potential oxidants in Figure 7.2: the free energies released by all organic compounds (normalized to one electron) are within a few percent of each other when the oxidant is strong [i.e., O_2 yields $\simeq 119\,kJ\,(mol\,e^-)^{-1}$], but their relative values decrease markedly, from formate to methane, when the oxidant is weak (i.e., SO_4^{2-}). The use of an average organic compound "CH_2O" $= \frac{1}{6}$glucose is thus appropriate when dealing with well-oxidized systems, but it can lead to relatively large errors in calculating free energies in reduced systems. Compare, for example, the energies released in the fermentation of methane from acetate and from glucose (Figure 7.2). In the first case, the energy released is only about $4\,kJ$ $(mol\,e^-)^{-1}$; in the second it is about four times larger.

Although thermodynamic calculations can only provide a very rough estimate of microbial energetics (and can be quite misleading), they do give insight into the workings of aquatic communities without requiring detailed understanding of the physiological processes.

Example 5. Inorganic Carbon Fixation by Nitrifiers Consider the chemoautotrophic bacteria of the genera *Nitrosomonas* and *Nitrobacter* which are specialized in oxidizing ammonium to nitrite and nitrite to nitrate, respectively. What is the minimum number of ammonium or nitrite molecules that must be oxidized by these organisms in order to fix 1 molecule of CO_2 into organic carbon? The answer is found by examining the energetics of the processes. From the data of Table 7.1 (and Table 7.3) we can write the appropriate oxidation and carbon fixation reactions:

TABLE 7.4 Intermediary Reactions during Methane Fermentation from Glucose

	ΔG_w° (kJ)	$\% \Delta G_w^\circ$ of Overall Total
1 Partial fermentation of glucose	−11.5	64.3
$0.042\,C_6H_{12}O_6 + 0.072\,HCO_3^- = 0.044\,CH_3COO^- + 0.016\,CH_3CH_2COO^-$		
$\qquad + 0.012\,CH_3CH_2CH_2COO^- + 0.096\,H_2$		
$\qquad + 0.139\,CO_2 + 0.043\,H_2O$		
2 Butyrate fermentation	−0.31	1.8
$0.012\,CH_3CH_2CH_2COO^- + 0.012\,HCO_3^- = 0.024\,CH_3COO^- + 0.006\,CH_4 + 0.006\,CO_2$		
3 Propionate fermentation	−0.39	2.2
$0.016\,CH_3CH_2COO^- + 0.008\,H_2O = 0.0016\,CH_3COO^- + 0.012\,CH_4 + 0.004\,CO_2$		
4 Acetate Fermentation	−2.52	14.1
$0.084\,CH_3COO^- + 0.0084\,H_2O = 0.084\,CH_4 + 0.084\,HCO_3^-$		
5 Hydrogen oxidation	−3.14	17.6
$0.096\,H_2 + 0.024\,CO_2 = 0.024\,CH_4 + 0.048\,H_2O$		
Overall $0.042\,C_6H_{12}O_6 = 0.126\,CH_4 + 0.126\,CO_2$	−17.18	100.0

Source: After McCarty 1972.[6]

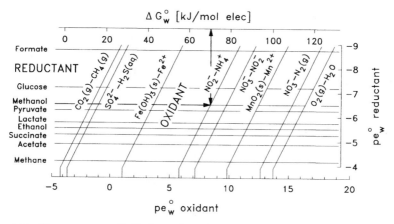

Figure 7.2   Free energy available from the oxidation of various organic substrates with various electron acceptors. The free energy released by a complete redox reaction (per mole of electrons transferred) is obtained from the intersection of the horizontal line corresponding to the reductant with the oblique line corresponding to the oxidant, as illustrated for the oxidation of methanol by nitrite. Data from Tables 7.1 and 7.2. Adapted from McCarty, 1972.[9] (N.B. $[Mn^{2+}] = [Fe^{2+}] = 1\,\mu M$).

Nitrosomonas   Ammonium oxidation:

$$\tfrac{1}{6}NH_4^+ + \tfrac{1}{4}O_2(g) = \tfrac{1}{6}NO_2^- + \tfrac{1}{3}H^+ + \tfrac{1}{6}H_2O$$

$$\Delta G_w^o = -45.3\,kJ\,mol^{-1} \tag{43}$$

$$\Delta G_1 = \text{energy available per N oxidized}$$

$$= 45.3 \times 6 = 272\,kJ$$

Carbon fixation with NH_4^+ as reductant:

$$\tfrac{1}{4}CO_2 + \tfrac{1}{6}NH_4^+ + \tfrac{1}{12}H_2O = \tfrac{1}{4}\text{``}CH_2O\text{''} + \tfrac{1}{6}NO_2^- + \tfrac{1}{3}H^+$$

$$\Delta G_w^o = +74.1\,kJ\,mol^{-1} \tag{44}$$

$$\Delta G_2 = \text{energy necessary per}$$
$$\text{inorganic C fixed}$$

$$= 74.1 \times 4 = 296\,kJ$$

$$\Delta G_2/\Delta G_1 \simeq 1.1$$

Thus if *Nitrosomonas* were perfectly efficient, they would be able to fix almost one CO_2 for each NH_4^+ oxidized.

Nitrobacter Nitrite oxidation:

$$\tfrac{1}{2}NO_2^- + \tfrac{1}{4}O_2(g) = \tfrac{1}{2}NO_3^-$$

$$\Delta G_w^o = -37.6\,kJ\,mol^{-1} \tag{45}$$

$$\Delta G_1 = 37.6 \times 2 = 75.2\,kJ$$

Carbon fixation with NO_2^- as reductant:

$$\tfrac{1}{4}CO_2 + \tfrac{1}{2}NO_2^- + \tfrac{1}{4}H_2O = \tfrac{1}{4}\text{“}CH_2O\text{”} + \tfrac{1}{2}NO_3^-$$

$$\Delta G_w^o = +81.8\,kJ\,mol^{-1} \tag{46}$$

$$\Delta G_2 = 81.8 \times 4 = 327\,kcal$$

$$\Delta G_2/\Delta G_1 \simeq 4.3$$

Plainly, *Nitrobacter* must work much harder than *Nitrosomonas* and must oxidize more than four NO_2^- for each CO_2 fixed.

Experimentally it is found that various strains of *Nitrosomonas* oxidize approximately 14–70 moles of nitrogen for every carbon assimilated; in *Nitrobacter* this ratio varies from 76 to 135.[10] For both types of organisms, the optimal energetic efficiency thus appears to be on the order of 5–10%, perhaps not an unexpected value considering the coupling of two biochemical pathways, each with a total energetic efficiency on the order of 30% or less.

On the basis of simple energetic calculations it is also possible to estimate the relative biomass yield from alternative heterotrophic processes or to decide which of two competing processes is more energetically favorable and hence more ecologically probable.

Example 6. Relative Biomass Yields from Aerobic and Denitrifying Bacteria
Consider two heterotrophic microbial populations, one composed of aerobes and the other one of denitrifiers, utilizing the same organic carbon source, "CH_2O." It is found from experiments that 80% of the organic carbon source is accounted for in the aerobe biomass. What is the expected yield for the denitrifiers? Are denitrifiers likely to outcompete aerobes in a mixed culture?

We assume that 20% of the organic carbon is utilized by the aerobes for various energetic sustenance and growth requirements: for every four organic carbons assimilated by the aerobes, one is burned, yielding 478 kJ. If the denitrifiers have approximately the same energetic requirement and the same efficiency as the aerobes, for every four organic carbons they assimilate they must expend 478 kJ and thus oxidize $478/(114 \times 4) = 1.05$ moles of organic carbon. The relative yield of the denitrifiers is thus $4/5.05 \simeq 79\%$, hardly distinguishable from the 80% of the aerobes. Energetically, both microbial populations are roughly equivalent and equally likely to succeed. This is because oxygen and nitrate are approximately equally good oxidants.

Note, however, that this result is pH dependent. For every increase by one pH unit above 7.0, the energy yield for the denitrifiers decreases by $5.69 \times 4/5 \simeq$ $4.6 \, \text{kJ mol}^{-1}$ of carbon oxidized and aerobic bacteria are thus increasingly favored energetically at high pH. Conversely, denitrifiers can actually have an energetic advantage at low pH (< 3.5).

Example 7. Competition among Sulfate Reducers and Methane Fermenters
The importance of the concentration dependent term in the free energy expression is exemplified by the question of competition between sulfate reducing and methanogenic bacteria.[11] From the data of Table 7.3, the sulfate reducers seem to obtain more energy per unit of substrate oxidized than the methane fermenters $[\Delta G_w^o = 20.5 \text{ vs. } 17.7 \, \text{kJ (mol e}^-)^{-1}]$. It is thus often assumed that the sulfate reducers should outcompete the methane fermenters as long as sulfate is available in the system. This result is of course unaffected by the exact nature of the organic substrate as long as suitable organisms of both types can use it. However, the two calculated energies are fairly similar and it is necessary to consider more precisely the actual chemistry of the system. Both processes yield the same energy when the reaction of oxidation of methane by sulfate is at equilibrium (substract one reaction from other; at equilibrium, $\Delta G = 0$, and thus $\Delta G_1 - \Delta G_2 = 0$):

$$\tfrac{1}{8}CH_4(g) + \tfrac{1}{8}SO_4^{2-} + \tfrac{1}{8}H^+ = \tfrac{1}{8}HS^- + \tfrac{1}{4}H_2O + \tfrac{1}{8}CO_2(g) \tag{47}$$

and from the appropriate half reactions in Table 7.1:

$$\log K = 4.25 - 2.86 = 1.39$$

The equilibrium relation

$$\frac{[HS^-]P_{CO_2}}{P_{CH_4}[SO_4^{2-}][H^+]} = 10^{11.1}$$

then describes the conditions under which both types of organisms can derive the same energy from the same organic substrate, (e.g., $[HS^-] \simeq 10^{-3} \, M$, $P_{CO_2} = 10^{-2} \, \text{atm}$; $P_{CH_4} = 10^{-6} \, \text{atm}$, $[SO_4^{2-}] = 10^{-2} \, M$, $[H^+] = 10^{-8} \, M$). Low sulfate, methane, and hydrogen ion concentrations concurrent with high sulfide and carbon dioxide concentrations will tend to favor methanogenesis over sulfate reduction and vice versa. Under some conditions, organisms can in fact derive energy from Reaction 47, oxidizing methane with sulfate; under other conditions the reaction is energy consuming.

* * *

There are many examples of ecological succession among microbes that can be rationalized on the basis of the redox functions of the various organisms: stepwise oxidation of reduced compounds promotes the sequential growth of specialized genera, as does the gradual exhaustion of the various oxidants. A case in point is the sequence of events leading to the formation of anoxic hypolimnetic waters: assemblages of aerobic organisms, denitrifiers, sulfate reducers, and methanogenic bacteria flourish one at a time as O_2, NO_3^-, and SO_4^{2-} are sequentially exhausted from the system. In soils and sediments where the oxides of manganese and iron are relatively abundant, manganese and iron reducers also appear after the disappearance of nitrate. Such a sequence is directly predicted by the relative energies of the corresponding oxidation reactions; as shown in Table 7.2 the pe_w^o's decrease in the order O_2/H_2O, NO_3^-/N_2, $MnO_2(s)/Mn^{2+}$, $Fe(OH)_3(s)/Fe^{2+}$, SO_4^{2-}/HS^-, and CH_2O/CH_4. The reason why the various organisms do not bloom simultaneously on the same organic substrate is not simply one of relative energetic advantage, however. In many cases there are specific biochemical processes that make it impossible for the organisms to coexist. For example, sulfate reduction is biochemically impossible in the presence of even traces of oxygen; thus aerobes and sulfate reducers must necessarily succeed each other.

From a macroscopic point of view, the energy fixed into reduced carbon by the initial process of photosynthesis is gradually utilized by the living biota in a multitude of serial and parallel interlocking processes. Step by step the energy is degraded into heat, except for a small amount of storage as refractory organic compounds and reduced minerals.

From a microscopic point of view, the energetic couplings among the various redox reactions are carried out by organisms using a few intracellular energy transfer compounds. Several of these compounds, NADP, NAD, and ferredoxin, transfer energy by undergoing changes in oxidation state and coupling with other appropriate redox couples to carry out complete redox reactions (see Table 7.5). However the principal compound for energy transfer in cells, ATP, does not undergo a redox reaction. Instead of an electron exchange, ATP transfers energy by exchanging phosphate groups with metabolic intermediates. In addition to the energetic scale provided by half redox reactions, biochemists thus also define an energetic scale by classifying "half" phosphorylation–dephosphorylation reactions in which inorganic phosphate is taken as a reactant.

A somewhat more mechanistic and biochemical approach to microbially mediated redox processes can be taken by coupling aquatic redox couples with the proper intracellular redox couples. For example, the light and dark reactions of photosynthesis can thus be separated:

Light reaction

$$2H_2O + 2NADP^+ \longrightarrow O_2 + 2H^+ + 2NADPH$$
$$\Delta G_w^o = +439 \, kJ \, mol^{-1} \tag{48}$$

Dark reaction

$$2H^+ + 2NADPH + CO_2 \longrightarrow 2NADP^+ + H_2O + CH_2O$$
$$\Delta G_w^o = +38.8 \, kJ \, mol^{-1} \tag{49}$$

TABLE 7.5   Some Cellular Energy Transfer Reactions

Half Redox Reactions (Reduction)	pe_w^o
$NAD^+ + 2H^+ + 2e^- = NADH + H^+$	-5.4
$NADP^+ + 2H^+ + 2e^- = NADPH + H^+$	-5.5
$2 \text{ Ferredoxin(Ox)} + 2e^- = 2 \text{ ferredoxin (Red)}$	-7.1
$\text{Ubiquinone} + 2H^+ + 2e^- = \text{ubiquinol}$	$+1.7$
$2 \text{ Cytochrome C (Ox)} + 2e^- = 2 \text{ cytochrome C (Red)}$	$+4.3$

Half Phosphate Exchange Reactions (Hydrolysis)	$\Delta G_w^o \text{ (kJ mol}^{-1})$
(Pi = inorganic phosphate)	
Phosphoenol pyruvate = pyruvate + Pi	-61.9
Phosphocreatinine = creatinine + Pi	-43.1
Acetylphosphate = acetate + Pi	-42.3
Adenosine triphosphate (ATP) = ADP + Pi	
$\quad$ 37°C, pH = 7.0, excess Mg^{2+}	-31
$\quad$ 25°C, pH = 7.4, $10^{-3} M$ Mg^{2+}	-37
$\quad$ 25°C, pH = 7.4, no Mg^{2+}	-40
Adenosine diphosphate (ADP) = AMP + Pi	-31
Glucose-1-phosphate = glucose + Pi	-21
Glucose-6-phosphate = glucose + Pi	-14
Glycerol-1-phosphate = glycerol + Pi	-9.2

The stoichiometry of these reactions is not exact and both reactions also involve ATP. Altogether the reduction of one carbon is found to require 2.2NADPH and 3ATP, plus 2ATP for auxiliary processes. In bacterial photosynthetic reactions where a reductant other than water is used as the electron donor, the light reaction requires much less energy:

$$2H_2S + 2NADP^+ \longrightarrow 2S^0 + 2H^+ + 2NADPH; \qquad \Delta G_w^o = +38.8 \text{ kJ mol}^{-1}$$

but the dark reaction is, of course, the same. Note that energetic calculations involving intracellular compounds are very much affected by the precise chemical composition of intracellular compartments. For example, ATP, NADPH, and NADH have weak acid–base properties and coordinate strongly with major cations (Table 7.5). Maintenance of H^+, Ca^{2+}, and Mg^{2+} gradients can be used to favor particular reactions, and ion "pumps" are an integral part of the energy transfer machinery of cells. Note also that the formation of high energy compounds such as NADPH and ATP during the light reaction can drive energy-consuming reactions other than the reduction of inorganic carbon. Rather than depending exclusively on the utilization (respiration) of carbohydrates, many cellular processes in plants are driven in whole or in part by direct coupling to the light reaction. This is, for example, the case for nitrate reduction, the energy for which is derived only in part from dark respiration, the symbolism of Table 7.3 notwithstanding.

3. PARTIAL REDOX EQUILIBRIUM CALCULATIONS:
pe CALCULATION IN COMPLEX SYSTEMS

Although many redox reactions in natural waters do not reach equilibrium, there are situations in which several redox couples are roughly at equilibrium with each other. To perform the corresponding equilibrium calculations, it is not always possible to consider that the pe is fixed by the "major" redox couple and simply to equate the pe of the "minor" couples as shown in Example 4. If the redox potential (pe) of the system results from a balance between several redox couples at similar concentrations, the objective of the calculation is in fact to obtain the equilibrium pe by considering the various redox equilibria simultaneously. We wish to examine here how this is done with the general methodology of Chapter 1, using tableaux and choosing appropriate components. For this purpose we use as an example the common situation of oxygen depletion in hypolimnetic waters.

3.1 Definition of the Problem and Choice of Components

Example 8. Anoxia in the Hypolimnion of a Lake   Consider the composition of the hypolimnion of a lake that becomes anoxic as a result of the decomposition of $0.5\,mM$ of organic carbon:

Recipe
$$\left.\begin{array}{l} [SO_4]_T = 2 \times 10^{-4}\,M \\ [Fe(III)]_T = 10^{-5}\,M \\ Alk = 5 \times 10^{-4}\,eq/L \\ [CO_2]_T = 10^{-3}\,M \\ [CH_2O]_T = 5 \times 10^{-4}\,M \\ [O_2]_T = 3 \times 10^{-4}\,M \end{array}\right\} \text{ before any redox reaction pH = 6.3}$$

As discussed previously, there is little point in considering the CH_2O/CO_2 and O_2/H_2O equilibria. If thermodynamics prevail, the reaction

$$CH_2O + O_2 \longrightarrow CO_2 + H_2O \tag{50}$$

must exhaust the oxygen of the system, and the reaction

$$CH_2O + \tfrac{1}{2}SO_4^{2-} + H^+ \longrightarrow CO_2 + \tfrac{1}{2}H_2S + H_2O \tag{51}$$

must proceed until elimination of the organic substrate (we do not consider here intermediate carbon or sulfur redox species). These reactions are important to define the mole balance constraints on the system but they do not result in interesting equilibrium relationships. Were we to write the mass law equations for the corresponding half redox reactions, we would end up calculating ridiculously small concentrations for CH_2O and O_2.

As is typical of such systems, the redox reactions for which we want to consider redox equilibrium are the $S(+VI)/S(-II)$ and the $Fe(III)/Fe(II)$ couples:

$$\tfrac{1}{8}SO_4^{2-} + \tfrac{5}{4}H^+ + e^- = \tfrac{1}{8}H_2S + \tfrac{1}{2}H_2O; \qquad pe^o = 5.12 \qquad (52)$$

$$Fe^{3+} + e^- = Fe^{2+}; \qquad\qquad pe^o = 13.0 \qquad (53)$$

The list of interesting species includes:

1. H^+ and OH^-
2. The usual carbonate species: $H_2CO_3^*$, HCO_3^-, CO_3^{2-}
3. SO_4^{2-} *and* H_2S, HS^-, S^{2-}
4. Fe^{3+}, $FeOH^{2+}$, $Fe(OH)_2^+$, $Fe(OH)_4^-$, *and* Fe^{2+}

There is also a possibility of precipitating $Fe(OH)_3(s)$, $FeCO_3(s)$, $Fe(OH)_2(s)$, or $FeS(s)$. Let us consider no solid initially.

Note We have not included O_2 as a species because this system contains an excess of reduced compounds (compare $[O_2]_T$ and $[CH_2O]_T$) and O_2 will be effectively exhausted. If, however, we had an excess of oxygen, it would be necessary to include O_2 as a species.

From the earlier discussion in this chapter, we include the electron in our choice of components for such a system. Without a good basis on which to decide what the principal components are, let us choose arbitrarily as a component set: H^+, e^-, $H_2CO_3^*$, SO_4^{2-}, and Fe^{2+} (see Tableau 7.1). The mole balance equations for H^+ and e^- are

$$\begin{aligned}
TOTH &= [H^+] - [OH^-] - [HCO_3^-] - 2[CO_3^{2-}] + 10[H_2S] + 9[HS^-] \\
&\quad + 8[S^{2-}] - x[Fe(OH)_x^{(3-x)+}] \\
&= -Alk + 4[CH_2O]_T - 4[O_2]_T = 3 \times 10^{-4}\,M \qquad (54)
\end{aligned}$$

$$\begin{aligned}
TOTe &= 8[H_2S] + 8[HS^-] + 8[S^{2-}] - [Fe^{3+}] - [Fe(OH)_x^{(3-x)+}] \\
&= -[Fe(III)]_T + 4[CH_2O]_T - 4[O_2]_T = 7.9 \times 10^{-4}\,M \qquad (55)
\end{aligned}$$

On inspection, it is clear that these two equations are not the most convenient or elegant ones that can be written. For example, taking their difference leads to an equation that does not contain the high stoichiometric coefficients for all the sulfide species nor the $(4[CH_2O]_T - 4[O_2]_T)$ term which dominates in both equations. This new equation would be obtained directly if H_2S, rather than e^-, were chosen as a component. In effect, what we are discovering here is an extension of our previous rules for choosing principal components: *instead of the electron, the principal components of the system include both the reduced and the oxidized species of the major redox couple.* The choice of the electron as a component, though conceptually pleasing and yielding straightforward stoichiometric expressions for all species, is numerically awkward. When the species of

TABLEAU 7.1

	H^+	e^-	$H_2CO_3^*$	SO_4^{2-}	Fe^{2+}	
Species						
H^+	1					
OH^-	-1					
$H_2CO_3^*$			1			
HCO_3^-	-1		1			
CO_3^{2-}	-2		1			
SO_4^{2-}				1		
H_2S	10	8		1		
HS^-	9	8		1		
S^{2-}	8	8		1		
Fe^{3+}		-1			1	
$Fe(OH)_x^{(3-x)+}$	$-x$	-1			1	
Fe^{2+}					1	
Recipe						
$[SO_4]_T$				1		$2 \times 10^{-4}\,M$
$[Fe(III)]_T$		-1			1	$10^{-5}\,M$
Alk	-1					$5 \times 10^{-4}\,M$
$[CO_2]_T$			1			$10^{-3}\,M$
$[CH_2O]_T$	$+4$	$+4$	1			$5 \times 10^{-4}\,M$
$[O_2]_T$	-4	-4				$3 \times 10^{-4}\,M$

the major redox couple are chosen as components, the mole balance equations can often be greatly simplified, one concentration being in large excess of all others.

3.2 pe Calculation: Solution of Example 8

As argued in Section 3.1, our choice of components should include S(+ VI) and S(− II) species. Since the oxidation of organic matter given in Reactions 50 and 51 is acid-producing (CO_2 and H_2S), we expect the equilibrium pH to be below the value of 6.3 which prevailed before any redox reaction. Thus H_2S and $H_2CO_3^*$ are the principal components for sulfide and carbonate. Choosing arbitrarily Fe^{2+} as our iron component then yields a complete component set: H^+, $H_2CO_3^*$, SO_4^{2-}, H_2S, and Fe^{2+} (see Tableau 7.2). [The stoichiometric coefficients are obtained by considering that $e^- = (H_2S)_{1/8}\,(SO_4^{2-})_{-1/8}\,(H^+)_{-5/4}\,(H_2O)_{1/2}$.]

$$
\begin{aligned}
TOTH &= [H^+] - [OH^-] - [HCO_3^-] - 2[CO_3^{2-}] - [HS^-] - 2[S^{2-}] \\
&\quad + \tfrac{5}{4}[Fe^{3+}] + (\tfrac{5}{4} - x)[Fe(OH)_x^{(3-x)+}] \\
&= + \tfrac{5}{4}[Fe(III)_T] - Alk - [CH_2O]_T + [O_2]_T \\
&= -6.88 \times 10^{-4}\,M
\end{aligned}
\tag{56}
$$

TABLEAU 7.2

	H^+	$H_2CO_3^*$	SO_4^{2-}	H_2S	Fe^{2+}
Species					
H^+	1				
OH^-	-1				
$H_2CO_3^*$		1			
HCO_3^-	-1	1			
CO_3^{2-}	-2	1			
SO_4^{2-}			1		
H_2S				1	
HS^-	-1			1	
S^{2-}	-2			1	
Fe^{3+}	$+\frac{5}{4}$		$+\frac{1}{8}$	$-\frac{1}{8}$	1
$Fe(OH)_x^{(3-x)+}$	$+\frac{5}{4}-x$		$+\frac{1}{8}$	$-\frac{1}{8}$	1
Fe^{2+}					1
Recipe					
$[SO_4^{2-}]_T$			1		$2 \times 10^{-4}\,M$
$[Fe(III)]_T$	$+\frac{5}{4}$		$+\frac{1}{8}$	$-\frac{1}{8}$	$1 \qquad 10^{-5}\,M$
Alk	-1				$5 \times 10^{-4}\,M$
$[CO_2]_T$		$+1$			$10^{-3}\,M$
$[CH_2O]_T$	-1	$+1$	$-\frac{1}{2}$	$+\frac{1}{2}$	$5 \times 10^{-4}\,M$
$[O_2]_T$	$+1$		$+\frac{1}{2}$	$-\frac{1}{2}$	$3 \times 10^{-4}\,M$

$$TOTH_2CO_3 = [H_2CO_3^*] + [HCO_3^-] + [CO_3^{2-}] = [CO_2]_T + [CH_2O]_T$$
$$= 1.5 \times 10^{-3}\,M \qquad (57)$$

$$TOTSO_4 = [SO_4^{2-}] + \tfrac{1}{8}[Fe^{3+}] + \tfrac{1}{8}[Fe(OH)_x^{(3-x)+}]$$
$$= [SO_4^{2-}]_T + \tfrac{1}{8}[Fe(III)]_T - \tfrac{1}{2}[CH_2O]_T + \tfrac{1}{2}[O_2]_T$$
$$= 1.01 \times 10^{-4}\,M \qquad (58)$$

$$TOTH_2S = [H_2S] + [HS^-] + [S^{2-}] - \tfrac{1}{8}[Fe^{3+}] - \tfrac{1}{8}[Fe(OH)_x^{(3-x)+}]$$
$$= -\tfrac{1}{8}[Fe(III)]_T + \tfrac{1}{2}[CH_2O]_T - \tfrac{1}{2}[O_2]_T$$
$$= 0.99 \times 10^{-4}\,M \qquad (59)$$

$$TOTFe = [Fe^{3+}] + [Fe(OH)_x^{(3-x)+}] + [Fe^{2+}] = [Fe(III)]_T$$
$$= 10^{-5}\,M \qquad (60)$$

The Fe terms are negligible to a first approximation in all but the last equation. The first equation is then the usual alkalinity expression and the others are straightforward expressions of conservation for CO_3^{2-}, SO_4^{2-}, S^{2-}, and Fe.

Since $TOTH_2CO_3$ is in excess of $TOTH_2S$ by a factor of 15, the sulfide terms in the $TOTH$ equation are relatively small, and the pH is roughly obtained in the usual manner for a pure carbonate system, by comparing Equations 56 and 57. (Since C_T is a bit more than twice the alkalinity, the pH should be slightly below 6.3.)

$$TOTH \simeq -[HCO_3^-] = -6.9 \times 10^{-4} = -10^{-3.16}$$

$$TOTH_2CO_3 \simeq [H_2CO_3^*] + [HCO_3^-] = 1.5 \times 10^{-3}$$

therefore

$$[H^+] = \frac{10^{-6.3}[H_2CO_3^*]}{[HCO_3^-]} \simeq \frac{10^{-6.3}(1.5 \times 10^{-3} - 6.9 \times 10^{-4})}{(6.9 \times 10^{-4})}$$

$$pH = 6.23$$

$$TOTSO_4 \simeq [SO_4^{2-}] = 10^{-4}\ M$$

$$TOTH_2S \simeq [H_2S] + [HS^-] = [H_2S](1 + 10^{-7} \times 10^{6.23}) = 10^{-4}\ M$$

$$[H_2S] = 10^{-4.07}\ M$$

$$[HS^-] = 10^{-4.84}\ M$$

Other minor carbon and sulfur species can then be obtained in a straightforward fashion.

To solve Equation 60, we need to write the constants corresponding to the Fe(III) species in the tableau. This is of course strictly equivalent to equating the pe of the Fe^{3+}/Fe^{2+} couple to that of SO_4^{2-}/H_2S (which is the pe of the system):

$$SO_4^{2-}/H_2S: \qquad\qquad pe = 5.12 - \tfrac{5}{4}pH + \tfrac{1}{8}\log\frac{[SO_4^{2-}]}{[H_2S]} = -2.66$$

$$Fe^{3+}/Fe^{2+}: \qquad\qquad -2.66 = 13.0 + \log\frac{[Fe^{3+}]}{[Fe^{2+}]}$$

therefore

$$[Fe^{2+}] = 10^{+15.66}[Fe^{3+}]$$

At a pH of 6.2, none of the hydrolysis species of Fe(III) have a sufficiently high formation constant to make them dominant over Fe^{2+} (see Chapter 5):

$$[FeOH^{2+}] = 10^{+11.8} \times 10^{-7.8}[Fe^{3+}] = 10^{-11.66}[Fe^{2+}] \ll [Fe^{2+}]$$

$$[Fe(OH)_2^+] = 10^{22.3} \times 10^{-15.6}[Fe^{3+}] = 10^{-8.96}[Fe^{2+}] \ll [Fe^{2+}]$$

$$[Fe(OH)_4^-] = 10^{34.4} \times 10^{-31.2}[Fe^{3+}] = 10^{-12.46}[Fe^{2+}] \ll [Fe^{2+}]$$

Even the solid am.Fe(OH)$_3$(s) cannot form

$$[Fe^{3+}] = 10^{-15.66}[Fe^{2+}] \leqslant 10^{-20.66}$$

$$[Fe^{3+}][OH^-]^3 \leqslant 10^{-20.66} \times 10^{-23.4} \ll K_s = 10^{-38.8}$$

$[Fe^{2+}]$ is thus the dominant term in Equation 60:

$$TOTFe \simeq [Fe^{2+}] = 10^{-5} M$$

We still have to ascertain whether FeCO$_3$(s) or FeS(s) precipitate:

$$[Fe^{2+}][CO_3^{2-}] = 10^{-5} \times \frac{10^{-10.3}}{10^{-6.23}} \times 10^{-3.16} = 10^{-12.23} < K_s = 10^{-10.7} M$$

$$[Fe^{2+}][S^{2-}] = 10^{-5} \times \frac{10^{-20.9} \times 10^{-4.07}}{10^{-12.46}} = 10^{-17.51} > K_s = 10^{-18.1} M$$

So in fact FeS(s) does precipitate. This affects the sulfide minimally since the total sulfide is 10 times in excess of the iron and it affects the alkalinity and the pH even less:

$$[Fe^{2+}] = \frac{10^{-18.1}}{[S^{2-}]} = \frac{10^{-18.1} \times 10^{-12.46}}{10^{-20.9} \times 10^{-4.07}} = 10^{-5.59} M$$

Three quarters of the iron is precipitated as FeS.

The calculations could be redone choosing FeS as a component instead of Fe^{2+}, but the results would be practically unchanged. The composition of this hypolimnetic water is thus characterized by the following chemistry:

$$pH = 6.23; \qquad pe = -2.66; \qquad [O_2] = 0 M$$

$$[H_2CO_3^*] = 10^{-3.09} M; \qquad [HCO_3^-] = 10^{-3.16} M$$

$$[H_2S] = 10^{-4.07} M; \qquad [HS^-] = 10^{-4.84} M$$

$$[SO_4^{2-}] = 10^{-4.0} M$$

$$[Fe^{2+}] = 10^{-5.59} M; \qquad [FeS.s] = 10^{-5.1} M$$

3.3 Redox in Sediments

As illustrated in Example 8, the decomposition of organic matter in environments that have restricted exchange with the atmosphere or oxygenated waters can result in the depletion of O_2 and the development of suboxic or anoxic conditions. These conditions often occur in sediments, flooded soils, and deep groundwaters. Characteristic profiles of dissolved inorganic species are observed in sediment pore waters. These profiles are paralleled by changes in the microbial communities with depth in the sediment and reflect the progression in terminal electron acceptors for microbial oxidation of organic carbon: O_2, NO_3^-, $MnO_2(s)$, $Fe(OH)_3(s)$, SO_4^{2-}, and organic C in fermentation reactions. Precipitation and dissolution reactions (e.g., calcite dissolution and precipitation of carbonate, phosphate, or sulfide minerals) also influence the chemical composition of sediment pore waters.

The pore water profiles from a lake sediment shown in Figure 7.3 illustrate some of these features.[14] Remineralization of organic N and P results in increases in ammonium and phosphate below the sediment–water interface. The increase in alkalinity is also a result of the decomposition of organic matter either directly (as organic N is remineralized to NH_4^+) or indirectly (as calcite dissolves in response to the release of CO_2 associated with remineralization of organic C). In this sediment, the profiles of the redox-active species are not clearly separated.

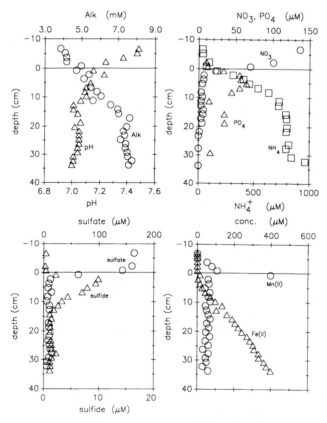

Figure 7.3   Concentrations of redox-active species in interstitial water as a function of depth in the sediment (cm) and in overlying water. The sediment–water interface is at a depth of 0 cm. Samples from Lake Greifen, Switzerland. Data from Wersin et al., 1991.[14]

Depletion of both NO_3^- and SO_4^{2-} and release of Mn^{2+} occurs very near the sediment–water interface; $MnO_2(s)$ is found only in the top centimeter of the sediments. The peak in sulfide is, however, slightly lower in the sediment profile than the peak in dissolved Mn^{2+}; the successive appearance of these species is in accord with the corresponding redox potentials. The decrease in the dissolved concentrations of Mn(II) and S($-$II) with depth are attributable to precipitation of rhodochrosite ($MnCO_3 \cdot s$) and iron sulfides. Precipitation of phosphate-containing minerals may account for the decrease in pore water phosphate. Vivianite, $Fe_3(PO_4)_2 \cdot 8H_2O$, was calculated to be supersaturated in the pore water, although this mineral could not be identified in the sediments. Equilibrium calculations, however, may not be strictly applicable in this system. The gradual increase in Fe(II) with depth may be attributed to slow kinetics of both precipitation of siderite ($FeCO_3$) and vivianite and dissolution of Fe(III) hydroxide.

Redox processes occurring at or near the sediment–water interface have important consequences for the geochemical cycling of metals, nutrients, and organic micropollutants. For example, reducing sediments can be a source of phosphate to overlying (anoxic) waters; this recycling of phosphorus can retard the restoration of eutrophic lakes even if external sources of phosphorus are reduced.[15] The residence time of organic micropollutants, such as PCBs, may also be increased by their recycling from bottom sediments.[16]

3.4 Redox in Wastewater Disposal Fields

Other interesting examples of pe calculations are found in aquatic systems where anoxic and oxic waters are mixed. A case in point is sewage or sludge discharge into fresh or marine waters where the total oxygen consumption is then approximated by a first-order rate constant that aggregates all oxidative biological processes. This constant is determined empirically in tests of biological oxygen demand (BOD). Slightly more sophisticated description of such waste disposal situations can include several oxidation kinetic parameters to represent the growth of subpopulations of aerobic microbes. In particular, one may include parameters for the "nitrogenous BOD" to reflect the oxidation of reduced nitrogen compounds (nitrification) which normally follows the oxidation of the nonrefractory organic carbon ("carbonaceous BOD"). One weakness in such empirical description of biological oxidation processes is the difficulty in matching the microbial community in the BOD test inoculum with that of the receiving waters.[8] Other kinetic parameters for oxygen addition by diffusion through the air–water interface and by algal photosynthesis can be superimposed on the oxygen consumption equation and included in a hydrodynamic transport model of the system. Because waste disposal systems are specifically designed to avoid anoxic conditions (and the BOD test is an aerobic test), their redox chemistry is in fact rarely interesting to the aquatic chemist: as seen in Example 3, the pe of such oxic systems is more sensitive to the pH than to the partial pressure of oxygen as long as oxygen is present in measurable concentration.

3.5 Redox in Hydrothermal Oceanic Vents

Leaving the murky waters of sewage disposal fields for the bottom of the blue ocean, a system that does provide an interesting case of redox calculations is that of the mixing of hydrothermal fluid from the ridge vents with deep ocean water. Although some of the important chemical processes in such a system are controlled by temperature ("pure" hydrothermal fluid is estimated to have a temperature of $350°C$), it is instructive to study a simple model system assuming uniform low temperature (our usual $25°C$, of course) for the mix.

Example 9. Mixture of Hydrothermal Fluid and Seawater* Concentrations for the various constituents of the two *end members* of the vent mixture, seawater and hydrothermal fluid, are given in Table 7.6. To keep the problem manageable, let us fix the ionic strength (say, $I = 0.5 M$) and ignore the chemistry of Na^+, Mg^{2+}, and Ca^{2+} (including the possible precipitation of $CaCO_3$), K^+, Cl^-, and H_2SiO_3 (which is considered a good conservative tracer in this system). We define h as the proportion of hydrothermal fluid in the mix and consider the composition of the system as h varies from 1 (hydrothermal fluid) to 0 (seawater). Because of the high temperature of hydrothermal fluid, the calculations at low dilution ($h > 0.1$) require difficult corrections. We thus limit our study to the

TABLE 7.6   Concentrations of Major Constituents in Seawater and Hydrothermal Fluid in Oceanic Vents[a,b]

		Seawater, $1 - h$ (2°C)	Hydrothermal Fluid, h (350°C)
Major cations	Na^+	466	527
	Mg^{2+}	52.7	0
	Ca^{2+}	10.3	25
	K^+	10.0	15
Major anions	Cl^-	542	594
	SO_4^{2-}	28.9	0
	Alk	2.45	−2.0
	C_T	2.38	9.0
Leached species	H_2SiO_3	0.16	18.0
Reduced species	H_2S	0	8.0
	Fe(II)	0	0.3
	Mn(II)	0	2
Oxygen	O_2	0.12	0

[a]*Note*: All concentrations are given in mM.
[b]The composition of the hydrothermal fluid was obtained from extrapolation of early data with lower temperature samples. More direct data suggest a higher alkalinity (ca. 0 to −0.5 mM).

*Specific conditions for this example were provided by R. E. McDuff.

acid–base and redox chemistry of the system at moderate $(0.1 > h > 0.01)$ and high $(h < 0.01)$ dilutions.

Before embarking on detailed calculations we note that over the range of interest to us the alkalinity of the mix is dominated by that of seawater. At relatively high values of h (moderate dilutions), the sulfide concentration from the hydrothermal fluid is still high $(10^{-3}\ M)$ and much in excess of the oxygen from seawater $(10^{-4}\ M)$. At some point, with decreasing values of h, the oxygen of seawater becomes important, while the sulfide is diluted and the system becomes oxic.

pH   Consider first moderate dilution factors, and assume that the pH is below 6. The principal components of the system are then: H^+, $H_2CO_3^*$, SO_4^{2-}, and H_2S (see Tableau 7.3). The mole balance equations are written:

$$
\begin{aligned}
TOTH &= [H^+] - [OH^-] - [HCO_3^-] - 2[CO_3^{2-}] + [O_2] \\
&\quad - [HS^-] - 2[S^{2-}] \\
&= 2h \times 10^{-3} + (1-h)(-2.45 + 0.12)10^{-3} \\
&= (4.33h - 2.33) \times 10^{-3}
\end{aligned}
\tag{61a}
$$

TABLEAU 7.3

	H^+	$H_2CO_3^*$	SO_4^{2-}	H_2S	
Species					
H^+	1				
OH^-	-1				
$H_2CO_3^*$		1			
HCO_3^-	-1	1			
CO_3^{2-}	-2	1			
O_2	1		$\frac{1}{2}$	$-\frac{1}{2}$	
SO_4^{2-}			1		
H_2S				1	
HS^-	-1			1	
S^{2-}	-2			1	
Hydrothermal fluid					
Alk	-1				-2×10^{-3}
$[CO_2]_T$		1			9×10^{-3}
$[H_2S]_T$				1	8×10^{-3}
Seawater					
Alk	-1				2.45×10^{-3}
$[O_2]_T$	1		$\frac{1}{2}$	$-\frac{1}{2}$	1.2×10^{-4}
$[CO_2]_T$		1			2.38×10^{-3}
$[SO_4]_T$			1		2.89×10^{-2}

$$TOTH_2CO_3 = [H_2CO_3^*] + [HCO_3^-] + [CO_3^{2-}]$$
$$= 9h \times 10^{-3} + (1-h)2.38 \times 10^{-3}$$
$$= (6.62h + 2.38) \times 10^{-3} \tag{62a}$$

$$TOTSO_4 = \tfrac{1}{2}[O_2] + [SO_4^{2-}]$$
$$= (1-h)(0.6 \times 10^{-4} + 2.89 \times 10^{-2})$$
$$= (2.9 - 2.9h) \times 10^{-2} \tag{63a}$$

$$TOTH_2S = -\tfrac{1}{2}[O_2] + [H_2S] + [HS^-] + [S^{2-}]$$
$$= 8h \times 10^{-3} - (1-h)0.6 \times 10^{-4}$$
$$= (8.06h - 0.06) \times 10^{-3} \tag{64a}$$

Using the usual approximations for a moderately low pH system, and expressing the appropriate mass laws (with apparent seawater constants—see chapter 6), one can simplify the equations to

$$TOTH = -[HCO_3^-] - [HS^-] = (4.33h - 2.33) \times 10^{-3} \tag{61b}$$

$$TOTH_2CO_3 = [H_2CO_3^*] + [HCO_3^-] = [HCO_3^-](1 + 10^{6.0}[H^+])$$
$$= (6.62h + 2.38) \times 10^{-3} \tag{62b}$$

$$TOTSO_4 = [SO_4^{2-}] = (2.9 - 2.9h) \times 10^{-2} \tag{63b}$$

$$TOTH_2S = [H_2S] + [HS^-] = [HS^-](1 + 10^{6.9}[H^+])$$
$$= (8.06h - 0.06) \times 10^{-3} \tag{64b}$$

Substitution of Equations 62b and 64b into 61b (and multiplication by 10^3) provides an implicit expression of $[H^+]$ as a function of h:

$$-\frac{6.62h + 2.38}{1 + 10^{6.0}[H^+]} - \frac{8.06h - 0.06}{1 + 10^{6.9}[H^+]} = 4.33h - 2.33 \tag{65}$$

This expression, which is plotted as line A in Figure 7.4, is valid as long as the oxygen concentration is small compared to that of sulfide. The relationship between $[O_2]$ and $[H_2S]$ is given by combination of the appropriate redox reactions:

$$\tfrac{1}{4}O_2(aq) + e^- + H^+ = \tfrac{1}{2}H_2O; \qquad pe^\circ = 21.45 \tag{66}$$

$$\tfrac{1}{8}SO_4^{2-} + e^- + \tfrac{5}{4}H^+ = \tfrac{1}{8}H_2S(aq) + \tfrac{1}{2}H_2O; \qquad pe^\circ = 5.12 \tag{67}$$

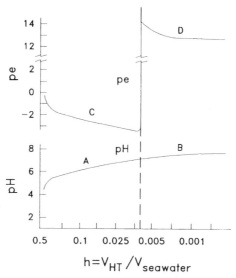

Figure 7.4 Variations of the calculated equilibrium composition of hydrothermal fluid as a function of its dilution (h) into seawater (Example 9). The letters labelling the graphs in the lower part of the figure correspond to equations in the text: A, B = pH; C, D = pe.

therefore

$$O_2(aq) + \tfrac{1}{2}H_2S(aq) = \tfrac{1}{2}SO_4^{2-} + H^+; \qquad \log K = +65.3 \qquad (68)$$

$$[O_2] = 10^{-65.3}[H^+][SO_4^{2-}]^{1/2}[H_2S]^{-1/2} \qquad (69)$$

In the range $0.1 > h > 0.01$, pH $\simeq 6.5$ (see Figure 7.4), and $[SO_4^{2-}] \simeq 10^{-1.5}$ (see Equation 63b),

$$[O_2][H_2S]^{1/2} \simeq 10^{-73} \qquad (70)$$

This very small value of the product of $[O_2]$ and $[H_2S]$ concentrations makes it possible for only one of them to be important at any time (this is of course a general result; see Section 2.2). Thus $[O_2]$ is the dominant term in Equation 64a for any negative value of the right-hand side, and $[H_2S]$ is the dominant term for any positive value:

$$TOTH_2S = -\tfrac{1}{2}[O_2] + [H_2S] + [HS^-] = (8.06h - 0.06) \times 10^{-3} \qquad (64c)$$

The value of the dilution parameter yielding $TOTH_2S = 0$,

$$h = \frac{0.06}{8.06} \simeq 7.4 \times 10^{-3} \qquad (71)$$

TABLEAU 7.4

	H^+	HCO_3^-	O_2	SO_4^{2-}	
Species					
H^+	1				
OH^-	-1				
$H_2CO_3^*$	1	1			
HCO_3^-		1			
CO_3^{2-}	-1	1			
O_2			1		
SO_4^{2-}				1	
H_2S	2		-2	1	
HS^-	1		-2	1	
S^{2-}			-2	1	
Hydrothermal fluid					
Alk	-1				-2×10^{-3}
$[CO_2]_T$	1	1			9×10^{-3}
$[H_2S]_T$	2		-2	1	8×10^{-3}
Seawater					
Alk	-1				2.45×10^{-3}
$[O_2]_T$			1		1.2×10^{-4}
$[CO_2]_T$	1	1			2.38×10^{-3}
$[SO_4]_T$				1	2.89×10^{-2}

corresponds to a critical dilution of the hydrothermal fluid, one at which the system undergoes a sudden transition from anoxic to oxic conditions.

Under oxic conditions ($h < 7.4 \times 10^{-3}$), the pH is obtained from the $TOTH$ equation using H^+, HCO_3^-, O_2, and SO_4^{2-} as the principal components (see Tableau 7.4):

$$TOTH = [H^+] - [OH^-] + [H_2CO_3^*] - [CO_3^{2-}] + 2[H_2S] + [HS^-]$$

$$= h(2 + 9 + 16)10^{-3} + (1 - h)(-2.45 + 2.38) \times 10^{-3}$$

$$= (27.1\,h - 0.07) \times 10^{-3} \qquad (72a)$$

The carbonate species concentrations are much larger than the others, leading to the simplified equation

$$TOTH = [H_2CO_3^*] - [CO_3^{2-}] = (27.1\,h - 0.07) \times 10^{-3} \qquad (72b)$$

(This equation can also be obtained from the previous simplified mole balances by the linear transformation $TOTH' = TOTH + TOTH_2CO_3 + 2 \times TOTH_2S$.)

The carbonate mole balance given by Equation 62a remains unchanged:

$$TOTHCO_3 = TOTH_2CO_3 = [H_2CO_3^*] + [HCO_3^-] + [CO_3^{2-}]$$
$$= (6.62\,h + 2.38) \times 10^{-3} \tag{62a}$$

Introducing the appropriate mass laws into Equations 62a and 72b and combining the equations yields an expression for $[H^+]$:

$$\frac{10^{6.0}[H^+] - 10^{-8.9}[H^+]^{-1}}{1 + 10^{6.0}[H^+] + 10^{-8.9}[H^+]^{-1}}(6.62\,h + 2.38) = 27.1\,h - 0.07 \tag{73}$$

This implicit function $[H^+]$ of h is plotted as line B on Figure 7.4. For $h = 7.4 \times 10^{-3}$ the pH change due to sulfide oxidation is so small (ca. 0.03) that line B is indistinguishable from line A. Note that the final seawater pH calculated in this example (7.45) is too low because we have neglected such species as borate which are important in the acid–base chemistry of seawater.

pe In the anoxic region, at moderate dilutions, the pe is given by the sulfate–sulfide ratio according to Reaction 67:

$$pe = 5.12 - \tfrac{5}{4}pH + \tfrac{1}{8}\log\frac{[SO_4^{2-}]}{[H_2S]} \tag{74}$$

Substituting Equations 63b and 64b yields the expression

$$pe = 5.12 - \tfrac{5}{4}pH + \tfrac{1}{8}\log(1-h)10^{-1.54} - \tfrac{1}{8}\log\frac{8.06\,h - 0.06}{1 + 10^{-6.9}[H^+]^{-1}}10^{-3}$$

therefore

$$pe = 5.19 - \tfrac{5}{4}pH + \tfrac{1}{8}\log\frac{(1-h)(1 + 10^{-6.9}[H^+]^{-1})}{h - 7.4 \times 10^{-3}} \tag{75}$$

(see line C in Figure 7.4). In the oxic region at high dilutions, the oxygen concentration controls the pe according to Reaction 66:

$$pe = 21.45 - pH + \tfrac{1}{4}\log[O_2]$$

The aqueous oxygen concentration is obtained by neglecting the sulfide species in Equation 64c (which results from Tableau 7.4 as well):

$$pe = 20.7 - pH + \tfrac{1}{4}\log(0.12 - 16.12\,h) \tag{76}$$

(see line D in Figure 7.4). Lines C and D exhibit a sharp discontinuity at $h = 7.4 \times 10^{-3}$, corresponding to a sudden change from a control of the redox potential by sulfide to a control by oxygen. Such a sudden change in the redox properties of hydrothermal waters has a dominant influence on the chemistry of many chemical species. For example, iron, which is relatively abundant as Fe(II) in hydrothermal fluid, precipitates as a sulfide at moderate dilutions and as Fe(III) oxide at high dilutions.

4. pe–pH DIAGRAMS

A common and convenient way to explore the redox chemistry of systems of known composition is to develop "pe–pH diagrams" (also known as E_H–pH or Pourbaix diagrams) which show the major species of the various components as a function of both pH and pe. Such diagrams display the pe–pH domains of stability of possible solids and the domains of dominance of soluble species.

Given a half redox reaction involving H^+, one can calculate the relative concentrations of the oxidant and the reductant as a function of pH and pe:

$$\text{Ox} + ne^- + m\text{H}^+ = \text{Red}; \quad \text{pe}^\circ \tag{77}$$

$$\text{pe} = \text{pe}^\circ - \frac{1}{n}\log\frac{[\text{Red}]}{[\text{Ox}]} + \frac{m}{n}\log\{\text{H}^+\}$$

$$n(\text{pe} - \text{pe}^\circ) + m\text{pH} = \log\frac{[\text{Ox}]}{[\text{Red}]} \tag{78}$$

When the expression on the left-hand side is less than zero, the reductant is the more important species and vice versa. The equation of the boundary ([Red] = [Ox]) is obtained by equating the left-hand side to zero:

$$n(\text{pe} - \text{pe}^\circ) + m\text{pH} = 0 \tag{79}$$

This expression defines a straight line of slope m/n on a pe–pH diagram. With pe as the ordinate, the oxidant is the dominant species above the line, the reductant below it. The difficulty in developing a pe–pH diagram lies in choosing the correct redox couple for which to write a half redox reaction. The reductant and the oxidant have to be the major species of the component under study in the pe–pH domain on each side of the dividing line. For example, depending on whether the pH is below or above 7, the S(+VI)/S(−II) couple should be represented by the SO_4^{2-}/H_2S or the SO_4^{2-}/HS^- reaction. Precipitation of sulfide by an excess of Fe(II) would make SO_4^{2-}/FeS the reaction to be studied. Short of repetitive trial and error procedures, the only solution is to develop through experience (i.e., through previous trial and error) an intuition for the probable major couples.

4.1 The Water, O_2, H_2 System

The chemistry of natural waters is limited by two redox boundaries. At these limits, water itself is oxidized to oxygen, at high pe, or reduced to hydrogen, at low pe. Most natural waters are oxic, their redox chemistry governed by the oxygen-water couple. However, oxidants other than O_2, such as H_2O_2, can also play an important role in these systems.

Example 10. pe–pH Diagram for the H_2O, O_2, H_2 System   The half redox reactions that express the oxidation of water to oxygen gas and its reduction to hydrogen gas provide two lines on a pe–pH diagram:

$$\tfrac{1}{4}O_2(g) + H^+ + e^- = \tfrac{1}{2}H_2O; \qquad pe^\circ = 20.75 \tag{80}$$

$$pe + pH = 20.75 + \tfrac{1}{4}\log P_{O_2} \tag{81}$$

and

$$H^+ + e^- = \tfrac{1}{2}H_2(g); \qquad pe^\circ = 0 \tag{82}$$

$$pe + pH = -\tfrac{1}{2}\log P_{H_2} \tag{83}$$

Certainly one atmosphere is an upper limit on the partial pressure of oxygen or hydrogen in natural waters and the equations

$$pe + pH = 20.75 \tag{84}$$

$$pe + pH = 0 \tag{85}$$

provide, respectively, upper and lower limits on a pe–pH diagram for natural waters (see Figure 7.5). Above the upper line water becomes an effective reductant (producing oxygen); below the lower line, water is an effective oxidant (producing hydrogen). Figure 7.5a shows how these limits vary with the partial pressures of the gases.

The possible oxidizing role of photochemically produced ozone and peroxide in natural waters is illustrated in Figure 7.5b, which provides the pe–pH diagrams corresponding to the reactions

$$\tfrac{1}{2}O_3(g) + H^+ + e^- = \tfrac{1}{2}O_2(g) + \tfrac{1}{2}H_2O; \qquad pe^\circ = 35.1 \tag{86}$$

$$\tfrac{1}{2}O_2(g) + H^+ + e^- = \tfrac{1}{2}H_2O_2; \qquad pe^\circ = 11.5 \tag{87}$$

$$\tfrac{1}{2}H_2O_2 + H^+ + e^- = H_2O; \qquad pe^\circ = 30.0 \tag{88}$$

(And also the acid–base reaction $H_2O_2 = H^+ + HO_2^-$; $pK = 11.6$.)

Ozone and peroxide are much stronger oxidants than oxygen in water

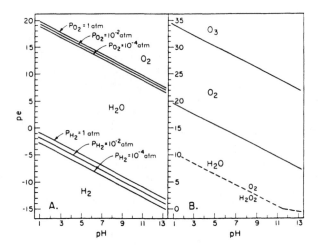

Figure 7.5 pe–pH diagrams for the O_2–H_2O–H_2 system (A) and the O_3–O_2–H_2O_2–H_2O system (B). In panel A the limits of the domain of H_2O stability are shown as a function of P_{O_2} and P_{H_2}. Within this domain O_2 acts as an oxidant (which it does not in the O_2 domain), and H_2 as a reductant (which it does not in the H_2 domain). The domain of stability of ozone, O_3, is shown in panel B. Hydrogen peroxide is never stable in the presence of oxygen and water; the dashed line in panel B indicates the region above which H_2O_2 is a better oxidant than O_2 in water.

producing, respectively, oxygen and water as the reduced species. For peroxide this may seem paradoxical since H_2O_2 is a product of the partial reduction of oxygen itself: the acquisition of the first two electrons (per mole of oxygen), from O_2 to H_2O_2, is less energetically favorable than the acquisition of the next two electrons, from H_2O_2 to H_2O. As a result there is no region of Figure 7.5*b* where H_2O_2 is the dominant species. The reaction

$$H_2O_2 = H_2O + \tfrac{1}{2}O_2(g)$$

has a negative standard free energy,

$$\Delta G° = -\frac{(30 - 11.5)}{0.175} = -106 \, \text{kJ} \, \text{mol}^{-1}$$

and tends to proceed to the right for measurable concentrations of the reactions. The line separating the regions where O_2 and H_2O_2 are the second most important species in the H_2O domain (Reaction 87) is shown on the graph. Above this line H_2O_2 is a better oxidant than O_2, and vice versa below the line. From a purely energetic point of view we can thus expect peroxide to be a more reactive oxidant than oxygen itself in oxic waters and the concentration of H_2O_2 in natural waters is accordingly low. The complete reduction of oxygen to water thus provides the major oxidizing reaction in natural waters and the O_2/H_2O couple effectively determines "the pe" of oxic waters—even if in many situations (including at the surface of platinum electrodes which are often used to measure "the pe") the acquisition of the last two electrons, from H_2O_2 to H_2O, is relatively slow.[17]

4.2 The Sulfate–Sulfide System

In anoxic waters, where oxygen is exhausted through oxidation of organic matter, sulfate becomes the primary oxidant. Although there exist important intermediate redox sulfur compounds (e.g., S^0), the reaction usually proceeds all the way to the formation of sulfides and can be written

$$\tfrac{1}{8}SO_4^{2-} + \tfrac{10}{8}H^+ + e^- = \tfrac{1}{8}H_2S + \tfrac{1}{2}H_2O; \qquad pe^o = 5.1 \tag{89}$$

A simplified sulfur system thus includes only sulfate and sulfide species.

Example 11. pe–pH Diagram for the S(VI)/S(−II) System In addition to Reaction 89 the pertinent reactions are

$$HSO_4^- = H^+ + SO_4^{2-}; \quad pK = 2.0 \tag{90}$$

$$H_2S = H^+ + HS^-; \quad pK = 7.0 \tag{91}$$

$$HS^- = H^+ + S^{2-}; \quad pK = 13.9 \tag{92}$$

Reactions 90–92 provide three vertical lines separating the various acid–base species on the pe–pH diagram (see Figure 7.6):

1. $HSO_4^- - SO_4^{2-}$; $pH = 2.0$
2. $H_2S - HS^-$; $pH = 7.0$
3. $HS^- - S^{2-}$; $pH = 13.9$

The separation of the sulfate and sulfide species then consists of four segments:

4. $HSO_4^- - H_2S$ for $pH < 2.0$
5. $SO_4^{2-} - H_2S$ for $2.0 < pH < 7.0$
6. $SO_4^{2-} - HS^-$ for $7.0 < pH < 13.9$
7. $SO_4^{2-} - S^{2-}$ for $13.9 < pH$

The corresponding reactions and pe–pH equations are obtained by combining

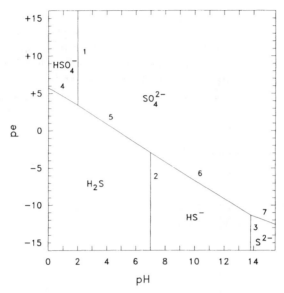

Figure 7.6 pe–pH diagram for the sulfate–sulfide system. See text for explanation of the numbered boundaries.

Reactions 89–92 and expressing the mass laws

4. $\frac{1}{8}HSO_4^- + \frac{9}{8}H^+ + e^- = \frac{1}{8}H_2S + \frac{1}{2}H_2O$; $pe^\circ = 4.85$
   $8pe + 9pH = 38.8$ for $[HSO_4^-] = [H_2S]$ (93)

5. $\frac{1}{8}SO_4^{2-} + \frac{10}{8}H^+ + e^- = \frac{1}{8}H_2S + \frac{1}{2}H_2O$; $pe^\circ = 5.1$
   $8pe + 10pH = 40.8$ for $[SO_4^{2-}] = [H_2S]$ (94)

6. $\frac{1}{8}SO_4^{2-} + \frac{9}{8}H^+ + e^- = \frac{1}{8}HS^- + \frac{1}{2}H_2O$; $pe^\circ = 4.23$
   $8pe + 9pH = 33.8$ for $[SO_4^{2-}] = [HS^-]$ (95)

7. $\frac{1}{8}SO_4^{2-} + H^+ + e^- = \frac{1}{8}S^{2-} + \frac{1}{2}H_2O$; $pe^\circ = 2.49$
   $pe + pH = 2.5$ for $[SO_4^{2-}] = [S^{2-}]$ (96)

The various equations are plotted in Figure 7.6.

In each of the regions of Figure 7.6 the free concentration of any sulfur species can be calculated from that of the dominant species (which is equal to the total sulfur concentration) by expressing the appropriate mass law. For example, the free sulfide concentration $[S^{2-}]$, which is important for assessing the solubility of metal sulfides (see Example 12), is given by

$$pS^{2-} = -19.9 + pS_T + 8pe + 8pH \text{ (in the } SO_4^{2-} \text{ region)}$$ (97)

$$pS^{2-} = 20.9 + pS_T - 2pH \text{ (in the } H_2S \text{ region)}$$ (98)

$$pS^{2-} = 13.9 + pS_T - pH \text{ (in the HS}^- \text{ region)} \tag{99}$$

If we consider the presence of hydrogen sulfide gas, the relevant redox reaction is obtained by combination of Reactions 89 and 100:

$$H_2S(g) = H_2S; \qquad pK_H = 1.0 \tag{100}$$

therefore

$$\tfrac{1}{8}SO_4^{2-} + \tfrac{10}{8}H^+ + e^- = \tfrac{1}{8}H_2S(g) + \tfrac{1}{2}H_2O; \qquad pe^o = 5.23$$

$$\log[SO_4^{2-}] = -41.8 + 8pe + 10pH + \log[P_{H_2S}] \tag{101}$$

This equation gives the sulfate concentration at redox equilibrium with any partial pressure of hydrogen sulfide.

*   *   *

Figure 7.6 displays the characteristic features of a pe–pH diagram. It is noteworthy that the domains of dominance of various species meet at common corners; typically three boundaries intersect at the same point. This is because the equations of the boundaries (which are the logarithmic expressions of mass laws) are linear combinations of each other. Very simply, the point where $[SO_4^{2-}] = [H_2S]$ *and* $[SO_4^{2-}] = [HS^-]$ is by necessity a point where $[H_2S] = [HS^-]$. This feature of pe–pH diagrams greatly facilitates the construction of the graphs and is useful for detecting errors. In redox systems that include high stoichiometric coefficients, this unique property of pe–pH diagrams is conserved only when the graph is strictly a *dominance diagram* rather than a *stability diagram*. Then boundaries represent the pH–pe condition where half the element of interest is in each of the two major species on either side of the line. To obtain the equations of these boundaries, the concentrations of the species themselves must thus be divided by the stoichiometric coefficient of the element, for example, $[HS^-] = \tfrac{1}{2}S_T$; $[S_2O_3^{2-}] = \tfrac{1}{4}S_T$; $[S_6^{2-}] = \tfrac{1}{12}S_T$. Note also that in this case the boundaries around the domain of a solid species correspond to the points where half of the solid is already precipitated, and not to the onset of precipitation. The diagram provides the domain of dominance of the solid, not its domain of stability. (The two are usually separated by only a fraction of a log unit.)

In the process of reducing sulfate or oxidizing sulfide, a number of compounds with intermediate oxidation states can be formed. As seen in Table 7.1, these include sulfite $[SO_3^{2-}]$, thiosulfate $[S_2O_3^{2-}]$, polysulfides $[S_n^{2-}]$, and solid sulfur in colloidal $[S_8^0 \cdot col]$ or orthorhombic $[S_8^0 \cdot ort]$ form. Although the production of these intermediate sulfur compounds is largely under biological control, partial chemical oxidation of sulfide by oxygen leads to the formation of elemental sulfur, polysulfides, and thiosulfate. Direct reaction between sulfide

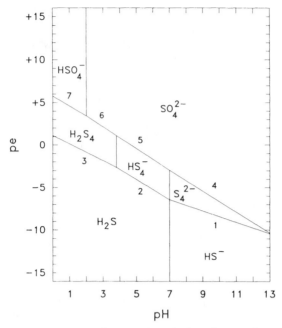

Figure 7.7   pe–pH diagram for sulfur: $S_T = 2 \times 10^{-2} M$. See text for explanation of the numbered boundaries.

and bacterially produced elemental sulfur also leads to the formation of polysulfides.[4,5,18]

The formation of intermediate sulfur species is illustrated in the pe–pH diagram of Figure 7.7, in which a high concentration of total sulfur is assumed, $S_T = 2 \times 10^{-2} M$, and all the sulfur species listed in Table 7.1 are considered. According to the diagram the tetrasulfide species is the only intermediate redox species to become dominant.

The equations of the various boundaries in Figure 7.7 are obtained by combination of the appropriate half redox (and acid–base) reactions of Table 7.1. The vertical boundaries are at the pKa's of the dominant species: 2.0 for SO_4^{2-}; 3.8 and 7.0 for S_4^{2-}; 7.0 for HS^-. The other boundaries are as follows:

1. $S_4^{2-} - HS^-$

$$\tfrac{1}{6}S_4^{2-} + \tfrac{2}{3}H^+ + e^- = \tfrac{2}{3}HS^-; \qquad pe^o = -2.66 \tag{102}$$

$$\tfrac{2}{3}pH + pe = -2.66 + \tfrac{2}{3}(2) - \tfrac{1}{6}(2.6) = -1.76$$

2. $HS_4^- - H_2S$

$$\tfrac{1}{6}HS_4^- + \tfrac{7}{6}H^+ + e^- = \tfrac{2}{3}H_2S; \qquad pe^o = 0.84 \tag{103}$$

$$\tfrac{7}{6}pH + pe = 0.84 + \tfrac{2}{3}(2) - \tfrac{1}{6}(2.6) = 1.74$$

3. $H_2S_4 - H_2S$

$$\tfrac{1}{6}H_2S_4 + H^+ + e^- = \tfrac{2}{3}H_2S; \qquad pe^\circ = 0.21 \tag{104}$$

$$pH + pe = 0.21 + \tfrac{2}{3}(2.0) - \tfrac{1}{6}(2.6) = 1.11$$

4. $SO_4^{2-} - S_4^{2-}$

$$\tfrac{2}{13}SO_4^{2-} + \tfrac{16}{13}H^+ + e^- = \tfrac{1}{26}S_4^{2-} + \tfrac{8}{13}H_2O; \qquad pe^\circ = 5.85 \tag{105}$$

$$\tfrac{16}{13}pH + pe = 5.85 + \tfrac{1}{26}(2.6) - \tfrac{2}{13}(2.0) = 5.65$$

5. $SO_4^{2-} - HS_4^-$

$$\tfrac{2}{13}SO_4^{2-} + \tfrac{33}{26}H^+ + e^- = \tfrac{1}{26}HS_4^- + \tfrac{8}{13}H_2O; \qquad pe^\circ = 6.12 \tag{106}$$

$$\tfrac{33}{26}pH + pe = 6.12 + \tfrac{1}{26}(2.6) - \tfrac{2}{13}(2.0) = 5.92$$

6. $SO_4^{2-} - H_2S_4$

$$\tfrac{2}{13}SO_4^{2-} + \tfrac{17}{13}H^+ + e^- = \tfrac{1}{26}H_2S_4 + \tfrac{8}{13}H_2O; \qquad pe^\circ = 6.27 \tag{107}$$

$$\tfrac{17}{13}pH + pe = 6.27 + \tfrac{1}{26}(2.6) - \tfrac{2}{13}(2.0) = 6.06$$

7. $HSO_4^- - H_2S_4$

$$\tfrac{2}{13}HSO_4^- + \tfrac{15}{13}H^+ + e^- = \tfrac{1}{26}H_2S_4 + \tfrac{8}{13}H_2O; \qquad pe^\circ = 5.96 \tag{108}$$

$$\tfrac{15}{13}pH + pe = 5.96 + \tfrac{1}{26}(2.6) - \tfrac{2}{13}(2.0) = 5.75$$

4.3 Metals in Carbonate and Sulfur-Bearing Waters

As we have seen in Chapter 5, many metals form carbonate, sulfide, and hydroxide complexes and solids in natural waters. Even if a metal has no interesting redox chemistry of its own, its speciation depends not only on the pH but also on the pe of the water because of the redox chemistry of sulfur with which it coordinates and precipitates. As a simple example let us consider the case of cadmium.

Example 12. pe–pH Diagram for Dominant Cadmium Species   Assume, for example,

$$[CO_2]_T = 10^{-2}\,M$$

$$[S]_T = 10^{-4}\,M$$

$$[Cd]_T = 2 \times 10^{-8}\,M$$

The major species of interest are Cd^{2+}, $CdCO_3(s)$, $Cd(OH)_2(s)$, and $CdS(s)$ ($pK_S = 13.7, 14.3,$ and 27.0, respectively.) For simplicity we consider only sulfate and sulfide as sulfur species (see Example 11) and denote the total concentrations by C_T, S_T, and Cd_T. In the high pe region where sulfide is not important, the vertical separations between the Cd^{2+}, $CdCO_3(s)$, and $Cd(OH)_2(s)$ regions are obtained in the manner studied in Chapter 5:

1. $Cd^{2+} - CdCO_3(s)$. The boundary is in the region where HCO_3^- is the

major carbonate species:

$$pCd^{2+} + pCO_3^{2-} = 13.7$$

$$pCO_3^{2-} = 10.3 + pHCO_3^- - pH$$

therefore $pCd^{2+} + pHCO_3^- = 3.4 + pH$

Equating $pCd^{2+} = p(Cd_T/2) = 8$ and $pHCO_3^- = pC_T = 2$ yields

$$pH = 6.6.$$

2. $CdCO_3(s) - Cd(OH)_2(s)$:

$$pCd^{2+} + pCO_3^{2-} = 13.7$$

$$pCd^{2+} + 2pOH^- = 14.3$$

$$pH + pOH^- = 14.0$$

therefore $pCO_3^{2-} + 2pH = 27.4$

The boundary is in the region where CO_3^{2-} is the major carbonate species:

$$pCO_3^{2-} = pC_T = 2 \text{ yields}$$

$$pH = 12.7.$$

It remains to define the boundaries between Cd^{2+}, $CdCO_3(s)$, $Cd(OH)(s)$, and $CdS(s)$. These are obtained most readily by considering the expressions of free sulfide concentrations in the various regions (see Example 11):

3. $Cd^{2+} - CdS(s)$ (in the SO_4^{2-} region);

$$pCd^{2+} + pS^{2-} = 27.0$$

$$pS^{2-} = -19.9 + pS_T + 8pe + 8pH$$

therefore $pCd^{2+} + pS_T + 8pe + 8pH = 46.9$

Equating $pCd^{2+} = p(Cd_T/2) = 8$ and $pS_T = 4$ yields

$$pe + pH = 4.36.$$

4. $Cd^{2+} - CdS$ (in the H_2S region):

$$pCd^{2+} + pS^{2-} = 27.0$$

$$pS^{2-} = 20.9 + pS_T - 2pH$$

therefore $pCd^{2+} + pS_T - 2pH = 6.1$
Equating $pCd^{2+} = p(Cd_T/2) = 8$ and $pS_T = 4$ yields

$$pH = 2.95$$

5. $CdCO_3(s) - CdS(s)$ (in the HCO_3^- and SO_4^{2-} region):

$$pCd^{2+} + pCO_3^{2-} = 13.7$$

$$pCd^{2+} + pS^{2-} = 27.0$$

$$pCO_3^{2-} = 10.3 + pHCO_3^- - pH$$

$$pS^{2-} = -19.9 + pS_T + 8pe + 8pH$$

therefore $pS_T - pHCO_3^- + 8pe + 9pH = 43.5$
Equating $pHCO_3^- = pC_T = 2$ and $pS_T = 4$ yields

$$8pe + 9pH = 41.5$$

6. $CdCO_3(s) = CdS(s)$ (in the CO_3^{2-} and SO_4^{2-} region).
Taking $pCO_3^{2-} = pC_T = 2$ in the equations listed above yields

$$pe + pH = 3.9.$$

7. $Cd(OH)_2(s) - CdS(s)$ (in the SO_4^{2-} region):

$$pCd^{2+} + 2pOH^- = 14.3$$

$$pCd^{2+} + pS^{2-} = 27.0$$

$$pS^{2-} = -19.9 + pS_T + 8pe + 8pH$$

$$pH + pOH^- = 14.0$$

therefore $pS_T + 8pe + 10pH = 60.6$
Equating $pS_T = 4$ yields

$$4pe + 5pH = 28.3.$$

The boundary lines 1–7 are plotted on the pe–pH diagram of Figure 7.8 which shows the domains of dominance of the various cadmium species.

This figure exhibits the characteristic features of many pe–pH diagrams for dominant metal species. At high pe's the metal is controlled from low to high pH by the free metal ion, the carbonate, and the hydroxide solids. In the neutral pH range, as the system becomes more reducing, the sulfide solid precipitates and replaces the free metal ion and the carbonate solid as the major cadmium species, before dissolved sulfide becomes dominant over sulfate. At low pe, if

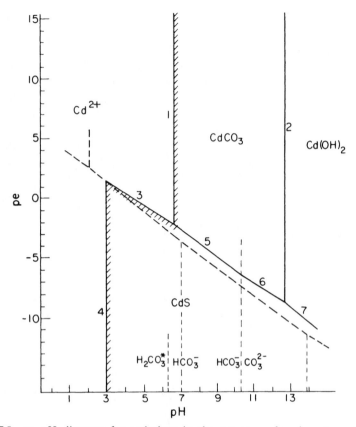

Figure 7.8 pe–pH diagram for cadmium in the presence of carbonate and sulfur: $[CO_2]_T = 10^{-2}\ M$; $[S]_T = 10^{-4}\ M$; $[Cd]_T = 2 \times 10^{-8}\ M$ (Example 12). Dashed lines are those of Figure 7.6.

the pH becomes low enough, the solid sulfide dissolves simply as a result of the decrease in free sulfide ion activity. With increasing pH, depending on the exact value of the pe, the carbonate and hydroxide solids can precipitate. In the diagram of Figure 7.8, however, the direct transition from sulfide to hydroxide solid occurs at too low a pe to be relevant to aquatic systems.

* * *

Although the equations are rarely difficult, the construction of such pe–pH diagrams for metal species can become quite time consuming. This is especially so when the metal itself is subject to redox reactions such as

1. Formation of oxides:

$$Fe_3O_4,\ MnO_2,\ PbO_2,\ \text{etc.}$$

2. Equilibrium among various soluble oxidation states:

$$Fe(II)/Fe(III), \quad Cu(II)/Cu(I), \quad \text{etc.}$$

3. Formation of the elemental state:

$$Cu^0, \quad Hg^0, \quad Ag^0, \quad \text{etc.}$$

For example, iron and manganese oxides, Fe_3O_4 and MnO_2, typically dominate the high pe and pH region of the diagrams, and elemental copper is stable in a region below the sulfide precipitation domain. In multimetal diagrams, even greater complications are introduced by the formation of mixed metal phases. At low pe's, in a system containing copper, iron, and sulfur, such solids as $CuFeS_2$ and Cu_3FeS_4 may control the speciation of the metals in large regions of the graph.

5. PHOTOCHEMICAL REACTIONS

So far in this chapter our attention has been focused principally on slow redox processes, usually mediated by microorganisms and often occurring near the bottom of the water column where organic debris accumulate. These redox processes depend ultimately on the biological utilization of solar energy in photosynthesis. Solar energy also generates a host of rapid redox reactions in surface waters through strictly abiotic processes. Photochemical reactions produce highly reactive radicals and unstable redox species which are important in some elemental cycles in the euphotic zone of natural waters. Although not all photochemical reactions are redox reactions, in aqueous solution most are, and the subject of aquatic photochemistry is conveniently approached in the framework of this chapter. In addition to the usual difficulties encountered in aquatic chemistry—very low concentrations, multiplicity of chemical constituents, presence of suspended colloidal phases—the study of photochemical processes in natural waters is complicated by the short lives of many important photochemical reactants.[1,19-23]

5.1 Input, Absorption, and Attenuation of Light in Natural Waters

The incident solar radiation reaching the earth's surface (Figure 7.9) has a spectrum encompassing near uv (300–400 nm), visible (400–700 nm), and infrared (> 700 nm) wavelengths. The infrared, which accounts for some 50% of the light energy, is absorbed and dissipated as heat in the surface waters (< 1 m). Some 90% of the rest penetrates into the water (10% is lost in reflection and backscattering), following a classical exponential attenuation curve according to the combined Beer–Lambert laws. The intensity of transmitted light at wavelength λ, $I_{t\lambda}$ (in photons cm^{-2} s^{-1}), is related to the intensity of

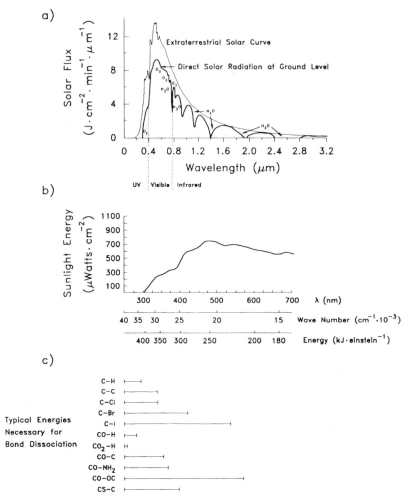

Figure 7.9 (a) Solar spectrum at the earth's surface. From Gates, 1962.[24] Comparison of the (b) uv–visible solar flux with (c) typical bond-breaking energies. From Zika, 1981.[20]

the incident light, $I_{0\lambda}$, by the expression

$$I_{t\lambda} = I_{0\lambda} 10^{-\alpha_\lambda l} \tag{109}$$

where α_λ is the wavelength-dependent attenuation coefficient of the medium and l is the pathlength. When light absorption is due to a solute rather than the medium, the concentration of the solute (C) and its wavelength-dependent extinction coefficient (ε_λ) must be included explicitly. Thus

$$I_{t\lambda} = I_{0\lambda}\, 10^{-\varepsilon_\lambda Cl} \tag{110}$$

The maximum penetration of this photon flux varies from a fraction of a meter in very turbid and colored water to some 100 m in clear ocean water. Light attenuation as a function of depth shows some variation with wavelength. Water, however, is relatively transparent to visible and near-uv light and thus the whole euphotic zone can be subjected to photochemical processes.

In photochemistry, it is the absorbed rather than the transmitted light that is of interest. The extent of light absorption may be expressed either as the reduction in light intensity

$$I_{0\lambda} - I_{t\lambda} = I_{0\lambda}[1 - 10^{-\alpha_\lambda l}] \tag{111}$$

or (more commonly) in terms of the *absorbance* A (or optical density) of the solution where

$$A = \log \frac{I_0}{I_t} = \alpha_\lambda l \tag{112}$$

If the concentration of the light-absorbing species, or *chromophore*, is included explicitly, a linear dependence of absorbance on concentration is obtained:

$$A = \varepsilon_\lambda Cl \tag{113}$$

Three principal aquatic constituents are responsible for the absorption of light. Water itself dominates light absorption in the ocean and clear water lakes.[27] Light absorption by dissolved organic matter (DOM) is due principally to the ill-defined colored humic material since identified aquatic organic solutes do not absorb in the relevant wavelength region.[19] (This may be partly tautological since photoreactive solutes produced in surface waters would be rapidly destroyed by the very large flux of light energy.) Finally, light is absorbed by living and dead suspended particles; their photoreactivity—aside from photosynthesis—is simply unknown and outside the realm of classical photochemistry. As shown by the great transparency of many saline waters, the inorganic solutes contribute little to the overall light absorption (in the visible range). Although absorption by trace constituents, particularly transition metals or organic micropollutants, plays an insignificant role in overall light attenuation, it may be very important in the chemistry of these species.

5.2 General Principles

Photochemical reactions are molecular transformations initiated by the absorption of light. These molecular transformations may involve only primary reactions of the chromophore or, in addition, secondary reactions with other molecules. Two of the fundamental rules pertaining to photochemical reactions are: (1) only the radiation *absorbed* by the molecule can lead to chemical changes and (2) each photon absorbed can activate *only one* molecule in the primary step of a photochemical process. Thus the absorption spectrum of a molecule defines its ability to reach electronically excited states, whose initial energies correspond to that of the absorbed photons.

The energy absorbed per mole, E, is given by the Plank equation:

$$E = \mathcal{N} h v = \frac{\mathcal{N} hc}{\lambda} \tag{114}$$

which relates the energy of light to its frequency, v, where h is Planck's constant. Alternatively, this expression can be written in terms of wavelength, λ, since the product of the wavelength of light and its frequency is a constant (c, the speed of light). Then, with λ in nanometers,

$$E = \frac{1.20 \times 10^5}{\lambda} \, \text{kJ mol}^{-1} \tag{115}$$

A molecule that absorbs at 520 nm must thus reach an excited state with an energy approximately 230 kJ mol^{-1} above the ground state, some large fraction of that energy being potentially available for photochemical reactions. Such high energy is reached by a negligible fraction of the molecules through purely thermal processes ($n_1/n_0 = \exp[-(E_1 - E_0)/RT] \simeq \exp[-230/2.5] \simeq 10^{-40}$ at room temperature according to the Boltzman distribution law), and it is also comparable to some bond and activation energies (cf. Figure 7.9b). This relatively large energy available from light absorption is a major reason why many chemical reactions that are unfavorable thermally are easily achieved photochemically.

The efficiency with which the absorbed light energy is utilized in photochemical processes is given by the *quantum yield*, defined as ratio of the number of molecules that photoreact to the number of photons absorbed at a given wavelength. The primary quantum yield (Φ_λ), which refers to an initial reaction following light absorption, is at the most equal to unity, and is usually less than one since processes other than chemical change may dissipate absorbed light quanta. The experimental quantum yield (ϕ_λ) can differ from the primary quantum yield if, for example, secondary thermal reactions such as free radical chain reactions are important. Because chain reactions can lead to more molecules of products than the initial number of reactant molecules, experimental quantum yields in such cases can be greater than one. The wavelength dependence of ϕ is important for photochemical redox reactions of metals and metal-organic complexes in natural waters[28] but is often neglected for reactions of organic micropollutants. In the latter case, it is necessary that ϕ be determined within the wavelength range of the light absorbed by the reacting species under natural conditions.[29]

Absorption of light results in a transition of a molecule, A, from its ground state to an electronically excited state, A^*. The molecule in the excited state can undergo several (photo)physical transformations. The electronic energy of the excited state can be dissipated as heat through vibrational relaxation processes of the molecule or through *quenching*, collision with another molecule or surface. The absorbed light energy can be reemitted as light (of lower energy than the absorbed light) with a transition of A^* back to the ground state; this reemission may be either fluorescence or phosphorescence, depending on the nature of the excited state from which the transition occurs.

Relaxation processes that dissipate absorbed energy as light or heat do not constitute photochemical reactions because no chemical transformation occurs. Chemical reactions provide an additional route for excited-state molecules to dissipate their excess energy. Both unimolecular (decomposition or rearrangement) and bimolecular reactions are observed. These are termed *direct* photochemical transformations when the chromophore itself is chemically changed. In many cases, the initial *primary photoproducts* of direct photochemical reactions are highly reactive intermediate species (termed *photoreactants* or *reactive transients*). Further thermal reactions of primary photoproducts are referred to as *indirect* photochemical transformations and ultimately produce *secondary photoproducts*. In a typical series of reactions

$$A \xrightarrow{h\nu} A^*$$

$$A^* \longrightarrow \text{primary photoproducts}$$

or

$$A^* + B \longrightarrow \text{primary photoproducts}$$

$$\text{primary photoproducts} \longrightarrow \text{secondary photoproducts}$$

Several steps may be required to obtain stable products. Indirect photochemical reactions may also be initiated if an excited-state molecule transfers its excess energy to another (ground-state) chemical species

$$A \xrightarrow{h\nu} A^*$$

$$A^* + B \longrightarrow A + B^*$$

In this case A^* acts as a *sensitizer*. The sensitized molecule, B^*, may then undergo further reactions. As we shall see, direct and indirect photochemical transformations cannot be entirely separated in natural waters. There is, however, some conceptual merit to the distinction.

5.3 Direct Photochemical Transformations

In natural waters, direct photochemical transformations can significantly affect the degradation of organic micropollutants and the redox cycling of transition metals. Direct photochemical processes are most likely to be important for those transition metals that have two stable oxidation states interconvertible by a one-electron transfer. The rate of a direct photochemical reaction can be estimated from the characteristics of the aquatic system, particularly the incident light intensity, and the properties of the chemical species undergoing the transformation, specifically its extinction coefficient and the quantum yield for the reaction.[23,25,30]

The rate of direct photolysis of a reactant X can be written

$$\frac{-d[X]}{dt} = k_{dp}[X] \tag{116}$$

The rate constant for direct photolysis, k_{dp}, depends on the quantum yield and the rate constant for light absorption by the reacting species, $k_{a\lambda}$:

$$k_{dp} = \sum_{\lambda} \phi_{\lambda} k_{a\lambda} \tag{117}$$

where the wavelength dependency of ϕ is included explicitly.[23,25,28]

The term $k_{a\lambda}$ represents the overlap between the light absorption spectrum of the reacting chemical species and the spectrum of the incident solar radiation in the aquatic environment. The product $k_{a\lambda}[X]$ is the average rate of light absorption by the reactant. It is equal to the product of the average rate of light absorption in a water column of depth D, $I_{a\lambda}$ (which is discussed in Sidebar 7.1), and the fraction of this light that is absorbed by the reactant X. If X does not contribute significantly to the overall light absorption in the water body, the fraction of light absorbed by X is approximately equal to $\varepsilon_{\lambda}[X]/\alpha_{\lambda}$. Then

$$k_{a\lambda} = \frac{\varepsilon_{\lambda} I_{a\lambda}}{j\alpha_{\lambda}} \tag{118}$$

where j is a constant introduced to reconcile units ($j = 6.02 \times 10^{20}$ for concentration in M and $I_{a\lambda}$ in photons·cm^{-3} s^{-1}).

In two limiting cases, fairly simple expressions for $I_{a\lambda}$, and thus $k_{a\lambda}$, can be obtained (see Sidebar 7.1 for detail). If the attenuation coefficient α_{λ} is large and most of the incident light is absorbed by the water over a relatively shallow depth, then the reaction rate constant is inversely proportional to the depth (D):

$$k_{a\lambda} = \frac{W_{\lambda} \varepsilon_{\lambda}}{j D \alpha_{\lambda}} \tag{119}$$

SIDEBAR 7.1

Solar Radiation in the Aquatic Environment

In the water column, the intensity of incident light is reduced as a result of light absorption both by the water and by some of its dissolved constituents (see Section 5.1). The intensity of the absorbed light is then equal to the difference between the incident and transmitted light intensities. For underwater solar radiation, the intensity of light absorbed in the water column by the medium over a depth interval D (or the average rate of absorption per unit volume), $I_{a\lambda}$ in photons·cm^{-3}s^{-1}, is described by the expression[25]

$$I_{a\lambda} = \frac{I_{d\lambda}(1 - 10^{-\alpha_\lambda l_d}) + I_{s\lambda}(1 - 10^{-\alpha_\lambda l_s})}{D} \tag{i}$$

where the incident light intensity ($I_{0\lambda}$ in Equations 109–112) is expressed explicitly in terms of the intensity of direct solar radiation ($I_{d\lambda}$) and sky radiation ($I_{s\lambda}$). Sky radiation is due to light scattering in the atmosphere. The pathlengths for direct and sky radiation are

$$l_d = D \sec \theta \tag{ii}$$

and

$$l_s = 1.2D \quad \text{(for water)} \tag{iii}$$

The terms W_λ and Z_λ correspond to $I_{a\lambda}$ under certain limiting conditions. If most of the incident light is absorbed by the medium (i.e., water),

$$I_{a\lambda} = \frac{I_{d\lambda} + I_{s\lambda}}{D} = \frac{W_\lambda}{D} \tag{iv}$$

where, by definition

$$W_\lambda = I_{d\lambda} + I_{s\lambda} \tag{v}$$

If only a small fraction ($< 5\%$) of the incident light is absorbed,

$$I_{a\lambda} = 2.303\alpha_\lambda (I_{d\lambda} l_d + I_{s\lambda} l_s) \tag{vi}$$

(continued)

Thus, with substitutions from Equations ii and iii,

$$I_{a\lambda} = 2.303\alpha_\lambda D(I_{d\lambda}\sec\theta + 1.2I_{s\lambda}) = 2.303\alpha_\lambda DZ_\lambda \qquad \text{(vii)}$$

where, again by definition,

$$Z_\lambda = I_{d\lambda}\sec\theta + 1.2I_{s\lambda} \qquad \text{(viii)}$$

Values for W_λ and Z_λ can then be used (under appropriate conditions) to replace $I_{a\lambda}$ in expressions for $k_{a\lambda}$.

In the other case, if only a small fraction ($< 5\%$) of the incident light is absorbed, then, to a first approximation, the reaction rate constant is independent of depth:

$$k_{a\lambda} = \frac{2.303\,\varepsilon_\lambda Z_\lambda}{j} \qquad (120)$$

The terms W_λ and Z_λ correspond to the intensity of light absorbed under the limiting conditions specified above and depend on such factors as latitude,

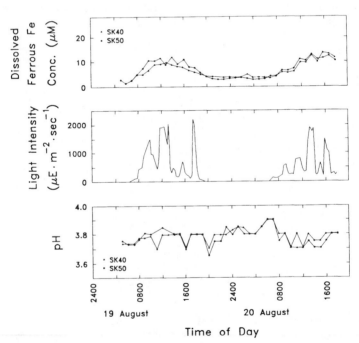

Figure 7.10 Diel photoredox cycle of iron in acidic stream waters. Ferrous iron concentration, light intensity, and pH as a function of time of day. From McKnight et al., 1988.[33]

season, and time of day. Values of W_λ and Z_λ for latitude 40°N are tabulated in reference 25 and values for other conditions can be obtained using the computer program GCSOLAR.[31] These parameters may then be used to calculate the half-life for a given chemical species with respect to removal by direct photolysis.

5.3.1 Direct Photochemical Reactions of Dissolved Metals in Natural Waters
Photoreduction of Fe(III) can be observed in low pH waters, where the reoxidation is slow.[32,33] The concentration of Fe(II) in acidic stream waters shows a diel cycle in which changes in concentration correlate with light intensity (Figure 7.10). Using the appropriate expression for the photochemical reaction rate constant as a function of light intensity, we may calculate the rate of Fe(II) production through direct photochemical reaction.

Example 13. Direct Photoreduction of Iron   The rate of direct photoreduction of dissolved Fe(III) in a thin surface layer of water can be calculated from the values of the ε_λ, ϕ_λ, and Z_λ given in Table 7.7. In this example we assume that

TABLE 7.7   Values of ε_λ and Z_λ for the Direct Photoreduction of FeOH^{2+} in Water[a]

λ (nm)	ε_λ	ϕ_λ^b	Z_λ ($\times 10^{-14}$) (photons·cm^{-2} s^{-1} for a 2.5-nm interval)	$\varepsilon_\lambda \phi_\lambda Z_\lambda$ ($\times 10^{-15}$)
297.5	2034	0.14	0.00716	0.204
300	2030	0.14	0.0240	0.681
302.5	2003	0.14	0.0723	2.03
305	1965	0.14	0.181	4.98
307.5	1920	0.14	0.305	8.20
310	1850	0.14	0.495	12.8
312.5	1798	0.14	0.717	18.0
315	1703	0.14	0.933	21.4
317.5	1623	0.13	1.15	23.9
320	1535	0.12	1.35	25.2
323	1430	0.11	2.52^c	41.0
330	1175	0.096	8.46^d	94.9
340	835	0.069	9.63	55.8
350	560	0.043	10.3	24.9
360	355	0.017	11.0	6.64

$$\Sigma \varepsilon_\lambda \phi_\lambda Z_\lambda = 341$$

[a] Midday, midsummer, latitude 40°N. References: Zepp and Cline (1977),[25] Zepp (1978),[29] and Faust and Hoigne (1990).[28]
[b] Experimental values of ϕ_λ obtained at 313 and 360 nm were used at these wavelengths. This value for ϕ_{313} was also used for $\lambda < 313$ nm. For intermediate wavelengths, values of ϕ_λ were calculated by linear interpolation between ϕ_{313} and ϕ_{360}.
[c] For a 3.75-nm interval.
[d] For 10-nm intervals.

$FeOH^{2+}$ is the predominant Fe(III) species in solution, a reasonable assumption for cloud- and fogwaters where the pH is often between 2.5 and 5. The rate of $FeOH^{2+}$ photoreduction by the reaction

$$FeOH^{2+} \xrightarrow{hv} Fe^{2+} + \cdot OH$$

is

$$\frac{-d[FeOH^{2+}]}{dt} = k_{dp}[FeOH^{2+}] \tag{121}$$

where

$$k_{dp} = \sum_{\lambda} \phi_{\lambda} k_{a\lambda} = \frac{2.303}{j} \sum_{\lambda} \phi_{\lambda} \varepsilon_{\lambda} Z_{\lambda} \tag{122}$$

From the values in Table 7.7,

$$k_{dp} = (3.41 \times 10^{17}) \left[\frac{2.303}{6.02 \times 10^{20}} \right] = 1.3 \times 10^{-3}\,\text{s}^{-1} \tag{123}$$

and the half-life for $FeOH^{2+}$ in sunlit water with respect to direct photo-reduction is

$$t_{1/2} = \frac{\ln 2}{1.3 \times 10^{-3}} = 530\,\text{s}$$

*   *   *

This example illustrates only a part of the complicated photoredox cycle of iron in natural waters. This cycle also includes the photochemical (and thermal) reduction of organically complexed Fe(III) and of Fe(III)-containing minerals as well as the reoxidation of Fe(II). Such photoreduction may also be taking place in surface waters at higher pH where the oxidation rate of ferrous iron is fast, thus maintaining Fe(II) at very low concentrations. The net effect of iron photoreduction is to accelerate the dissolution of Fe(III) minerals and the dissociation of Fe(III) chelates. Thus photochemical processes enhance the availability of iron for uptake by microorganisms[34] (see Chapter 6). Also the reoxidation of Fe(II) to Fe(III) and its precipitation as a fresh hydrous oxide may maintain iron in highly disperse colloidal form with high solubility, adsorption capacity, and affinity for aquatic solutes.

Formation of organic complexes, either in solution or at mineral surfaces, may significantly enhance the photochemical reactivity of metals in natural waters. Organic complexation of Cu(II) is important in natural waters and many

Cu(II) complexes with ligands having carboxylate and amino functional groups are photoreactive.[35] Light absorption by such complexes can result in reduction of the metal and oxidation of the ligand. Thus

$$Cu(II)L \xrightarrow{hv} Cu(II)L^*$$

$$Cu(II)L^* \longrightarrow Cu(I) + L^+$$

Degradation of the organic ligand generally accompanies this oxidation. Photochemical degradation of Fe(III)EDTA complexes may be one of the most important removal processes in natural waters for EDTA, which is relatively resistant to microbial degradation.[36] The depth profile of Cu(I) in seawater exhibits a surface maximum, consistent with a photochemical reduction mechanism (note that indirect photochemical processes may also contribute to this phenomenon). And up to 15% of the total copper in surface seawater may be present as Cu(I), whose re-oxidation to Cu(II) is inhibited by chloride complexation.[35]

5.3.2 Direct Photochemical Reactions of Metal Oxides in Natural Waters

Photochemical reactions involving solid phases are most often important for metals, such as iron and manganese, that are relatively insoluble in their higher oxidation states. The effect of photochemical reduction on manganese cycling in seawater is particularly noticeable because the oxidation rate of Mn^{2+} is relatively slow and because the two principal oxidation states of manganese in aquatic systems, Mn(II) and Mn(IV), are easily identified as the dissolved and particulate fractions of the metal, Mn^{2+} and MnO_2. The depth profiles of manganese in seawater exhibit a pronounced Mn(II) maximum and Mn(IV) minimum at the surface.[37-40] (A secondary Mn(II) maximum and Mn(IV) minimum are also often observed at the oxygen minimum where the reoxidation rate should be the slowest.) Solubilization of MnO_2 suspensions in seawater exposed to light is enhanced in the presence of humic substances.[38,41] Although the mechanism of manganese photoreduction is unknown, it is likely to involve the $+III$ oxidation state of the metal; the formal oxidation state of manganese in aquatic oxides is in fact between 2.5 and 4.

The mechanisms of photochemical reductive dissolution of minerals and the role of organic substances in this process are the subject of active research. These processes may result in the solubilization of metal oxides (as described above), the decomposition of organic compounds,[42,43] and production of transient, reactive species, such as $\cdot OH$ or H_2O_2 (see Section 5.4). Several crystalline iron oxides, including hematite, goethite, and lepidocrocite, are semiconductors. Light absorption by the bulk mineral itself occurs in the near-uv/visible region and induces or enhances electron transfer reactions at the mineral–water interface.[44] The characterization of solids as semiconductors is based on the energy gap (or band gap) between filled and vacant electronic energy levels of the material; the band gap for semiconductors is intermediate between that of insulators and metals. Light absorption (in the near-uv/visible region) by a semiconductor can result in promotion of an electron from the valence band to the conduction band. A vacancy, or hole, is left in the valence band. In this case, organic substrates adsorbed at the mineral–water interface may act merely as electron donors. That is, an electron-hole pair is generated by light absorption by the mineral and the hole is scavenged by the organic substrate, which is oxidized in the process. Alternatively, an organic ligand may form an inner-sphere surface complex that, itself, absorbs light. Thus either the oxide or the surface complex may be the chromophore. This distinction is significant because surface complexes are likely to absorb light further into the visible region than would the semiconducting oxide.[45] Thus the efficiency of light of decreasing energy (increasing λ) in promoting reductive dissolution of oxides depends on the mechanism of the photochemical process.

5.4   Indirect Photochemical Transformations

In natural waters the direct photochemical reactions described above may be of less importance than indirect photochemical transformations. Dissolved organic matter (DOM) is often the chromophore in natural waters and acts as a sensitizer to produce reactive transients. The most common of these are singlet oxygen, peroxide, and the free radical species superoxide, hydroxyl radical, and organic peroxides (see Table 7.8). Recent measurements indicate that approximately $0.1–60\,\mathrm{n}M$ of reactive radicals can be produced in seawater per minute of full sun illumination, with highest production rates observed in organic-rich estuarine waters.[46] A general scheme for the photochemical formation of reactive transients in natural waters is shown in Figure 7.11. In both the figure and table, a distinction is made between singlet and triplet excited states (which

TABLE 7.8   Photochemically Produced Reactive Transients in Surface Waters

O_2^-	Superoxide ion radical
HO_2	Hydroperoxyl radical
$ROO\cdot$	Peroxyl radical
$\cdot OH$	Hydroxyl radical
CO_3^-	Carbonate radical
Br_2^-	Dibromide radical
H_2O_2	Hydrogen peroxide
e_{aq}^-	Solvated or hydrated or aqueous electron
$^3DOM*$	"Triplet" excited state of DOM (the term "triplet" is used here only as a qualitative description of molecular properties, e.g., relative lifetime or reactivity)
$^1O_2$	Singlet oxygen

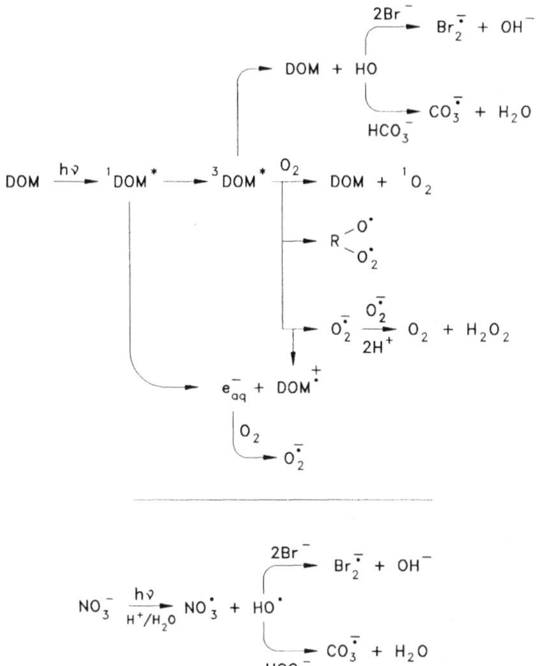

Figure 7.11 Schematic representation of pathways for the formation and transformations of photoreactants in natural waters. Adapted from Haag and Mill, 1990.[47]

differ in their net electron spin). This distinction is significant because, for quantum-mechanical reasons, the triplet excited state is often longer-lived than the singlet excited state (with respect to nonreactive relaxation) and has a correspondingly greater probability of encountering another reactive molecule in solution. Note that oxygen is unusual in having a triplet ground state and a singlet excited state ($^1O_2$). Reactive transients may also be produced through photochemical reactions involving dissolved transition metals[28,48] or metal oxides.[43,49,50] The importance of certain reactions may differ in freshwater and seawater. For example, the reaction of ·OH with Br^- is only significant in seawater.

Sources and sinks of photoreactants in natural waters are listed in Table 7.9. The constant (and generally rapid) turnover of reactive transients in illuminated waters results in steady-state concentrations of the various species ranging from 10^{-18} M for the hydroxyl radical (·OH) in open ocean surface waters to micromolar concentrations of H_2O_2, a relatively stable species.[21,22,47] Generally, the very low steady-state concentrations of the reactive transients are not directly measurable. They are estimated from the rate of reaction of added probe compounds which are chosen to react selectively with the transients of interest.[1,23]

TABLE 7.9 Reactive Transients in Surface Waters: Sinks, Fluxes, and Midday, Surface Concentrations

Transient	Sinks[a]	k_{sink}	Formation Rate ($M \cdot s^{-1}$)	Loss Rate (s^{-1})	Concentration (M)
^{3}DOM	$k_q[O_2]$	$2 \times 10^9\ M^{-1}s^{-1\,d}$	$(3\text{–}300) \times 10^{-9}$	5×10^5	$(1\text{–}5) \times 10^{-13\,d}$
$^1O_2$	$k_q(H_2O)$	$2.5 \times 10^5\ s^{-1\,e}$	$(3\text{–}300) \times 10^{-9}$	2.5×10^5	$10^{-14}\text{–}10^{-13\,b,d,f}$
					$10^{-14}\text{–}10^{-12\,c}$
					$10^{-14}\text{–}10^{-10\,c,g}$
ROO·	$k_t[\text{ROO·}]^2$	?	$10^{-11}\text{–}10^{-10}$	$0.1\text{–}1?$	
	$k_r[\text{DOM}]$	?			
·OH	$k_r[\text{DOM}]$	$1.3 \times 10^9\ M^{-1}s^{-1\,h}$	$(3\text{–}300) \times 10^{-12\,b}$	$10^{6\,b}$	$10^{-18}\text{–}10^{-17\,a,i}$
	$k_t[\text{Br}^-]^b$	$2.5 \times 10^4\ L\,mg^{-1}s^{-1\,j}$	$10^{-11}\text{–}10^{-10\,c}$	$(0.2\text{–}2) \times 10^5$	$(2\text{–}6) \times 10^{-16\,c,j}$
	$k_r[\text{DOM}]^c$		$(5\text{–}10) \times 10^{-11}$	$(0.5\text{–}1.5) \times 10^7$	$(1\text{–}2) \times 10^{-17\,c,k}$
e_{aq}^-	$k_r[O_2]$	$2 \times 10^{10}\ M^{-1}s^{-1\,h}$			
	$k_r[\text{NO}_3^-]$	$1 \times 10^{10}\ M^{-1}s^{-1\,h}$			
O_2^-	$k_t[O_2^-]^2$	$6 \times 10^{12}\ [\text{H}^+]\ M^{-1}s^{-1\,l}$	$10^{-11}\text{–}10^{-7}$	$10^{-3}\text{–}1$	$10^{-9}\text{–}10^{-8\,m}$
	$k_r[\text{DOM}]$	?			
CO_3^-	$k_r[\text{DOM}]$	$40\ L\,mg^{-1}s^{-1\,n}$	$10^{-11}\text{–}10^{-10\,c}$	$20\text{–}1000^c$	$10^{-14}\text{–}10^{-13\,c,n}$
Br_2^-	k_r	$2.5 \times 10^3\ s^{-1\,o}$	$(3\text{–}300) \times 10^{-12\,h}$	$2.5 \times 10^{3\,b}$	$10^{-15}\text{–}10^{-14\,b,i,o}$

Source: From Werner and Haag (1990).[47]

[a]Transient reacts with species indicated by brackets or parentheses. Rate constant subscript indicates type of reaction: q = quenching; t = termination of two radicals; r = other reactions

[b]Value of rate constant in seawater.

[c]Value of rate constant in freshwater.

[d]Zepp et al. (1985).[51]

[e]Rodgers and Snowden (1982).[52]

[f]Haag and Hoigné (1986).[53]

[g]Faust and Hoigné (1987).[54]

[h]Buxton et al. (1988).[55]

[i]Mopper and Zhou (1990).[56]

[j]Haag and Hoigné (1985).[57]

[k]Zepp et al. (1987).[58]

[l]Bielski et al. (1985).[59]

[m]Petasne and Zika (1987).[60]

[n]Larson et al. (1988)[61]

[o]Zafiriou et al. (1987).[62]

Some species may undergo both direct and indirect photochemical transformations. The direct photolysis of organic micropollutants and metal–organic complexes has already been mentioned; these types of compounds also react with photochemically generated reactive transients. Indirect photochemical processes, specifically reaction with ·OH, may also play a key role in the transformations and degradation of DOM.[56] Empirically derived rates of DOM removal by reaction with ·OH (together with data on typical DOM concentrations) have been used to calculate an average residence time for oceanic DOM. The calculated value of approximately 40,000 years assumes that the indirect photochemical reaction with ·OH is the only removal mechanism for DOM. Thus, this calculation will overestimate the DOM residence time if other mechanisms (e.g., direct photolysis) also contribute to DOM removal, and, indeed, the calculated residence time is about six to seven times the age of deepsea DOM determined by radiocarbon dating. This result suggests that, although removal mechanisms other than reaction with ·OH cannot be neglected, the indirect photochemical degradation of DOM is likely to contribute significantly to the removal of oceanic DOM.[56] Reaction of DOM with ·OH results in the transformation of biologically refractory organic matter to low molecular weight compounds which can be utilized by microorganisms.[63] Thus such photochemical processes may have significant implications for microbial ecology.

$$*\qquad*\qquad*$$

Overall, (nonphotosynthetic) photochemical reactions in oxygenated natural waters are ultimately oxidizing, and one of their principal impacts is probably the formation and degradation of refractory organic compounds. In some instances, these processes involve the concomitant reduction of transition metals. Depending on the reoxidation rate, thermodynamically unstable reduced metallic species may thus be maintained at significant concentrations in euphotic waters. This interaction between light, trace metals, and trace organic compounds (one could add particles) is probably a major controlling factor in the geochemistry and geobiology of surface waters.

6. REDOX KINETICS

Thus far, our discussion has focused predominantly on the energetics of redox processes or on the effects of dominant redox couples, such as O_2/H_2O or SO_4^{2-}/H_2S on trace metal speciation. This focus has allowed us to develop an equilibrium framework for redox chemistry in natural waters. But it has, at the same time, drawn our attention away from the dynamic nature of redox reactions. For instance, the results of Example 8 indicate that only Fe(II), either

Fe^{2+} or FeS(s), should be present in anoxic hypolimnetic waters, whereas Fe(III) hydroxides have been reported to persist even in anoxic lake sediments, most probably because of slow reductive dissolution kinetics.[14] Significant concentrations of trace metals may be maintained in nonequilibrium oxidation states because of photochemical or microbial processes. As discussed in Section 5.3, photochemical reactions produce Fe(II), Mn(II), and Cu(I) in oxic surface waters. The presence of reduced Se and As species in oxygenated waters has been attributed to microbial reduction.[64] Ambient concentrations of these nonequilibrium species result from the balance between the rates of their formation and elimination. Redox kinetics can also influence the fate and transport of organic micropollutants, such as chlorinated pesticides or solvents, in anoxic groundwaters or sediments.

Redox kinetics in natural waters are difficult to describe systematically for several reasons. The reactions may occur either in solution or at solid–water interfaces; in heterogeneous redox reactions, the solid may be either a reactant or a catalyst. Many important redox reactions in natural waters are microbially mediated and thus reaction rates cannot be predicted from chemical parameters alone. Even for strictly abiotic redox reactions, quantitative, a priori prediction of reaction rates is often difficult. Nonetheless, we may make some generalizations as to the form of the rate laws applicable to homogeneous or heterogeneous redox reactions and some estimates of the rate constants for such reactions, either from theory or based on experimental data.

6.1 Rate Laws for Redox Reactions

The rates of homogeneous redox reactions are generally dependent on the concentrations of both reductant and oxidant. Thus, for the general reaction

$$Red_1 + Ox_2 \longrightarrow Ox_1 + Red_2$$

the rate law would be of the form:

$$rate = k[Red_1][Ox_2] \tag{124}$$

Such a rate law is indeed followed in the oxidation of Fe(II) by O_2, a reaction particularly relevant to natural waters, thus

$$rate = \frac{-d[Fe(II)]}{dt} = k'[Fe(II)][O_2] \tag{125}$$

For a fixed partial pressure of O_2, the reaction is pseudo first order in $[Fe(II)]$ and the rate law may be written as

$$rate = \frac{-d[Fe(II)]}{dt} = k_{obs}[Fe(II)] \tag{126}$$

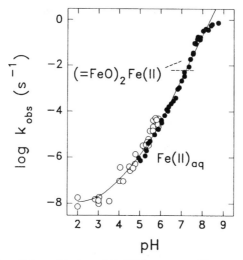

Figure 7.12 Kinetics of the oxidation of Fe(II) by oxygen ($P_{O_2} = 1$ atm). Logarithm of the observed rate constant for rate $= -d[\mathrm{Fe(II)}]/dt = k_{obs}[\mathrm{Fe(II)}]$ as a function of pH. Solid line calculated as described in the text. Dashed lines indicate oxidation of adsorbed Fe(II). From Wehrli, 1990.[65]

The dependence of k_{obs} on pH, shown in Figure 7.12, arises from the contribution of hydrolyzed iron species to the overall reaction[65] where

$$k_{obs}[\mathrm{Fe(II)}] = k_0[\mathrm{Fe^{2+}}] + k_1[\mathrm{FeOH^+}] + k_2[\mathrm{Fe(OH)_2}] \qquad (127)$$

The solid line shown with the data in Figure 7.12 is calculated by explicitly including the dependence of Fe(II) speciation on pH. If the concentration of each of the iron species is expressed as a fraction of the total Fe(II), then

$$k_{obs} = k_0 \frac{[\mathrm{Fe^{2+}}]}{[\mathrm{Fe(II)}]} + k_1 \frac{[\mathrm{FeOH^+}]}{[\mathrm{Fe(II)}]} + k_2 \frac{[\mathrm{Fe(OH)_2}]}{[\mathrm{Fe(II)}]} \qquad (128)$$

As shown in Examples 1 and 2 of Chapter 6, each of these fractions can be written as explicit functions of $[\mathrm{H^+}]$, thus

$$k_{obs} = \left(k_0 + k_1 \frac{K_1 K_w}{[\mathrm{H^+}]} + k_2 \frac{\beta_2 K_w^2}{[\mathrm{H^+}]^2} \right) \left(1 + \frac{K_1 K_w}{[\mathrm{H^+}]} + \frac{\beta_2 K_w^2}{[\mathrm{H^+}]^2} \right)^{-1} \qquad (129)$$

where K_1 and β_2 are the formation constants for the mono- and dihydroxy species (in terms of $[\mathrm{OH^-}]$) and K_w is the dissociation constant for water. Values of the rate constants (k_0, k_1, and k_2) are given in Table 7.10.

TABLE 7.10   Rate Constants^a for Reaction of Fe(II) Species with O_2

Fe(II) Species^b	Pseudo First-Order Rate Constant^c (s^{-1})	Second-Order Rate Constant^d ($M^{-1}s^{-1}$)
Fe^{2+}	1.0×10^{-8}	7.9×10^{-6}
$FeOH^+$	3.2×10^{-2}	25
$Fe(OH)_2$	1.0×10^4	7.9×10^6
$\equiv Fe(II)$	6.3×10^{-3}	5.0

^{a}From Wehrli, 1990.[65]
^{b}The species $\equiv Fe(II)$ represents Fe(II) adsorbed on the surface of goethite.
^{c}For constant $P_{O_2} = 1$ atm.
^{d}Derived from corresponding pseudo first-order rate constants. Concentration of O_2 calculated with Henry's constant of $10^{-2.9}\,M \cdot atm^{-1}$.

Iron speciation may also be incorporated into the rate law in Equation 125, in which $[O_2]$ is explicitly included. Then

$$\text{rate} = \frac{-d[\text{Fe(II)}]}{dt} = k'[\text{Fe(II)}][O_2]$$

$$= k'_0[Fe^{2+}][O_2] + k'_1[FeOH^+][O_2] + k'_2[Fe(OH)_2][O_2] \qquad (130)$$

where each second-order rate constant (in $M^{-1}s^{-1}$) is obtained by dividing the corresponding pseudo first-order rate constant (in s^{-1}) by the O_2 concentration, for example, $k' = k_{obs}/[O_2]$. Values of the second-order rate constants are also given in Table 7.10.

In this case, formation of hydroxy complexes increases the rate of Fe(II) oxygenation ($k_0 \ll k_1 \ll k_2$). In contrast, inorganic complexation of Fe(II) by Cl^- and SO_4^{2-} results in a significant decrease in the oxidation rate. Thus Fe(II) oxidation is relatively slow in seawater compared to freshwater.[66,67]

The Fe(II) oxygenation reaction can also occur as a heterogeneous reaction. At high pH, Fe(III) formed in the reaction precipitates as hydrous ferric oxide. Autocatalysis is observed because of the reaction of adsorbed Fe(II) with O_2. To account for this reaction, the terms $k_3 [\equiv Fe(II)]$ or $k'_3[\equiv Fe(II)][O_2]$, where $\equiv Fe(II)$ indicates adsorbed Fe(II), must be included in the rate laws given by Equations 126, 127, and 130, respectively. The estimated rate constants for reaction of Fe(II) sorbed on goethite[65] are roughly comparable to the rate constants for reaction of the monohydroxy species (see Table 7.10).

This example illustrates the general form of rate laws for redox reactions and the importance of speciation in redox reactions of trace metals, but leaves us with the question of why some redox reactions are more facile than others—for example, why the various Fe(II) species should react with O_2 at different rates.

Estimating (or even rationalizing) the rates of redox reactions requires some understanding of the mechanisms of these reactions.

6.2 Redox Reaction Mechanisms

Redox reactions[68-70] involve the transfer of one or more electrons between reductant and oxidant. For redox couples such as Fe(II)/Fe(III) or Cu(I)/Cu(II), only one-equivalent reactions are possible. When the stable oxidation states of the redox couple differ by two electrons, for example, Mn(II)/Mn(IV), the reactants are formally two-equivalent redox agents. Redox reactions are referred to as *complementary* reactions when both oxidant and reductant are either one-electron or two-equivalent redox agents and as *noncomplementary* when a one-electron redox agent reacts with a two-electron redox agent. It is a useful empirical observation that noncomplementary redox reactions are often slow compared to complementary reactions.

A more sophisticated rationalization of the rates of redox reactions involves the application of molecular orbital theory (MOT), described briefly in Chapter 6. Although a detailed discussion of this approach is beyond the scope of this text, the basic principles may be simply stated. In a redox reaction, electron transfer must occur from the highest occupied molecular orbital (HOMO) of the reductant to the lowest unoccupied molecular orbital (LUMO) of the oxidant. For this electron transfer to be facile, the reductant HOMO and the oxidant LUMO must be of similar energies and have compatible symmetry. Then, if the bonds made or broken in the electron transfer step are consistent with the end products of the redox reaction, the reaction is termed "symmetry-allowed" and is expected to be rapid. The observed kinetics of some important redox reactions in natural waters, such as pyrite oxidation and Fe(II) and Mn(II) oxygenation, have been explained on the basis of MOT considerations.[71,72]

Redox reactions may also be characterized as either inner- or outer-sphere reactions. The distinction between inner- and outer-sphere redox reactions depends on whether the electronic interactions are promoted by chemical bonds between reductant and oxidant. In the reaction of metal complexes, this corresponds to outer-sphere reactions in which the coordination spheres of the metals are essentially intact and inner-sphere reactions in which there is interpenetration of the coordination spheres of the metals in the transition state. The additional interaction between reductant and oxidant in the inner-sphere mechanism allows electron transfer to proceed through a more efficient, lower-energy pathway than in the corresponding outer-sphere mechanism (i.e., for the same reductant/oxidant pair). Rate laws for inner- and outer-sphere redox reactions have the same form and the reaction mechanisms cannot be differentiated on this basis.

6.2.1 *Outer-Sphere Electron Transfer*
Outer-sphere electron transfer reactions[73-75] are the most amenable to quantitative prediction of rate constants. The outer-sphere reactions consists of several steps: (1) formation of the precursor

complex, (2) electron transfer, and (3) dissociation of the successor complex

$$\text{Step 1} \quad A + B \rightleftharpoons A, B$$

$$\text{Step 2} \quad A, B \overset{ET}{\rightleftharpoons} A^-, B^+$$

$$\text{Step 3} \quad A^-, B^+ \rightleftharpoons A^- + B^+$$

As discussed in Chapter 3, reaction rate constants (or free energies of activation, $\Delta G^{\ddagger}$) can be related to equilibrium constants, that is, to the driving force for the reaction (ΔG°). A quantitative relation between $\Delta G^{\ddagger}$ and ΔG° for outer-sphere electron transfer reactions is given by the theory of Marcus.[76] It includes the contributions of the following factors to $\Delta G^{\ddagger}$: (1) the loss of translational and rotational free energy on formation of the precursor complex, (2) the work (arising from electrostatic interactions) required to bring the reactants together in the precursor complex and to separate the products from the successor complex, (3) the internal reorganization of the reactants, and (4) solvent reorganization. The contributions of these factors result in a finite $\Delta G^{\ddagger}$ (referred to as λ by Marcus) even when the electron transfer involves no net chemical reaction, $\Delta G^{\circ} = 0$.

A useful simplification of the Marcus theory is possible when comparing electron transfer reactions with no net chemical reaction (*self-exchange*) for two redox agents, that is,

$$Ox_1 + {}^*Red_1 \rightleftharpoons Red_1 + {}^*Ox_1$$

$$Ox_2 + {}^*Red_2 \rightleftharpoons Red_2 + {}^*Ox_2$$

(where the asterisk indicates an isotopically labeled species) with the electron transfer between the two redox agents (*cross-reaction*), that is,

$$Ox_1 + Red_2 \rightleftharpoons Red_1 + Ox_2$$

In this case, the factors contributing to the term λ (described above) in the cross reaction (denoted by the subscript "12") and the self-exchange reactions (denoted by the subscripts "11" and "22") effectively balance out and

$$\lambda_{12} = \tfrac{1}{2}(\lambda_{11} + \lambda_{22}) \tag{131}$$

Then the rate constant for the cross reaction (k_{12}) can be related to the rate constants for the self-exchange reactions (k_{11} and k_{22}), the equilibrium constant for cross exchange (K_{12}) and a steric and statistical factor f by the expression

$$k_{12} = (k_{11}k_{22}K_{12}f)^{1/2} \tag{132}$$

where
$$\log f = \frac{(\log K_{12})^2}{4\log(k_{11}k_{22}/z^2)} \quad (133)$$

and z is the encounter rate constant (ca. $10^{11}\,M^{-1}\,s^{-1}$). (Note that a term for the contribution of the work required to bring the reactants together in the self- and cross-exchange reactions can be, but is usually not, included explicitly in the expression for k_{12}.)

The utility of this expression may be illustrated by comparing the reactions of Fe(II) complexes with the oxidant Ce(IV), that is,

$$Ce^{IV} + Ce^{III} \overset{k_{11}}{\rightleftharpoons} Ce^{III} + Ce^{IV}$$

$$Fe^{II}L + Fe^{III}L \overset{k_{22}}{\rightleftharpoons} Fe^{III}L + Fe^{II}L$$

$$Ce^{IV} + Fe^{II}L \overset{k_{12}}{\rightleftharpoons} Ce^{III} + Fe^{III}L$$

For a series of different ligands L, the variation in k_{12} will be due chiefly to the variation in the equilibrium constant for the $Fe^{III}L/Fe^{II}L$ couple. This equilibrium constant can be directly related to the standard redox potential $E°$ ($E° = -\Delta G°/nF$). The correlation between k_{12} and the redox potentials for $Fe^{II}L/Fe^{III}L$ couples is illustrated in Figure 7.13.

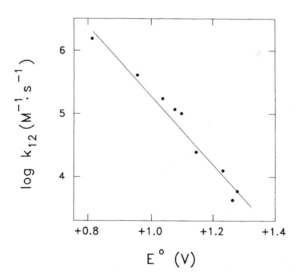

Figure 7.13 Correlation between rate constants and equilibrium constants for redox reactions based on Marcus theory. The logarithm of rate constants for the oxidation of a series of Fe(II) complexes by Ce(IV) is plotted as a function of the standard redox potential of the Fe(III)/Fe(II) couples of the respective complexes. From Burgess, 1988.[75]

TABLE 7.11 Comparison of Calculated and Observed Rate Constants for Outer Sphere Cross Reactions[a]

Reaction	Observed	Calculated
$IrCl_6^{2-} + W(CN)_8^{4-}$	6.1×10^7	6.1×10^7
$IrCl_c^{2-} + Fe(CN)_6^{4-}$	3.8×10^5	7×10^5
$Mo(CN)_8^{3-} + W(CN)_8^{4-}$	5.0×10^6	4.8×10^6
$Fe(CN)_6^{4-} + MnO_4^-$	1.3×10^4	5×10^3
$V(H_2O)_6^{2+} + Ru(NH_3)_6^{3+}$	1.5×10^3	4.2×10^3
$Ru(en)_3^{2+} + Fe(H_2O)_6^{3+}$	8.4×10^4	4.2×10^5
$Fe(H_2O)_6^{2+} + Mn(H_2O)_6^{3+}$	1.5×10^4	3×10^4

Sample Calculation[b]

$$Fe(CN)_6^{4-} + Mo(CN)_8^{3-} \rightleftharpoons Fe(CN)_6^{3-} + Mo(CN)_8^{4-}; K_{12} = 100$$

Fe self-exchange $\quad k_{11} = 740 \, M^{-1} s^{-1}$

Mo self-exchange $\quad k_{22} = 3.0 \times 10^4 \, M^{-1} s^{-1}$

$$\log f = \frac{(\log K_{12})^2}{4 \log(k_{11} k_{22}/z^2)}$$

$$z = 10^{11} \, M^{-1} s^{-1}; f = 0.86$$

$$k = (k_{11} k_{22} K_{12} f)^{1/2} = 4.4 \times 10^4$$

$$\text{obs } k_{12} = 3 \times 10^4 \, M^{-1} s^{-1}$$

[a] From Atwood (1985).[73]
[b] From Cotton et al. (1984).[69]

As shown in Table 7.11, calculations based on the Marcus cross relation often agree well with observed rate constants. Marcus theory describes a specific application (i.e., to redox reactions) of the more general linear free energy relationships between kinetic and thermodynamic constants discussed in Chapter 3.

One important aspect of this discussion is that the complexation of metals by different ligands has a significant effect on the overall driving force for electron transfer and thus on the rate of redox reactions. The stabilization of different oxidation states of a metal by various ligands plays a crucial role in biochemical electron transfer reactions.

In some cases, however, the observed rate constants for redox reactions are very significantly underestimated by calculations for outer-sphere reactions. Enhancement of the rate of electron transfer is due to increased interaction between reductant and oxidant, that is, to an inner-sphere reaction mechanism.[74]

6.2.2 Inner-Sphere Electron Transfer Very often, inner-sphere redox reactions between metal cations or complexes involve the formation of a bridged intermediate. When the overall reaction involves transfer of the bridging ligand between the reactants, an inner-sphere reaction mechanism can sometimes be demonstrated unambiguously. Inner-sphere pathways have been proven both for homogeneous and heterogeneous reactions. For the homogeneous reaction

$$Cr^{2+} + Co(NH_3)_5Cl^{2+} \xrightleftharpoons{H^+} CrCl^{2+} + Co^{2+} + 5NH_4^+$$

formation of the bridged complex must precede electron transfer because of the inertness of Cr(III) toward ligand substitution.[77] For the heterogeneous reactions

$$U_{aq}^{4+} + MO_2 \rightarrow UO_2^{2+} + M^{2+}$$

(where M = Pb or Mn) isotopic tracer experiments demonstrate that both of the oxygens in the product UO_2^{2+} derive from the solid via a bridged intermediate.[78] (Note: In contrast, the outer-sphere mechanism can only be conclusively demonstrated in the rare situation when rates of electron transfer exceed rates of ligand substitution for both reactants.)

We may write a mechanism for inner-sphere reactions with the same general steps as for outer-sphere reactions: formation of the precursor complex, electron transfer, and dissociation of the successor complex. For inner-sphere reactions, however, the first and last steps involve not only diffusion of the reactants toward and products away from each other and some distortion of the reacting species in the transition state (as is the case for outer-sphere reactions) but also bond making and bond breaking. Any one of the three steps can be rate limiting. Prediction of rates of inner-sphere electron reactions is extremely difficult because of the strong electronic coupling of reductant and oxidant.

6.3   Redox Reactions at the Mineral–Water Interface

Both inner- and outer-sphere electron transfer reactions can occur at the mineral–water interface as well as in solution. In heterogeneous reactions, the solid may be either a reactant or a catalyst in the redox reaction. In the first case, reduced minerals [such as pyrite, iron(II)-containing silicates, or uraninite] may be oxidized or oxidized minerals [such as Fe(III) or Mn(III, IV) oxides] reduced. In the second case, redox reactions may be accelerated by the association of one of the reactants with the mineral surface. In either case, the kinetics of redox reactions are constrained by a particular characteristic of surface reactions. That is, once the surface becomes saturated with some chemical species and its maximum surface concentration is attained, further increase in the dissolved concentration of that species cannot increase its rate of reaction at the surface.

6.3.1 Minerals as Redox Reactants Redox reactions at the mineral–water interface in which the solid itself is oxidized or reduced ofteń result in the dissolution of the solid because of the variation of metal solubilities with oxidation state. For a given mineral, rates of redox-enhanced dissolution can be significantly greater than rates of proton- or ligand-promoted dissolution discussed in Chapter 5. The acceleration of Fe(III) oxide dissolution in the presence of the reductant ascorbate is shown in Figure 7.14. As for nonreductive dissolution, the rates of redox-enhanced mineral dissolution are generally considered to be surface controlled; the plateau in the dissolution rate observed at high total ascorbate concentrations (Figure 7.14b) is consistent with a linear dependence of of the rate on the surface concentration of the reductant (Figure 7.14c).

As with other surface-controlled dissolution reactions, we may expect a direct dependence of the redox-enhanced dissolution rate on the surface area of the reacting mineral; such a dependence has been proposed for the dissolution of magnetite in anoxic marine sediments.[80] The dependence of the rate on the *dissolved* concentration of the reductant (or oxidant) may be expected to vary from first-order at low surface coverage of the mineral to zero-order when the surface is saturated. The extent of surface coverage is likely to vary greatly among different types of environments.

6.3.2 Minerals as Catalysts Redox reactions can also occur at the mineral–water interface in which the mineral itself is not oxidized or reduced but, rather, catalyzes the electron transfer reaction. As discussed above, the oxygenation of Fe^{2+} is accelerated both by hydrolysis and by complex formation at oxide surfaces. Similar effects are observed[81–83] for oxygenation of Mn^{2+} and VO^{2+}.

Surface catalysis can also be important in photochemical processes. Formation of H_2O_2 in illuminated suspensions of TiO_2, ZnO, and desert sand with acetate as an electron donor has been demonstrated.[50] Photochemical degradation of chlorinated phenols is also catalyzed by TiO_2; the rate of degradation depends on the adsorbed concentration of the phenol and is saturated at high concentrations as is characteristic for surface-controlled processes.[84]

6.4 Microbial Redox Reactions

As has been discussed throughout this chapter, many redox transformations in natural waters are microbially mediated. Once again, the example of metal oxygenation may serve to illustrate some of the important features of microbial redox kinetics. Figure 7.15 shows the dramatic effect of addition of "Mn-bacteria," that is, settleable material collected from hypolimnetic waters, on the rate of Mn(II) oxidation. The dependence of manganese oxidation rates in seawater on dissolved manganese concentration has been interpreted in terms of Michaelis–Menten kinetics for enzymatic processes (cf. Chapters 3 and 6). Both of these observations, however, are also consistent with (abiotic) surface catalysis. An

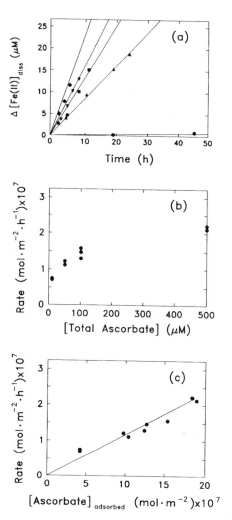

Figure 7.14 Reductive dissolution of hematite by ascorbate (at pH 3). (*a*) Dissolved concentration of Fe(II) as a function of time for initial ascorbate concentrations: (●) 0, (▲) 10, (■) 50, (▼) 100, and (◆) 500 μM. The notation $\Delta[Fe(II)]_{diss}$ indicates the correction of the observed [Fe(II)] for initial, rapid release. (*b*) Dissolution rate vs. total ascorbate concentrations showing saturation at high total ascorbate. (*c*) Dissolution rate vs. adsorbed ascorbate showing first-order dependence on surface ascorbate concentration. Adapted from Sulzberger et al., 1989.[79]

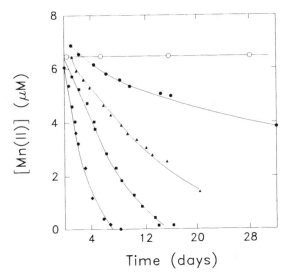

Figure 7.15 Effect of "Mn-bacteria" on Mn(II) oxygenation kinetics. Decreasing dissolved Mn(II) concentration is observed over time in lake water (pH 7.5) spiked with Mn^{2+} to which varying amounts of a suspension of "Mn-bacteria"(settleable material collected from hypolimnetic waters) were added: (○) filtered control; amounts of "Mn-bacteria" added increased in the order (●) < (▲) < (■) < (◆). From Diem and Stumm, 1984.[85]

active role of microorganisms in Mn(II) oxidation has been claimed based on the effect of microbial inhibitors[86] and the observed photoinhibition of oxidation in surface waters.[40]

* * *

In part, the difficulty in describing redox kinetics in natural waters is a result of the ubiquity of these processes. Redox reactions can transform dissolved or adsorbed species or solid phases. They can be abiotic or microbially mediated and may be induced thermally or photochemically. Clearly, we must disentangle these processes conceptually, even if they are co-occurring under natural conditions, in order to rationalize observed redox kinetics. Although quantitative predictions of the rates of redox reactions in natural waters are not yet possible, we may attempt to understand the dependence of redox reaction rates on the concentrations of reductant and oxidant and the factors affecting the magnitude of the rate constants for redox reactions.

REFERENCES

1. O. C. Zafiriou, N. V. Blough, E. Micinski, B. Dister, D. Kieber, and J. Moffett, *Mar. Chem.*, **30**, 45 (1990).

2. L. C. Sillén and A. E. Martell, *Stability Constants of Metal Ion Complexes*, Special Publication 17, Chemical Society, London, 1964.

3. W. M. Latimer, *The Oxidation States of the Elements and Their Potentials in Aqueous Solutions*, 2nd ed., Prentice-Hall, Englewood Cliffs, NJ, 1952.

4. J. Boulègue and G. Michard, *J. Francais d'Hydrologie*, **9**, 27 (1978).

5. J. Boulègue, *CR Acad. Sci., Paris*, **283**, 591 (1976).

6. G. Schwarzenbach and A. Fischer, *Helv. Chim. Acta*, **43**, 1365 (1960).

7. J. A. Raven, *Energetics and Transport in Aquatic Plants*, MBL Lectures in Biology, Vol. 4, Liss, New York, 1984.

8. A. F. Gaudy and E. T. Gaudy, *Microbiology for Environmental Scientists and Engineers*, McGraw-Hill, New York, 1980.

9. P. L. McCarty, in *Water Pollution Microbiology*, R. Mitchell, Ed., Wiley, New York, 1972.

10. M. Alexander, *Introduction to Soil Microbiology*, 2nd ed., Wiley, New York, 1965.

11. R. W. Howarth, *Biogeochemistry*, **1**, 5 (1984).

12. A. L. Lehninger, *Biochemistry*, 2nd ed., Worth, New York, 1975.

13. *CRC Handbook of Biochemistry*, 2nd ed., H. A. Sober, Ed., Chemical Rubber Co., Cleveland, OH, 1970.

14. P. Wersin, P. Höhener, and W. Stumm, *Chem. Geol.*, **90**, 233 (1991).

15. R. Gächter and D. Imboden, in *Chemical Processes in Lakes*, W. Stumm, Ed., Wiley-Interscience, New York, 1985, Chapter 16.

16. D. E. Armstrong, J. P. Hurley, D. L. Swackhammer, M. M. Shafer, in *Sources and Fates of Aquatic Pollutants*, R. A. Hites and S. J. Eisenreich, Eds., ACS Advances in Chemistry Series No. 216, American Chemical Society, Washington, DC, 1987, Chapter 15.

17. W. Stumm and J. J. Morgan, *Aquatic Chemistry*, 2nd ed., Wiley, New York, 1981, pp. 461–462.

18. J. Boulègue and G. Michard, in *Chemical Modeling in Aqueous Systems*, E. A. Jenne, Ed., ACS Symposium Series 93, American Chemical Society, Washington, DC, 1979.

19. O. C. Zafiriou, *Mar. Chem.*, **5**, 497 (1977).

20. R. G. Zika, in *Marine Organic Chemistry*, E. K. Duursma and R. Dawson, Eds., Elsevier, Amsterdam, 1981.

21. W. J. Cooper, R. G. Zika, R. G. Petasne, and A. M. Fischer, in *Aquatic Humic Substances*, I. H. Suffet and P. MacCarthy, Eds., ACS Advances in Chemistry Series No. 219, American Chemical Society, Washington, DC, 1989, pp. 333–362.

22. J. Hoigné, B. C. Faust, W. R. Haag, F. E. Scully, Jr., and R. G. Zepp, in *Aquatic Humic Substances*, I. H. Suffet and P. MacCarthy, eds., ACS Advances in Chemistry Series No. 219, American Chemical Society, Washington, DC, 1989, pp. 353–381.

23. J. Hoigné, in *Aquatic Chemical Kinetics*, W. Stumm, Ed., Wiley-Interscience, New York, 1990, pp. 43–70.

24. D. M. Gates, *Energy Exchange in the Biosphere*, Harper & Row Biological Monographs, New York, 1962.

25. R. G. Zepp and D. M. Cline, *Environ. Sci. Technol.*, **11**, 359 (1977).

26. R. C. Smith and J. E. Tyler, *Photochem. Photobiol. Rev.*, **1**, 117 (1976).

27. S. Q. Duntley, *J. Opt. Soc. Am.*, **53**, 214 (1963).

28. B. C. Faust and J. Hoigné, *Atmos. Environ.*, **24A**, 79 (1990).

29. R. G. Zepp, *Environ. Sci. Technol.*, **12**, 327 (1978).

30. A. Leifer, *The Kinetics of Environmental Aquatic Photochemistry*, American Chemical Society, Washington, DC, 1988S, 304 pp.

31. GCSOLAR may be obtained from: The Center for Exposure Assessment Modeling, Environmental Research Laboratory, U. S. EPA, Athens, GA.

32. R. Collienne, *Limnol. Oceanogr.*, **28**, (1983).

33. D. M. McKnight, B. A. Kimball, K. E. Bencala, *Science*, **240**, 637 (1988).

34. M. A. Anderson and F. M. M. Morel, *Limnol. Oceanogr.*, **27**, 789 (1982).

35. J. W. Moffett and R. G. Zika, in *Photochemistry of Environmental Aquatic Systems*, R. G. Zika and W. J. Cooper, Eds., ACS Symposium Series No. 327, American Chemical Society, Washington, DC, 1987, pp. 117–129.

36. J. H. Carey and C. H. Langford, *Can. J. Chem.*, **51**, 3665 (1973).

37. J. H. Martin and G. A. Knauer, *Earth Planet. Sci. Lett.*, **51**, 266 (1980).

38. W. G. Sunda, S. A. Huntsman, and G. R. Harvey, *Nature*, **301**, 234 (1983).

39. W. G. Sunda and S. A. Huntsman, *Deep-Sea Res.*, **35**, 1297 (1988).

40. W. G. Sunda and S. A. Huntsman, *Limnol. Oceanogr.*, **35**, 325 (1990).

41. T. D. Waite, I. C. Wrigley, and R. Szymczak, *Environ. Sci. Technol.*, **22**, 778 (1988).

42. K. M. Cunningham, M. C. Goldberg, and E. R. Weiner, *Environ. Sci. Technol.*, **22**, 1090 (1988).

43. C. Siffert and B. Sulzberger, *Langmuir*, **7**, 1627 (1991).

44. P. Pichat and M. A. Fox, in *Photoinduced Electron Transfer*, M. A. Fox and M. Channon, Eds., Elsevier, Amsterdam, 1988, pp. 242–302.

45. B. Sulzberger, in *Aquatic Chemical Kinetics*, W. Stumm, Ed., Wiley-Interscience, New York, 1990, pp. 401–429.

46. O. C. Zafiriou and B. Dister, *J. Geophys. Res.*, **96**, 4939 (1991).

47. W. R. Haag and T. Mill in *Effects of Solar Ultraviolet Radiation on Biogeochemical Dynamics in Aquatic Environments*, N. V. Blough and R. G. Zepp, Eds., WHOI Technical Report, WHOI-90-09, Woods Hole, MA, 1999, p. 82–88.

48. Y. Zuo and J. Hoigné, *Environ. Sci. Technol.*, **26**, 1014 (1992).

49. T. D. Waite and F. M. Morel, *Environ. Sci. Technol.*, **18**, 860 (1984).

50. C. Kormann, D. W. Bahnemann, and M. R. Hoffmann, *Environ. Sci. Technol.*, **22**, 798 (1988).

51. R. G. Zepp, P. F. Schlotzhauer, and R. M. Zink, *Environ. Sci. Technol.* **19**, 74 (1985).

52. M. A. J. Rodgers and P. J. Snowden, *J. AM. Chem. Soc.*, **104**, 5541 (1982).

53. W. R. Haag and J. Hoigné, *Environ. Sci. Technol.*, **20**, 341 (1986).

54. B. C. Faust and J. Hoigné, *Environ. Sci. Technol.*, **21**, 957 (1987).

55. G. V. Buxton, C. L. Greenstock, W. P. Helman, and A. B. Ross, *J. Phys. Chem. Ref. Data*, **17**, 513 (1988).

56. K. Mopper and X. Zhou, *Science*, **250**, 661 (1990).

57. W. R. Haag and J. Hoigné, *Chemosphere*, **14**, 1659 (1985).

58. R. G. Zepp, Y. I. Skurlatov, and J. T. Pierce, in *Photochemistry of Environmental Aquatic Systems*, R. G. Zika and W. J. Cooper, Eds., ACS Symposium Series No. 327, American Chemical Society, Washington, DC, 1987, Chapter 16.

59. B. H. J. Bielski, D. R. Cabelli, R. L. Arudi, and A. B. Ross, *J. Phys. Chem. Ref. Data*, **14**, 1941 (1985).

60. R. G. Petasne and R. G. Zika, *Nature*, **325**, 516 (1987).

61. R. A. Larson, L. L. Hunt, and D. W. Blankenship, *Environ. Sci. Technol.*, **21**, 492 (1977).

62. O. C. Zafiriou, M. B. True, and E. Hayon, in *Photochemistry of Environmental Aquatic Systems*, R. G. Zika and W. J. Cooper, Eds., ACS Symposium Series No. 327, American Chemical Society, Washington, DC, 1987, Chapters 7 and 8.

63. D. J. Kieber, J. McDaniel, and K. Mopper, *Nature*, **341**, 637 (1989).

64. J. J. Wrench and C. I. Measures, *Nature*, **299**, 431 (1982).

65. B. Wehrli, in *Aquatic Chemical Kinetics*, W. Stumm Ed., Wiley-Interscience, New York, 1990, pp. 311–336.

66. F. J. Millero, *Geochim. Cosmochim. Acta*, **49**, 547 (1985).

67. F. J. Millero, S. Sotolongo, and M. Izaguirre, *Geochim. Cosmochim. Acta*, **51**, 793 (1987).

68. E. A. Cotton and W. Wilkinson, *Advanced Inorganic Chemistry*, 3d ed., Wiley-Interscience, New York, 1972, pp. 672–680.

69. E. A. Cotton, G. Wilkinson, and P. L. Gans, *Basic Inorganic Chemistry*, 2nd ed., Wiley, New York, 1984, 708 pp.

70. J. E. Huheey, *Inorganic Chemistry*, 3d ed., Harper International, New York, 1983, pp. 559–564.

71. G. W. Luther, III, *Geochim. Cosmochim. Acta*, **51**, 3193 (1987).

72. G. W. Luther, III in *Aquatic Chemical Kinetics*, W. Stumm, Ed., Wiley-Interscience, New York, 1990, pp. 173–198.

73. J. D. Atwood, *Inorganic and Organometallic Reaction Mechanisms*, Brooks/Cole, Monterey, CA, 1985, 322 pp.

74. J. J. Zuckerman, Ed., *Inorganic Reactions and Methods*, Vol. 15, VCH Publishers, FL, 1976, pp. 1–174.

75. J. Burgess, *Ions in Solution*, Ellis Horwood, Chichester, 1988, 191 pp.

76. R. A. Marcus, in *The Nature of Seawater*, E. D. Goldberg, Ed., Dahlem Conference Proceedings, 1975, pp. 477–504.

77. H. Taube, H. Myers, and R. L. Rich, *J. Chem. Soc.*, **75**, 4118 (1953).

78. G. Gordon and H. Taube, *Inorg. Chem.*, **1**, 69 (1962).

79. B. Sulzberger, D. Suter, C. Siffert, S. Banwart, *Mar. Chem.*, **28**, 127 (1989).
80. D. E. Canfield and R. A. Berner, *Geochim. Cosmochim. Acta*, **51**, 645 (1987).
81. W. Sung and J. J. Morgan, *Geochim. Cosmochim. Acta*, **45**, 2377 (1981).
82. S. H. R. Davies and J. J. Morgan, *J. Coll. Int. Sci.*, **129**, 63 (1989).
83. B. Wehrli and W. Stumm, *Geochim. Cosmochim. Acta*, **53**, 69 (1989).
84. H. Al-Ekabi, N. Serpone, E. Pellizzetti, C. Minero, M. A. Fox, and R. B. Draper, *Langmuir*, **5**, 250 (1989).
85. D. Diem and W. Stumm, *Geochim. Cosmochim. Acta*, **48**, 1571 (1984).
86. J. W. Moffett, *Nature*, **345**, 421 (1990).

PROBLEMS

7.1 Given $[Mn]_T = [Co]_T = 10^{-6}\, M$, what are the principal oxidation states of manganese and cobalt in oxic waters at $pH = 7$? What concentration of a complexing ligand ($K_{Mn}^{app} = 10^{11}$; $K_{Co}^{app} = 10^{12}$) is necessary to keep Mn(II) and Co(II) in solution? Answer the same questions at $pH = 9$.

7.2 Denitrification can lead to the formation of NO(g) or N_2O(g) instead of N_2(g).
a. At $pH = 7$, under what partial pressure of NO or N_2O in the atmosphere are these processes favorable compared to N_2 evolution?
b. At $pH = 10$?
c. At $pH = 5$?

7.3 Suppose you were a chemoautotroph trying to make a living out of oxidizing Fe(II) to Fe(III) with oxygen, in a solution saturated with $FeCO_3$(s) ($pH = 8$; $C_T = 10^{-3}\, M$) and am.Fe(OH)$_3$(s). Assuming 10% overall efficiency, how much iron do you need to oxidize in order to fix 1 mole of inorganic carbon. (Assume that Fe^{2+} is the reductant in C fixation.) Discuss the effects of pH and Fe(II) or Fe(III) chelation on your survival.

7.4 An aquifer is polluted with methanol. Upon exhaustion of the oxygen and the nitrate, hematite becomes the principal electron acceptor for methanol oxidation by the bacterium *Shewanella putrefaciens*. Write the appropriate complete redox reactions. Calculate the free energy of the reaction assuming $P_{CO_2} = 10^{-1.5}$ atm, precipitation of $FeCO_3$(s), and $pH = 7.3$. Under the same conditions, calculate the free energy of the reaction as a function of pH. Making appropriate assumptions, can you identify a pH range where denitrification may be less favorable or sulfate reduction more favorable than iron reduction? Discuss.

7.5 During a storm a reduced swamp doubles its volume and reaches complete acid–base and redox equilibrium. (Assume organic carbon formation and utilization to be negligible.) Given the initial conditions

$$pH = 8.0$$
$$Alk = 10^{-3} M$$
$$[SO_4^{2-}]_T = 4 \times 10^{-3} M$$
$$[H_2S]_T = 4 \times 10^{-4} M$$

and assuming rainwater to be slightly polluted (Alk $= -10^{-4} M$) and saturated with atmospheric CO_2 and O_2 ($[H_2CO_3^*] \simeq 10^{-5} M$; $[O_2.aq] \simeq 3 \times 10^{-4} M$), what is the final composition (Alk, pH, pe) of the swamp?

7.6 With the data of Tables 7.1 and 6.3 prepare a pe–pH diagram for chromium. What are the principal stable forms of chromium in natural waters?

7.7 **a.** Given the data of Tables 7.1, 6.3, and 5.4, prepare a pe–pH diagram for iron in an oxidized carbonated system ($[CO_3]_T = 10^{-3} M$; pe > 0); the major species to consider are

$$Fe(OH)_x^{(3-x)+} \qquad Fe^{2+}$$
$$\alpha\text{-}Fe_2O_3(s) \qquad FeCO_3(s)$$
$$Fe_3O_4(s) \qquad Fe(OH)_2(s).$$

b. Prepare a pe–pH diagram for iron in reduced systems (pe < 0), given $S_T = 10^{-4} M$, $[CO_3^{2-}]_T = 10^{-3} M$. Do not consider elemental sulfur, sulfite, thiosulfate, polysulfides, or pyrite; major species to consider are Fe^{2+}, FeS(s), $FeCO_3$(s), and $Fe(OH)_2$(s).

c. Complete the diagram by determining the boundaries between parts a and b.

7.8 An anoxic freshwater is characterized by

$$Na_T = Ca_T = Mg_T = K_T = Cl_T = [SO_4]_T = C_T = [S\cdot -II]_T = 10^{-3} M$$

a. Given pH $= 8.0$, calculate the alkalinity, the carbonate, and the sulfide species and the partial pressures of CO_2 and H_2S.

b. Calculate the pe and the partial pressure of O_2.

c. Is methane fermentation likely to take place in this system?

d. Write the complete redox reactions for methane fermentation of methanol (CH_3OH) and formaldehyde (CH_2O). What are the qualitative effects of such fermentation on alkalinity and pH?

e. Using the buffer capacity of the system (considered closed and isolated from gas phases), calculate the pH change due to fermentation of $10^{-4}\,M$ CH_2O.

f. Considering that $10^{-3}\,M$ of O_2 are introduced into the system and react rapidly with the sulfide, calculate the new composition of the system: pe, pH, alkalinity, carbonate, and sulfide species (no gas exchange).

CHAPTER 8

REACTIONS ON SOLID SURFACES

As noted in Chapter 5, some 80% of the material transported by rivers and discharged into the oceans is in the form of suspended particles, and only some 20% in the form of dissolved species. Acid–base and precipitation–dissolution reactions control the exogenic cycle of the major cations and anions. In contrast the geochemistry of most trace constituents is controlled by their reaction with solid surfaces. For example, simple chemical models for the residence time of many trace elements in the oceans are based on particle sedimentation rates and on the partitioning of species between the dissolved and particulate fractions.[1,2] (Other, biological, models, applicable to "bioactive" elements, emphasize the uptake of trace elements by living microorganisms.) In most cases this partitioning can be shown to be controlled not by the precipitation of a pure solid phase, but by the adsorption of the element on the surface of a separate mineral. The extent of this adsorption is often the only chemical parameter entering such geochemical models.

The topic of solute adsorption on suspended solids in aquatic systems is one that has undergone important conceptual evolution over the past two decades. Physical processes, particularly the electrostatic interactions between surfaces and solutes, used to receive most of the attention. The main focus is now on the chemical reactions between solutes and functional groups at surfaces.[3,4] As argued by Stumm and Morgan (1981),[5] we have now witnessed a return to the even more ancient *chemical* theory of surface processes. The new theory, embodied in the *surface complexation model*, describes surface adsorption reactions as consisting both of a chemical bonding of solutes to surface atoms and of an electrostatic interaction between ions and charged surfaces.

By considering that the reactions between solutes and functional groups on solid surfaces are effectively coordination reactions that obey mass law equations, we may apply the concepts and the mathematics of complexation processes to adsorption. It is a convenient way to introduce the topic, focusing on the nature and coordinative properties of the surface groups found on aquatic particles. The importance of adsorption in the speciation of trace aquatic constituents is illustrated in numerical examples showing how both metals and ligands—sometimes interacting with each other—can be controlled by adsorption on particles such as those of aluminum oxide.

A major difference between complexation reactions among (small) solutes and coordination reactions on solid surfaces stems from the long-range nature of electrostatic interactions and the physical proximity of adsorption sites. We have seen how long-range electrostatic forces affect the interactions between metal ions and polyelectrolytes such as humic acids. We have also indicated how these effects can be accounted for by separating the total free energy of coordination into a purely chemical term and an electrostatic term. There lies the main difficulty in applying complexation concepts to adsorption processes; no universal equilibrium constant can be simply defined and various adsorption models differ principally by the manner in which the electrostatic interaction term is calculated. This subject is addressed here, starting with a study of the charging mechanism for aquatic particles and the presentation of the classical Gouy–Chapman theory of the double layer to calculate the surface potential. Superposition of the resulting coulombic energy of interaction with purely chemical energy thus leads to a simple one-layer adsorption model. This is applied to a study of the acid–base and ligand and metal adsorbing properties of hydrous ferric oxide, a ubiquitous adsorbent in natural waters.[6]

Finally, a few related topics such as ion exchange, adsorption–desorption kinetics, and particle coagulation are briefly introduced.

1. AQUATIC PARTICLES (NATURAL HYDROSOLS)

Suspended particles in natural waters comprise both inorganic solids and dead or live organic matter, the proportion of each varying widely in time and place. Inorganic solids account for most of the total suspended matter transported by rivers and consist principally of alumino-silicates derived from physical erosion and partial weathering of continental rock. They constitute the red clays that dominate the composition of much of the ocean sediments. Other inorganic solids such as oxides and carbonates (e.g., FeO_x, MnO_x, $CaCO_3$, and even SiO_2) are in large part precipitated in situ and account for most of the total mass deposited on the ocean floor. Only a fraction of this mass is accumulated in the deep sea sediments, however, because of partial redissolution of the major "organic"* constituents, SiO_2 and $CaCO_3$. In the euphotic zone of lakes and oceans, organisms—principally bacteria, microalgae, and zooplankton— typically contribute a major portion ($> 90\%$) of the total mass of suspended particles, most of it (ca. 75%) in the form of dead remains and fragments.[7,8] In the deep ocean where suspended solids are chiefly inorganic, the major part of the suspended organic matter originates from bacteria and zooplankton.

The concentration range of suspended particles in natural waters spans some seven orders of magnitude, from 0.01 mg L^{-1} in the deep ocean to $50,000$ mg L^{-1} in particularly turbid rivers and estuaries. Concentrations of a few milligrams per liter are characteristic of productive surface waters; eutrophic lakes may contain up to a few hundred milligrams per liter of suspended solids.

The size range of aquatic particles is also very large, from 10^{-9} m for colloidal

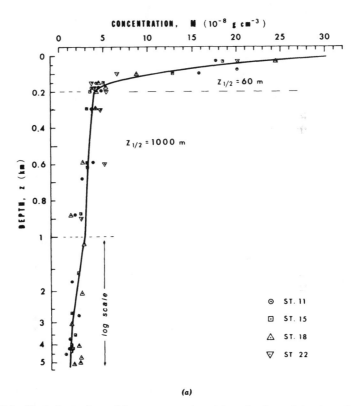

(a)

Figure 8.1 Variation of particle concentration (a) and of particle size distribution (b) as a function of depth in the Pacific Ocean. Note the constancy of the slope of the logarithmic size distribution (the exponent of the lower law) with depth as the particle concentration varies by a factor of more than 10. (Newer data indicate lower deep ocean concentrations, on the order of 10^{-9} g cm^{-3}.) From Lerman, 1979.[9]

*Recall that when referring to sediments, oceanographers use the adjective "organic" to designate any material derived from organisms, including SiO_2 and $CaCO_3$. Recall also that in the same context the adjective detrital refers not to fecal or decomposing organic matter but to solid material of continental origin, mostly alumino-silicates.

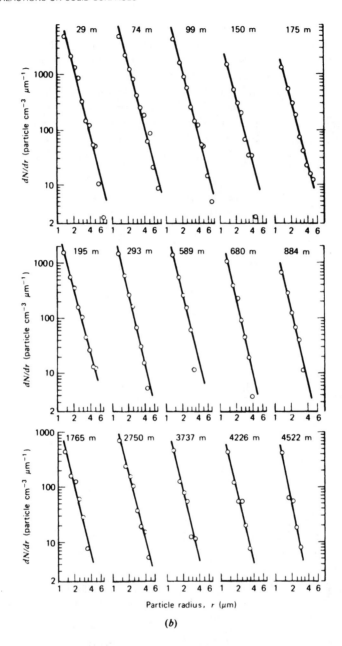

Particle radius, r (μm)

(b)

oxides and humic substances to 10^{-2} m for large zooplankton or fecal aggregates. (We are not considering here a wide assortment of important, large particles such as sargassum weed, marine mammals, icebergs, ships, flotsam, tar balls, and bottles with desperate messages.) The particle size distribution usually follows a power law function of the form[9]

$$n(r) = \frac{dN}{dr} = Ar^{-p} \tag{1}$$

where A and p are constants, and dN is the number of particles per fluid volume with radii between r and $r + dr$. The total number of particles is thus given by the integral

$$N = \int_{r_{min}}^{r_{max}} n \, dr$$

In aquatic systems the exponent p of such a power law ranges from 2 to 5, most of the data being well approximated by an exponent close to 4 in the size range $1-100 \, \mu m$. This size distribution is observed at all depths in the oceans, even though the total particle concentration decreases sharply with depth in the top mixed layer (Figure 8.1). A high power distribution can be predicted by considering a steady state between aggregation of small particles and settling of larger aggregates[11] (see Section 7).

On the basis of the size distribution one can calculate the total number, area, and volume of particles in any size class. When the exponent of the power law is 4, the number of particles of radius r is proportional to r^{-3}. Thus the small particles clearly dominate numerically. They also contribute most of the area available for adsorption (proportionally to r^{-1}). In contrast, the total particle volume, and thus approximately the mass, are distributed uniformly over the various size classes.

2. COORDINATIVE PROPERTIES OF SURFACES

As we shall see in this section, the surfaces of aquatic particles contain functional groups whose acid–base and other coordinative properties are similar to (and can be correlated with) those of their counterparts in dissolved complexes. What, then, makes the coordinative behavior of surfaces different and so difficult to study in comparison with inorganic and organic complexing weak acids such as those we examined in Chapter 6? The major answer to this question lies simply in the geometric restrictions imposed by the solid nature of surfaces. Before examining the surface chemistry of specific types of solids let us briefly mention the effects of these geometric restrictions that are common to all solids.

2.1 General Considerations

First of all, and as mentioned earlier, the obligatory proximity of the various functional groups on a surface makes them susceptible to long-range coulombic (i.e., electrostatic) interactions from other neighboring groups. (Recall that this is also the case for polyelectrolytes such as humic acids.) Consider, for example, a surface composed of identical negatively charged weak acid groups. Upon acidimetric titration the tendency of these groups to coordinate a proton decreases as the increasingly positive charge of the overall surface, resulting from the coordination of H^+, contributes a decreasing attractive electrostatic energy of bonding to each remaining acid group. The net result is a continuously decreasing equilibrium constant for the protonation reaction, a situation that renders difficult both the analysis of experimental data and the calculation of model systems.

The geometry of solid-solution interfaces creates other complications. The necessary exclusion of water molecules from the solid volume may necessitate a greater degree of dehydration for coordinated species than in similar complexes in solution. Owing to these and other steric effects, and to the influence of neighboring atoms, chemically similar groups in different geometrical configurations may exhibit very different affinities for solutes. Superimposed on the continuously varying coulombic effect, the chemically reactive groups at solid surfaces may then also exhibit an intrinsic variability in their coordinative properties.

Finally, solid surfaces in solution exhibit adsorptive properties in the absence of any specific chemical affinity for solutes. Obviously, electrostatically charged surfaces—whatever the origin of the charge—must attract ions of opposite charge. In addition, the mere exclusion of water molecules from the solid volume makes it energetically favorable for hydrophobic molecules to be situated at the solid–water interface. (The total free energy of the system includes a term for the surface tension which is decreased by organic adsorption.) These two adsorptive processes based on purely electrostatic and hydrophobic effects must be integrated into a general thermodynamic description of the coordinative properties of solid surfaces in solution.

2.2 Adsorption Isotherms

The adsorption of a solute C on a solid X is traditionally quantified by an *adsorption density* parameter Γ (moles per gram) representing the number of moles of C adsorbed per unit mass of X. Other dimensions such as grams per gram, moles per mole , and moles per square meter are of course possible and equivalent. Various theoretical and empirical mathematical expressions for Γ as a function of medium chemistry, chiefly the concentration of C, can be formulated. They are known as adsorption isotherms because they are strictly applicable to systems at constant temperature—a condition much more critical in the gaseous systems for which the formulae were originally developed.

Consider first the simplest case in which the adsorption process can be represented as a coordination reaction with 1:1 stoichiometry:

$$\equiv X + C = \equiv XC \qquad (2)$$

where $\equiv X$ is now taken to represent an adsorptive surface site on the solid. If a constant standard free energy of adsorption, ΔG^o_{ads}, can be assigned to this reaction, the mass law is written

$$\frac{\{\equiv XC\}}{\{\equiv X\}\{C\}} = K_{ads} = \exp\left[-\frac{\Delta G^o_{ads}}{RT}\right] \tag{3}$$

Making the further assumptions that (1) the activities of the surface species are proportional to their concentrations* (a key assumption throughout this chapter), and (2) the number of adsorption sites, $[\equiv X]_T$, is fixed, we can combine this mass law with the mole balance for surface sites:

$$TOT \equiv X = [\equiv X] + [\equiv XC] = [\equiv X]_T \tag{4}$$

to yield the expression

$$[\equiv XC] = [\equiv X]_T \frac{K_{ads}[C]}{1 + K_{ads}[C]} \tag{5}$$

Defining $\Gamma = [\equiv XC]$/solid mass and $\Gamma_{max} = [\equiv X]_T$/solid mass and substituting in Equation 5 yield the well-known expression of the *Langmuir isotherm*:

$$\Gamma = \Gamma_{max}\frac{K_{ads}[C]}{1 + K_{ads}[C]} \tag{6}$$

This hyperbolic expression is widely used to interpret adsorption data that demonstrate a saturation in surface coverage (Figure 8.2).

Although the great majority of adsorption data do exhibit a decrease in incremental adsorption at increasing solute concentrations, not all show the clear maximum of the Langmuir isotherm. To generalize the issue, let us write the adsorption reaction as a simple change in the nature of the species C:

$$C = C_{ads}; \quad \Delta G^{o\prime}$$

The free energy of the sorption reaction can thus be expressed as a function of the dissolved and absorbed concentrations (both expressed in moles per liter of solution):

$$\Delta G = \Delta G^{o\prime} + RT \ln \frac{[C_{ads}]}{[C]} \tag{7}$$

---

*Surface species concentrations can be expressed in moles per liter of solution, per gram of solid, per square meter of solid surface, or per mole of solid. For convenience we choose here the first of these possibilities (moles per liter) which provides a consistent set of units for all aqueous and adsorbed species.

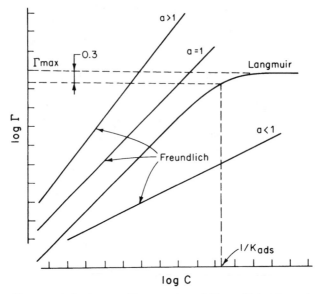

Figure 8.2 Characteristic shapes of Langmuir and Freundlich adsorption isotherms.

In order for the ratio of surface to bulk concentrations, $[C_{ads}]/[C]$, to decrease at increasing concentrations, the adsorption reaction must somehow become less favorable, $\Delta G^{o\prime}$ must increase. If we compare Equation 7 with that corresponding to the Langmuir equation,

$$\Delta G = \Delta G^o + RT \ln \frac{[\equiv XC]}{[\equiv X][C]} \tag{8}$$

we see that the Langmuir equation (in which C_{ads} is represented by $\equiv XC$) provides for an increase in the effective free energy of adsorption $\Delta G^{o\prime}$ through the decrease in the activity of free surface sites $[\equiv X]$ as the surface becomes saturated:

$$\Delta G^{o\prime} = \Delta G^o - RT \ln [\equiv X] = \Delta G^o - RT \ln([\equiv X]_T - [C_{ads}]) \tag{9}$$

As $[C_{ads}]$ approaches $[\equiv X]_T$, $\Delta G^{o\prime}$ goes to infinity and further adsorption becomes energetically impossible. There are obviously many alternative formulations to account for an increase in $\Delta G^{o\prime}$ as the adsorption reaction proceeds. Some are purely empirical, some are based on theories; all result in an isotherm expression—typically attached to a Germanic name—relating the concentration sorbed at the surface to that in the bulk solution.

Perhaps the simplest approach is to consider that the free energy of adsorption increases linearly with the log of the concentration of adsorbate. It is then not necessary to consider that the solid surface ever becomes saturated:

$$\Delta G^{o\prime} = \Delta G^{o} + \alpha R T \ln [C_{ads}] \tag{10}$$

At equilibrium

$$\Delta G = 0 = \Delta G^{o} + (1 + \alpha) R T \ln[C_{ads}] - R T \ln [C]$$

therefore

$$\ln [C_{ads}] = -\frac{\Delta G^{o}}{(1 + \alpha) R T} + \frac{1}{1 + \alpha} \ln [C] \tag{11}$$

Defining $a = 1/(1 + \alpha)$ and $\ln A = -a \Delta G^{o}/RT$, we obtain the exponential expression

$$[C_{ads}] = A C^{a} \tag{12}$$

Once divided by the solid mass to obtain the adsorption density, this expression is known as the Küster or Freundlich isotherm:

$$\Gamma = A' C^{a} \tag{13}$$

which accommodates many data sets over some range of concentrations.

Of the many other simple formulations that correspond to an increase in the free energy of adsorption with surface coverage, there is particular interest in those that yield a continuum between adsorption and precipitation ($\Delta G^{o\prime}$ varies from ΔG^{o}_{ads} to ΔG^{o}_{ppn}). The famous Brunauer, Emmett, and Teller (BET) isotherm for gases corresponds to such a formulation; by considering adsorption at multiple layers over the surface, it provides a continuum between monolayer gas adsorption on a solid and ultimate condensation of the gas.[12] Similar formulations for solute adsorption on suspended solids are possible. For example, one can consider, at the surface of a solid, the formation of a solid solution (see Section 6.2) whose composition varies continuously between that of the original solid (the adsorbent) and a pure precipitate of the adsorbate.[13] Such a model is useful in describing adsorption at high surface coverage.

2.3 The Complexation Model for Adsorption

The basic adsorption model used throughout this chapter is a straightforward generalization of the Langmuir formulation: instead of considering that only a single 1:1 complex can be formed at the surface of the solid, we allow for the presence of any number of such surface complexes with any reasonable stoichiometry. As in the Langmuir derivation, the fundamental assumption (hence the word "model") is that one may apply mass laws to the corresponding complexation reactions, using activities of surface species that are proportional to their concentrations.

We thus consider that aquatic surfaces contain functional groups with well-defined coordinative properties (surface sites), that for each type of surface site a total concentration $\equiv X_T$ can be defined, and that a standard free energy of adsorption ΔG°_{ads} can be assigned to each adsorption reaction. To account separately for the electrostatic adsorption term, the free energy of adsorption can be taken as the sum of an intrinsic and a coulombic free energy:

$$\Delta G^\circ_{ads} = \Delta G^\circ_{int} + \Delta G^\circ_{coul} \tag{14}$$

The corresponding adsorption constant is of course given as the product of two constants:

$$K_{ads} = K_{int} \cdot K_{coul} \tag{15}$$

with the usual formula for each constant:

$$K = \exp\left(-\frac{\Delta G^\circ}{RT}\right)$$

For the adsorption reaction

$$\equiv X + C = \equiv XC$$

the mass law is then written

$$\frac{[\equiv XC]}{[\equiv X][C]} = K_{ads} = K_{int} K_{coul} \tag{16}$$

In the case of a polydentate surface complex, the formulation of the mass law presents a minor difficulty. Consider, for example, a general adsorption reaction in which a solute C reacts with x surface sites $\equiv X$:

$$x \equiv X + C = \equiv X_x C \tag{17}$$

One may be tempted to write the mass law

$$\frac{[\equiv X_x C]}{[\equiv X]^x [C]} = K_{ads} \tag{18}$$

However, the exponent x in this expression has doubtful validity. At a naive level, it expresses that the probability of finding x sites together for C to react with is proportional to the xth power of the site concentrations. This is valid for "independent" molecules in solution, but it is certainly not true of fixed sites on a surface. A more satisfying approach is to consider that C is reacting with a single multidentate surface species $\equiv X_x$. At low surface coverage the concentration of such a species is simply given by $[\equiv X]/x$. As long as x is not too large (typically $x \leqslant 3$) and the surface is not close to saturation, an exponent of 1 is more satisfactory in the mass law expression:

$$\frac{[\equiv X_xC]}{[C][\equiv X]/x} = K \tag{19}$$

The coefficient $1/x$ can of course be included in the equilibrium constant (as are the activity coefficients) leading to the expression

$$\frac{[\equiv X_xC]}{[\equiv X][C]} = K_{ads} = K_{int} K_{coul} \tag{20}$$

To describe adsorption processes in aquatic systems on the basis of the surface complexation model, it is then necessary and sufficient to define the stoichiometry and the intrinsic constants of the various surface species and to provide an expression for the coulombic term. To make matters simple, we shall consider initially only the intrinsic adsorption term, effectively ignoring the variable electrostatic interaction. This is equivalent to studying the adsorption process under conditions where the surface is (approximately) uncharged or the coulombic term is taken to be constant and included in the overall adsorption constant. The energetics and the stoichiometry of adsorption can then be studied without the complication of unspecific long-range coulombic interactions.

As we have noted for other chemical processes, over a certain range of conditions often only one reaction dominates the chemical behavior of a complex system. When this is true for adsorption processes—a situation often encountered in dilute solutions at fixed pH—the surface complexation model simply degenerates to the Langmuir formulation, a simple and satisfactory result in view of the wide applicability of the Langmuir isotherm to experimental data.

3. ADSORPTION OF METALS AND LIGANDS ON AQUATIC PARTICLES

3.1 Oxide Surfaces

The bulk of the recent literature on adsorption in aquatic systems deals with metal oxides, principally those of silicon, aluminum, iron, and manganese. At the surface of a solid not all the coordinative possibilities of the metal or the ligand atoms constituting the solid can be satisfied. If the metal is in excess of the ligand, the surface must exhibit metal-like coordination properties; conversely if the ligand is in excess, the surface must behave like a ligand toward

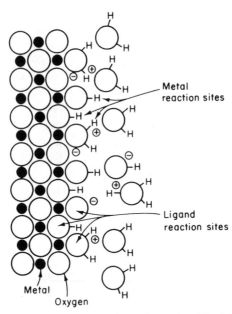

Figure 8.3 A schematic model of the surface of metal oxides illustrating the sites of metal and ligand reactions. The surface oxygen atoms, hydroxyl groups, and water molecules act as exchangeable ligands, giving the surface both ligand- and metal-like properties. Modified from Schindler, 1981.[14]

solutes at the interface. For dispersed suspensions of an oxide in water, both situations are true simultaneously because the medium itself is effectively an infinite source of exchangeable ligands. The metal atoms closest to the surface can always bind the appropriate number of hydroxyl groups from the solution to fulfill their coordination requirements (Figure 8.3). The surface can thus be seen either as ligand rich, if one considers the surface layer of hydroxyl groups as belonging to the solid, or metal rich if the surface hydroxyl groups are seen as adsorbed solution species. Indeed, metal oxides are observed to adsorb both metals and ligands from solution.

The most commonly accepted symbolism is to represent surface groups of metal oxides by a general hydrolyzed species $\equiv X\text{OH}$. Acid–base reactions are then written by considering the loss or gain of a proton rather than that of a hydroxide ion, as is more usual for the soluble hydroxide species:

$$\equiv X\text{OH} = \text{H}^+ + \equiv X\text{O}^-$$

$$\equiv X\text{OH} + \text{H}^+ = \equiv X\text{OH}_2^+$$

Replacing H^+ by other metals provides the symbolism for metal adsorption reactions and ligand adsorption is represented by an exchange with the hydroxyl groups as illustrated in Figure 8.4. Several types of species may be formed in this manner including ion pairs with major bulk electrolytes, surface chelates of metal ions (since the surface is clearly multidentate), and ternary surface complexes in which metal–ligand combinations are adsorbed at the surface.

One should be careful to distinguish between the physical reality of these various species—a reality that is difficult to ascertain experimentally—and their appropriateness in representing experimental data in the framework of a particular adsorption model. For example, one may argue whether or not ion pairs such as $\equiv XO.Na$ or $\equiv XOH_2.Cl$ do or do not actually exist—and if they do exist, whether they are inner or outer sphere complexes as suggested by their name. The issue is more conceptual than experimental. The undeniable experimental fact is that the nature and the concentration of the background electrolyte affect the adsorption behavior of oxides, including their acid–base properties. As is the case for solutions, but without as much experimental backing, one may wish to account for such effects by either ion pair formation or activity coefficient corrections (included in an effective equilibrium constant). From a practical point of view, the advantage of more universal applicability of the ion pair model is largely offset by its greater complexity and its greater data requirement compared with the effective constant approach. The degree of hydration of adsorbed ions is of course of great theoretical importance in understanding the nature of the interactions between solid surfaces, solutes, and water.

The choice of appropriate surface species such as $\equiv XOM^+$, $(\equiv XO)_2M^\circ$, or $\equiv XOMOH^\circ$, is governed either by available surface spectroscopic data (which are still rarely unambiguous) or by the necessity to match experimental partitioning data describing the effects of pH and adsorbate concentrations on adsorption. Since pH is also the principal parameter determining surface charge, the choice of the species and the values of the corresponding intrinsic constants are largely dependent on the way the coulombic interaction is evaluated. To illustrate the effect of adsorbate concentration and pH on adsorption let us now consider the case of adsorption of a metal on a metal oxide.

Example 1. Adsorption of Lead on Alumina   Consider a solution containing a background electrolyte (say 0.1 M NaCl) and ad hoc concentrations of strong acid and base to adjust the pH in the range 4–7.

The solution contains a suspension of alumina (γ-Al$_2$O$_3$) and a low concentration of a lead salt. The total number of adsorption sites on alumina ($[\equiv X]_T$, expressed here in moles per liter) is obtained by multiplying the solid mass with the specific site density (approximately $1.3 \times 10^{-4}\,mol\,g^{-1}$ for γ-Al$_2$O$_3$).* The

---

*For simplicity, we assume in this example that all the acid sites on the alumina surface have the same affinity for lead sorption. In fact, as discussed in Example 5 for iron hydroxide, oxides usually have high- and low-affinity sites for metal sorption. At very low sorption density, when even the high-affinity sites are undersaturated, it is only the product $K_1^{app}L_T$ that matters, not the individual values of the sorption constant and the surface ligand concentration.

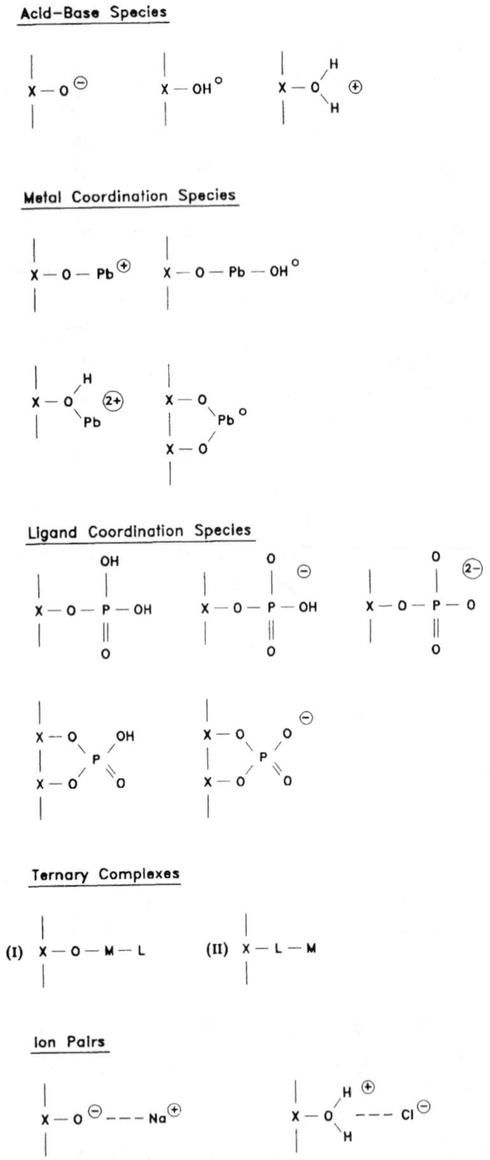

Figure 8.4 Some possible coordination species that may form between solutes and an oxide surface. Direct evidence for the existence of a particular surface species is generally lacking, and "reasonable" stoichiometries are thus chosen simply to fit available adsorption data.

total concentration of lead in the system, Pb_T, is assumed to be so low that the surface is not close to saturation ($Pb_T \ll \equiv X_T$). The surface reactions to be considered are

$$\equiv XOH_2^+ = H^+ + \equiv XOH; \qquad K_{a1}^{app} = 10^{-6.0} \tag{21}$$

$$\equiv XOH = H^+ + \equiv XO^-; \qquad K_{a2}^{app} = 10^{-7.7} \tag{22}$$

$$\equiv XOPb^+ = \equiv XO^- + Pb^{2+}; \qquad K_1^{app} = 10^{-6.1} \tag{23}$$

These constants have been chosen from the literature[15] as being applicable in this type of background electrolyte and in the pH range 5–6, which is of most interest to us. They are *not* universal constants and depend not only on the pH (see Section 5) and on the background electrolyte, but also on the mode of preparation of the aluminum oxide. Since we are not considering here explicitly any electrostatic interactions between the surface and the solutes, the constants are adjusted to reflect some average degree of electrostatic repulsion.

In addition to surface reactions, we have to consider the usual reactions in solution, in particular, the hydrolysis of lead:

$$Pb^{2+} + H_2O = PbOH^+ + H^+; \qquad K_{OH} = 10^{-7.7} \tag{24}$$

other hydrolysis species are not important below pH 7.

Choosing H^+, Pb^{2+}, and $\equiv XOH_2^+$ as principal components (see Tableau 8.1), we obtain the following mole balance and mass law equations:

$$TOT \equiv XOH_2 = [\equiv XOH_2^+] + [\equiv XOH] + [\equiv XO^-] + [\equiv XOPb^+]$$
$$= [\equiv X]_T \tag{25}$$

$$TOTPb = [Pb^{2+}] + [PbOH^+] + [\equiv XOPb^+] = Pb_T \tag{26}$$

$$[\equiv XOH] = 10^{-6.0}[H^+]^{-1}[\equiv XOH_2^+] \tag{27}$$

TABLEAU 8.1

	H^+	Pb^{2+}	$\equiv XOH_2^+$
H^+	1		
OH^-	-1		
Pb^{2+}		1	
$PbOH^+$	-1	1	
$\equiv XOH_2^+$			1
$\equiv XOH$	-1		1
$\equiv XO^-$	-2		1
$\equiv XOPb^+$	-2	1	1

$$[\equiv XO^-] = 10^{-13.7}[H^+]^{-2}[\equiv XOH_2^+] \tag{28}$$

$$[PbOH^+] = 10^{-7.7}[H^+]^{-1}[Pb^{2+}] \tag{29}$$

$$[\equiv XOPb^+] = 10^{-7.6}[H^+]^{-2}[\equiv XOH_2^+][Pb^{2+}] \tag{30}$$

Neglecting the last two terms of Equation 25, which must be relatively small, and substituting Equation 27 yields

$$[\equiv XOH_2^+] = [\equiv X]_T(1 + 10^{-6.0}[H^+]^{-1})^{-1} \tag{31}$$

Introducing Equations 29 and 30 in Equation 26 and solving for $[Pb^{2+}]$,

$$[Pb^{2+}] = Pb_T(1 + 10^{-7.7}[H^+]^{-1} + 10^{-7.6}[H^+]^{-2}[\equiv XOH_2^+])^{-1} \tag{32}$$

Combining Equations 30 and 32,

$$[\equiv XOPb^+] = \frac{10^{-7.6}Pb_T[\equiv XOH_2^+]}{[H^+]^2 + 10^{-7.7}[H^+] + 10^{-7.6}[\equiv XOH_2^+]} \tag{33}$$

Finally an expression for the fraction of lead adsorbed can be derived by introducing Equation 31:

$$\text{Percent lead adsorbed} = \frac{[\equiv XOPb^+]}{Pb_T}$$

$$= \frac{[\equiv X]_T}{10^{7.6}[H^+]^2 + 10^{1.6}[H^+] + 10^{-6.1} + [\equiv X]_T} \tag{34}$$

The principal features of this formula are illustrated in Figure 8.5. For a fixed adsorbent concentration the percentage of lead adsorbed increases sharply

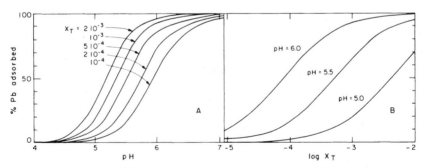

Figure 8.5 Adsorption of lead on alumina as calculated in Example 1: variation in lead adsorption (A) as a function of pH for various surface site concentrations, and (B) as a function of surface site concentration at various pH's. Note the effect of surface site concentration on the "adsorption edge" in panel A.

above a critical pH (ca. 5–6 for the conditions of the example), and this "adsorption edge" occurs at lower pH's with increasing solid concentrations (Figure 8.5a). At constant pH, the percentage is simply a hyperbolic function of the total adsorbent concentration as shown in Figure 8.5b. Because this example is calculated for low adsorption densities ($[\equiv X]_T \gg Pb_T$), the percentage of adsorbed metal is independent of the total metal concentration; otherwise a typical Langmuir isotherm would be calculated at constant pH.

*   *   *

Example 1 illustrates the three principal features of metal adsorption on oxides:

(a) Adsorption increases sharply with pH because of the decreasing competition with H^+ for the surface ligand (this is similar to the effect of pH on complexation in solution).

(b) The adsorption edge (i.e., the pH at which the percent of sorbed metal increases sharply) is displaced to lower pH's with decreasing adsorbate–adsorbent ratios.

(c) Metal adsorption is significant even below the first pK_a of the surface acid groups. The dominant surface species is $\equiv XOH_2^+$ and the surface is positively charged when adsorption commences. Thus adsorption takes place *against* an electrostatic repulsion. The extent of this electrostatic repulsion is of course important to assess; in our example it is accounted for by the choice of the effective adsorption constant.

Example 2. Adsorption of Phosphate on Alumina As a contrast to Example 1, consider the adsorption of phosphate on aluminum hydroxide. The pH range of interest is somewhat higher than for lead adsorption (ca. 5–9 pH) and, as will be seen later, the effective acidity constants are consequently larger:

$$\equiv XOH_2^+ = H^+ + \equiv XOH; \qquad K_{a1}^{app} = 10^{-6.8} \tag{35}$$

$$\equiv XOH = H^+ + \equiv XO^-; \qquad K_{a2}^{app} = 10^{-8.7} \tag{36}$$

In order for this model system to exhibit the appropriate adsorption properties, several phosphate surface species need to be considered:[15]

$$\equiv XOH_2PO_3^0 + H_2O = \equiv XOH_2^+ + H_2PO_4^-; \qquad K^{app} = 10^{-4.5} \tag{37}$$

$$\equiv XOHPO_3^- + H^+ + H_2O = \equiv XOH_2^+ + H_2PO_4^-; \qquad K^{app} = 10^{+1.3} \tag{38}$$

These surface reactions and their apparent equilibrium constants are chosen so as to obtain the same general dependency of adsorption on concentration and pH as is observed experimentally.

Finally, the acid–base properties of phosphate have to be considered. In the

TABLEAU 8.2

	H^+	$H_2PO_4^-$	$\equiv XOH_2^+$
H^+	1		
OH^-	-1		
$H_2PO_4^-$		1	
HPO_4^{2-}	-1	1	
$\equiv XOH_2^+$			1
$\equiv XOH$	-1		1
$\equiv XO^-$	-2		1
$\equiv XOH_2PO_3$		1	1
$\equiv XOHPO_3^-$	-1	1	1

pH range 5–9 only the second acidity constant is important:

$$H_2PO_4^- = H^+ + HPO_4^{2-}; \qquad K_a = 10^{-7.0} \tag{39}$$

With $H_2PO_4^-$ and $\equiv XOH_2^+$ as principal components (see Tableau 8.2), the two relevant mole balance equations are written

$$TOT \equiv XOH_2 = [\equiv XOH_2^+] + [\equiv XOH] + [\equiv XO^-] + [\equiv XOH_2PO_3]$$
$$+ [\equiv XOHPO_3^-] = [\equiv X]_T \tag{40}$$

$$TOTH_2PO_4 = [H_2PO_4^-] + [HPO_4^{2-}] + [\equiv XOH_2PO_3]$$
$$+ [\equiv XOHPO_3^-] = P_T \tag{41}$$

In the concentration ranges where $[\equiv X]_T$ and P_T are similar—as is typical of most experimental studies of anion adsorption which, in contrast to metal adsorption, are usually performed at high surface coverage—no term in these two equations can be neglected and the algebra of the problem is a bit unwieldy. Nonetheless, since there is no high exponent in any of the mass laws, for any given set of conditions (pH, $[\equiv X]_T$, P_T), the problem is amenable to an explicit solution. Let us define the parameters:

$$a = 1 + \frac{10^{-7}}{[H^+]}$$

$$b = 10^{4.5} + \frac{10^{-1.3}}{[H^+]}$$

$$c = 1 + \frac{10^{-6.8}}{[H^+]} + \frac{10^{-15.5}}{[H^+]^2}$$

Then, through judicious substitutions in Equations 35–39, we find that the

$H_2PO_4^-$ concentration is given by the solution of the quadratic:

$$abx^2 + [ac + b([\equiv X]_T - P_T)]x - P_T = 0 \tag{42}$$

For any set of conditions, we can solve Equation 42 for $x(= [H_2PO_4^-])$ and obtain the adsorbed phosphate concentration by substituting into Equation 43:

$$P_{ads} = \frac{b[H_2PO_4^-]}{c + b[H_2PO_4^-]}[\equiv X]_T \tag{43}$$

The principal feature of the results shown in Figure 8.6 for various total phosphate and concentrations (fixed $[\equiv X]_T$) is a decrease in phosphate adsorption with pH. This is due to the acid–base properties of the surface and the ligand. For example, at a pH just above 7, the principal surface reaction consumes one

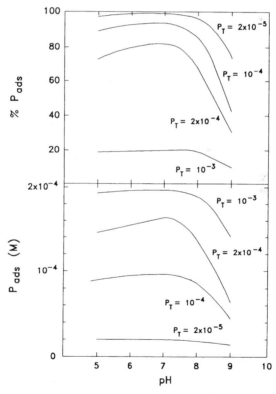

Figure 8.6 Adsorption of phosphate on alumina as calculated in Example 2 for $[\equiv X]_T = 2 \times 10^{-4} M$. Note the general decrease in adsorption at high pH (in contrast to Figure 8.5) and the complex effects of the total phosphate concentration, P_T, on the pH dependency of the fractional P adsorbed.

proton:

$$\equiv XOH + HPO_4^{2-} + H^+ = \equiv XOHPO_3^- + H_2O$$

At low pH, the adsorption often reaches a maximum or even decreases, owing to either exhaustion of the phosphate from solution (e.g., $[\equiv X]_T = 2 \times 10^{-4} M$, $P_T = 10^{-4}$) or to a proton release according to the reaction

$$\equiv XOH_2^+ + H_2PO_4^- = \equiv XOHPO_3^- + H^+ + H_2O$$

In this situation, the position of the adsorption maximum is controlled by the relative acidities of the surface hydroxyl group, the ligand in solution, and the ligand bound to the surface.

*   *   *

The characteristics of adsorption illustrated in Example 2 are general for all types of ligands and oxide surfaces. For example, Figure 8.7 shows how the

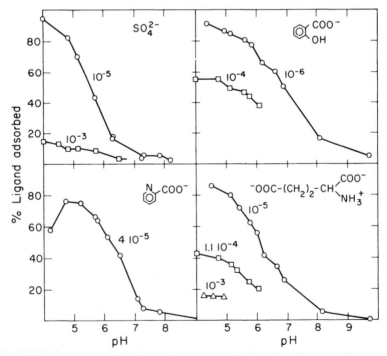

Figure 8.7   Adsorption of sulfate, salicylate, picolinate and glutamate on amorphous iron oxide. $[Fe(OH)_3]_T = 10^{-3} M$, except $6.9 \times 10^{-4} M$ for picolinate. From Davis and Leckie, 1978.[16]

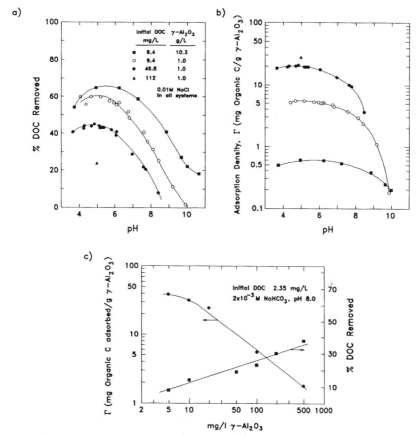

Figure 8.8 Adsorption of natural organic matter on alumina: (*a*) percent DOC adsorbed as a function of pH; (*b*) adsorption density (DOC adsorbed per mass of solid) as a function of pH; (*c*) percent DOC adsorbed and adsorption density as a function of solid concentration. From Davis, 1984.[30]

adsorption of sulfate, salicylate, picolinate, and glutamate on amorphous iron oxide all exhibit pronounced maxima at low pH.[16] Adsorption of natural organic material on γ-Al_2O_3 (Figure 8.8) also shows the same characteristic pH dependency and can be modeled quantitatively by assuming the same type of surface complexes with the surface aluminum ion.[17]

If we consider the adsorption of natural organic matter as a function of suspended solid concentration (Figure 8.8*b*), we find that even at the relatively high pH of 8, the typical DOC concentration of a few milligrams per liter is sufficient to saturate the available inorganic surface on a few milligrams per liter of suspended solid. (Note that the value of 40 mg org·C g^{-1} Al_2O_3 is consistent with a site density of about 10^{-4} mol g^{-1} Al_2O_3 and a concentration of functional groups of about 10 meq g^{-1} org·C. Thus, in this case approximately one quarter of the organic functional groups would have reacted with the surface groups.) This result suggests that whereas only a fraction of the organic matter in aquatic systems might be adsorbed on suspended solids, the solid surface may be entirely covered by organic material and hence exhibit physical and chemical properties characteristic of organic matter.

3.2 Carbonate and Sulfide Surfaces

For oxide minerals, the chemistry of the surface is dominated by the chemistry of the surface hydroxyl groups. This is not necessarily the case for other minerals that contain another ligand such as carbonate or sulfide. The weak acid properties of that ligand at the surface and its metal binding properties can be important in determining the surface properties of the mineral.

The acid–base chemistry for the surface of calcite, for example, has been modeled[17] both as the hydrolysis of surface calcium,

$$\equiv Ca^+ + H_2O = \equiv CaOH + H^+$$

and as the protonation of the surface carbonate,

$$\equiv CO_3^- + H^+ = \equiv HCO_3$$

Metal adsorption on carbonates is usually thought to occur by coordination with the carbonate ligand. This adsorption does not exhibit "Langmuirian" saturation and is often modeled as a surface precipitation–solid solution reaction[18,19] (see Section 6.2). Studies of Cd^{2+} adsorption on calcareous aquifer sand suggest that carbonate minerals may be responsible for significant metal adsorption on solids and sediments.[21] In contrast, adsorption of trace metals on calcium carbonate in the water column appears to be insignificant and the settling of calcium carbonate particles does not contribute to the flux of trace metals to the sediments.[22]

Similarly, in the case of sulfide minerals—whose study is complicated by the difficulty in preventing oxidation of the surface—both hydrolysis of the constituent metal and protonation of the sulfhydryl surface group contribute to the chemical properties of the surface. For example, the surface chemistry of CdS[22,24] can be modeled as

$$\equiv Cd^+ + H_2O = \equiv CdOH + H^+$$

$$\equiv S^- + H^+ = \equiv HS$$

Trace metals may be incorporated into sulfide minerals by adsorption, precipitation, or lattice exchange. In this last process, the constituent metal of the sulfide mineral is replaced, at the surface, by another metal whose corresponding sulfide is less soluble. The limited studies of metal adsorption on sulfide minerals suggest that adsorption is dominated by the surface hydroxyl groups rather than the surface sulfhydryl groups. The surface interactions of metals with sulfide minerals are likely to have a marked influence on the fate and transport of metals in anoxic environments.[25]

3.3 Organic Surfaces

Minerals are important constituents of the suspended material in natural waters and they provide reactive surfaces for the adsorption of trace solutes. Organic surfaces also contribute to the adsorption of trace constituents in aquatic systems,[2,26,27] either in the form of organic particles proper (live organisms, dead remains, condensed humic material), or in the form of organic coatings on inorganic particles such as oxides and clay minerals. The presence of organic coatings is evidenced by the negative surface charge of most particles in (oxygenated) natural waters.[26,28] In the absence of organic coatings, most oxides (excepting silica and manganese oxides) would have a positive surface charge at ambient pH. In environments where fresh hydrous oxide phases are continuously precipitated, for example, rivers receiving significant inputs of iron-rich acid mine drainage waters, positively charged particles have indeed been observed.[29] The correspondence between sorption of natural organic matter (NOM) on oxides and surface charge (or electrophoretic mobility) is shown in Figure 8.9. Adsorbed organic matter, by determining surface charge, can control the stability of suspended particles in natural waters (see Section 7.3). It can also of course affect the adsorption of solutes.

In the case of organic particles proper, the surface coordination of metals by surface ligands such as carboxyl, phenolic, or sulfhydryl groups is conceptually similar to the equivalent process in solution; the essential difference is that brought about the long-range coulombic interactions, a subject treated in the next section. The situation is quite different for organic coatings on inorganic surfaces, for the organic ligands themselves may react reversibly with the surface. The system, inorganic surface–organic ligand–metal ion ($\equiv X - Y - M$), may thus yield four general types of surface reactions:

(i) Metal adsorption:

$$\equiv X + M = \equiv X M$$

(ii) Ligand adsorption:

$$\equiv X + Y = \equiv X Y$$

(iii) Ligand coordination to the adsorbed metal ion (i.e., formation of ternary surface complex I):

$$\equiv X M + Y = \equiv X M Y$$

(iv) Metal complexation by the adsorbed ligand (i.e., formation of ternary

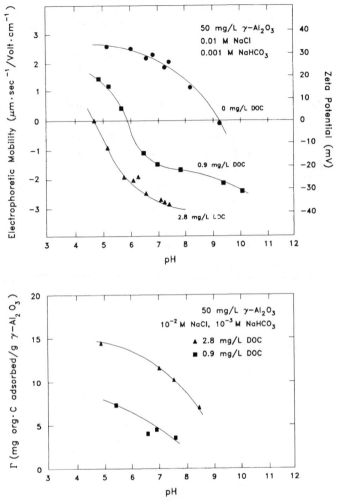

Figure 8.9 Adsorption of humic material (Suwannee River fulvic acid) on alumina and its effect on surface charge. From Davis, 1982.[17]

surface complex II):

$$\equiv XY + M = \equiv XYM$$

In addition, the metal and the organic ligand may react in solution according to the usual complexation reactions:

$$M + Y = MY$$

For each of these five types of reactions several different species may be formed with various stoichiometries and acid–base chemistries.

The net metal adsorption on inorganic surfaces can be increased by the presence of natural organic compounds which may bind the metal to the surface by forming ternary surface complexes (II). It may also be decreased through competition between the ligand and the metal for surface sites or, more likely, through competition between the adsorbing surface and the dissolved organic ligand for the metal. In model systems both increases and decreases of metal sorption in the presence of organic ligands are in fact observed, as is the implicit third one: no effect on metal adsorption.

The adsorption of copper on amorphous iron oxide is unaffected by sulfate or salicylate, prevented by the presence of picolinic acid, and increased at low pH by the presence of glutamate (Figure 8.10).[16] These results can be rationalized quantitatively as shown in the following example.

Example 3. Coadsorption of Copper and Organic Ligands on Amorphous Iron Oxide   Consider the principal copper complexation reactions for systems containing both an iron oxide suspension and organic ligands (mixed constants corrected for $I = 0.1\ M$):

$$Cu^{2+} + \equiv FeOH^\circ = \equiv FeOCu^+ + H^+; \qquad {}^*K_{ads} = 10^{0.7}$$

(The effective constant ${}^*K_{ads}$ includes both the activity coefficients of Cu^{2+} and the coulombic term applicable at the mid-range pH of 5.5.)

$$Cu^{2+} + SO_4^{2-} = CuSO_4; \qquad\qquad K = 10^{1.5}$$

$$Cu^{2+} + HSal^- = CuSal + H^+; \qquad {}^*K' = 10^{-2.8}$$

$$Cu^{2+} + HPic = CuPic^+ + H^+; \qquad {}^*K' = 10^{+2.7}\ (below\ pH = 5.4)$$

$$Cu^{2+} + HGlut^- = CuGlut + H^+; \qquad {}^*K' = 10^{-1.7}\ (above\ pH = 4.4)$$

With the ligand concentrations present in these experiments $[\equiv FeOH^\circ]_T = 5 \times 10^{-3} \times Fe_T = 5 \times 10^{-6}\ M$; $[SO_4^{2-}]_T = 10^{-3}\ M$; $[HSal^-]_T = 10^{-3}\ M$; $[HPic]_T = 10^{-4.4}\ M$; $[HGlut^-]_T = 10^{-4}\ M$, we expect the iron oxide surface to outcompete sulfate and salicylate but not picolinate for complexation of the cupric ion. This is indeed observed in Figure 8.10. Note that the concentration of copper ($10^{-6}\ M$) is smaller than that of the ligands and thus cannot affect ligand speciation. Although sulfate, salicylate, and picolinate are largely adsorbed at low pH, their functionalities apparently do not allow formation of type II ternary surface complexes ($\equiv XYM$). The acid functional groups of these molecules are bound to the surface hydroxyl groups. On the contrary, glutamate has coordinative functional groups at both ends of the molecule and formation of a type (II) ternary surface complex is evidenced by the increased copper adsorption at low pH. At higher pH, when glutamate desorbs, its affinity for

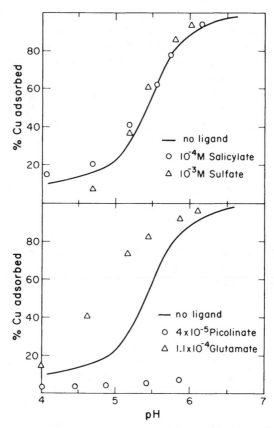

Figure 8.10   Adsorption of copper on amorphous iron oxide ($Fe_T = 10^{-3} M$) in the presence of various ligands (shown in Figure 8.7). The decrease in copper adsorption in the presence of picolinate is explained by the formation of the nonadsorbable Cu-picolinate complex in solution. The increase in copper adsorption in the presence of glutamate is explained as the formation of a type II ternary complex, $\equiv X - L - M$, on the surface. From Davis and Leckie, 1978.[16]

copper is not quite high enough to outcompete the oxide surface. The results of Figure 8.10 provide no evidence for or against formation of type I ternary surface complexes ($\equiv XMY$) whose existence is supported by other data.

* * *

In the absence of sufficient information to develop a meaningful theoretical model for the effect of natural organic material on the adsorption of metal ions in natural water, consider the data of Davis on the adsorption of copper on

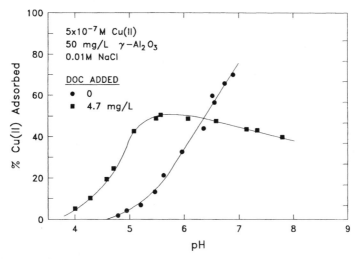

Figure 8.11   Adsorption of copper on alumina in the presence of natural organic matter. The addition of dissolved organic carbon to the system enhances adsorption at low pH (presumably due to type II ternary surface complexes) and decreases it at high pH (presumably due to formation of nonadsorbed complexes). From Davis, 1983.[30]

γ-Al$_2$O$_3$[30] (Figure 8.11). At low pH the presence of natural organic matter drastically increases the copper adsorption while it actually decreases it at higher pH. These observed effects can be rationalized by the formation of type (II) ternary surface complexes at low pH where the organic is largely adsorbed (see Figure 8.8) and by complexation of copper by the desorbed organic ligand at high pH. (The same effect might have been observed with glutamate in Figure 8.10 if the ligand concentration had been higher.)

A study of copper complexation with the same natural organic matter in solution showed very high affinity of the ligand for copper (similar to Example 7 in Chapter 6), such that above pH = 5, all the copper should be complexed (note the excess of ligand over metal in this example: $Cu_T = 5 \times 10^{-7} M$ vs. $L_{1T} + L_{2T} \simeq 7 \times 10^{-6} M$; see Table 6.12). The initial increase in copper adsorption with pH in the presence of the organic matter is thus interpreted as the increase in complexation of the metal by the adsorbed ligand whose degree of protonation decreases (resulting in an increase in the effective complexation constant). To account quantitatively for the eventual plateau in adsorption it is necessary to estimate the relative contributions of surface reactions (i), (iii), and (iv) to the overall adsorption process. This is not possible in the absence of further information since the presence of the metal must affect the ligand adsorption via formation of type (II) ternary surface complexes and, vice versa, the presence of the ligand must affect the metal adsorption via formation of type (I) complexes.

4. ELECTROSTATIC INTERACTIONS ON SURFACES

In this chapter we have repeatedly pointed out the importance of long-range electrostatic interactions in adsorption reactions. To obtain·a quantitative description of such interactions—which are most responsible for making coordination at surfaces different from coordination among solutes—we start with a system in which only electrostatic forces are at play, considering a charged surface immersed in an ideal (nonreactive) electrolyte solution. By using the principle of additivity of energies, we can then superimpose the results with any chosen set of surface complexation reactions, thus obtaining a complete adsorption model that considers both coulombic and chemical interactions. The calculation of the electrostatic interactions—the Gouy–Chapman theory—and its superposition on the chemical interactions—the Stern theory—are formally and conceptually similar to the Debye–Hückel theory of nonideal behavior of electrolytes and to the polyelectrolyte theory presented in the context of metal binding to humic acids. The same long-range electrostatic interactions are considered; the same mixture of electrostatic (Poisson) and thermodynamic (Boltzmann) effects are used for the derivation; the same approximations of uniform charge distributions on surfaces and continuous charge distribution in the medium are involved.

For simplicity we consider a one-dimensional system consisting of an inert electrolyte solution neighboring an infinite flat surface. This geometry is generally applicable to particles in solution because the distances over which coulombic forces from the charged surfaces affect ions in solution ($1/\kappa$, see discussion below) are usually much shorter than the radii of the particles. The distribution of co- and counterions near the flat surface is governed by the Poisson–Boltzmann equation which, unlike the situation encountered for charged spheres, has an explicit solution (given in Sidebar 8.1). This solution was derived around 1910 by Gouy and Chapman,[31] more than a decade before the similar derivation by Debye and Hückel. The result is the spatial distribution of ions near the surface which involves a characteristic distance $1/\kappa$, the so-called double layer thickness ("the Debye length") defined by

$$\kappa^2 = 2\frac{F^2}{\varepsilon\varepsilon_0 RT}I \tag{44}$$

$$\frac{1}{\kappa} = 0.30I^{-1/2} \qquad (\kappa^{-1} \text{ in nm and } I \text{ in } M) \tag{45}$$

where F is the Faraday constant, I is the ionic strength of the bulk solution, ε is the relative dielectric constant of water, ε_0 is the permittivity of vacuum, and RT has its usual meaning. As shown in Figure 8.12, this distance $1/\kappa$ is characteristic of the extent of the electrostatic influence of the surface in the solution and determines how far from the surface the concentration of counterions (ions of

SIDEBAR 8.1

Solutions to the Poisson–Boltzmann Equation:
III. GOUY–CHAPMAN THEORY (1910–1913)

Our main objective is to obtain the relation between surface charge σ and surface potential Ψ_0 for an infinite flat plate immersed in a solution considered to be a continuous dielectric medium. In addition, the derivation provides the distribution of ions and of electric potential Ψ near the surface. For simplicity we consider a 1:1 electrolyte (NaCl) at a bulk concentration I (ionic strength). The one-dimensional Poisson–Boltzmann equation is written:

$$\frac{d^2\Psi}{dx^2} = \frac{-FI}{\varepsilon\varepsilon_0}(e^{-F\Psi/RT} - e^{+F\Psi/RT}) \qquad (i)$$

It is easily integrated by multiplying both sides by $2\,d\Psi/dx$. With the boundary conditions in the bulk solution $(x = \infty)\Psi = d\Psi/dx = 0$, we obtain

$$\left(\frac{d\Psi}{dx}\right)^2 = \frac{2FI}{\varepsilon\varepsilon_0}\frac{RT}{F}(e^{-F\Psi/RT} + e^{+F\Psi/RT} - 2) \qquad (ii)$$

$$= \frac{8RT}{\varepsilon\varepsilon_0}I \cdot \sinh^2\frac{F\Psi}{2RT} \qquad (iii)$$

we take the square root (with the necessary negative sign):

$$\frac{d\Psi}{dx} = -\left(\frac{8RTI}{\varepsilon\varepsilon_0}\right)^{1/2}\sinh\frac{F\Psi}{2RT} \qquad (iv)$$

With an appropriate change of variable this differential equation can be integrated further (into a cumbersome formula) with the boundary condition $\Psi = \Psi_0$ at $x = 0$. From the distribution of Ψ with distance, one can deduce the distribution of co- and counterions:

$$[Na^+] = Ie^{-F\Psi/RT}$$
$$[Cl^-] = Ie^{+F\Psi/RT} \qquad (v)$$

(continued)

A much simplified solution is obtained in the case of small potentials ($F\Psi/RT < 1$) for which (iv) can be integrated to yield

$$\Psi = \Psi_0 e^{-\kappa x} \tag{vi}$$

(κ is defined in the text).

The surface charge can be obtained by calculating the excess counter charge in solution:

$$\sigma = -F \int_0^x ([Na^+] - [Cl^-])dx = \varepsilon\varepsilon_0 \int_0^x \frac{d^2\Psi}{dx^2}dx \tag{vii}$$

$$= -\varepsilon\varepsilon_0 \left(\frac{d\Psi}{dx}\right)_{x=0} \tag{viii}$$

Thus the surface charge–surface potential relation can be obtained (without the small potential approximation) from (iv) and (viii):

$$\sigma = (8\varepsilon\varepsilon_0 RTI)^{1/2} \sinh F\Psi_0/2RT \tag{ix}$$

For small potentials, the formula can be simplified as shown in the text.

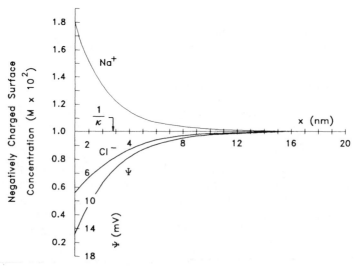

Figure 8.12   Theoretical distributions of ion concentrations and electrical potential near a charged surface. (Calculated for the small potential approximation.) $1/\kappa$ is the thickness of the double layer.

opposite charge) is increased compared to its bulk value and that of co-ions (ions of like charge) is decreased. In dilute solution, the double layer thickness can be relatively large ($1/\kappa = 3.0$ nm for $I = 10^{-2}$ M), whereas it is "compressed" in high ionic strength systems ($1/\kappa = 0.4$ nm for $I = 0.5$ M).

The Gouy–Chapman theory also yields the relation between the surface charge, σ, and the surface potential, Ψ_0:

$$\sigma = (8\,RT\,\varepsilon\varepsilon_0 I)^{1/2} \sinh\left(z\Psi_0 F/2\,RT\right) \tag{46}$$

where z is the valence of the symmetrical background electrolyte. At 25°C, this equation becomes

$$\sigma = 0.1174\,I^{1/2} \sinh\left(z\Psi_0 \times 19.5\right) \tag{47}$$

For small potentials, the surface charge is simply proportional to the surface potential:

$$\sigma = \varepsilon\varepsilon_0 \kappa \Psi_0 \tag{48}$$

$$\sigma = 2.3\,I^{1/2}\Psi_0 \qquad (\sigma \text{ in coulombs m}^{-2};\ I \text{ in } M;\ \Psi_0 \text{ in } V) \tag{49}$$

The coefficient of proportionality between σ and Ψ_0 has the dimensions of a capacitance:

$$C_{GC} = \varepsilon\varepsilon_0 \kappa = 2.3\,I^{1/2} \qquad (C_{GC} \text{ in farads m}^{-2} \text{ and } I \text{ in } M) \tag{50}$$

This Gouy–Chapman capacitance is on the order of 1–0.03 farad m^{-2}, and varies with the square root of the ionic strength of the solution ($I = 10^{-1}$–10^{-4} M). For high values of the potential ($\Psi_0 > 25$ mV), where the hyperbolic sine relation obtains, the apparent capacitance σ/Ψ_0 of the system increases with the surface charge and potential.

4.1 Inclusion of the Coulombic Term in Surface Complexation Calculations

In the previous section we saw how purely electrostatic effects result in the adsorption (positive and negative) of ions in the "diffuse layer" (also called the Gouy layer) near a charged surface. In practice such an effect is relatively unimportant because ions are affected strictly on the basis of their charge and major electrolyte ions thus contribute practically all the counter charge excess near the surface. In other words, unless surface potentials are very high and available surface areas very large, electrostatic adsorption in the diffuse layer results simply in a minute spatial rearrangement of the major ions.

What is vastly more interesting to us is the influence of the electrostatic effect on the specific (i.e., "chemical") adsorption of trace solutes. This problem is apparently easily solved by considering that the total adsorption energy is the sum of the chemical (intrinsic) and the electrostatic (coulombic) energy terms as originally defined by the complexation model (see Section 2.3)

$$\Delta G^o_{ads} = \Delta G^o_{int} + \Delta G^o_{coul} \tag{51}$$

Recall from Example 1 that adsorption of cations on oxides usually takes place against an electrostatic repulsion; the electrostatic contribution to the total energy of adsorption can thus often be negative ($\Delta G_{coul} > 0$).

A simple way to understand how the coulombic term is calculated is to consider the electrostatic work involved in transporting the reacting solute from the bulk solution to the surface and the dissolved products from the surface to the bulk solution:

$$\Delta G^o_{coul} = F \Delta Z \Psi_0 \tag{52}$$

where Ψ_0 is the surface potential (compared to a reference potential of 0 in the bulk solution) and $F \Delta Z$ is the change in the molar charge of the surface species due to the adsorption reaction. For example,

$$\equiv XOH_2^+ = 2H^+ + \equiv XO^-; \qquad \Delta Z = -2; \quad \Delta G^o_{coul} = -2F\Psi_0$$

$$\equiv XOPb^+ + H^+ = \equiv XOH + Pb^{2+}; \qquad \Delta Z = -1; \quad \Delta G^o_{coul} = -F\Psi_0$$

$$\equiv XOHPO_3^- + H^+ + H_2O = \equiv XOH_2^+ + H_2PO_4^-;$$
$$\Delta Z = +2; \quad \Delta G^o_{coul} = +2F\Psi_0$$

What we have done is to consider that an adsorption reaction such as

$$\equiv XOPb^+ + H^+ = \equiv XOH + Pb^{2+}; \qquad \Delta Z = -1; \qquad \Delta G^o_{coul} = -F\Psi_0$$

is effectively the sum of three reactions:

1. First the reactant in solution (H^+) moves to the surface:

$$H^+_{bulk} = H^+_{surface}; \qquad \Delta G^o_1 = \Delta G^o_{coul,1} = +F\Psi_0$$

2. Then the reaction itself proceeds at the surface:

$$\equiv XOPb^+ + H^+_{surface} = \equiv XOH + Pb^{2+}_{surface}; \qquad \Delta G^o_2 = \Delta G^o_{int}$$

3. Finally, the dissolved product moves to the bulk solution:

$$Pb^{2+}_{surface} = Pb^{2+}_{bulk}; \qquad \Delta G^o_3 = \Delta G^o_{coul,2} = -2F\Psi_0$$

and

$$\Delta G^\circ_{coul} = \Delta G^\circ_{coul,1} + \Delta G^\circ_{coul,2} = (1 - 2)F\Psi_0$$

The idea of adding purely chemical and coulombic free energies to calculate the specific adsorption of solutes on surfaces was developed by Stern, who extended the Gouy–Chapman theory in 1924.[32] (Sidebar 8.2 shows how the coulombic term is formally an activity coefficient for charged surface species.) The additional key idea of the surface complexation model is to consider that the surface charge itself is entirely due to the chemical reactions at the surface. Thus the consideration of long-range coulombic interactions introduces an additional unknown—the surface potential Ψ_0—into the equilibrium calculations (in addition to the usual species concentration as seen in Examples 1 and 2), and the surface complexation model provides the necessary additional equation from the charge-potential relation of the Gouy–Chapman theory (for a 1:1 electrolyte):

$$\sigma = \alpha \sinh(F\Psi_0/2RT) = \frac{\alpha}{2}[\exp(F\Psi_0/2RT) - \exp(-F\Psi_0/2RT)] \quad (53)$$

where the coefficient α is a function of ionic strength (Equation 46), and the

SIDEBAR 8.2

Coulombic Term and Activity Coefficients

For didactic simplicity, we have derived the expression of the coulombic term by calculating the electrostatic energy necessary to bring bulk solutes to the adsorbing surface. This derivation does not clearly delineate the extrathermodynamic assumptions involved in the surface complexation model and hides the true nature of the coulombic term as an activity coefficient for surface species.

Consider the reaction

$$XOPb^+ + H^+ = XOH + Pb^{2+}; \quad K$$

The mass law for this reaction can always be written

$$\frac{[XOH][Pb^{2+}]}{[XOPb^+][H^+]} = K^{int} \cdot \frac{\gamma_{XOPb^+} \cdot \gamma_{H^+}}{\gamma_{XOH} \cdot \gamma_{Pb^{2+}}}$$

(continued)

in which all nonideal behavior is accounted for explicitly by the activity coefficients γ. The activity coefficients of the dissolved species H^+ and Pb^{2+} have their usual meaning and can be obtained from Debye–Hückel theory or some empirical formula.

To calculate the activity coefficients for the surface species we choose a condition of zero surface charge and potential for their reference states. A mean surface potential Ψ_0 then contributes a free energy of $+F\Psi_0$ to the positively charged surface species $XOPb^+$. Thus,

$$\frac{\gamma_{XOPb^+}}{\gamma_{XOH}} = e^{F\Psi_0/RT} \cdot \frac{\gamma'_{XOPb^+}}{\gamma'_{XOH}}$$

The remaining activity coefficients γ'_{XOPb^+} and γ'_{XOH} account for all other nonideal contributions to the activities of the surface species, including for example local variation in the surface potential around the mean value Ψ_0. We assume, of course, that the ratio of these two coefficients can be taken as unity. Thus the first extrathermodynamic assumption involved in the surface complexation model is that all nonideal behavior of the surface species can be attributed to the effects of a mean surface potential Ψ_0. The second extrathermodynamic assumption is that Ψ_0 can be calculated from the surface charge according to the Gouy–Chapman theory (or one of its variations).

surface charge σ, given by

$$\sigma = \frac{F}{AS}([\equiv XOH_2^+] - [\equiv XO^-] + [\equiv XOPb^{2+}] - [\equiv XOHPO_3^-] + \cdots)$$

$$(54)$$

The coefficient F/AS (F = Faraday constant; A = specific surface area in square meters per gram; S = solid concentration in grams per liter) converts the concentration units (M) of the surface species into the appropriate surface charge density $(C\,m^{-2})$.

The following three examples demonstrate how the additional equation and additional unknown can be conveniently incorporated in the tableau format. They also illustrate the acid–base properties of hydrous ferric oxide—HFO, an important adsorbent in natural waters—as well as its affinity for both cations and anions.

Example 4. Acid–Base Chemistry of HFO* Consider a 90-mg L^{-1} suspension of hydrous ferric oxide (HFO) in a background electrolyte of $10^{-2}\,M$ whose

---

*Examples 4–6 follow closely similar examples in Dzombak and Morel, 1990.[6]

TABLE 8.1 Physical–Chemical Characteristics of Hydrous Ferric Oxide (HFO)

- Molecular weight: 1 mol Fe/90 g HFO
- Specific surface area $= A = 600 \, \text{m}^2 \, \text{g}^{-1}$
- Site concentrations
  Low-affinity sites $[\equiv \text{Fe}^w\text{OH}]_T = 0.2 \, \text{mol/mol Fe} \, (\text{H}^+, \text{anions, cations})$
  High-affinity sites $[\equiv \text{Fe}^s\text{OH}]_T = 0.005 \, \text{mol/mol Fe (cations)}$

pH can be adjusted by appropriate addition of strong acid or strong base. The physical characteristics of HFO are given in Table 8.1 and its intrinsic adsorption constants in Table 8.2. What is the surface speciation of HFO at a given pH—say pH = 6.0—and how much base or acid must have been added to the suspension to obtain this pH?

Recipe $[\equiv \text{FeOH}]_T = 0.09 \, \text{g HFO L}^{-1} \times [1 \, \text{mol Fe}/90 \, \text{g HFO}]$

$$\times \, [0.2 \, \text{mol site/mol Fe}] = 2 \times 10^{-4} \, M$$

$[\text{HA}]_T = ?$ (assuming we are on the acid side of the titration)

Species $\text{H}^+, \text{OH}^-, \text{A}^-, \text{B}^+, \equiv \text{FeOH}_2^+, \equiv \text{FeOH}, \equiv \text{FeO}^-$

Reactions $\equiv \text{FeOH}_2^+ = \equiv \text{FeOH} + \text{H}^+; \qquad K_1 = K_{a1}^{int} \cdot P^{-1}$ (55)

$\equiv \text{FeOH} = \equiv \text{FeO}^- + \text{H}^+; \qquad K_2 = K_{a2}^{int} \cdot P^{-1}$ (56)

The parameter P, representing the exponential of the dimensionless potential,

$$P = \exp(-F\Psi_0/RT)$$ (57)

is introduced for convenience as a principal variable of the system (rather than Ψ_0). Choosing H^+, $\equiv \text{FeOH}$, and P (and implicity A^-) as principal components yields Tableau 8.3, in which the coefficients of the "P column" are simply the charge numbers of the surface species. Thus the mass law equations for the

TABLEAU 8.3

	H^+	$\equiv \text{FeOH}$	P
H^+	1		
OH^-	-1		
$\equiv \text{FeOH}_2^+$	1	1	1
$\equiv \text{FeOH}$		1	
$\equiv \text{FeO}^-$	-1	1	-1
$[\equiv \text{FeOH}]_T$		1	
HA_T	1		

TABLE 8.2 Intrinsic Sorption Constants for Hydrous Ferric Oxide

Reactions			Adsorbing species	$\log K_1^{int}$	$\log K_2^{int}$	$\log K_3^{int}$	$\log K_4^{int}$
Acidity Constants							
$\equiv FeOH_2^+$	$= \equiv FeOH^0 + H^+$;	K_{a1}	H^+	-7.29	-8.93		
$\equiv FeOH^0$	$= \equiv FeO^- + H^+$;	K_{a2}					
Alkaline Earth Cations							
$\equiv Fe^sOH^0 + M^{2+}$	$= \equiv Fe^sOHM^{2+}$;	K_1	Ca^{2+}	4.97	-5.85		
$\equiv Fe^wOH^0 + M^{2+}$	$= \equiv Fe^wOM^+ + H^+$;	K_2	Sr^{2+}	5.01	-6.58	-17.60	
$\equiv Fe^wOH^0 + M^{2+} + H_2O$	$= \equiv Fe^wOMOH^0 + 2H^+$;	K_3	Ba^{2+}	5.46			
Transition and Post-Transition Metal Cations							
$\equiv Fe^sOH^0 + Ag^+$	$= \equiv Fe^sOAg^0 + H^+$;	K_1	Ag^+	-1.72			
$\equiv Fe^sOH^0 + M^{2+}$	$= \equiv Fe^sOM^+ + H^+$;	K_1	Co^{2+}	-0.46	-3.01		
$\equiv Fe^wOH^0 + M^{2+}$	$= \equiv Fe^wOM^+ + H^+$;	K_2	Ni^{2+}	0.37			
			Cd^{2+}	0.47	-2.90		
			Zn^{2+}	0.99	-1.99		
			Cu^{2+}	2.89			
			Pb^{2+}	4.65			
			Hg^{2+}	7.76			

Trivalent Cations

$\equiv Fe^sOH^0 + Cr^{3+} + H_2O \;=\; \equiv Fe^sOCrOH^+ + H^+$; K_1

Species	K_1
Cr^{3+}	2.06

Trivalent Anions

$\equiv FeOH^0 + A^{3-} + 3H^+ = \equiv FeH_2A^0 + H_2O$; K_1
$\equiv FeOH^0 + A^{3-} + 2H^+ = \equiv FeHA^- + H_2O$; K_2
$\equiv FeOH^0 + A^{3-} + H^+ = \equiv FeA^{2-} + H_2O$; K_3
$\equiv FeOH^0 + A^{3-} = \equiv FeOHA^{3-}$; K_4

Species	K_1	K_2	K_3	K_4
PO_4^{3-}	31.29	25.39		
AsO_4^{3-}	29.31	23.51		
VO_4^{3-}		17.72	10.58	13.57

Arsenite and Borate

$\equiv FeOH^0 + H_3A^0 = \equiv FeH_2A^0 + H_2O$; K_1

Species	K_1
H_3AsO_3	5.41
H_3BO_3	0.62

Divalent Anions

$\equiv FeOH^0 + A^{2-} + H^+ = \equiv FeA^- + H_2O$; K_2
$\equiv FeOH^0 + A^{2-} = \equiv FeOHA^{2-}$; K_3

Species	K_2	K_3
SO_4^{2-}	7.78	0.79
SeO_4^{2-}	7.73	0.80
SeO_3^{2-}	12.69	5.17
$S_2O_3^{2-}$		0.49
CrO_4^{2-}	10.85	

The superscripts s and w are used to denote strong and weak sorbing sites on hydrous ferric oxide.

surface acid base reactions are obtained by using the corresponding exponent for the component P and the corresponding intrinsic constants:

$$[\equiv FeOH_2^+] = [H^+] [\equiv FeOH] \cdot P^1 \cdot [K_{a1}^{int}]^{-1} \tag{58}$$

$$[\equiv FeO^-] = [H^+]^{-1} [\equiv FeOH] \cdot P^{-1} K_{a2}^{int} \tag{59}$$

Since the total concentration of acid is not specified, the $TOTH$ equation is not useful. The mole balance equation for the surface species is straightforward:

$$TOT \equiv FeOH = [\equiv FeOH_2^+] + [\equiv FeOH] + [\equiv FeO^-] = [\equiv FeOH]_T$$
$$= 10^{-3.7} \tag{60}$$

Using the coefficients of the tableau, the "$TOTP$" equation yields, within the F/AS factor, the surface charge of the HFO:

$$TOTP = [\equiv FeOH_2^+] - [\equiv FeO^-] = \frac{AS}{F} \sigma \tag{61}$$

Thus the right-hand side given by Equation 61 can be expressed as a function of P:

$$TOTP = [\equiv FeOH_2^+] - [\equiv FeO^-] = \frac{AS}{F} \alpha \sinh \frac{F\Psi_0}{2RT}$$
$$= \frac{AS}{F} \cdot \alpha \cdot \frac{P^{-1/2} - P^{1/2}}{2} \tag{62}$$

One should recall that the coefficient α is a function of ionic strength:

$$\alpha = 0.1174 \, I^{1/2}$$

At $I = 10^{-2} \, M$: $\alpha = 0.01174 \, C \, m^{-2}$.

$$\frac{AS}{F} \alpha = \frac{600 \, m^2 \, g^{-1} \times 0.09 \, g \, L^{-1}}{96,500 \, C \, mol^{-1}} \times 1.174 \times 10^{-2} \, C \, m^{-2}$$
$$= 6.57 \times 10^{-4} \, M$$

For any given value of H^+, the unknowns are $[\equiv FeOH_2^+]$, $[\equiv FeOH]$, $[\equiv FeO^-]$, and P and they can be calculated from Equations 58–62. In practice the solution to these four simultaneous equations requires a numerical scheme. Computer solutions for various ionic strengths are shown in the form of log C–pH diagrams in Figure 8.13.

Because they are useful in this and other examples, the values of P for HFO

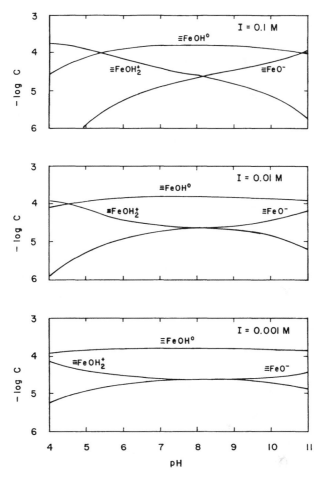

Figure 8.13 Calculated surface speciation as a function of pH and ionic strength (1:1 electrolyte) for a $90\,\text{mg}\,\text{L}^{-1}$ (TOTFe $= 10^{-3}\,M$) suspension of HFO.

(in the absence of specific adsorbates other than H^+, and for a 1:1 electrolyte) are given in Table 8.3 as a function of pH and ionic strength.

For example, at a pH of 6 and $I = 0.01$,

$$\log P = -1.85$$

We need only introduce $[H^+] = 10^{-6}$ and $P = 10^{-1.85}$ in Equations 58–60 to obtain the acid–base speciation of the surface:

$$[\equiv\text{FeOH}_2^+] = 10^{-6} \cdot 10^{-1.85} \cdot 10^{+7.29}[\equiv\text{FeOH}]$$
$$= 10^{-0.56}[\equiv\text{FeOH}]$$

TABLE 8.3 Coulombic Correction Factor as a Function of pH and Ionic Strength (HFO, Low Sorption Density, 1:1 Electrolyte)

$$\log P = \log[\exp(-F\Psi_0/RT)] \text{ for } I =$$

pH	1.0	0.5	0.1	0.05	0.01	0.005	0.001	0.0005	0.0001
4.00	−1.58	−1.85	−2.45	−2.68	−3.11	−3.27	−3.56	−3.66	−3.86
4.20	−1.57	−1.84	−2.41	−2.62	−3.01	−3.16	−3.43	−3.53	−3.70
4.40	−1.55	−1.81	−2.35	−2.54	−2.91	−3.05	−3.29	−3.38	−3.53
4.60	−1.54	−1.78	−2.28	−2.46	−2.80	−2.93	−3.15	−3.23	−3.36
4.80	−1.51	−1.74	−2.20	−2.37	−2.68	−2.80	−3.00	−3.08	−3.19
5.00	−1.47	−1.69	−2.12	−2.27	−2.56	−2.67	−2.85	−2.92	−3.01
5.20	−1.43	−1.63	−2.02	−2.16	−2.43	−2.53	−2.69	−2.75	−2.83
5.40	−1.37	−1.56	−1.92	−2.05	−2.29	−2.39	−2.53	−2.58	−2.65
5.60	−1.31	−1.48	−1.81	−1.93	−2.15	−2.24	−2.36	−2.40	−2.46
5.80	−1.23	−1.39	−1.69	−1.80	−2.01	−2.08	−2.19	−2.22	−2.27
6.00	−1.15	−1.29	−1.57	−1.67	−1.85	−1.92	−2.01	−2.04	−2.08
6.20	−1.06	−1.18	−1.44	−1.53	−1.70	−1.75	−1.83	−1.85	−1.88
6.40	−0.96	−1.07	−1.31	−1.39	−1.53	−1.58	−1.64	−1.66	−1.69
6.60	−0.86	−0.96	−1.17	−1.24	−1.36	−1.41	−1.46	−1.47	−1.49
6.80	−0.75	−0.84	−1.03	−1.09	−1.19	−1.23	−1.27	−1.28	−1.30
7.00	−0.64	−0.72	−0.88	−0.93	−1.02	−1.05	−1.07	−1.09	−1.10
7.20	−0.53	−0.60	−0.73	−0.77	−0.84	−0.86	−0.88	−0.89	−0.90

7.40	−0.41	−0.47	−0.57	−0.60	−0.65	−0.68	−0.69	−0.70	−0.70
7.60	−0.30	−0.34	−0.41	−0.44	−0.47	−0.49	−0.49	−0.50	−0.51
7.80	−0.18	−0.21	−0.25	−0.27	−0.29	−0.30	−0.30	−0.31	−0.31
8.00	−0.06	−0.07	−0.09	−0.09	−0.10	−0.11	−0.10	−0.11	−0.11
8.20	0.05	0.06	0.07	0.08	0.09	0.08	0.09	0.09	0.09
8.40	0.17	0.19	0.24	0.25	0.28	0.27	0.29	0.29	0.29
8.60	0.29	0.33	0.40	0.42	0.46	0.46	0.48	0.48	0.49
8.80	0.40	0.46	0.56	0.59	0.65	0.65	0.68	0.68	0.69
9.00	0.52	0.58	0.71	0.76	0.83	0.84	0.87	0.87	0.88
9.20	0.63	0.71	0.87	0.92	1.01	1.02	1.07	1.07	1.08
9.40	0.74	0.83	1.01	1.08	1.18	1.21	1.26	1.26	1.28
9.60	0.85	0.95	1.16	1.23	1.36	1.39	1.45	1.46	1.47
9.80	0.95	1.06	1.30	1.38	1.52	1.56	1.63	1.65	1.67
10.00	1.05	1.17	1.43	1.52	1.69	1.73	1.82	1.83	1.86
10.20	1.14	1.28	1.56	1.66	1.85	1.90	2.00	2.02	2.06
10.40	1.22	1.38	1.68	1.79	2.00	2.06	2.18	2.20	2.25
10.60	1.30	1.47	1.80	1.92	2.15	2.22	2.35	2.38	2.44
10.80	1.37	1.55	1.91	2.04	2.29	2.37	2.52	2.56	2.63
11.00	1.42	1.62	2.01	2.15	2.42	2.51	2.68	2.73	2.81

$$[\equiv FeO^-] = 10^{+6} \times 10^{+1.85} \times 10^{-8.93}[\equiv FeOH]$$

$$= 10^{-1.08}[\equiv FeOH]$$

$$10^{-0.56}[\equiv FeOH] + [\equiv FeOH] + 10^{-1.08}[\equiv FeOH] = 10^{-3.7}$$

$$[\equiv FeOH] = 10^{-3.83}$$

$$[\equiv FeOH_2^+] = 10^{-4.39}$$

$$[\equiv FeO^-] = 10^{-4.78}$$

The concentration of acid that must be added to a neutral HFO suspension (obtained, for example, by a careful stoichiometric addition of base to a ferric salt solution) is given by the *TOTH* equation:

$$TOTH = [H^+] - [OH^-] + [\equiv FeOH_2^+] - [\equiv FeO^-] = HA_T$$

$$\therefore HA_T = 2.5 \times 10^{-5} M \tag{63}$$

The log C–pH diagrams of Figure 8.13 present two interesting characteristics. First, the neutral surface species $\equiv FeOH$ dominates the surface speciation over a very wide pH range, much beyond the intrinsic pK_a's, because of repulsive coulombic forces. At low pH as the pH is decreased, the increasing positive charge of the surface inhibits its further protonation, and at high pH as the pH is increased, the increasing negative charge of the surface inhibits its further deprotonation. Second, the pH range of dominance of the neutral surface species decreases with increasing ionic strength. This results from the shielding of the coulombic interactions by the bulk electrolyte ions; the repulsion of like charges decreases as the ionic strength increases. As a result, at high ionic strength, the proton exchange reactions at the surface resemble more the reactions of dissolved weak acids, unaffected by coulombic interactions.

The calculation of Example 4 is effectively the reverse of the procedure by which the acid–base properties of HFO (or any solid) are determined in the first place: the net protonation of the surface is obtained from the concentration of acid added, corrected by the hydrogen and hydroxide ion concentrations:

$$[\equiv FeOH_2^+] - [\equiv FeO^-] = HA_T - [H^+] + [OH^-]$$

Thus the experimental data consist of graphs of surface charge (within the factor AS/F) vs. pH as shown in Figure 8.14c. An important feature of these graphs is the point of zero charge. The ZPC (= pH at which the net surface charge is zero) is independent of ionic strength and can be obtained by experimental means other than acid–base titration (such as electrophoretic mobility). Through the Gouy–Chapman theory $\Psi_0 = \Psi_0(\sigma)$ (Figure 8.41b), these data can be transformed to provide the coulombic term ($\log P = -2.3F\Psi_0/RT$) as a function of pH (Figure 8.14a). On this basis, one can deduce the intrinsic acidity constants of the surface. An interesting feature of the Ψ_0 (or $\log P$) vs. pH graphs in

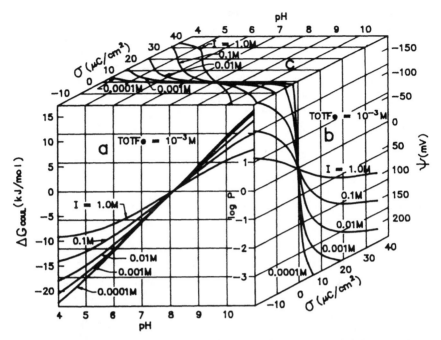

Figure 8.14 Relationship between pH, surface potential, Ψ_0 (or coulombic term, log P, or coulombic free energy, ΔG_{coul}), and surface charge density, σ (or surface protonation) for various ionic strengths of a 1:1 electrolyte: (a) dependence of the coulombic term and surface potential on solution pH; note the near-Nerstian beavior at low ionic strength; (b) Ψ_0 versus σ; these curves correspond to the Gouy–Chapman theory; (c) σ versus pH; these are the curves obtained experimentally. From Dzombak and Morel, 1990.[6]

Figure 8.14a is that the low ionic strength curves are already linear and have a slope of approximately 59 mv. Thus at low ionic strength the pH determines the surface potential at near-Nernstian fashion.

Example 5. Adsorption of H_2CrO_4 on HFO To the system of the previous example, let us now add 0.1 μM H_2CrO_4. What is the extent of chromate adsorption at pH = 6? How does the chromate adsorption vary with pH?

To the species of Example 3, we add the dissolved and surface chromate species $HCrO_4^-$, CrO_4^{2-}, and $\equiv FeCrO_4^-$ and the corresponding reactions:

$$HCrO_4^- = H^+ + CrO_4^{2-}; \qquad \log K = -6.38 \qquad (64)$$

$$H^+ + CrO_4^{2-} + \equiv FeOH = \equiv FeCrO_4^- + H_2O; \qquad \log K = 10.67 - \log P \quad (65)$$

The constants have been corrected for ionic strength effects. With the component choice: H^+, $HCrO_4^-$, $\equiv FeOH$, and P, we obtain Tableau 8.4. Assuming the pH

TABLEAU 8.4

	H^+	$HCrO_4^-$	$\equiv FeOH$	P
H^+	1			
OH^-	-1			
CrO_4^{2-}	-1	1		
$HCrO_4^-$		1		
$\equiv FeOH_2^+$	1		1	1
$\equiv FeOH$			1	
$\equiv FeO^-$	-1		1	-1
$\equiv FeCrO_4^-$		1	1	-1
$[\equiv FeOH]_T$			1	
$[H_2CrO_4]_T$	1	1		
HA_T	1			

given, we ignore the $TOTH$ equation and write the other three mole balances:

$$TOTHCrO_4 = [HCrO_4^-] + [CrO_4^{2-}] + [\equiv FeCrO_4^-] = 10^{-7} \tag{66}$$

$$TOT \equiv FeOH = [\equiv FeOH_2^+] + [\equiv FeOH] + [\equiv FeO^-] + [\equiv FeCrO_4^-]$$
$$= 10^{-3.7} \tag{67}$$

$$TOTP = [\equiv FeOH_2^+] - [\equiv FeO^-] - [\equiv FeCrO_4^-]$$
$$= 6.57 \times 10^{-4}\frac{P^{1/2} - P^{-1/2}}{2} \tag{68}$$

Again the solution of this equilibrium problem is best obtained by computer; the concentration of adsorbed chromate as a function of pH is shown in Figure 8.15.

Because the chromate concentration is much smaller than the concentration of adsorbing sites; however, the adsorption of chromate has no significant effect on the surface speciation and hence on the surface charge or the coulombic forces. Thus the surface speciation is as calculated in Example 4 and we can again use the results of Table 8.3 to obtain the value of log $P = -1.85$, for $I = 10^{-2} M$ and pH $= 6$.

The mass laws for the chromate species are then written:

$$[CrO_4^{2-} = 10^{-6.38}[HCrO_4^-]/10^{-6} = 10^{-0.38}[HCrO_4^-]$$
$$[\equiv FeCrO_4^-] = 10^{4.29} \times 10^{+1.85}[HCrO_4^-][\equiv FeOH]$$
$$= 10^{6.14}[HCrO_4^-][\equiv FeOH]$$

Introducing $[\equiv FeOH] = 10^{-3.83}$ and substituting in the mole balance equation

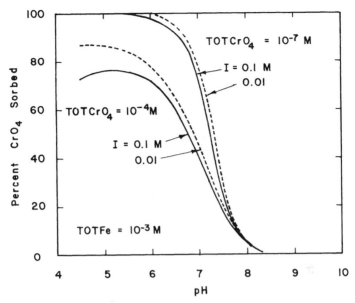

Figure 8.15 Calculated pH edges for chromate sorption on HFO ($90\,mg\,L^{-1}$) at two chromate loadings and ionic strengths (1:1 electrolyte). From Dzombak and Morel (1990).[6]

66, we obtain

$$[HCrO_4^-] + 10^{-0.38}[HCrO_4^-] + 10^{2.31}[HCrO_4^-] = 10^{-7}\,M$$

Thus the chromate is indeed entirely adsorbed,

$$[\equiv FeCrO_4^-] = 10^{-7}\,M$$

the dissolved species concentrations being less than 1% of the total,

$$[HCrO_4^-] = 10^{-9.31}\,M$$
$$[CrO_4^{2-}] = 10^{-9.69}\,M$$

The decrease in sorption with increasing pH shown in Figure 8.15 is characteristic of anion sorption on oxides. As in Example 2 it is due to competition by OH^- for the surface sites (surface deprotonation) and also now by the attendant decrease in coulombic attraction.

Example 6. Adsorption of Zn^{2+} on HFO Consider again the system of Example 4 to which we now add $0.1\,\mu M\ Zn^{2+}$. As seen in Table 8.2, cations are adsorbed at both high-affinity surface sites $[\equiv Fe^sOH]$, which are only 2.5% of the total ($0.005\,mol/Fe$), and low-affinity $[\equiv Fe^wOH]$ sites. At this low adsorbate to adsorbent ratio, we need consider only the adsorption at high-affinity sites:

$$Zn^{2+} + \equiv Fe^sOH = \equiv Fe^sOZn^+ + H^+; \qquad \log K = 0.99 + \log P \qquad (69)$$

Again, the adsorption constant has been corrected for ionic strength effects. With H^+, Zn^{2+}, $\equiv Fe^sOH$, $\equiv Fe^wOH$, and P as components, we obtain Tableau 8.5:

$$TOTZn = [Zn^{2+}] + [\equiv Fe^sOZn^+] = 10^{-7} \qquad (70)$$

$$TOT \equiv Fe^sOH = [Fe^sOH_2^+] + [\equiv Fe^sOH] + [\equiv Fe^sO^-]$$
$$+ [\equiv Fe^sOZn^+] = 10^{-5.3} \qquad (71)$$

$$TOT \equiv Fe^wOH = [Fe^wOH_2^+] + [\equiv Fe^wOH] + [\equiv Fe^wO^-]$$
$$= [Fe^wOH]_T = 10^{-3.7} \qquad (72)$$

$$TOTP = [\equiv Fe^wOH_2^+] - [\equiv Fe^wO^-] + [\equiv Fe^sOH_2^+]$$
$$- [\equiv Fe^sO^-] + [\equiv Fe^sOZn^+]$$
$$= 6.57 \cdot 10^{-4} \frac{P^{1/2} - P^{-1/2}}{2} \qquad (73)$$

The adsorbed zinc concentration as a function of pH obtained by numerical solution of these equations is shown in Figure 8.16.

As in the previous example, the adsorbate–adsorbent ratio is low, and the surface charge is negligibly affected by zinc sorption. The problem is thus again amenable to a simple solution, using the values of P tabulated in Table 8.3.

TABLEAU 8.5

	H^+	Zn^{2+}	$\equiv Fe^sOH$	$\equiv Fe^wOH$	P
H^+	1				
OH^-	-1				
Zn^{2+}		1			
$\equiv Fe^wOH_2^+$	1			1	1
$\equiv Fe^wOH$				1	
$\equiv Fe^wO^-$	-1			1	-1
$\equiv Fe^sOH_2^+$	1		1		1
$\equiv Fe^sOH$			1		
$\equiv Fe^sO^-$	-1		1		-1
$\equiv Fe^sOZn^+$	-1	1	1		1
Zn_T		1			
$[\equiv Fe^sOH]_T$			1		
$[\equiv Fe^wOH]_T$				1	

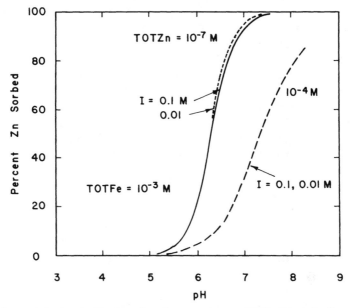

Figure 8.16 Calculated pH edges for zinc sorption on HFO $(90 \, mg \, L^{-1})$ at two zinc loadings and ionic strengths (1:1 electrolyte). From Dzombak and Morel (1990).[6]

For example, at pH = 6.0, log $P = -1.85$

$$[\equiv Fe^{s}OZn^{+}] = 10^{-1.04}[Zn^{2+}][\equiv Fe^{s}OH][H^{+}]^{-1} \tag{74}$$

At pH 6, the acid-base speciation of the surface is as calculated in Example 4:

$$[\equiv Fe^{s}OH] = 10^{-0.13}[\equiv Fe^{s}OH]_{T} = 10^{-5.63}$$

Thus

$$[\equiv Fe^{s}OZn^{+}] = 10^{-0.67}[Zn^{2+}]$$

Introducing this result into the zinc mole balance (Equation 70):

$$[Zn^{2+}] + 10^{-0.67}[Zn^{2+}] = 10^{-7}$$

$$[Zn^{2+}] = 10^{-7.08} \, M$$

$$[\equiv Fe^{s}OZn^{+}] = 10^{-7.75} \, M$$

The zinc is thus 20% adsorbed at this pH as can be seen in Figure 8.16. To obtain the complete figure, the calculation can be done manually for a series of pH's.

An interesting aspect of metal adsorption on oxides concerns the effect of ionic strength. Contrary to what one might expect in view of the large effect of ionic strength on the acid–base chemistry of oxide surfaces, changing the ionic strength has no measurable influence on metal adsorption. This is because changes in ionic strength have two opposite effects on the adsorption reaction: a higher ionic strength decreases the electrostatic repulsion of the positively charged surface (shielding) and, at the same time, decreases the activity of the metal ion in solution. These two effects cancel each other almost perfectly. For example, the effective constant for reaction 69 at a pH of 6.2 near the midpoint of the pH edge (Figure 8.16) is calculated as

$$\log K = 0.99 + \log \gamma_2 + \log P$$
$$= -1.89, \; -1.89, \; -1.90 \quad \text{for } I = 0.1, \, 0.01, \, 0.001 \; M$$

5. PARTITIONING OF SOLUTES INTO BULK PARTICULATE PHASES

A number of aquatic processes that remove solutes from solution and incorporate them into particles are no more adequately described by surface adsorption then they are by precipitation of pure solid phases. Aside from the important phenomenon of uptake by living organisms, these processes, which can be described by the generic term "sorption," involve the partitioning of solutes between the solution and the whole of a particulate phase. The particle–solution interface is not considered to control such processes, which are dominated by the bulk properties of the particles themselves. The word particle here signifies a phase that is hydrodynamically and thermodynamically distinct from the bulk solution; it may be a localized region of an aqueous solution, a solid, or an organic phase, as we now discuss briefly in the cases of ion exchange, solid solution, and organic film solvation.

5.1 Ion Exchange

The general model of an ion exchanger consists of a porous lattice containing fixed charges.[33] Hydrated clays, in which isomorphous replacement of silicon by aluminum or of aluminum by magnesium results in a net negative charge, are the most common aquatic particles approximating this model. The water included in the lattice constitutes an aqueous solution distinct from the bulk solution and in which electroneutrality is maintained by inclusion of an excess of counter ions from the bulk solution.

Using the subscript x to refer to a cation exchanger, we can write, for example, the exchange reaction for sodium and calcium:

$$2Na_x^+ + Ca^{2+} = Ca_x^{2+} + 2Na^+ \tag{75}$$

As is the case for surface adsorption, the energy that attracts ions in the exchanger can be purely electrostatic or partly chemical. Even when the energy of attraction is purely electrostatic, high specificity may result from a matching of the pore or mesh size of the exchanger with that of the hydrated ions. On the basis of a largely electrostatic interaction, a preference for multivalent ions is generally expected and observed in ion exchangers. For example, if we assume that ion partitioning is caused purely by coulombic interactions and can be accounted for by a net potential difference $\Delta\Psi_0$ between the exchanger and the bulk solution, we can write

$$Na_x^+ = Na^+; \qquad \frac{\{Na_x^+\}}{\{Na^+\}} = \exp\left[-\frac{F\Delta\Psi_0}{RT}\right] \tag{76}$$

$$Ca_x^{2+} = Ca^{2+}; \qquad \frac{\{Ca_x^{2+}\}}{\{Ca^{2+}\}} = \exp\left[-\frac{2F\Delta\Psi_0}{RT}\right] \tag{77}$$

Such equations describe the partitioning of ions in relatively dilute solutions on both sides of a semipermeable membrane—the Donnan equilibrium—and they explain qualitatively the role of electrical charge in ion exchange. However, the equations are of little quantitative use for most ion exchangers because activity coefficients are all but unknown in the ion exchange phase, a medium of effective ionic strength in excess of $0.3\,M$ in which some of the charges are fixed. The same difficulty applies to the use of the equilibrium constant for Reaction 75:

$$K = \frac{\{Ca_x^{2+}\}\{Na^+\}^2}{\{Na_x^+\}^2\{Ca^{2+}\}} \tag{78}$$

The corresponding "selectivity coefficient"

$$^c K = \frac{[Ca_x^{2+}][Na^+]^2}{[Na_x^+]^2[Ca^{2+}]} \tag{79}$$

is far from constant over a range of sodium and calcium concentration and there is no simple functionality to relate it to the equilibrium constant, K.

Although a number of theories have been developed to predict or at least rationalize the behavior of ion exchangers, empirical information on the ion exchange properties of clays is probably most useful to aquatic chemists. In classical ion exchange literature this empirical information usually takes the form of an ion exchange isotherm in which the mole fraction of a given ion in the exchanger is plotted against the mole fraction of the same ion in solution. More directly pertinent to our general formulation of equilibrium is to plot the selectivity coefficient as shown in Figure 8.17 for the exchange of monovalent ions in a bentonite clay.[34] The selectivity sequence, $Cs^+ > Rb^+ > K^+ > Na^+ > Li^+$, is opposite to that of the hydrated radii. A selectivity for smaller hydrated ions is

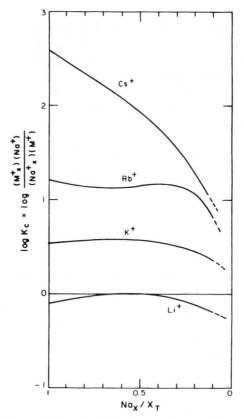

Figure 8.17 Selectivity coefficients for the exchange of monovalent cations in a bentonite clay. The smaller hydrated ions are exchanged preferentially over the larger ones, in accordance with a principally electrostatic interaction energy. Adapted from Gast, 1969.[34]

expected on the basis of a principally electrostatic sorption energy. Note that a strong preference for retention of potassium over sodium in clays is consistent with the relative sodium impoverishment in sedimentary rock compared with igneous rock (see Chapter 5).

5.2   Solid Solution

When a solid phase precipitates in a multicomponent system, foreign ions, metals or ligands, are often incorporated into the solid matrix. Although the first step in such a process may be that of surface adsorption (as it is for precipitation of the pure solid), the actual inclusion of a foreign ion in the bulk of the solid phase is restricted to ions whose valences and interatomic bonding distances are a close match for those of the precipitating ions. For example, in aquatic systems, Sr^{2+},

Mg^{2+}, and Mn^{2+} are often incorporated in calcium carbonate.[35] Many metal oxides and sulfides show varying degrees of enrichment in other trace metals, as is conspicuously exemplified in ferromanganese nodules.

A thermodynamic description of this process is obtained by considering the solid phase as a solution of one pure solid in the other and attributing to each of these solids an activity that is not unity. Consider, for example, the substitution of Mn^{2+} for Ca^{2+} in calcite:

$$Mn^{2+} + CaCO_3(s) = MnCO_3(s) + Ca^{2+} \tag{80}$$

The apparent equilibrium constant for this reaction—now called a distribution coefficient—expressed as

$$D = \frac{\{MnCO_3 \cdot s\}}{\{CaCO_3 \cdot s\}} \times \frac{\{Ca^{2+}\}}{\{Mn^{2+}\}} \tag{81}$$

is equal to the ratio of the solubility products of calcium and manganese carbonates:

$$K_{CaCO_3} = \frac{\{Ca^{2+}\}\{CO_3^{2-}\}}{\{CaCO_3 \cdot s\}} \tag{82}$$

$$K_{MnCO_3} = \frac{\{Mn^{2+}\}\{CO_3^{2-}\}}{\{MnCO_3 \cdot s\}} \tag{83}$$

therefore

$$D = \frac{K_{CaCO_3}}{K_{MnCO_3}} \simeq \frac{10^{-8.3}}{10^{-10.4}} = 10^{2.1} \tag{84}$$

The difficulty with this approach of course is to evaluate the activities of the solids in the solid solution. This can be done empirically from the equilibrium partitioning of Mn^{2+} and Ca^{2+}, thus making the whole approach an empirical rather than a predictive one. If we instead make the gross assumption that the solid solution is ideal and equate the activities of the solids to their mole fractions, we obtain a qualitative description of the role of solid solutions in the solubility of ions as shown in the following Example.

Example 7. Solid Solutions of Carbonates Consider the addition of manganese ion to a system defined by a fixed free carbonate ion concentration and a given total concentration of calcium; $[CO_3^{2-}] = 10^{-5.3} M$; $TOTCa = 10^{-2} M$. By defining the total solid concentration,

$$S_T = [CaCO_3 \cdot s] + [MnCO_3 \cdot s] \tag{85}$$

the activities of the individual solids in the ideal solution can be written:

$$\{CaCO_3 \cdot s\} = \frac{[CaCO_3 \cdot s]}{S_T} = \frac{[Ca^{2+}][CO_3^{2-}]}{K_{CaCO_3}} = 10^3 [Ca^{2+}] \tag{86}$$

$$\{MnCO_3 \cdot s\} = \frac{[MnCO_3 \cdot s]}{S_T} = \frac{[Mn^{2+}][CO_3^{2-}]}{K_{MnCO_3}} = 10^{5.1} [Mn^{2+}] \tag{87}$$

The three mole balance expressions

$$TOTCa = [Ca^{2+}] + [CaCO_3 \cdot s] = [Ca^{2+}](1 + 10^3 S_T) = 10^{-2} \tag{88}$$

$$TOTMn = [Mn^{2+}] + [MnCO_3 \cdot s] = [Mn^{2+}](1 + 10^{5.1} S_T) \tag{89}$$

$$S_T = [CaCO_3 \cdot s] + [MnCO_3 \cdot s] = S_T(10^3 [Ca^{2+}] + 10^{5.1} [Mn^{2+}]) \tag{90}$$

complete the mathematical description of the system. Algebraic manipulation of these equations leads to the approximate solution:

$$[Mn^{2+}] = \frac{TOTMn}{TOTMn + TOTCa} \times \frac{K_{MnCO_3}}{[CO_3^{2-}]} \tag{91}$$

which is plotted in Figure 8.18. The principal feature of this result—and the

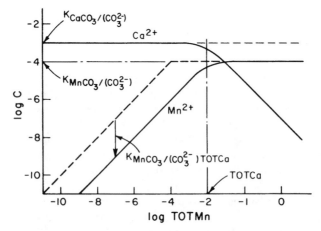

Figure 8.18 Variations in dissolved calcium and manganese ion concentrations as a function of total manganese in the presence of an ideal $CaCO_3$–$MnCO_3$ solid solution ($[CO_3^{2-}]$ is fixed at $10^{-5.3}$ M). Note the large decrease in dissolved manganese compared to total manganese at low concentrations.

major effect of solid solution formation—is the decrease in the dissolved manganese $[Mn^{2+}]$ at low total manganese concentration in the presence of calcite. With the assumption of ideality of the solid solution, this decrease is a factor of 100. Such a large effect is due in part to the greater insolubility of $MnCO_3$ compared to $CaCO_3$, and in part to the large concentration of solid compared

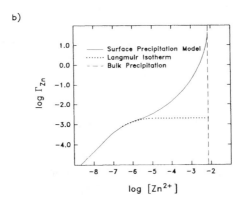

Figure 8.19 The surface precipitation model. (*a*) Schematic of surface reactions showing surface complexation and subsequent formation of a surface solid solution. (*b*) Generalized isotherm showing how formation of a surface solid solution provides a continuous transition between surface complexation and precipitation of a new solid phase. From Farley et al. (1985)[13].

to dissolved calcium, as seen in the formula

$$TOTMn = [Mn^{2+}]\left[1 + \frac{K_{CaCO_3}}{K_{MnCO_3}} \times \frac{S_T}{[Ca^{2+}]}\right] \qquad (92)$$

Note that a decrease in the solubility of calcium due to a decrease in the activity of $CaCO_3(s)$ is only seen when the total manganese is in excess of the total calcium, not a very probable circumstance in natural waters. However, such a decrease in the dissolved concentrations of a precipitating metal due to solid solution with another may occur in the case of metal oxides, when the metals are both present at trace concentrations.

* * *

Often a solid solution may result from an adsorption reaction near the point of surface saturation and it may involve only a fraction of the bulk solid. This situation can be described by a surface precipitation model as schematized in Figure 8.19 for the case of zinc adsorbing on ferric hydroxide. This model provides a continuum between the adsorption of zinc at low site densities and the precipitation of a pure $Zn(OH)_2(s)$ phase on the iron hydroxide. In the transition, the surface phase is treated as a solid solution of $Fe(OH)_3(s)$ and $Zn(OH)_2(s)$. In this model Fe^{3+} is also allowed to adsorb on the surface and adsorption of Fe^{3+} on $Fe(OH)_3(s)$ is formally and mathematically equivalent to precipitation of $Fe(OH)_3(s)$.

5.3 Organic Film Solvation

Many trace organic pollutants are hydrophobic and their fate in aquatic systems is largely governed by their partitioning among gas, water, and solid phases. A widely accepted and reasonably successful predictive model for the water–solid partitioning of such hydrophobic trace compounds is to consider a two-phase equilibrium between the water and what is imagined to be an organic film coating the sediment or soil particles.[36,37] Three major empirical observations are at the core of such a simple model:

1. Sorption isotherms for hydrophobic compounds are typically linear over the range of concentration of interest.
2. For a given compound, sorption is proportional to the total organic content of the solid (the coefficient of proportionality is somewhat affected by the size class of the sediments).
3. For a given sediment, sorption is proportional to the octanol–water partition coefficient of the organic compound. (Note also that octanol–water partition coefficients are inversely correlated with water solubilities.)

Although these observations may support other models, the image of a bulk organic phase in which the hydrophobic compound is dissolved—effectively a solvent extraction procedure—is particularly simple and appealing even if it does not have a strong conceptual or theoretical backing. The general fomula for such organic film solvation of trace compounds is written

$$C_x = a \times K_{ow} \times f_{ow} \times C \qquad (93)$$

where C_x is the concentration of the compound in the solid phase, a is an empirical coefficient of proportionality ($a \simeq 0.63$) representing the solvent properties of condensed aquatic organic matter compared to octanol, K_{ow} is the octanol water partition coefficient, f_{oc} is the fraction of organic carbon in the solid, and C is the concentration of the compound in water. The parameters a, K_{ow}, and f_{oc} are dimensionless and both C_x and C are expressed in the same units: C_x is in grams or moles per kilogram of dry solid and C in grams or moles per kilogram of solution. The wide applicability of this formula and its small data requirements make it a valuable tool to predict the fate of organic pollutants in aquatic systems.

Complicated effects of natural organic matter on the sorption (and reactivity) of hydrophobic organic micropollutants have been observed. Thus organic micropollutants may interact with both sorbed and dissolved humic substance; the former interaction serves to increase partitioning of the pollutant to the solid phase and the latter, by competition, to decrease it.[38-40] The reactivity of organic micropollutants (i.e., in transformation reactions) may also be altered in the presence of natural sediments. The association of organic micropollutants with humic material (either dissolved or sorbed on sediment) inhibits at least some degradation reactions, particularly base-catalyzed hydrolysis.[41,42]

6. KINETIC CONSIDERATIONS

In the introduction to this chapter we noted that the geochemical fate of trace elements and compounds was often controlled by (ad)sorption on a solid surface and eventual settling and incorporation in sediments. This overall phenomenon includes not only the two obvious steps of sorption and settling, but also the process of particle aggregation through which colloidal material ($< 10 \, \mu$m) is transformed into particles ($> 10 \, \mu$m) which have non-negligible settling velocities. Any of the three processes, sorption, aggregation, or settling, may control the overall rate of elimination of a conservative chemical compound from the water column. In this section we discuss each of these briefly.

6.1 Adsorption Kinetics

Most of the experimental adsorption data that are shown or alluded to in this chapter were obtained with equilibration times on the order of minutes to hours and were usually observed to be reversible on the same time scale. When the experiments were continued for longer times, often a second, slower, and typically irreversible adsorption process was observed with a characteristic equilibration time on the order of days to months. This second process is usually thought to correspond to a slow change in the nature of the solid itself (adsorption data are often obtained with metastable solids) or to a diffusion of the adsorbate toward the interior of the solid which may be porous. Indeed, many theoretical treatments of adsorption kinetics focus on diffusion and often consider a two-step process: diffusion to the particle in the boundary layer and diffusion within the particle.[43] The expression for boundary layer diffusion has the usual form introduced in Chapters 4 and 5; that for pore diffusion is complicated by pore geometry and by reaction of the sorbate with the pore surface (retardation). Other general rate equations simply describe the kinetics of the adsorption process as a bimolecular reversible chemical reaction:

$$\equiv X + C \rightleftharpoons \, \equiv XC \tag{94}$$

$$\frac{d[\equiv XC]}{dt} = k_f[\equiv X][C] - k_b[\equiv XC] \tag{95}$$

In this formulation which is consistent with a Langmuirian description of adsorption equilibrium, the law of microscopic reversibility is usually considered to hold and only one independent kinetic constant must be obtained:

$$\frac{d[\equiv XC]}{dt} = k\left([\equiv X][C] - \frac{[\equiv XC]}{K_{ads}}\right) \tag{96}$$

As noted in Chapter 5, reaction and diffusion control of solute–solid reactions are often difficult to distinguish from each other and applicability of a given rate law is rarely sufficient proof of the underlying mechanism.

6.2 Effects of Adsorption on the Kinetics of Surface Reactions

Reactions occurring at mineral surfaces commonly exhibit saturation kinetics, that is, a plateau in the rate of reaction is observed at high (dissolved) reactant concentrations. This effect has been noted both for dissolution reactions (Chapter 5) and for surface-catalyzed redox reactions (Chapter 7). The origin of such saturation kinetics should now be clear. If the rate-limiting step of a reaction involves a surface species, then the reaction rate must depend on the surface concentration of that species. A parallel between the dependence of surface concentration and of reaction rate on the dissolved concentration of the reacting species is thus expected. The shift from first-order dependence of surface concentration on dissolved concentration when the surface is far from saturation to zero-order dependence once the surface is saturated is the basis for the common observation of (apparent) fractional-order dependence of reaction rates on dissolved concentration over some limited concentration range (cf. Figure 5.15).

In general, a simple Langmuirian model has been applied to describe observed dissolution kinetics for proton-promoted, ligand-promoted, and reductive dissolution of oxides.[44,45] Although dissolution reactions are considered to proceed in parallel at surface sites with varying reactivity, a constant distribution of sites is assumed. Thus, the contribution of different site types to the overall reaction rate cannot be distinguished. Electrostatic effects are not explicitly considered. Detachment of a surface metal atom (destabilized by formation of a surface complex, protonation of neighboring hydroxyl groups, or reduction or oxidation to a more labile oxidation state) from the lattice is generally taken to be the rate-limiting step in oxide dissolution. Rapid and reversible adsorption is assumed, as is rapid regeneration of the mineral surface; this latter assumption is supported by direct spectroscopic evidence. Although these models are quite effective in describing bulk dissolution kinetics, they rely on a vastly oversimplified picture of the mineral surface. Modern techniques for examining the microtopography and atomic structure of mineral surfaces (particularly scanning tunneling and atomic force microscopy) demonstrate the remarkable heterogeneity and dynamic nature of mineral surfaces.[46] Further development of models for surface-controlled dissolution should integrate this improved picture of atomic-scale surface structure with macroscopic observations of dissolution kinetics.

6.3 Settling

The settling velocity of aquatic particles is generally assumed to follow Stokes law:

$$V_s = \frac{g}{18\mu}(\rho_s - \rho)d^2 \tag{97}$$

where V_s is the settling velocity; g is the acceleration due to gravity; μ is the dynamic viscosity of the water; ρ_s is the density of the particle; ρ is the density of water; and d is the diameter of the particle assumed to be spherical. The density of aquatic particles is not always well known, in most cases it is slightly in excess of that of water and increasing with the inorganic content of the particle. Let us focus our attention on particle size to which, because of the square term, the settling velocity is most sensitive. The most important notion to be conveyed here is that a particle diameter is really not an inherent, conservative property of a given particle. In the size range below $10\,\mu m$, particles continuously aggregate, and the balance of aggregation and settling processes results in a characteristic size distribution that is a property of the suspension, not that of the individual particles. Thus in order to describe quantitatively the fate of particles (and of their load of sorbed material), we must understand the mechanisms of particle aggregation in natural waters, coagulation, and ingestion by organisms.

6.4   Particle Aggregation

From a thermodynamic point of view, suspensions of small particles are unstable since the total surface energy is greater in the dispersed state than it is in the aggregated state.* The classical treatment of the energetics of particle interactions, the DLVO theory (Derjaguin, Landau, Verwey, and Overbeek), considers the balance of short-range attractive Van der Waals forces and the longer-range repulsive electrostatic forces.[47] Because the Van der Waals forces vary with the inverse of the fourth power of the interparticle distance (x^{-4}), and the electrostatic forces with x^{-2}, a plot of interaction energy versus x yields the typical graph of Figure 8.20. When the particles are very close together, attractive forces dominate and the particles are coagulated. When the particles are farther apart, electrostatic repulsion creates an energy barrier that prevents coagulation and "stabilizes" the suspension. Thus the stability of aquatic particles (i.e., their resistance to coagulation and hence sedimentation) depends chiefly on the factors that determine surface charge—the extent of organic coatings and pH—and on the ionic strength which modulates the extent of coulombic repulsion.[49] The large effect that increasing ionic strength has on lowering the coagulation energy barrier, as illustrated in Figure 8.20, is simply due to compression of the double layer and decrease in the electrostatic repulsion between particles. The transport of colloidal iron and aluminum in rivers and their deposition within estuaries is strongly influenced by sorbed humic substances. In freshwaters, colloids are stabilized by sorbed organics which impart to them a large negative charge; flocculation and coagulation occur in estuaries where the ionic strength and the concentration of Ca^{2+} and Mg^{2+} (which may partly neutralize the negative surface charge) are higher.[50-52]

There are three principal mechanisms by which particles can overcome their electrostatic energy barrier, collide, and coagulate:[53] Brownian motion, as determined by the thermal energy of the suspension (perikinetic coagulation); laminar or turbulent fluid shear controlled by the kinetic energy of the fluid (orthokinetic coagulation); and differential settling which depends on the gravitational (and hence kinetic) energy of the particles. These processes are very much dependent on the sizes of the coagulating particles as well as on the temperature and mixing regime of the water. The particle size distribution observed in aquatic systems (see Figure 8.1 and Table 8.1) can be rationalized as a steady state between coagulation of the smaller particles and settling of the larger ones, and values of the exponent of the power law near $p = 4$ are predicted by such theory. In situations where coagulation is the step limiting the overall sedimentation process, the rate of elimination of particles from the water column may be approximately dependent on the square of the particle concentration (second order process) rather than being proportional to the particle concentration, as it is when settling is the rate-limiting step. In shallow waters, which are often well mixed vertically,

---

*Surface energies are in fact sufficiently large to affect the solubility of solid phases (cf. Chapter 5, Section 7.2).

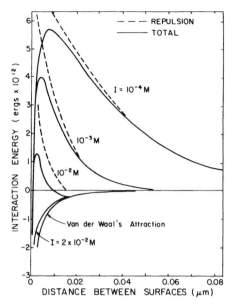

Figure 8.20 Variation in the energy of attraction (Van der Waals) and repulsion (electrostatic) as a function of distance between two charged spheres ($d = 0.2 \mu$m, $\Psi_0 = 50$ mV) in a medium of low ionic strength ($2 \times 10^{-2} M$ to $10^{-4} M$). From Kasper, 1971.[48]

the particle may be eliminated from the water column chiefly through coagulation with particles at the surface of the sediments (filtration by the "fluff layer").[54,55]

Although the subject of particle ingestion by aquatic organisms is beyond the scope of this chapter, it should be strongly emphasized that zooplankton feeding, shedding of outer layers, and feces packaging are generally considered to be important, perhaps dominant, factors in the sedimentation of aquatic particles, particularly in the sea.[7,8,56] Although zooplankton certainly exhibit some taste and discrimination in their feeding habits, one may hope that, on a larger scale, their particle aggregating activities are describable by coagulation equations. Although this "biological coagulation" mechanism may sometimes be more effective than orthokinetic or perikinetic coagulation and promote a faster sedimentation rate, in some places over some size range it may result in the same type of particle size distribution in the water column. Again we find that the composition of natural waters results from an interplay among physical, chemical, and biological processes. The key is to find which are dominant, and when.

REFERENCES

1. P. W. Schindler, *Thalassia Yugoslavia*, **11**, 101 (1975).
2. L. Balistrieri, P. G. Brewer, and J. W. Murray, *Deep-Sea Res.*, **28A**, 101 (1981).

3. M. C. Kavanaugh and J. O. Leckie, Eds., *Particulates in Water*, Advances in Chemistry Series 183, American Chemical Society, Washington, DC, 1980.

4. W. Stumm, *Chemistry of the Solid-Water Interface*, Wiley-Interscience, New York, 1992.

5. W. Stumm and J. J. Morgan, Aquatic Chemistry, 2nd ed., Wiley, New York, 1981, pp. 600–601.

6. D. A. Dzombak and F. M. M. Morel, *Surface Complexation Modeling*, Wiley-Interscience, New York, 1990.

7. R. J. Gibbs, Ed., *Suspended Solids in Water*, Plenum, New York, 1974.

8. J. K. B. Bishop, J. E. Edmond, D. R. Ketten, M. P. Bacon, and W. B. Silker, *Deep Sea Res.*, **24**, 511 (1977).

9. A. Lerman, *Geochemical Processes, Water and Sediments Environments*, Wiley, New York, 1979.

10. A. Lerman, K. L. Calder, and P. R. Betzer, *Earth Planet. Sci. Letters*, **37**, 61 (1977).

11. J. R. Hunt, in *Particulates in Water*, M. C. Kavanaugh and J. O. Leckie, Eds., Advances in Chemistry Series 183, American Chemical Society, Washington, DC, 1980.

12. A. W. Adamson, *Physical Chemistry of Surfaces*, 2nd ed., Wiley, New York, 1967.

13. K. J. Farley, D. A. Dzombak, and F. M. M. Morel, *J. Colloid Interface Sci.*, **106**, 226 (1985).

14. P. W. Schindler, in *Adsorption of Inorganics at Solid–Liquid Interfaces*, M. A. Anderson and A. J. Rubin, Eds., Ann Arbor Science, Ann Arbor, MI, 1981.

15. H. Hohl and W. Stumm, *J. Colloid Int. Sci.*, **55**, 281 (1976).

16. J. A. Davis and J. O. Leckie, *Environ. Sci. Technol.*, **12**, 1309 (1978).

17. J. A. Davis, *Geochim. Cosmochim. Acta*, **46**, 2381 (1982).

18. P. Somasundaran and G. E. Agar, *J. Colloid Int. Sci.*, **24**, 433 (1967).

19. R. N. J. Comans and J. J. Middleburg, *Geochim. Cosmochim. Acta*, **51**, 2587 (1987).

20. J. A. Davis, C. C. Fuller, and A. D. Cook, *Geochim. Cosmochim. Acta*, **51**, 1477 (1987).

21. C. C. Fuller and J. A. Davis, *Geochim. Cosmochim. Acta*, **51**, 1491 (1987).

22. L. Sigg, in *Aquatic Surface Chemistry*, W. Stumm, Ed., Wiley-Interscience, New York, 1987, pp. 319–349.

23. S. W. Park and C. P. Huang, *J. Colloid Interface Sci.*, **117**, 431 (1987).

24. S. W. Park and C. P. Huang, *J. Colloid Interface Sci.*, **128**, 245 (1989).

25. J. A. Davis and D. B. Kent, in *Mineral–Water Interface Geochemistry: Reviews in Mineralogy*, Vol. 23, M. F. Hochella, Jr. and A. F. White, Eds. Mineralogical Society of America, Washington, DC, 1990, pp. 177–260.

26. K. A. Hunter and P. S. Liss, *Nature*, **282**, 823 (1979).

27. E. Tipping and D. Cooke, *Geochim. Cosmochim. Acta*, **46**, 75 (1982).

28. R. Beckett and N. P. Le, *Colloids and Surfaces*, **44**, 35 (1990).

29. P. P. Newton and P. S. Liss, *Limnol. & Oceanog.* **32**, 1267 (1982).

30. J. A. Davis, *Geochim. Cosmochim. Acta*, 48, 679 (1984).

31. D. C. Grahame, *Chem. Rev.*, **41**, 441 (1947).

32. O. Stern, *Z. Elektrochem.*, **30**, 508 (1924).

33. F. Hellferich, *Ion Exchange*, McGraw-Hill, New York, 1962.

34. R. G. Gast, *Soil Sci. Soc. Am. Proc.*, **33**, 37 (1969).

35. W. Stumm and J. J. Morgan, *Aquatic Chemistry*, 2nd ed., Wiley, New York, 1981, pp. 287–291.

36. C. T. Chiou, L. J. Peters, and V. H. Freed, *Science*, **206**, 831 (1979).

37. S. W. Karickhoff, B. J. Brown, and T. A. Scott, *Water Res.*, **13**, 241 (1979).

38. G. Caron and I. H. Suffet, in *Aquatic Humic Substances: Influence on Fate and Treatment of Pollutants*, I. H. Suffet and P. MacCarthy, Eds., Advances in Chemistry Series 219, American Chemical Society, Washington, DC, 1989, pp. 117–130.

39. D. E. Kile and C. T. Chiou, in *Aquatic Humic Substances: Influence on Fate and Treatment of Pollutants* I. H. Suffet and P. MacCarthy, Eds., Advances in Chemistry Series 219, American Chemical Society, Washington, DC, 1989, pp. 131–157.

40. L. L. Henry, I. H. Suffet, and S. L. Friant, in *Aquatic Humic Substances: Influence on Fate and Treatment of Pollutants*, I. H. Suffet and P. MacCarthy, Eds., in 219, American Chemical Society, Washington, DC, 1989, pp. 159–171.

41. R. G. Zepp and N. L. Wolfe, in *Aquatic Surface Chemistry*, W. Stumm, Ed., Wiley-Interscience, New York, 1987, pp. 423–455.

42. D. L. Macalady, P. G. Tratnyek, and N. L. Wolfe, in *Aquatic Humic Substances: Influence on Fate and Treatment of Pollutants*, I. H. Suffet and P. MacCarthy, Eds., in 219, American Chemical Society, Washington, DC, 1989, pp. 323–332.

43. S. C. Wu and P. M. Gschwend, *Environ. Sci. Technol.*, **20**, 717 (1986).

44. G. Furrer and W. Stumm, *Geochim. Cosmochim. Acta*, **50**, 1847 (1986).

45. B. Sulzberger, D. Suter, C. Siffert, and S. Banwart, *Mar. Chem.*, **28**, 127 (1989).

46. M. F. Hochella, Jr., in *Mineral–Water Interface Geochemistry, Reviews in Mineralogy*, Vol. 23, M. F. Hochella, Jr. and A. F. White, Eds., Mineralogical Society of America, Washington, DC, 1990, pp. 87–132.

47. W. Stumm and J. J. Morgan, *Aquatic Chemistry*, 2nd ed., Wiley, New York, 1981, p. 656.

48. D. Kasper, "Theoretical and Experimental Investigations of the Flocculation of Charged Particles in Aqueous Solutions by Polyelectrolytes of Opposite Charge," Ph.D. Dissertation, California Institute of Technology, Pasadena, CA, 1971.

49. L. Liang and J. J. Morgan, *Aquatic Sci.*, **52**, 32 (1990).

50. E. R. Sholkovitz, *Geochim. Cosmochim. Acta*, **40**, 831 (1976).

51. E. A. Boyle, J. M. Edmond, and E. R. Sholkovitz, *Geochim. Cosmochim. Acta*, **41**, 1313 (1977).

52. E. R. Sholkovitz and D. Copland, *Geochim. Cosmochim. Acta*, **45**, 181–189 (1981).

53. S. Friedlander, *Smoke, Dust and Haze: Fundamentals of Aerosol Behavior*, Wiley, New York, 1977.

54. K. A. Newman, F. M. M. Morel, K. D. Stolzenbach, *Environ. Sci. Technol.*, **24**, 506 (1990).

55. K. D. Stolzenbach, K. N. Newman, C. Wong, *J. Geophys. Res.*, **97**, C11, 17889 (1992).

56. M. M. Mullin, in *Particulates in Water*, M.C. Kavanaugh and J. O. Leckie, Eds., Advances in Chemistry Series 183, American Chemical Society, Washington, DC, 1980.

PROBLEMS

8.1 **a.** Assuming a solid density $\simeq 1$, and a power law size distribution with an exponent of -4 between radii of 0.1 and $10\,\mu m$, what is the total surface area of $5\,mg\,L^{-1}$ of suspended solids?

 b. Using a cross-sectional area $\simeq 10^{-15}\,cm^2$ for typical ionic groups, estimate an upper limit on the adsorption capacity $[\equiv X]_T$ of such a suspension. Compare your result with the values given in Example 1 and Example 3 (specific site density and specific surface area per solid mass).

8.2 Consider a solid surface characterized by the reactions

$$\equiv XOH_2^+ = H^+ + \equiv XOH; \qquad pK_{a_1} = 5.0$$

$$\equiv XOH = H^+ + \equiv XO^-; \qquad pK_{a_2} = 7.0$$

$$\equiv XO^- + Cu^{2+} = \equiv XOCu^+; \qquad pK = -8.0$$

and for which electrostatic effects are considered unimportant.

 a. Calculate the adsorption isotherm at pH $= 7$, $X_T = 10^{-6}\,M$.

 b. Assuming no other solution species than Cu^{2+}, calculate the fraction of copper adsorbed as a function of pH ($[\equiv X]_T = 10^{-6}\,M, Cu_T = 10^{-8}\,M$).

 c. Assuming the same inorganic species and the same water composition as in Example 3 of Chapter 6, calculate the speciation of copper at pH $= 7$.

8.3 **a.** Calculate the double-layer thickness in a solution of $5 \times 10^{-4}\,M$ NaCl.

 b. In such a medium what is the charge density of a surface at a potential of $-20\,mV$?

 c. Calculate the effects of the adsorption in the diffuse layer on the bulk ion composition (Na^+ and Cl^-) if the solid is present at the high experimental density of $10\,g\,L^{-1}$ (specific surface area $= 100\,m^2\,g^{-1}$).

 d. Assuming that the surface charge has been developed by removal of acidic protons from the surface and that the solution is not buffered at all, what is the pH of the solution?

 e. What is the pH at the surface?

8.4 Following the method illustrated in Example 6, calculate the extent of sorption of $10^{-8}\,M$ Ni on $180\,mg\,L^{-1}$ HFO as a function of pH.

8.5 Following the method illustrated in Example 5, calculate the extent of sorption of $10^{-7}\,M$ arsenate on $180\,mg\,L^{-1}$ HFO as a function of pH.

8.6 Recalculate the speciations of copper and nickel in the model freshwater and seawater of Examples 3 and 4 in Chapter 6, considering that the water contains $45\,mg\,L^{-1}$ of HFO in suspension.

INDEX